3rd International Conference of Indonesia Society for Remote Sensing (ICOIRS 2017)

IOP Conference Series: Earth and Environmental Science
Volume 165

Semarang, Indonesia
31 October – 1 November 2017

ISBN: 978-1-5108-6582-2
ISSN: 1755-1307

Printed from e-media with permission by:

Curran Associates, Inc.
57 Morehouse Lane
Red Hook, NY 12571

Printed by Curran Associates, Inc. (2018)

For permission requests, please contact the Institute of Physics
at the address below.

Institute of Physics
Dirac House, Temple Back
Bristol BS1 6BE UK

Phone: 44 1 17 929 7481
Fax: 44 1 17 920 0979

techtracking@iop.org

Additional copies of this publication are available from:

Curran Associates, Inc.
57 Morehouse Lane
Red Hook, NY 12571 USA
Phone: 845-758-0400
Fax: 845-758-2633
Email: curran@proceedings.com
Web: www.proceedings.com

Table of contents

Volume 165

3rd International Conference of Indonesia Society for Remote Sensing (ICOIRS 2017)

31 October to 1 November 2017, Semarang, Indonesia

Accepted papers received: 12 June 2018
Published online: 3 July 2018

Preface

Papers

Agriculture

Coastal and Marine

Geodesy

GIS (Geographic Information System)

Hazard

Land Cover and Land Use

Plantation Analysis and Development

Remote Sensing and Photogrammetry

Urban Geoplanning and Development

Unmanned Aerial Vehicle

Water Resources

On behalf of the local organizing committee, I would like to welcome you all to The 3rd International Conference of Indonesia Society for Remote Sensing (ICOIRS 2017) – PIT MAPIN XXIV will be held at Semarang on 31 October – 1 November 2017. This Program is part of the 60th Anniversary of Diponegoro University in 2017. The Conference is organized by collaboration of Indonesian Society for Remote Sensing (ISRS/MAPIN), PUI – PKMBRP (Center of Excellent for Science and Technology – Center for Coastal Mitigation and Rehabilitation Studies) Diponegoro University, PUI – PPPJ (Center of Excellent for Science and Technology – Center for Utilization of Remote Sensing) Indonesian National Institute of Aeronautics and Space, BIG (Geospatial Information Agency) and Oceanography Department, Faculty of Fisheries and Marine Science, Diponegoro University under the theme "Global Cooperation for Conserving The Earth". This program is supported by International Society for Photogrammetry and Remote Sensing (ISPRS), Faculty of Engineering, such as Geodesy, Urban & Regional Planning, Geology Department and Center for Spatial Data Infrastructure Development (PPIDS) of Diponegoro University. Over 150 papers and 300 participants are expected in this conference.

The event will bring together researchers, policy makers, and practitioners from developed and developing countries to share insights into the challenges and opportunities of remote sensing technology and its application in solving the problems of Indonesia and South East Asia countries. It will showcase cutting-edge research from around the South East Asia, focusing on themes of equity and risk, learning, capacity building, methodology, and possibly investment approaches in remote sensing. It will explore practical adaptation policies, approaches and share strategies for decision making to support global cooperation for conserving the earth.

ICOIRS 2017 (PIT MAPIN XXIV) would not have been possible without the many people who have worked with us in planning and organizing the technical program, exhibition, and social events. We thanks to MAPIN for its tremendous guidance and support on organizing the event; special thanks to the local organizing committee members who have all work extremely hard to make this conference happen.

Finally, I would like to ask all of you to attend this exciting conference by presenting papers, discussing technical issues, exchanging ideas, and enjoying the city of Yogyakarta together. We are looking forward to see you all at ICOIRS 2017 in Semarang.

IOP Conf. Series: Earth and Environmental Science **165** (2018) 011001 doi:10.1088/1755-1315/165/1/011001

Dr. Muhammad Helmi, S.Si., M.Si. (PUI-PKMBRP UNDIP)., ICOIRS 2017 (PIT MAPIN XXII) Chairman

IOP Conf. Series: Earth and Environmental Science **165** (2018) 011002 doi:10.1088/1755-1315/165/1/011002

Peer review statement

All papers published in this volume of *IOP Conference Series: Earth and Environmental Science* have been peer reviewed through processes administered by the proceedings Editors. Reviews were conducted by expert referees to the professional and scientific standards expected of a proceedings journal published by IOP Publishing.

IOP Conf. Series: Earth and Environmental Science **165** (2018) 012001 doi:10.1088/1755-1315/165/1/012001

Innovation on Geolocation and Pattern Recognition for Paddy Growth Stages Reporting in Indonesia

Agustan [1*], Swasetyo Yulianto [1], Lena Sumargana [1], Heri Sadmono [1], Fauziah Alhasanah [1]

[1]Center of Technology for Regional Resources - Agency for the Assessment and Application of Technology (PTPSW - BPP Teknologi), Jakarta - Indonesia

agustan@bppt.go.id

Abstract. Information on paddy growth stages is important for rice yield prediction. By utilizing statistical approach such as Area Frame Sampling method, data on paddy growth stages in the certain area may be used to estimate its harvest potential. The method of area frame sampling is based on segment observations determined from a stratified random sampling. Normally, information of paddy growth stages is obtained from terrestrial and remote sensing method. Since the life cycle of paddy is around 120 days, temporal resolution for observation becomes the main consideration. There is a need to develop a robust location-based reporting system for paddy growth stage. This system is designed by controlling the observer to report paddy condition on a determined location, called segment; the observer must report from the center of each segment. This paper discusses the innovation on the use of mobile phone for geolocation and pattern recognition to collect paddy growth stage data. The GPS on the mobile phone is explored for geolocation whereas the camera on the mobile phone is utilized to capture the paddy images. This information is then sent to the server for automatic pattern recognition. The statistical method with pre-processing, feature extraction, classification, feature selection and learning were applied on pattern recognition. Testing on geolocation was conducted in the Java area since May 2017 and installed for 2,356 observers. It is found that 73% of observers successfully reported the paddy growth stage from the center of each segment, which is locked in 10-meter accuracy, and 27% reported not from the center of segment due to field conditions.It is also found that a combination of GPS and mobile network or assisted-GPS can speed up the positioning.

1. Introduction

Indonesia has established the Law No. 18/2012 for national food security policy, which then followed by the Government Regulation No. 17/2015. These legal bases mandate the institution responsible for providing information about food, that cover food information system, area planted, area harvested, productivity, land use, and machinery.

In Indonesia, rice is the most staple food compared to corn, sago, and cassava [1], therefore paddy becomes one of the priority crops to monitor. Paddy field in Indonesia is commonly found throughout the country with a total area around 8 million hectares. The most productive paddy field is located in Java, Sumatra and Sulawesi islands with different varieties. Most paddy varieties have 120 days life cycle and its growth stage is classified into three main stages: vegetative, reproductive and ripening [2].

Paddy growth stage is an important parameter for rice crop monitoring [3], [4], that by knowing the area and distribution of paddy growth stage, the time and quantity of harvest can be estimated.

Recently, two methods have been developed for paddy growth stage monitoring which is satellite-based remote sensing and statistic-based terrestrial observation. Satellite-based remote sensing method has the advantage of area coverage and repeats observations (temporal resolution), whereas statistic-based terrestrial observation method has the advantage of actual paddy condition. Both methods are assessed to improve paddy reporting mechanism in Indonesia that conducted on the monthly, quarterly and yearly basis.

In Indonesia, statistic-based terrestrial observation method collects two basic information of paddy crop: complete report of harvested area and amount of productivity with field survey. In data acquisition method based on statistical approach, Area Frame Sampling (AFS) was developed to improve the data quality of the harvested area and was used by Statistics Indonesia, a government agency responsible for statistical data. AFS may be defined as a listing of land areas from which the sample is selected as the basis for identifying all the statistical units to be enumerated [5]. Since AFS is linked with locational-based [6] reporting mechanism, it is important to develop a system that can handle the reporting directly from the field in a simple way using a compact and mobile device. Compact and mobile device means the integration of required and functional sensors such as communication, positioning, and camera in one and handy device, which is found in mobile phones [7]. This paper describes the innovation on utilizing android based mobile phones by developing geolocation for data and image acquisition, and pattern recognition as a control system in the main server to support the implementation of AFS in paddy growth stages reporting in Indonesia.

2. Data and Method

The goal of this research is to improve the data quality in paddy growth stages reporting in Indonesia, therefore the AFS method is the foundation of this research. The ability of smartphone in data transfer through the internet connection combined with positioning and image capturing functions is assessed and developed in one integrated reporting system. Due to the compatibility and availability of smartphone, this research exploits Android-based smartphone by developing a system with Java and Eclipse programming languages [8], whereas MySQL, PHP and Python script are developed for database server application. The system has consisted of two components: operational and office. The operational component is developed for the observer and installed on their smartphone to collect and report data. The office component is developed to receive, classify and archive data then disseminate the information.

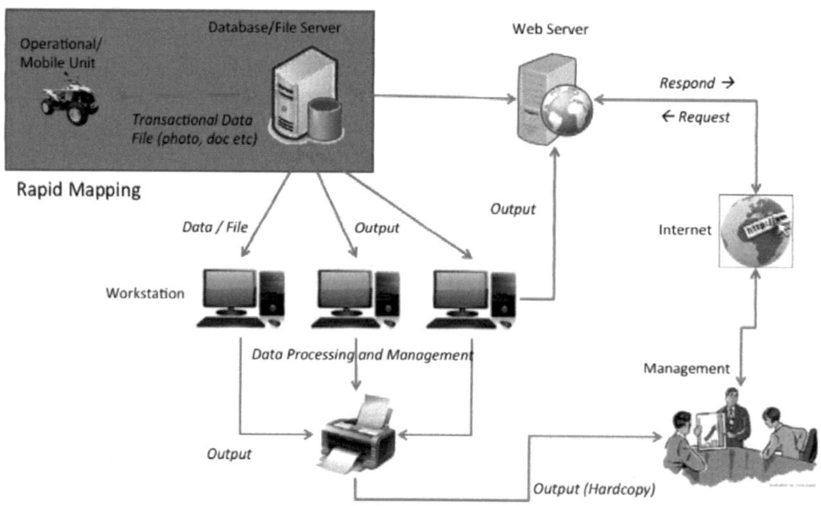

Figure 1. Basic data observation, communication and reporting workflow

2.1. Operational component with geolocation function

The operational system is developed to combine the geolocation or navigational function with several features including accommodating segment with "google map image" as background or base map, determining location based on GPS and assisted-GPS [9] facility, acquiring image by embedded camera, and sending data to server as reporting through data/internet connection.

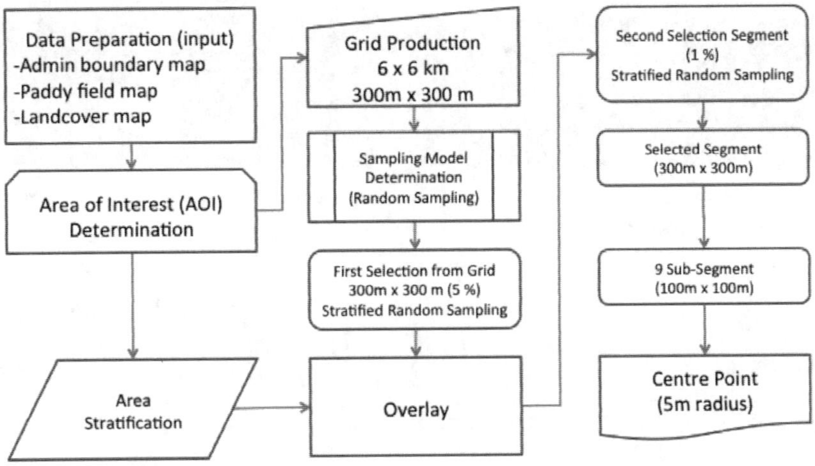

Figure 2. Diagram of coordinates determination based on AFS Method

AFS method adopts the locational-based system, which means data reported must contain the real condition from a pre-determined location. To ensure that the observer reports from the correct position, this system enables reporting paddy growth stage only from coordinates in the center of the sub-segment as a final product from AFS. This means that the observer must find the pre-determined coordinates using the navigation function in the system, and after reaching the exact location, functions for image capturing and data reporting are then activated automatically.

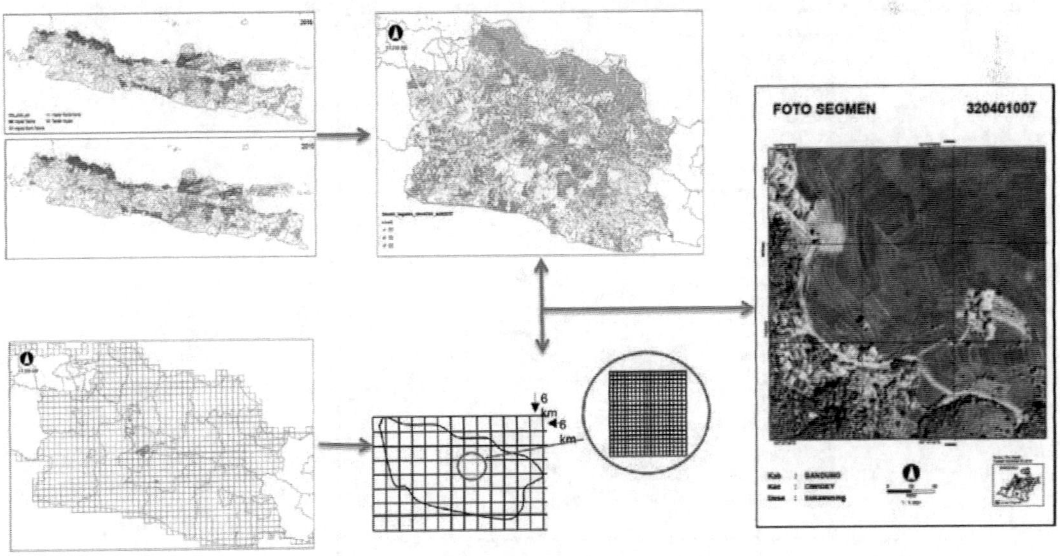

Figure 3. Diagram of Coordinates Determination based on AFS Method

2.2. Office component with pattern recognition function

The office system applied to pattern recognition to distinguish paddy growth stage that works in the database server. Various paddy images are taken from the field by utilizing image capture feature that is built into the system. This feature is enhanced with image compression capability hence the image size is standardized into approximately 500 kilobytes.

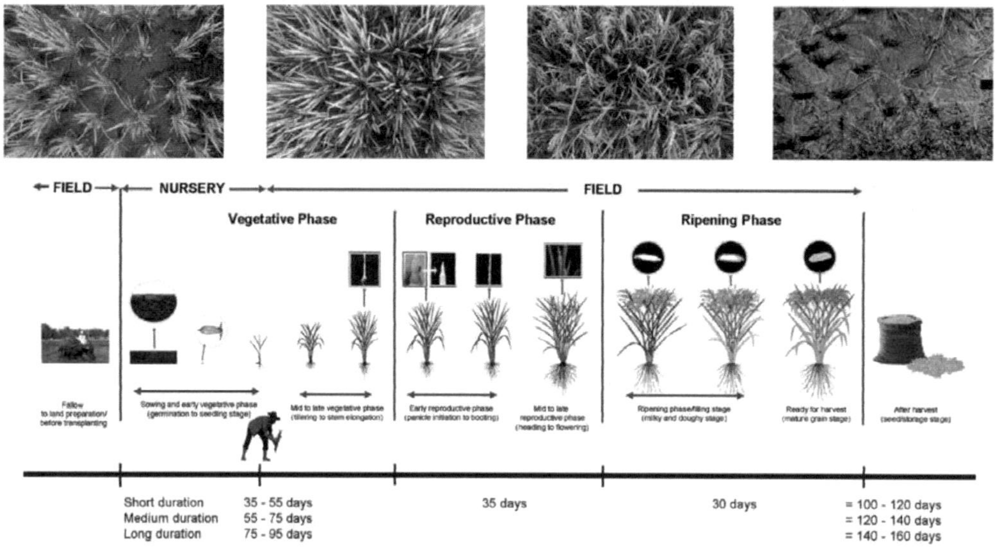

Figure 4. Captured images from field and paddy phenological stage (modified from IRRI-Rice Knowledge Bank)

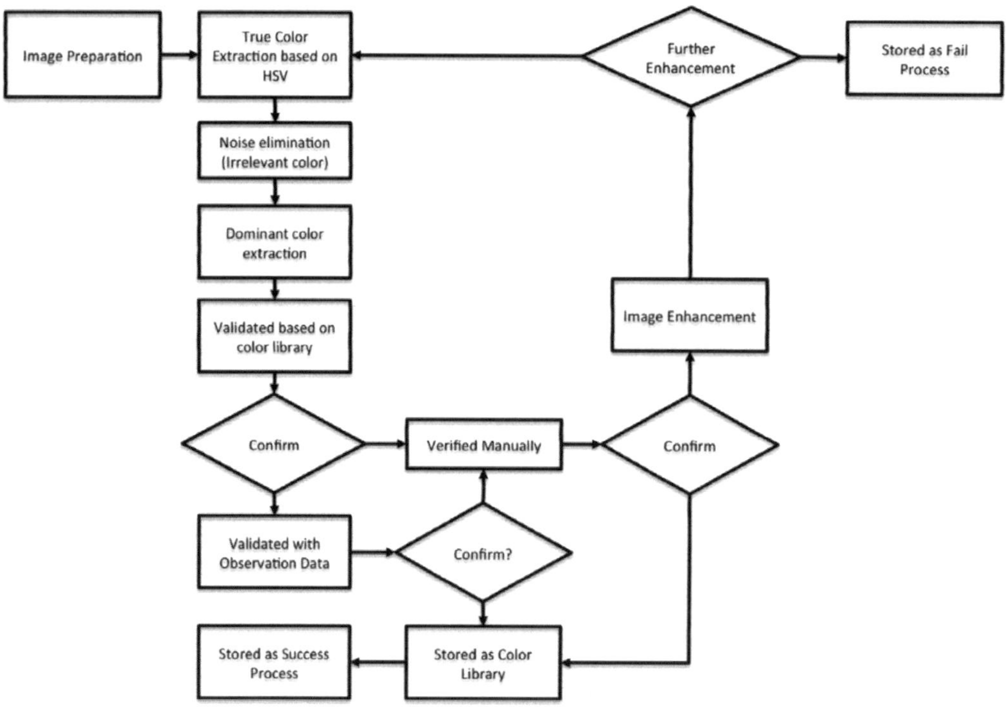

Figure 5. Strategy for Image Processing

Paddy images received and stored in the server are classified into phenological stage (Figure 4). A digital number of color is extracted based on Hue-Saturation-Value (HSV) color space and grouped by K-means clustering. Then, these data are discriminated based on decision trees algorithm [10]. The first step is classified as brown and not brown. Image with 90% brown, 5% green, 5% others are classified as land preparation; image with 60% brown, 20% dark blue or black, 10% green, 10% others are classified into early vegetative (V1); image with 20% brown, 60% green, 10% yellow, 10% others is classified into late vegetative (V2); image with 40% green, 40% yellow, 10% brown, 10% others is classified into generative; whereas image with 75% yellow, 15% green, 10% others are classified into ripening or harvest.

3. Result and Discussion

Testing was conducted and installed for 2,356 observers, and it is found that 73% of observers successfully reported the paddy growth stage from the centre of segment, which is locked in 10 meter accuracy, whereas the other 27% reported not from the centre of segment due to field conditions., It is also found that a combination of GPS and mobile network or assisted-GPS can speed up the positioning.

For the office component, 200 images were captured and stored in the server, which was processed with the developed algorithm. The output of this program is a text file that contains location and file name input, segment code, observation value, result value, similarity matching, ranks of the dominant color, rank of pixel number. Since the classification is related to physical features of paddy growth stage, therefore the output is presented without land preparation class. Figure 6 illustrates the process of image segmentation based on HSV value to extract the true and dominant color for each image.

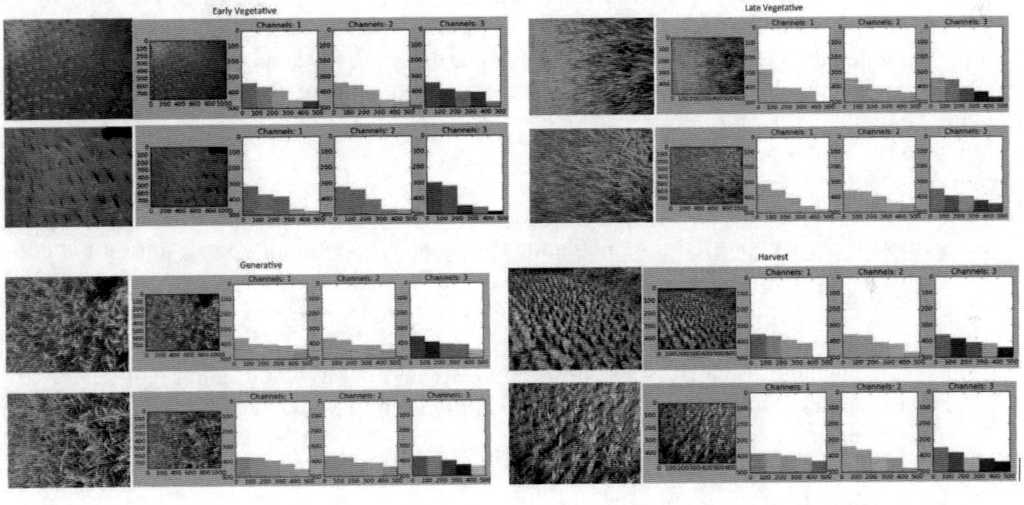

Figure 6. Illustration of Pattern Recognition for Paddy Phenological Stage Based on HSV

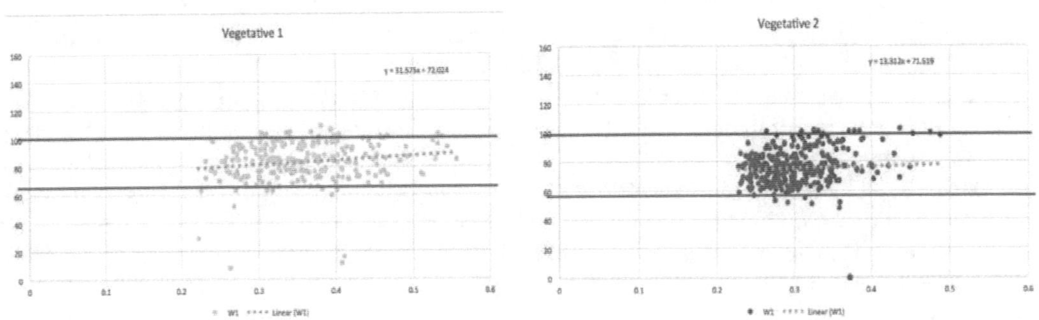

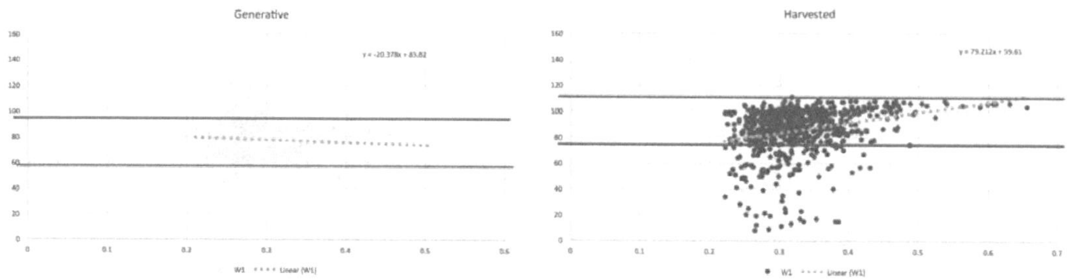

Figure 7. The Spectrum of Dominant Color Related to Paddy Phenological Stage

Figure 6 illustrates the color spectrum distribution based on pattern recognition developed using Python script. The two parallel red lines in Figure 7 represent the threshold of lower and upper limit of color digital number to be classified into phenological stages. By implementing the rules into the data, the accuracy of this system is around 60%. It is found that for vegetative phase the images can be segmented easily, however for generative and harvest phases, the images are more difficult to classify.

4. Conclusions

In urban areas or areas with well-developed communication infrastructure, assisted-GPS mode reaches the segment center faster. However, in areas with no communication signals, assisted-GPS mode displays unsettled coordinates, hence it often causes confusion. Smartphone-based positioning is useful for ensuring the quality of data reported for AFS method. The automatic image classification based on HSV image segmentation still needs improvement since the accuracy is only 60%. The similarity of the prime color of paddy growth stage and non-standardized camera used may be the cause these difficulties.

References

[1] Hartini T N S, Padmawati R S, Lindholm L, Surjono A and Winkvist A 2005 The importance of eating rice. Changing food habits among pregnant Indonesian women during the economic crisis, *Soc. Sci. Med.*, **61** (2005), pp. 199–210.

[2] Yoshida S 1981 *Fundamental of Rice Crop Science*, The International Rice Institute, Los Banos, Laguna, Philippines, 269 p.

[3] Shao Y, Xiangtao F, Liua H, Xiaoa J, Rossc S, Briscod B, Brownc R and Staplese G 2001 Rice monitoring and production estimation using multitemporal RADARSAT, *Remote Sensing of Environment*, **76**, pp. 310-325.

[4] Tan C P, Koay J Y, Ewe H T, Chuah H T and Bahari S 2006 Applications of remote sensing in the monitoring of rice crops, *IEM Journal*, Vol. **67**, No. 4, pp.1-11.

[5] Gallego F J 2015 Area Sampling Frames for Agricultural and Environmental Statistics, EUR 27595 EN, doi:10.2788/88253

[6] Lin C Y and Hung M T 2014 A location-based personal task reminder for mobile users, *Personal Ubiquitous Computing*, vol. **18**, pp. 303-314.

[7] Casey E and Turnbull B 2011 *Digital evidence on mobile devices*. In Eoghan Casey, Digital Evidence and Computer Crime. Third Edition. Forensic Science, Computers, and the Internet, Academic Pres.

[8] Dexter M 2007 *Eclipse and Java for Total Beginners Companion Tutorial*, Licensed under the Educational Community License version 1.0, 45 p.

[9] LaMance J, DeSalas J and Järvinen J 2002 Assisted GPS: A Low-Infrastructure Approach. GPS World. March 1, 2002.

[10] Jiang L, Wang W, Yang X, Xie N and Cheng Y 2011 Classification Methods of Remote Sensing Image Based on Decision Tree Technologies. In: Li D., Liu Y., Chen Y. (eds) Computer and

Computing Technologies in Agriculture IV. CCTA 2010. *IFIP Advances in Information and Communication Technology*, vol **344**. Springer, Berlin, Heidelberg

Acknowledgments
Ministry of Research, Technology and Higher Education – Directorate General of Research and Development for INSINAS-IRPI 2017 funding scheme; Statistics Indonesia for supports, observers and facilities

IOP Conf. Series: Earth and Environmental Science **165** (2018) 012002 doi:10.1088/1755-1315/165/1/012002

Rice Productivity Prediction Model Design Based On Linear Regression of Spectral Value Using NDVI and LSWI Combination On Landsat-8 Imagery

Yudo Prasetyo[1*], Abdi Sukmono[1], Kurnia Wisnu Aziz[1] and Bernardinus Joko Prakosta Santu Aji[1]

[1]Department of Geodetic Engineering, Faculty of Engineering, Diponegoro University Semarang, Indonesia

yudo.prasetyo@ft.undip.ac.id

Abstract. Rice is one of the most important agriculture crops in many countries, and it is a primary food source for more than three billion people worldwide. The global rice consumption is projected to be 873 million tonnes in 2030. Based on national socio-economic survey (susenas) data from Indonesia Statistic Agency (BPS), Indonesia's population of rice consumption 2016 reached 78,816 kg per capita per year. In the recent decades, two major issues like population growth (in particular in the major rice producing or consuming countries) and climate change put enormous pressure on the global food demand and its production. Almost all countries made the issue of food as a major issue. Now, Indonesian government wants to realize Indonesia became self-sufficient of rice in 2017. Some food security programs implemented to achieve it. One of them is predicting or estimation rice production and consumption. Prediction or estimation rice productivity before harvest is crucial, especially in regions characterised by climatic uncertainties. It is also enables governments to put in a good strategic contingency plans for the redistribution of food during times of famine. Satellite remote sensing has been widely applied and recognised as a powerful and effective tool for identifying agriculture crops. An important goal of agricultural remote sensing research is to spectrally estimate crop variables related to crop conditions, which can subsequently be entered into crop simulation and production models. In this research will be use a global and medium resolution of satellite remote sensing imagery (Landsat 8 ETM+) with 2016-2017 acquisition year in order to estimation of rice productivity. The rice productivity prediction modeling methods will be build in this research based on regression and spectral pattern analysis of NDVI (and LSWI (Land Surface Water Index in multi temporal satellite imagery. In addition, this prediction model will be tested against fact the level of productivity in the rice field. We will take a proportional field sampling test using cluster sampling test and random spectral value checking in order to measure accuracy level of rice productivity prediction model. The results of this research indicate that NDVI and LSWI algorithm is the best combination of linear regression in estimation of rice productivity level with coefficient of determination equal to 0.639. Validation of regression model equation to data of Agriculture Agency of Demak has difference or RMSE 8,394 Quintal/ Ha

1. Introduction

Rice is one of the most important agriculture crops in many countries, and it is a primary food source for more than three billion people worldwide [1]. The global rice consumption is projected to be 873 million tonnes in 2030[2]. In the recent decades, two major issues like population growth (in particular in the major rice producing/consuming countries) [3] and climate change put enormous pressure on the global food demand and its production [4]. Almost all countries made the issue of food as a major issue. Various food security programs promoted in order to maintain food sovereignty.

Indonesia is one of producer and consumer of rice. Based on national socio-economic survey (SUSENAS) data from Indonesia Statistic Agency (BPS), Indonesia's population of rice consumption 2016 reached 78,816 kg per capita per year [5]. It is assessed by the Ministry of Agriculture is fairly high consumption. In fact, more than world rice consumption is only 60 kg per capita per year. In Asia, Indonesia's rice consumption is the highest. Indonesian rice production to meet domestic demand for rice in the range of 1980 to 2008. After that the needs of Indonesian rice deficit forced the government to import rice from abroad. Now, Indonesian government wants to realize Indonesia became self-sufficient of rice in 2017. Some food security programs implemented to achieve it. One of them is predicting or estimation rice production and consumption.

Prediction or estimation rice productivity before harvest is crucial, especially in regions characterised by climatic uncertainties. Forecasting enables planners and decision makers to predict how much to import in the case of a shortfall or, optionally, to export in the case of a surplus. Prediction and estimation also enables governments to put in place strategic contingency plans for the redistribution of food during times of famine. Therefore, the monitoring of crop development, crop growth and early productivity prediction are generally of great importance [6]. Indonesian territory is large enough. This makes an early estimate of rice production is very important.

Most common and widely used methods for estimating rice production are the use of agricultural statistical data acquired through field visits and interviewing the farmers. The methodology for mapping area under rice cultivation is basically done through annual/seasonal sample surveys based on a number of sample clusters that are constituted all over the country for measuring cultivated area during the crop growing season. Each cluster is visited many times and areas are recorded by the field staffs, checked, and then processed by regional statistical officers. Despite its invaluable ability for understanding historical trends in rice area, this method is extremely tedious, time-consuming, less precise, costly, inconsistent, too generalized and labour-intensive [7]. In most countries, the data become available too late for the appropriate actions to be taken to avert food shortage [6]. In addition, this method unable to provide information in timely manner as well as unable to delineate detailed geographical spatial distribution of areas under rice cultivation.

Satellite remote sensing has been widely applied and is recognised as a powerful and effective tool for identifying agriculture crops [8]. An important goal of agricultural remote sensing research is to spectrally estimate crop variables related to crop conditions, which can subsequently be entered into crop simulation and production models. Some research use global and medium resolution of satellite remote sensing imagery in estimation of rice productivity, such as MODIS, NOAA and Landsat ETM+. Nuarsa (2012) [9] use of remote sensing in rice productivity in Tabanan, Bali. In this research. This research develops estimation of rice productivity in the next generation of Landsat satellite imagery that Landsat-8 satellite imagery. Beside that, the modelling methods will be used in this research based on regression and spectral pattern analysis of NDVI and LSWI multi temporal

Based on the above introductory description as for the formulation of the problem of this research are:

1. How the right model for rice productivity estimation before harvesting through Landsat-8 satellite imagery?
2. How the productivity of rice in study area (Demak Regency) at 2017?

2. Methodology

2.1. Description of Study Area

The study was conducted in Demak of the Province of Central Java, Indonesia. Centred at latitude 6° 43' 26"–7° 09' 43" and longitude 110° 27' 58 "-110° 48' 47" and is located approximately 25 km east of Semarang City (Figure 1). The longest distance from West to East is along 49 km and from North to South along the 41 km [5].

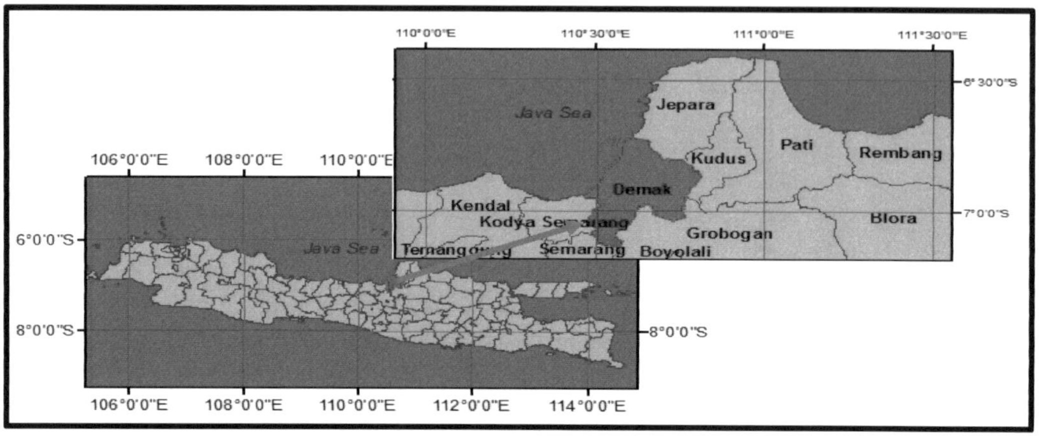

Figure 1. Location of study area

The Demak Regency was selected as the study area because Demak is the one of central area for rice production in Central Java. Based on BPS (Indonesian Central Statistics Agency) data 2016, rice production of Demak Regency in 2015 reached 653547 tons. It was 3th rice production in Central Java after Sukoharjo Regency and Cilacap Regency. Beside that, Demak is located near from Semarang City. This makes Demak regency as food stock or food buffer zone for Semarang City [5]. Demak has twice the rice planting season, the rainy growing season and growing season during the dry season. Rice planting patterns simultaneously both for wetland paddy nor dry land. In Demak Almost 95 percent of paddy field are wetland paddy [10].

The harvested area of paddy of Demak in 2015 reached 98.618 hectares. If compared with previous year, it increased 2.01 percent. In 2015, paddy production was 653547 tons dried paddy. That increased 15.11 percent, compared with previous year. In 2015, the productivity of paddy was about 66.27 quintal per hectare, it increased 12.75 percent compared with 2014 as 58.73 quintal per hectare [10].

2.2. Data Collection

Data collection includes spatial data collection and data collection non spatial. Spatial data includes Landsat 8 satellite image data (obtained via download from glovis.usgs.gov), Administration Map of Demak (obtained from Development Planning Agency of Demak/BAPPEDA), Indonesian Topographical map (RBI) of Demak with scale 1:25.000 (obtained from Indonesian Spatial Agency/BIG). Non spatial data include rice productivity data (obtained from the Department of Agriculture), Biophysical data parameters (growth stage) obtained from rice planting calendar.

2.3. Method

Method of this study was followed the flowchart in Figure 2. It was divided in to four step, that were image preprocessing step, image processing and model development step, statistical testing and validation step, and model aplication step.

The principle method of this research was prediction of rice productivity using pattern value of NDVI, EVI and LSWI of rice plant before harvest in vegetative and generative growth stage. Because both in vegetative and generative growth stage were important phase of rice to growth. In this study, three images Landsat-8 multi temporal, i.e. the image on May 29, 2015 as time 1, the image on June 14, 2015, as time 2 and the image on June 30, 2015 as time 3 were used to make a Model rice productivity. The models aqcuired using linear regression betwen Index (NDVI, EVI, LSWI) multi temporal (time 1, time2 and time 3) and rice productivity. Then this model were tested using statistical testing to choose the best model. It was validated using independent rice productivity, with the requirement of RMSE is 10 quintal/ha. After that the model was aplicated to predict of rice productivity on rice planting II, 2017.

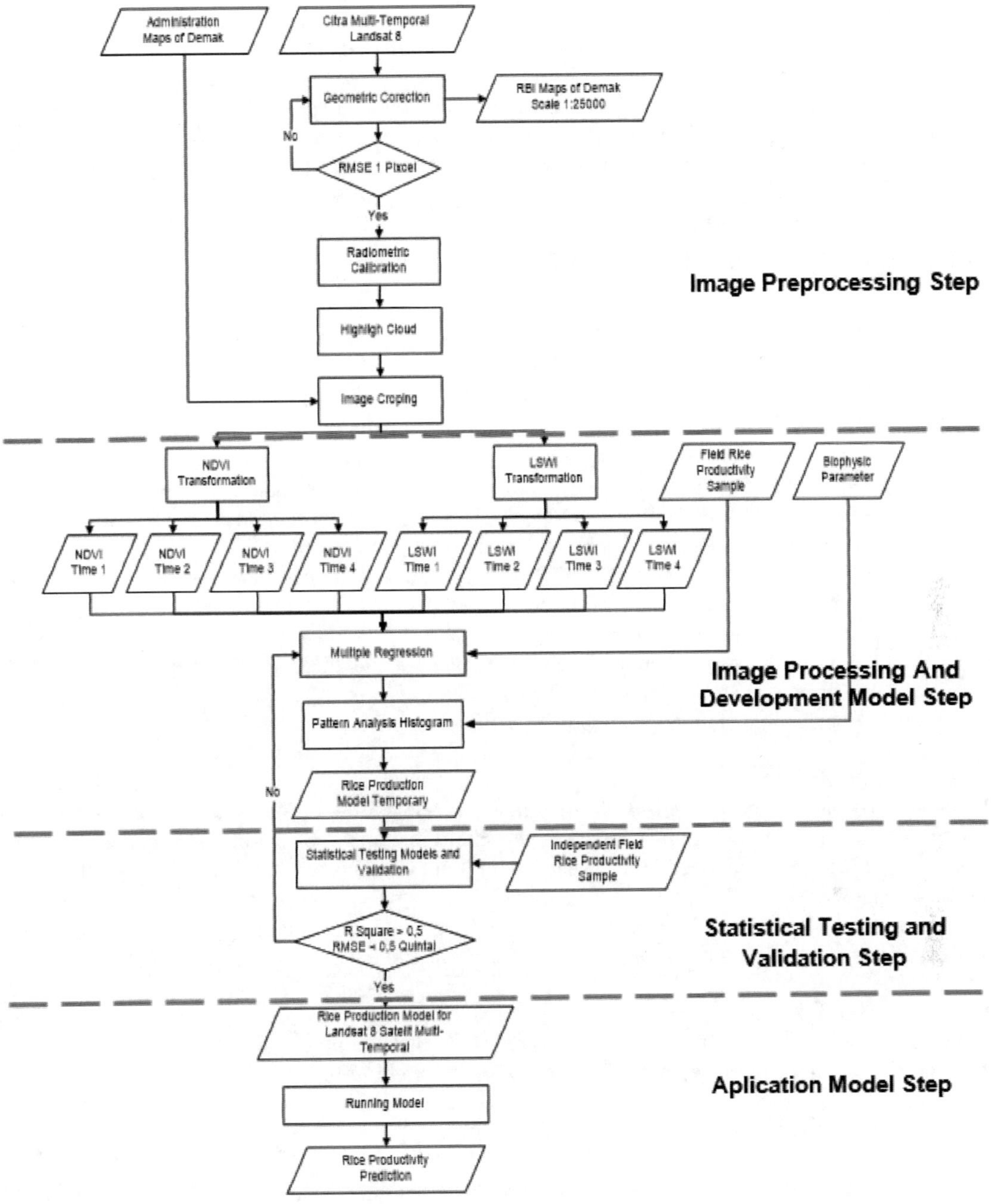

Figure 2. Flowhchart of the research

2.3.1 Preprocessing Image Method

a) Geometric Corection

Geometric correction geo-referenced image to real world coordinates. The geometric correction method of the image used image to map rectification with reference Indonesian Topographical map (RBI) of Demak, scale 1:25,000. Ground control point (GCP) on this research taken as many as 16 points of each image. Root mean square error (RMSE) in this process were required maximum 1 pixel.

b) Radiometric correction

This stage corrected image using the conversion of digital number to Top of Atmosphere (TOA) Reflectance. OLI band data of Landsat 8 can be converted to TOA planetary reflectance using reflectance rescaling coefficients provided in the product metadata file (MTL file). MTL is the format of metadata file for Landsat 8 imagery. The following equation is used to convert digital number (DN) or pixel values to TOA reflectance for operational land imager (OLI) data as follows [11]:

$$\rho\lambda' = M\rho Qcal + A\rho \quad (1)$$

TOA reflectance with a correction for the sun angle is then [6]:

$$\rho\lambda = \frac{\rho\lambda'}{\cos(\theta SZ)} = \frac{\rho\lambda'}{\sin(\theta SE)} \quad (2)$$

where:

$\rho\lambda'$ = TOA planetary reflectance, without correction for solar angle. Note that $\rho\lambda'$ does not contain a correction for the sun angle

M_ρ = Band-specific multiplicative rescaling factor from the metadata (REFLECTANCE_MULT_BAND_x, where x is the band number)

A_ρ = Band-specific additive rescaling factor from the metadata (REFLECTANCE_ADD_BAND_x, where x is the band number)

Q_{cal} = Quantized and calibrated standard product pixel values (DN)

$\rho\lambda$ = TOA planetary reflectance

θSE = Local sun elevation angle. The scene center sun elevation angle in degrees is provided in the metadata (SUN_ELEVATION)

θ_{SZ} = Local solar zenith angle; $\theta_{SZ} = 90° - \theta_{SE}$

c) Highlight cloud

Clouds and shadows in the image is a problem that can't be avoided. This is an obstacle in the use of medium resolution imagery such as Landsat 8, especially in equatorial regions. This causes the loss of some of the information contained in the images. Highlight Cloud is used to discriminate or distinguish clouds with land cover [12]. Because of the clouds in the application or interpretation of land use is considered as interference to other land cover reflectance values, so that should be eliminated in the analysis. In this study, the clouds separated by digitization after sharpening cloud value.

d) Image Cropping

Cropping images do reduce the file size of the image so that the processing of data to be lighter and faster. It are made in accordance with the required area or region of interest (ROI). Moreover, in this study uses data multi-temporal for classification process, so cropping of the image is needed in order to accelerate the image processing [13].

2.3.2 Image Processing and Model Development Method. In this research, the value of reflectance TOA are transformed to other spectral value using transformation spectral. Transformation method by using Normalized Difference Vegetation Index (NDVI) Enhanced Vegetation Index (EVI) and Land Surface Water Index (LSWI) will be used in each image. The results of image transformation at several different times are combined for a single image with three band (NDVI time1, NDVI time 2, NDVI time 3) and second image (LSWI time 1, LSWI time 2, LSWI time 3). The models aqcuired using linear regression betwen Index (NDVI, EVI, LSWI) multi temporal (time 1, time2 and time 3) and rice productivity in 50 point. The best model combination was chosen based on the highest value of coeficient determination in several models.

2.3.3 Statistical Testing and Validation Method. Several statistical testing were used to test the model. The statstical testing used were Clasical assumtion test (normality, multicolinearity, heteroscedasticity and autocorrelation) and F test. Beside that, models that have been formed need to be validated to determine the quality of the model. Validation the model using 15 point of rice productivity independent data. So it is known that the difference between the estimation results with the data of Agriculture Department of Demak Regency and RMSE.

3. Result and Discussion

3.1 Development Model of Rice Productivity Estimtion

The model development of the relationship between independent variables (EVI, NDVI LSWI multi temporal before harvest) and the dependent variable of rice production using multiple linear regression method. Number after index vegetation showed time 1, time 2 and time 3 (example :NDVI 1 as NDVI time 1). In the regression process is also done a classical assumption tested to ensure the right statistical rules. After that the model was chosen by observing the highest coefficient of determination. This coefficient shows how strong the model is able to represent the bound variables of the independent variable. The model feasibility tested was used as a test tool to ensure that this model is truly feasible for estimating rice production using multi-temporal landsat-8 satellite image data.

Based on the relationship between rice productivity variables with NDVI transformation value, EVI and LSWI at multitemporal 3 time before harvest period can be made linear regression model as follows in Table 1.

Table 1. The Regression Result of Rice Productivity Estimation Model

No	Parameters	Equation of Regression	(R^2)
1	**NDVI** (NDVI 1, NDVI 2, NDVI 3)	$Y = 99{,}030 - 236{,}143X_1 + 37{,}624X_2 - 56{,}722X_3$	0.38
2	**EVI** (EVI 1, EVI 2, EVI 3)	$Y = 80{,}253 - 22{,}57X_1 + 22{,}684X_2 - 19{,}463X_3$	0.21
3	**LSWI** (LSWI 1, LSWI 2, LSWI 3)	$Y = 68{,}461 + 8{,}133X_1 - 23{,}064X_2 + 28{,}567X_3$	0.07
4	**NDVI and LSWI** (NDVI 1, NDVI 2, NDVI 3, LSWI 1, LSWI 2, LSWI 3)	$Y = 101{,}027 - 9{,}185X_1 + 16{,}334X_2 - 72{,}067X_3 - 38{,}408X_4 - 54{,}432X_5 + 45{,}716X_6$	0.639
5	**EVI and LSWI** (EVI 1, EVI 2, EVI 3, LSWI 1, LSWI 2, LSWI 3)	$Y = 93{,}929 - 20{,}114X_1 + 16{,}867X_2 - 26{,}437X_3 + 1{,}135X_4 - 44{,}631X_5 + 24{,}95X_6$	0.413

Based on the regression result can be seen that the highest coefficient of determination resulted from the linear regression between rice productivity as the dependent variable and the value of NDVI and LSWI as independent variables. The coefficient of determination approach 0.639. This model has also tested the classical assumptions and required all assumptions. Therefore, in this research will be analyzed further is the relationship between rice productivity variables with the value of NDVI and LSWI. Based on regression, the equation model are:

$$Y = 101{,}027 - 9{,}185X_1 + 16{,}334X_2 - 72{,}067X_3 + 38{,}408X_4 - 54{,}432X_5 + 45{,}716X_6 \quad (3)$$

Where :

Y : The rice productivity (quintal/ha)
X_1 : The value NDVI time 1
X_2 : The value NDVI time 2
X_3 : The value NDVI time 3
X_4 : The value LSWI time 1
X_5 : The value LSWI time 2
X_6 : The value LSWI time 3

3.2 Statistical Testing Result

3.2.1 Classical Assumtion Test Result. The classical assumption test is a test of statistical assumptions that must be required in multiple linear least squares regression (MLS). A classical assumption test is needed to determine whether the linear regression estimation results is done from the error. In this study the classical assumption test performed is the normality test, multicolinearity test, autocorrelation test and heteroscedasticity test. The linear regression estimation model must satisfy the test so that the resulting equation is of good quality. The result of classical assumption test can be seen in Table 2.

Table 2. The summary result of classical Assumtion Test

Classical Assumtion Test	Tested Method	Value	Requirement	Conclusion
Normality	*Komolgorov-Smirnov* Method	Sig value parameters in range 0.087-0.2	sig value> 0.05	Required
Heteroscedasticity	Glejser method	Sig value parameters in range 0.099-0.857	sig value> 0.05	Required
Multicolinearity	Variance Inflection Factor (VIF) Method	VIF value parameters in range 1.692-4.236	VIF value < 10	Required
Autocorrelation	Durbin Watson method	Durbin Watson value 2.152	Durbin-Watson count is greater than 1.784 and smaller than 2.216	Required

3.2.2 F-Test Result. The model simultaneous test or model feasibility test or more popularly referred to as F test is a step to identify the estimated regression model is feasible or not. Worthy (reliable) here means a model that is feasible to be used to explain the effect of independent variables on the dependent variable. The test name is referred to as the F test, as it follows the following F distribution whose testing criteria are like One Way Anova.

Table 3. The Result of F Test

	Model	*Sum of Squares*	*Df*	*Mean Square*	*F*	*Sig.*
1	Regression	1186,129	6	197,688	15,457	.000[b]
	Residual	549,961	43	12,790		
	Total	1736,090	49			

Taking conclusions on the F test is done by observing the sig value of the regression model made. In table 3, it can be seen that the model sig value is 0.000. It shows that the value of sig model meets the first requirement of sig <0,05. So Ho is rejected and Ha accepted. Therefore it can be concluded that the resulting linear regression model is feasible to be used to predict the dependent variable.

3.3 Validation

Multiple linear regression models generated to predict rice productivity need to be validated to find out how the quality of the model when applied. To perform such validation by entering the value of free variables are NDVI and LSWI on the model. Furthermore, based on the results of the calculation can be known the difference between the results of calculations with data from the Department of Agriculture Demak Regency. And can be known RMSE (Root Mean Square Error) from the results of the calculation data. The result calculation in 15 sample data there is total difference of rice productivity equal to 45,742 quintal / Ha between data from Agricultural Service and result of model calculation. In the 15 samples used it has an RMSE value of 8.394 quintal / Ha. RMSE value 8,394 quintal / Ha is still categorized as a medium error. The error rate is between 10-15%. This could be due to a sample in the form of secondary data that have no coordinates so that the location is not accurate. Therefore, further research is needed which can take samples of rice productivity directly in the field. The sampling must have the exact coordinates.

3.4 Aplication Model

The model was reliable to estimate the productivity because after compared with the rice productivity 2016 from Department of Agriculture [14], it has different 10% error. The error can be caused from the sample from the secunder data statistic. The research [15] can result best model until minimal error 4,2 % with the same parameter (LSWI and EVI). It used modis data. [15] used rice productivity from in situ or primary data. Although this research has not yet reached the best model but with a 10% error can still be tolerated for initial estimation of rice productivity.

The resulting model is applied to the image in 2017. The objective of the implementation is to know the estimation of rice productivity in Demak Regency in 2017, planting season II (May-August). The productivity estimate will then be explained and visualized to obtain the distribution of the productivity value and the extent of each class.

Estimation of rice productivity in 2017 is done by entering the value of NDVI and LSWI (multi temporal three different time) in 2017 on the equation of multiple linear regression model that has been produced. The values of NDVI and LSWI are applied to the equations of multiple linear regression models that have been made. The result of model implementation resulted that the smallest rice productivity estimation in 2017 is 28,246 quintal/ ha. While for estimation of the biggest rice productivity is 155,461 quintal/ ha. The average on the estimate of rice productivity in 2017 amounted to 78.904 quintal / ha.Based on the result, a class can be used to determine the level of rice productivity. In this study, 3 classes were created to differentiate the level of rice productivity in Demak Regency by 2017. The calculation of the range for each class can be done by reducing the largest and smallest value in a data and dividing it by the desired number of classes. Range of rice productivity level in 2017 can be seen in table 4.

Table 4. Classification Rice Productivity level In 2017 based statistical method

Level of Rice Productivity	Range level productivity (Quintal/Ha)
Low Productivity	28,246 - 70,651
Medium Productivity	70,652 - 113,056
High Productivity	113,057 - 155,461

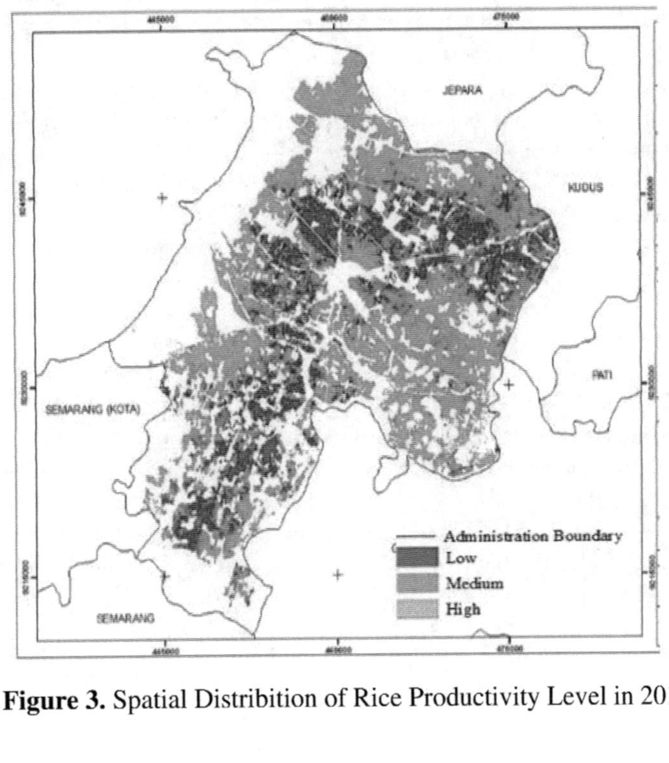

Figure 3. Spatial Distribibition of Rice Productivity Level in 2017

Based on Table 4, the distribution of productivity levels in 2017 can be visualized as shown in Figure 3. Based on figure 3 can be known the area of each class. In 2017, Demak District has a total area of rice fields of 47632.847 hectares of rice productivity estimates. The area consists of low rice productivity level of 8255,778 Ha, the level of medium rice productivity is 39010,413 Ha and high rice productivity level of 366,655 Ha.

4. Conclusion

This research resulted model of rice productivity estimation before harvesting using Landsat-8 with parameter NDVI and LSWI. The model was reliable to estimate the productivity of rice with determinant coefficient of 0.639 and average error 10%. So from the model that is able to show that the value of NDVI and LSWI affect the value of rice productivity as much as 63.9% while the rest of 36.1% is caused by other factors. Other factors could be the existence of pest and fertilizer application which is not discussed in this research. This model can be aplied to predict the productivity of rice In Demak at 2017. The model could show that Demak Regency has a total area of rice fields of 47632.847 hectares of rice. The area consists of low rice productivity level of 8255,778 Ha, the level of medium rice productivity is 39010,413 Ha and high rice productivity level of 366,655 Ha.

References

[1] Khush G S 2005 What it will take to feed 5 billion rice consumers in 2030. *International Journal of Molecular Biology*, Vol **59**, 1–6.
[2] Purevdorj M and Kubo M 2005 The future of rice production, consumption and seaborne trade: Synthetic prediction method. *Journal of Food Distribution Research*, **36** (1): 250-259.
[3] GeoHive—Historic, Current and Future Population: Asia. Available online: http://www.geohive.com/earth/his_proj_asia.aspx. (Accessed on 4 April 2017).
[4 Fifth Assessment Synthesis Report; IPCC: Geneva, Switzerland, 2014; p. 116. Available online: http://www.ipcc.ch/pdf/assessment-report/ar5/syr/SYR_AR5_LONGERREPORT.pdf. (Accessed on 4 April 2017).
[5] BPS (Center Statistic of Indonesia) 2014 *Demak in Figure 2014*. Demak: BPS.
[6] Sawasawa Haig L S 2003 Crop Yield Estimation: Integrating RS, GIS, and Management Factor. A case study of Birkoor and Kortigiri Mandals, Nizamabad District India. Thesis of Master Degree in ITC Enchede: Netherland.
[7] Prasad A K, Chai L, Singh R P and Kafatos M 2006 Crop yield estimation model for Iowa using remote sensing and surface parameters. *International Journal of Applied earth observation and geoinformation*, Vol **8**: 26-33.
[8] Mekong Delta Using ENVISAT/ASAR Dual Polarization Data. *IEEE Trans. Geoscience and Remote Sensing*, 47, pp. 517 –526.
[9] Nuarsa I W, Nishio F and Hongo C 2012 Rice Yield Estimation Use Landsat ETM+ and Field Observation. *Journal of Agriculture Science*, Vol **4**, No 3, 45-56.
[10] Department of Agriculture 2015 *Conditions of Land Agriculture in Demak Th. 2014-2015*. Demak: Department of Agriculture.
[11] USGS 2013 Using the USGS Landsat 8 Product. http://Landsat.usgs.gov. 10 April 2015.
[12] Putra E H 2011 *Remote Sensing Using ERMapper*. Yogyakarta: Graha Ilmu.
[13] Sukmono Abdi and Ardiansyah 2017 Identification of rice field using Multi-Temporal NDVI and PCA method on Landsat 8 (Case Study: Demak, Central Java). *IOP Earth and Environmental Science* Vol 54 paper 012001.
[14] Department of Agriculture 2017 Rice Productivity in 2016. Agricultural Agency: Demak
[15] Wijesingha J S J, N L Deshapriya, L Samarakoon 2015 Rice Crop Monitoring And Yield Assesment With Modis 250m Gridded Vegetation Product : A Case Study in Sa Kaeo Province Thailand. *The International Archives of The Photogrammetry, Remote Sensing and Spatial Information Science*, Volume XL-7/W3.

IOP Conf. Series: Earth and Environmental Science **165** (2018) 012002 doi:10.1088/1755-1315/165/1/012002

Acknowledgments
We extend our thanks to remote sensing and photogrammetry laboratory Diponegoro University for all their help in data processing equipment and thanks to Agricultural Agency of Demak Regency for all their help in suporting data.

IOP Conf. Series: Earth and Environmental Science **165** (2018) 012003 doi:10.1088/1755-1315/165/1/012003

Study of Spatial Variation of Ocean Surface Waves by Using SAR Images in Coastal Area

A Sulaiman[1*], Donysius B Sencaki[1], Awaludin[1] and A Darmawan[1]

[1]Pusat Teknologi Pengembangan Sumberdaya Wilayah-BPPT, Geostech 820, Kawasan Puspiptek Serpong, Tangerang Selatan, Indonesia

albertus.sulaiman@bppt.go.id

Abstract. A temporal pattern or characteristic of the ocean surface wave can be learned from bouy observation. In terms of spatial variations, the use of Synthetic Aperture Radar (SAR) image is considered to be the most efficient as SAR is not weather dependent. This paper aimed to study the pattern of ocean waves with study area in Jakarta coast by using SAR image. We focused on spatial variation so that we emphasized in the study of the wave pattern in the area close to the coastline and away from the coast. The low pass filter was applied to reduce the noise and the discontinuity in wave images can be removed as well. The two dimensional Fast Fourier Transform and the dispersion relation were applied to get the wave spectrum and then the amplitude of ocean wave was successfully obtained. By comparing the spectral ocean wave model, the pattern of ocean waves can be understood. The results showed that in the area close to the coastline, the random wave behavior was not dominant, where a single wave pattern as a result of the nonlinear process or the wave-wave interaction appeared. Meanwhile, in the offshore area, the random effect was very dominant.

1. Introduction

The growth of coastal town will give the change of shoreline so that the detailed information about the pattern of ocean waves is needed. The shoreline dynamic is mostly influenced by ocean waves. Therefore, the knowledge of ocean wave climate and dynamics are indispensable for developing shoreline management plan. The information of ocean wave can be obtained by using wave gauge measurement. This measurement gives detailed information temporally but weak spatially [1]. This weakness can be handled by distributing many wave gauges onto the ocean but this is not a cost-effective way. The use of radar satellite for observing ocean surface wave fills this weakness.

The spatial and temporal variation of ocean surface waves can be retrieved from SAR images. In this image, mesoscale phenomena can be observed i.e. waves with the periods between 3s~30s (wind sea to swell). The SAR image consists of two-dimensional electromagnetic backscatter of the ocean surface. The wave backscatter energy from SAR image depends on wavelength, polarization, geometry, and roughness of ocean surface [2]. The wave provides the change of ocean surface so that following the fundamental principle, by knowing roughness generated from SAR image, the information of ocean surface wave can be obtained.

Currently, the knowledge of extraction of surface wave from SAR image was already developed [3,4]. In this paper, we used similar manner to investigate the ocean surface wave dynamics in Jakarta Bay. Jakarta is the largest coastal city in Indonesia that is growing rapidly so that the information of ocean wave is very important. The wave information of wave climate usually is obtained by wave gauge measurement which has good temporal variation but less on spatial variation. This lack may be equipped with the satellite data. The satellite data that is often used for the wave information is Topex-Poseidon but it had a low resolution so that not adequate for Jakarta bay. The satellite data most adequate to this proposed is SAR. We used the SAR images to see the behavior of the wave dynamics for both offshore areas and the onshore area in Jakarta Bay. The paper is organized as follow, the data and data processing are presented in sec-2. The data analysis and discussion are described in sec-3. The paper is ended by a conclusion

2. Method

2.1. Data

The study area was in Jakarta bay that is located in the north coast of Java with latitude around 106°50E and longitude 05°09S. The bay is relatively shallow and flat, with an average depth about 15m. It has many coral islands, reef-flats and related shallow-platforms occur in this bay. We used ALOS-Palsar II with HH polarization image on January, 26 2015 that is depicted in Figure 1. The data was provided by JAXA under ALOS Research Announcement Project number PI417002 entitled "Sea Surface Roughness Identification based on Synthetic Aperture Radar Data in Indonesian Waters". The study area are represented in white box for both offshore and close to shoreline.

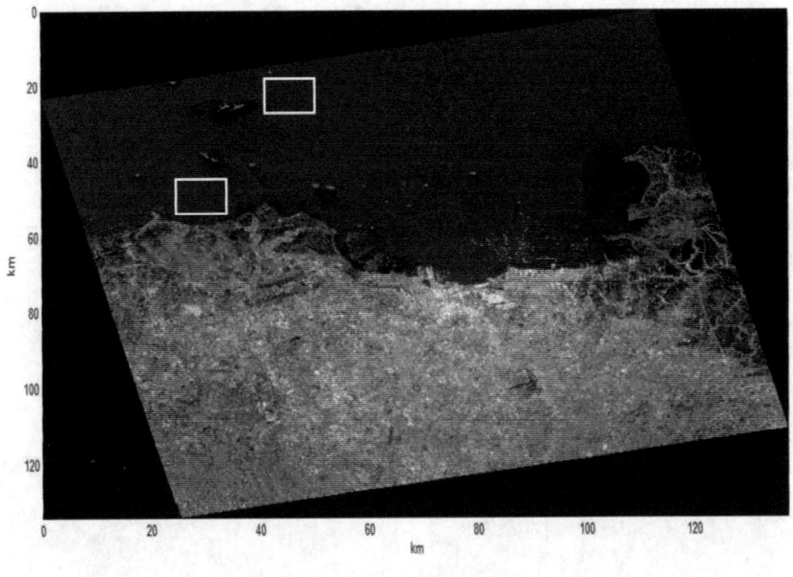

Figure 1. SAR images of ALOS Palsar II with HH polarization on January, 26 2015.

2.2. Data Processing

The Data processing chain for ScanSAR mode (wide-swath dual polarization, 1.5 level) was started by converting the ALOS-2 CEOS data format to EAM-DIMAP data format. To reduce the speckle noise after radiometric calibration then the Refined Lee filter was applied [5]. The result was a normalized radar cross section (sigma naught $\sigma 0$) in terms of decibel (dB) value that is depicted in Figure 1. The first step was to capture the study areas in the offshore region and close to shoreline region (see the white box in Figure 1). The statistical behavior of surface wave was investigated from backscatter distribution of sigma naught along with the wave crest. Then the probability function (pdf) can be calculated.

A single SAR image contains a lot of kind of ocean surface waves that are described in the form of its wavelength. The characteristic of each wave can be seen from the spectrum of the wave numbers. But the SAR image spectrum does not represent the real ocean surface displacement spectrum. The relation between them is expressed by the modulation transfer function T(k) (MTF) where k= . The relation is usually not linear and difficult to obtain. From one SAR image (radar cross section σo), we can observe the directional wave spectrums in which the ocean wavelength and direction are evaluated. It means that from SAR image, we evaluate the wave number spectra rather than frequency spectra which are intensively observed by wave gauge. The 2D Fast Fourier Transform (FFT) is used to get wave number spectrum S(kx,ky) that represents the wave energy. The relation between SAR spectrum and ocean surface wave spectrum Ψ(kx,ky) is given by [4],

$$S(k_x,k_y) = T(\vec{k})\Psi(k_x,k_y) \tag{1}$$

In two dimensional, the wave has the energy and the direction that is represented in term of directional spectrum $\Psi(f,\theta)$. This relation is between spatial wave spectrum and temporal wave spectrum [3]

$$\Psi(f,\theta) = \Psi(k_x,k_y)\frac{4\pi k\sqrt{gk\tanh(kh)}}{gk\tanh(kh) + gkh\sec h^2(kh)} \tag{2}$$

where g is the acceleration of gravity and h is water depth. The 2D FFT is applied to get dominant wavelength before applying the averaging filter. The band pass filter is constructed based on the equation (2). We use MTF linear Equation (1) with T(k)=105 to get ocean wavenumber spectrum. Finally, the inverse Fourier transform is applied to get the wave amplitude.

3. Result and Discussion

We studied two interest areas i.e. offshore and close to the shoreline. They are depicted in Figure 2 and Figure 3. The wave that is observed by SAR image is usually generated by wind force that propagates from offshore and dissipated by the change of bathymetry and wave breaking. The previous measurement showed that the wave characteristics of Jakarta bay were determined by the wind velocity [6]. Most of the wave had 1-10 dm for wave height, 1-8 sec for wave periods and 20-30 m for the wavelength [6]. When the waves propagate close to the coast then the wave direction is perpendicular to the shoreline. The behavior of radar cross-section of the offshore and near coastline region is depicted in Figure 2 and Figure 3.

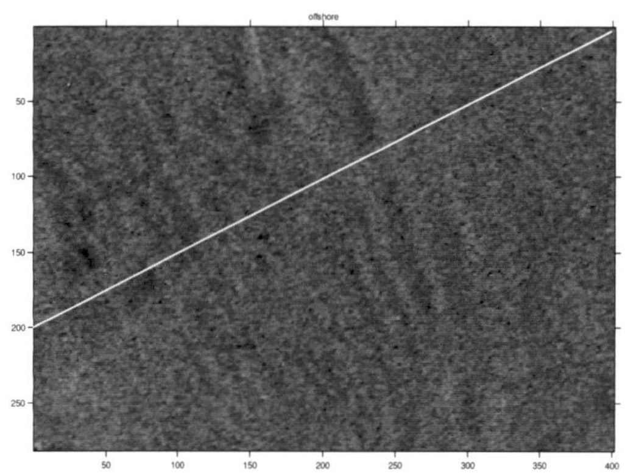

a)

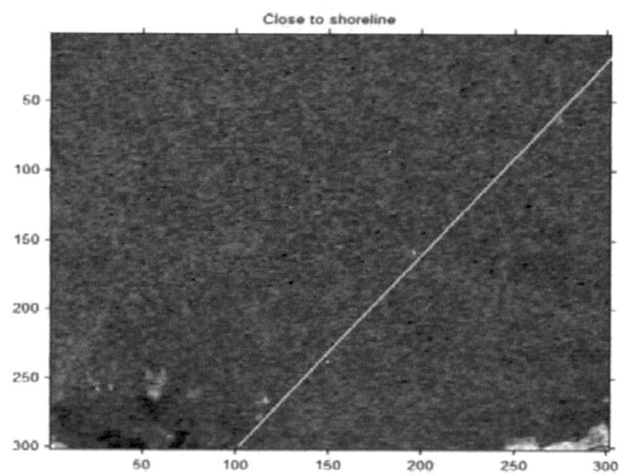

d)

IOP Conf. Series: Earth and Environmental Science **165** (2018) 012003 doi:10.1088/1755-1315/165/1/012003

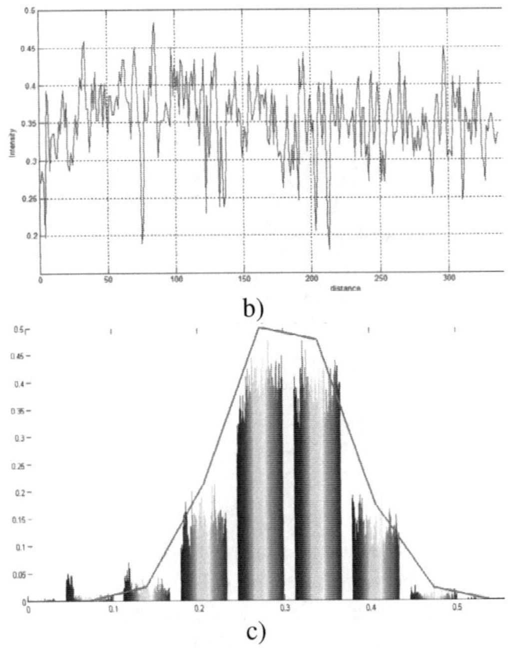

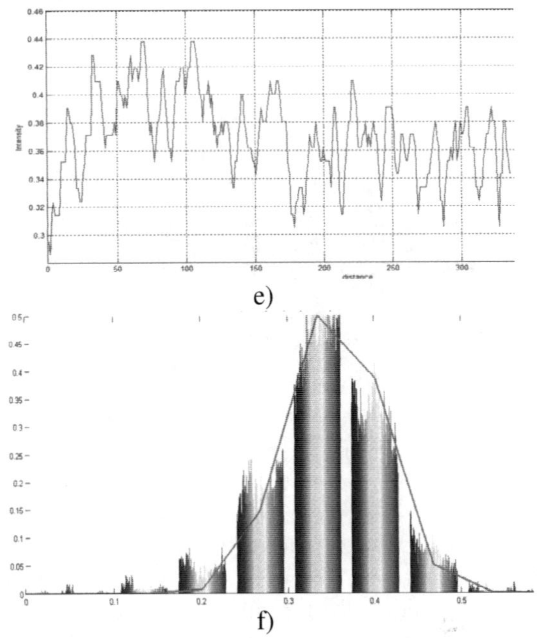

Figure 2. a) The capture of offshore region corresponds to white box in Fig.1. b) The backscatter value of white line and c) represents the probability density function of backscatter distribution of white line.

Figure 3. d) The cropping of onshore region corresponds to white box in Fig.1. e) The backscatter value of white line and f) represents the probability density function of backscatter distribution of white line.

For the offshore region the radar cross section had pdf in the form of $\rho=1/(2\pi\sigma)\exp(-(x-\mu)2/\sigma2)$ where the mean of deviation standard σ was about 0.0038 and the peak displacement $\mu = 0.31$ m-1. This form was a Gaussian distribution. In the other hand, for the region close to the shoreline the pdf had the form of $\rho=1/(2\pi\sigma)\exp(-(x-\mu)2/\sigma2)-0.01$ where the mean of deviation standard σ was about 0.0046 and the peak displacement was $\mu = 0.34$ m-1.

The 2D FFT and the wave amplitude of offshore are depicted in Figure 4 where area close to the shoreline is depicted in Figure 5.

.

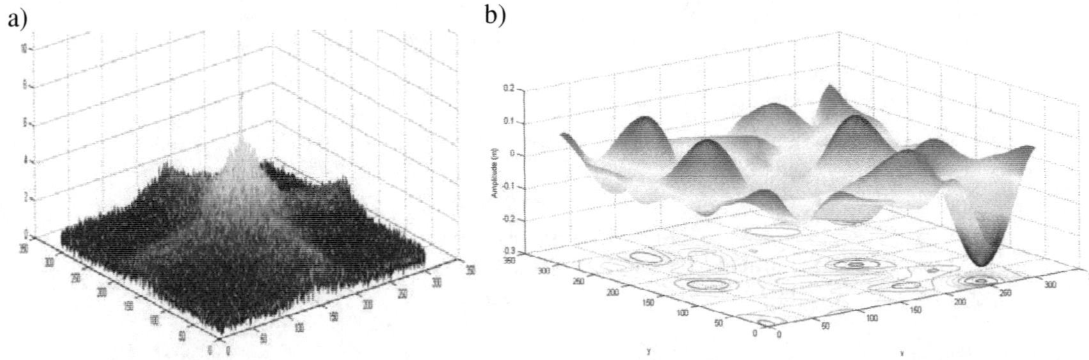

Figure 4. a) The 2D FFT for an offshore area where a horizontal axis is a wave number and the vertical axis is energy, b) The wave amplitude as result of 2D IFFT with linear MTF where horizontal axis in m and the vertical axis in da.

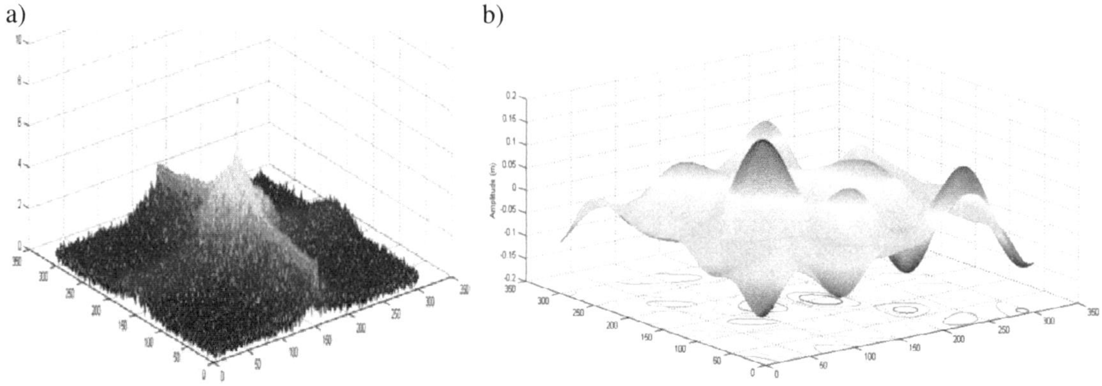

Figure 5. a) The 2D FFT for near shoreline area where horizontal axis is a wavenumber and the vertical axis is energy, b) The wave amplitude as result of 2D IFFT with linear MTF where horizontal axis in *m* and vertical axis in *da*.

The results showed that for an offshore region the wave fields had average amplitude about 0.85 m and it had a random distribution. The significant wave high measurement based on Altimetry satellite and modeling show that for inter-island waters of Indonesia (such as Jakarta bay) had less than 1.2 meters wave height in January, which increased in February and March and reaches about 1.5 meters height [7,8]. It can be inferred that the wave fields had the wavelengths which a bit random but the amplitude is relatively uniform. This is consistent with the radar cross section that has Gaussian pdf. Contrasts with the offshore, for the nearshore the wave fields had amplitude with the average 0.8 m and appeared to tend to be a single wave. The Figure 5 b) also shows that there are two single waves with the amplitude is about 1m among the smaller wave high.

4. Conclusion

The spatial variation of ocean surface wave fields from SAR images in Jakarta Bay was investigated. We used ALOS-Palsar II images with HH polarization images on January, 26 2015 in this study. By using linear MTF then the ocean surface wave fields were obtained. The result showed that the area closes to the coastline, the wave appeared to be a single wave but in the offshore region the wave has random wavelength and the amplitude tend to uniform.
2017.

References

[1] Apel J R 1988 *Principles of Ocean Physics*, (New York: Academic-Press).
[2] Alpers W 2011, *IEEE Trans on Geosci and Remote Sensing*. **49** 3 1114.
[3] Kuo Y Y *et al* 1999 , *Ocean Engineering*. **26** 1125
[4] Kanevsky M B 2009 *Radar imaging of the ocean waves,* (Amsterdam: Elsevier)
[5] Veci L *et al* 2014 *Proc. Int. IEEE International Geoscience and Remote Sensing Symposium (IGARSS)* p 1
[6] Anonymous 1975. *Laporan hasil penelitian terhadap arus, gelombang, sedimentasi den pasang surut di perairan pantai Marunda*, (Jakarta: Lembaga Oseanologi Nasional).
[7] Kurniawan R and Khotimah M K 2015, *Journal for Tech and Sci*, **26** 1
[8] Atmoko R A *et al* 2015, *Proc. Int. Seminar on Science and Technology*, G502-131.

Acknowledgments
This work was supported by Laboratorium Geomatika PTPSW-BPPT for fiscal year

Mangrove Leaf Area Index Estimation Using Sentinel 2A Imagery in Teluk Ratai, Pesawaran Lampung

Anwar Juniansah[1], Galih Candra Tama[1], Kadek Risna Febriani[1], Muamar Nandika Baharain[1], Tito Kanekaputra[1*], Yuniarsita Setyo Wulandari[1] and Muhammad Kamal[1]

[1]Geographic Information Science, Faculty of Geography, Gadjah Mada University Yogyakarta

titoputra98@yahoo.com

Abstract. Mangroves are estuary dominant vegetation in tropic and subtropic area which have an important role to balance the coastal ecosystem. One of the main mangrove biophysics parameters is leaf area index, which is defined as an unitless quantity from the area of one side of the leaf on each unit of ground surface area. LAI measurement using satellite imagery is more efficient than direct measurement because it covers the isolated area in the mangrove forest. This paper discussion focus on mangrove leaf area index estimation model comparison to obtain the best estimation model based on accuracy test value. The imagery used in this research is Sentinel 2A with 10 meters resolution and generic vegetation index (NDVI) compared with nongeneric index vegetation (EVI). Normalized vegetation index is chosen because its sensitivity to chlorophyll tissue of the leaves beside, Enhanced Vegetation Index is chosen because its sensitivity to vegetation canopy structure. The LAI measurement result in Teluk Ratai, Pesawaran, Lampung showing values range from 0.37 until 1.39. The correlation analysis result showing an adequate strong relationship between NDVI and EVI with LAI field measurement value. The correlation value between NDVI and the field measurement value of LAI is 0.779 also the correlation value between EVI and the field measurement value of LAI is 0.762. Based on those value both of vegetation index have a strong relationship with the field measurement result of LAI. From the standard error estimation value, the LAI estimation model accuracy using NDVI is 79.8% and 78.78% using EVI. Visual comparison also done by compare vegetation density pattern in Sentinel 2A with estimation classes in NDVI and EVI model. The best model to estimate leaf area index mangrove is EVI model based on visual comparison, accuracy test, and saturation effect from NDVI.

Keyword: Leaf Area Index, Estimation, Accuracy, Model, Comparison

1. Introduction

Mangrove forest is commonly estuary vegetation on tropical and subtropical coastal area. Mangrove has important roles on balancing the coastal ecosystems because of its capacity to absorb carbon dioxide (CO_2) around the atmosphere and change it into sediment form [1]. Those capacity are strongly affected by their biophysics characteristic, such as primary production, total of biomass, nitrogen cycle, and vegetation evapotranspiration. One of the main biophysics parameter is leaf area index (LAI) that can be define as vast on a leaf compare with the surface [2].

Leaves are important organ of a vegetation because in this part all of gasses exchange process to the atmosphere occur. Therefore, leaf area index is an important key on ecology process modeling such as photosynthesis, carbon dioxide absorption and gas exchange cycle on the atmosphere [3]. For instance, measuring leaf area index on mangrove vegetation are important to estimate the energy exchange between land and atmosphere also for another ecological process modeling.

Data acquisition of mangrove leaf area index through field measurement usually need more time, energy, and costly because the accessibility to mangrove area are difficult but the accuracy of this method usually higher than indirect method. LAI measurement using satellite imagery is more efficient than field measurement especially for large mangrove area because of its capacity to cover a wide area and reduce field measurement activity (direct method) [4]. Based on the problems above researcher interested to measure the accuracy level of leaf area index estimation using generic vegetation index and non-generic vegetation index. Generic vegetation index only affected by the vegetation density value, whereas non-generic vegetation index is affected by another factor.

Satellite imagery that used in this research is Sentinel-2A with 10 meters spatial resolution. The vegetation index are Normalized Difference Vegetation Index (NDVI) as the generic index and Enhanced Vegetation Index (EVI) as the non-generic index. NDVI is used because it is chlorophyll sensitive and only use two spectral bands; near infrared and red band. Advantage of using NDVI is the rationing concept that able to reduce multiplicative noise (cloud shadow, atmosphere attenuation, topographic variation). In other hand, EVI is used because of its sensitivity to respond canopy structure variation [5]

The research location is in Padang Cermin Village and Sanggi Village, Padang Cermin Sub District, Pesawaran District, Lampung Province. The study sites are located at the Way Sabu River Estuary that empties into Ratai Bay with coordinates 105°9'38.984" BT - 105°10'58.957" BT and 5°34'43.172" LS - 5°36'3.543" LS.

Based on the figure 1 the research location is broadly part of Lampung Bay with calm water conditions which tend to have high sedimentation rate. The study area has a tropical climate, in 2008 with average rainfall in Pesawaran District ranging from 161.8 mm/month, and the average number of rain days 13.1 days/month. The mean intermediate temperature is 22.9°C - 32.4°C. The average relative humidity interval is between 56.8% to 93.1%.

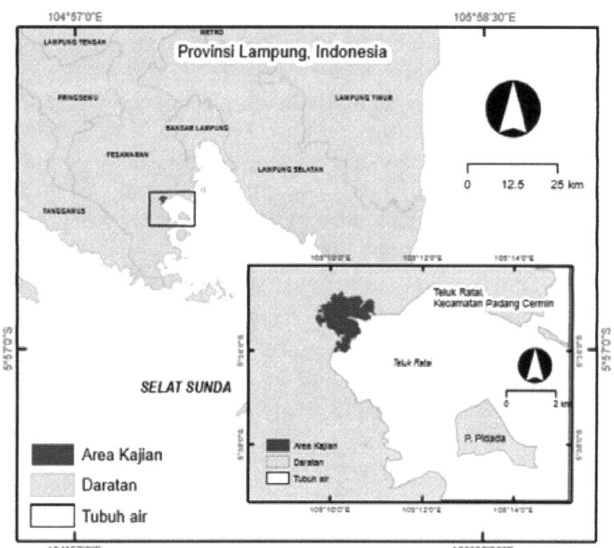

Figure 1. Map of Lampung Bay

IOP Conf. Series: Earth and Environmental Science **165** (2018) 012004 doi:10.1088/1755-1315/165/1/012004

2. Data and Method

2.1. Data

2.1.1. Sentinel-2A image of Ratai Bay study area Pesawaran District Lampung Province 10 meter resolution, recording September 10, 2016.

2.1.2. Image of Google Earth study area of Ratai Bay of Pesawaran Regency of Lampung Province.

2.2. Method

2.2.1. Flow Diagram of Reasearch

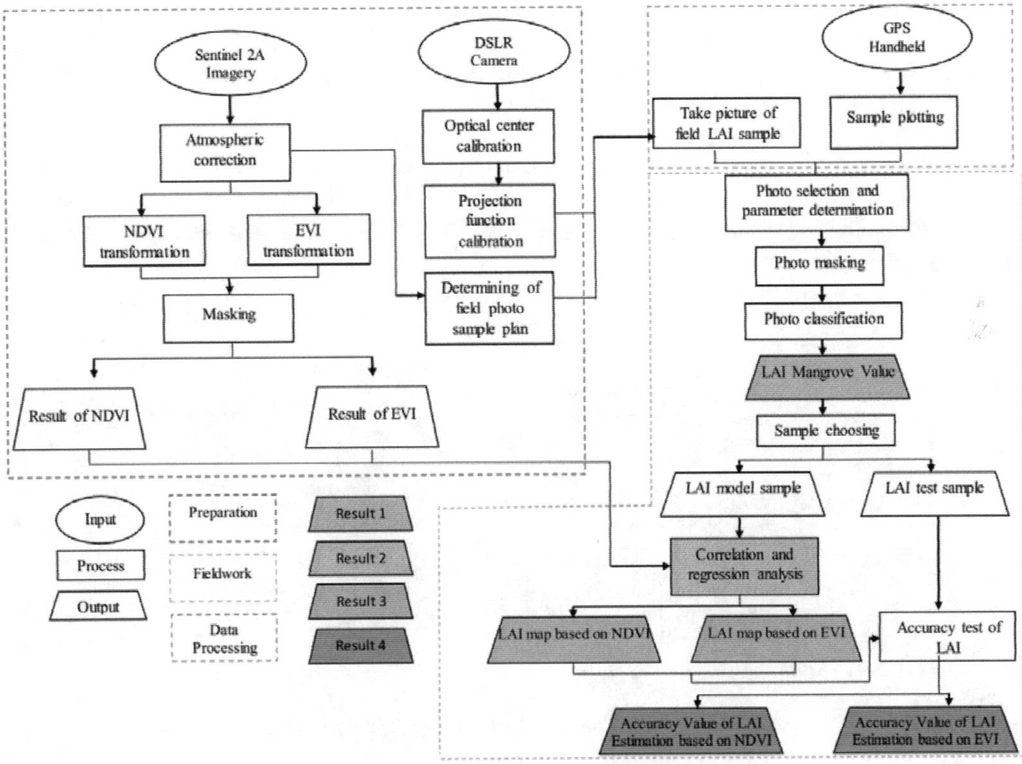

Figure 2. Flow Diagram Of Research

2.2.2. Pre-Processing Data. Pre-processing data phase is a step before conducting the research and this is an extraction of information from Sentinel-2A imagery. The phase includes literature review which is the preparation phase in collecting information regarding the research methods, theoretical basis to the analysis techniques that needs to be performed in this research. The references referred to were literature in the field of remote sensing, leaf area index (LAI), and vegetation index. Subsequently, image correction which was conducted through the improvement of image quality. This correction was an atmospheric correction. In this research, the data used was Sentinel-2A image with 1C radiometrically corrected level (reflectance at sensor). Conversion was performed by using ENVI 5.3 software.

Transformation of the imagery was a phase following the correction. This imagery of Ratai Bay mangrove area in LAI analysis has used two image transformations. The two transformations are utilized based on their effects, i.e., generic in the form of Normalized Difference Vegetation Index (NDVI) and non-generic in the form of Enhanced Vegetation Index (EVI). The phase of selecting the research area has used the imagery masking method. It was conducted by using ENVI 5.3 software, with an input in a form of layer masking. This layer masking is created by conducting visual interpretation of the mangrove area.

The final step was camera calibration. The aim of camera calibration was to set the camera and hemisphere lens prior to Leaf area index (LAI) sampling. Camera calibration performed includes several steps such as optical center calibration and projection function calibration. Optical center calibration was conducted to determine the coordinate from the center point during the shoot. In this stage, hemisphere lens is perforated at some point to determine the pixel coordinates. Projection function calibration was based on pixel value from optical center calibration. This calibration was conducted by two shoots with different coverage area. The aim was to determine the minimum and maximum distance.

2.2.3. Field Data Collecting. Leaf area index (LAI) sampling of the mangrove area was conducted by using purposive random sampling due to the consideration on the difficulty of access to the mangrove area in Ratai Bay, Pesawaran District. The difficulty of access to mangrove area is affected by the sea tides. The location of mangroves also located in several major river estuaries, high mangrove density, and so forth. The distribution of sampling in Sentinel-2A imagery was placed for sampling area with the size of 15m x 15m. In the area, the leaf area index (LAI) shooting was conducted at the midpoint and elsewhere in the sampling area. In addition, in this area of 15m x 15m, plotting has been carried out with the aim to coordinate the measurement of leaf area index (LAI) obtained and to be compared with image transformation data.

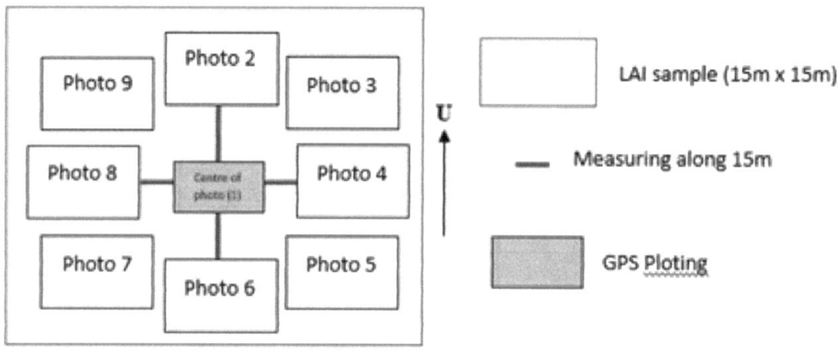

Figure 3. Scheme of taking a picture in the field

2.2.4. Field Data Processing. The sample measurement process of mangrove LAI was conducted through several steps that are shown in the following photos. Photo was taken sequentially from the north at each sample point which are assumed for ESU conformity (Elementary Sampling Unit) in the field. The lighting conditions in each photo should be the same, therefore photos were selected before being processed with Can-Eye software. The number of photos taken were amounted to 9 pieces of photos for ESU characterization [6]. Photos used are photos with upward conditions, i.e., photos which were taken from the bottom to the top.

The process parameters performed should already have optical calibration and projection on the camera lens as they are required to process Angular resolution and FAPAR computation. The parameters were username, image size, optical center & projection function, Circle of Interest (COI), Subsample factor, Angular resolution, Fcover, and FAPAR. Photo masking needs to be performed in order to remove non-vegetation objects and sky background. In advance to masking, gamma factor correction was conducted to clarify the appearance of vegetation and non-vegetation.

Class definition used was no mixed pixel. This option merely classified the pixels based on one class. The class used was green vegetation. Prior to classification, it is necessary to determine the effective leaf zone in the photograph. Therefore, the classification was merely conducted based on leaf area and its canopy.

2.2.5. Statistical Analysis of Data. The processing of field data and Sentinel 2A imagery in this research has also used statistical analysis. Statistical analysis conducted includes correlation and regression analysis, to determine the strength and the form of correlation between dependent variable and independent variable. Regression analysis also yielded the value of determinant coefficient (R^2) which stated the magnitude of the independent variable affecting the dependent variable.

The vegetation index values (NDVI and EVI) was determined as the independent variable, whereas the field LAI value considered as the dependent variable. The regression equation of the two variables was used to construct the LAI estimation model, with 15 samples of the LAI model.

Model accuracy test was conducted by using 10 test samples. Accuracy test was performed by using standard error estimate method, 1:1 plot, and visual comparison. Through SEE, the percentage of estimated accuracy was obtained and the 1:1 plot provided the information on the accuracy of the estimation result (over/underestimate).

3. Result and Discussion

Field measurement result from 25 samples in Ratai Bay indicated the range value of LAI between 0.37—1.39. The LAI field measurement result was strongly affected by the photo exposure, therefore the over- and under-exposure photos were not chosen to be analyzed. This samples were then divided into two, 15 samples as model and 10 as the validation samples.

The distribution of mangrove species in Ratai Bay identified was various in types, among others are *Rhizophora apiculata*, *Rhizophora mucronata*, *Avicennia lanata*, *Xylocarpus moluccensis*, and *Bruguiera cylindrical*. Based on the mangrove pattern analysis, it was known that mangrove in Ratai Bay tends to avoid the natural mangrove distribution pattern. For example, *Rhizophora apiculata* and *Bruguiera cylindrical* which supposed to be located in the middle part of mangrove forest and tends to head towards the mainland, but in Ratai Bay they were mostly found in the coastline heading to the beach.

NDVI value (derived from Sentinel 2A) were in the range of 0.48—0.84 (n=25, mean=0.75, median=0.79, SD=0.09). The normality of NDVI data was tested with Kolmogorov Smirnov and the dnmax value was 0.179. The EVI value (also derived from Sentinel 2A) were in the range of 2.65—3.23 (n=25, mean=3.12, median=3.17, SD=0.13). The normality of EVI data was also tested with Kolmogorov Smirnov, the dnmax value obtained was 0.187. The threshold value in critical testing table was 0.338, based on this value it was known that both, NDVI and EVI data, were distributed normally.

Correlation analysis indicated strong correlation between NDVI and EVI with the LAI field measurement result. Correlation value between NDVI and EVI was 0.983391 therefore they were strongly correlated. Correlation value between NDVI and LAI field measurement result was 0.779, EVI and LAI field measurement result was 0.762.

LAI estimation modeling was conducted by linear regression, where the result was shown in Table 1. The linear regression equation used was based on equation (1).

$$y = \mathbf{a}x + \boldsymbol{b} \tag{1}$$

The dependent variable was from LAI field measurement data model samples and the independent variable was from vegetation index (in this case NDVI and EVI).

Table 1. The result of regression equation

Vegetation index	Regression equation result	R^2
NDVI	y = 2.1472x-0.658	0.60
EVI	y = 1.3397x + 3.222	0.58

The result of equation indicated that 60% of LAI value from field measurement result can be explained by NDVI and 58% of LAI value from field measurement result can be explained by EVI.

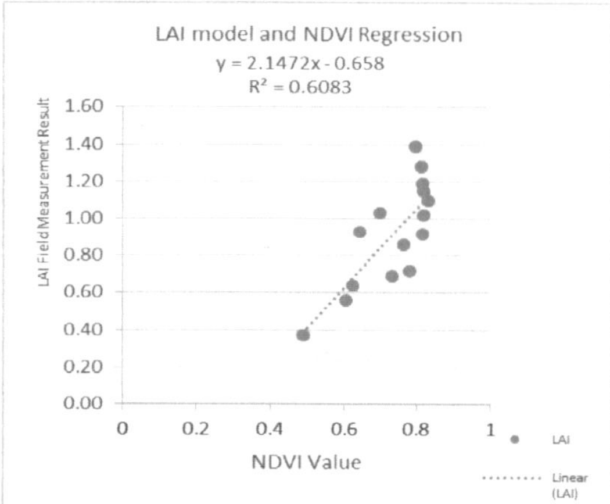

Figure 4. Regression of LAI model and NDVI

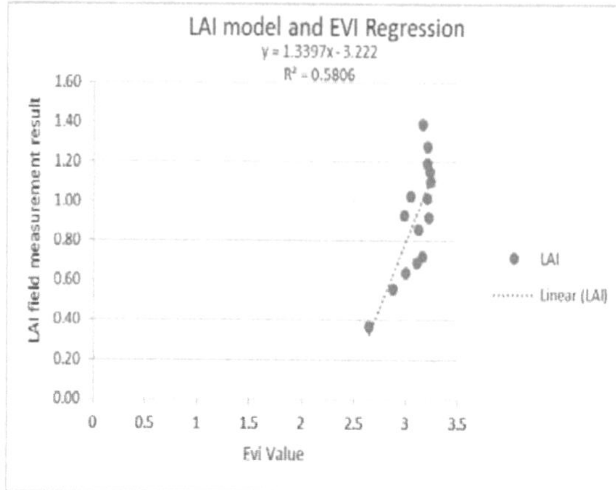

Figure 5. Regression of LAI model and EVI

Accuracy of the estimation result was tested with Standard Error Estimate (SEE) and 1:1 plot by using field measurement data. The SEE obtained a result with the maximum accuracy value of 79.8% with NDVI and 78.78% with EVI. The difference between NDVI and EVI estimation model was not significant, but NDVI model was considered to be better than EVI model statistically. This can be observed by the statistics summary in Table 2. SEE Calculation. The models were also tested with 1:1 plot to determine the accuracy of the LAI estimation.

Graphic 1:1 plot from the estimation result by using NDVI and LAI Validation samples variable indicated that the estimation result was overestimate. Similar with estimation result with EVI model, overestimate result was obtained. This can be observed from the point distribution which were mainly located in the upper part of 1:1 plot line (Figure 6. 1:1 Plot). The point referred to the estimation of LAI value by using each model.

Table 2. Standart Error Estimate Calculation

	EVI	NDVI
SEM	0.395591813	0.357134937
SEM	0.049448977	0.044641867
Mean	0.90000000000	0.90000000000
Standard deviation	0.239165215	0.239165215
Confidence Level 95%	0.148233412	0.148233412
upper range	1.04823341207	1.04823341207
lower range	0.75176658793	0.75176658793
maximum error	29.57982733	28.1052955
minimum error	21.21390676	20.15640968
maximum accuracy	78.78609324	79.84359032
minimum accuracy	70.42017267	71.8947045

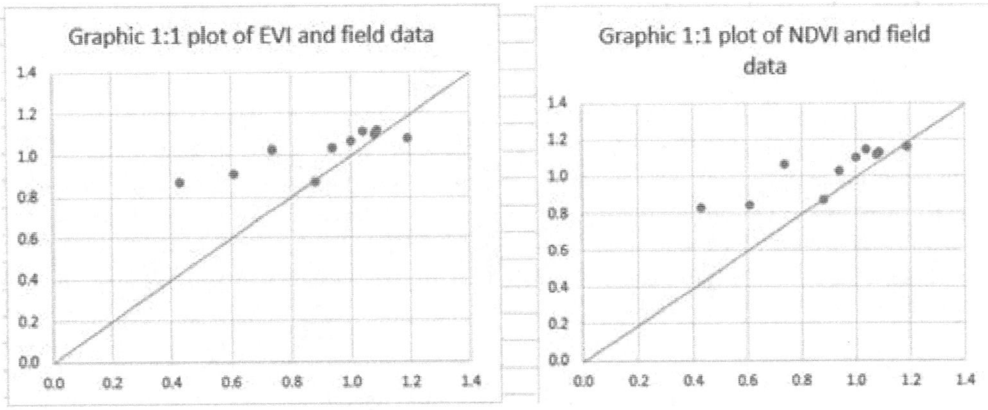

Figure 6. 1 :1 Plot between EVI with LAI model and NDVI with LAI model.

Overestimate can be caused by the photo exposure which was overly obscure, therefore Can-Eye software indicated many leaf structures than it should be. Furthermore, inequality of samples distribution also affects this estimation result. It is difficult to obtain prevalent sample distribution in mangrove area because the condition of mangrove forest is hard to access.

The accuracy of LAI estimation that have been tested statistically was subsequently compared to the visual analysis. This visual analysis was conducted by comparing the mangrove density in Sentinel 2A imagery (recorded in 21st September 2016) with the estimation result by using NDVI and EVI index (shown in Figure 7). Based on this visual analysis, estimation pattern with EVI model is more similar with mangrove density in imagery.

By observing the mangrove density distribution pattern, it can be seen in the Sentinel imagery that mostly, high mangrove density has no strict boundary and was not massive in nature like the pattern of LAI estimation by using NDVI model. For example, mangrove in the middle-northern pattern of the river indicated that high-density mangrove was not massive and it has no strict boundary. Different with the estimation by using NDVI, in that location, the mangrove seems to be highly massive and have a strict boundary.

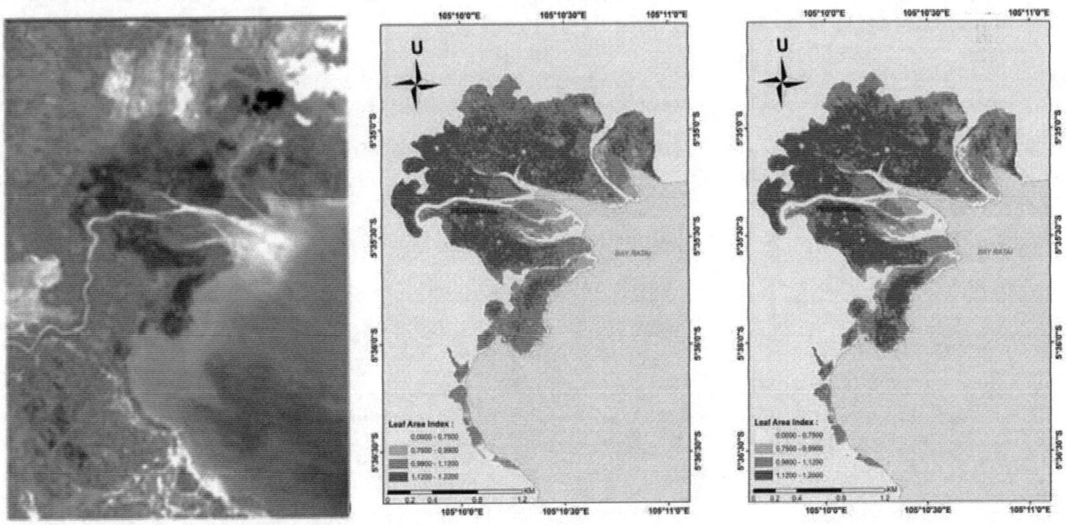

Figure 7. Comparison between mangrove density in Sentinel-2A, and LAI estimation result using NDVI and EVI

The comparison by using density value approach considered to be quite representative. Based on the data analysis by using Can-Eye and statistical analysis, there were strong correlation between the canopy mangrove densities with LAI, moreover, it is impossible to obtain the LAI value by visual interpretation in satellite imagery.

Statistically the NDVI value was considered better than EVI to represent the LAI value, but in the high mangrove canopy density, the NDVI model has undergone saturations. Because of the saturation, NDVI lacks of sensitivity to differentiate the LAI value in high density of mangrove. In addition, the inequality field measurement samples also caused the samples to be less representative. Therefore, the EVI model was considered to be more representative to estimate the mangrove LAI value in Ratai Bay because the mangrove was highly dense.

4. Conclusion

The best model to estimate LAI mangrove was by using EVI vegetation index because it was considered to be more compatible with the mangrove density pattern in imagery and mangrove real condition in Ratai Bay. The NDVI vegetation index was considered inappropriate due to the saturation that occurred in the high-density mangrove.

References

[1] Candra E D, Hartono and Wicaksono P 2016 Above Ground Carbon Stock Estimates of Mangrove Forest Using Worldview-2 Imagery in Teluk Benoa, Bali. *IOP Conf. Ser.: Earth and Env. Sci.* 47 1—11

[2] Green P, Edmund, Peter J, Mumby, Alaishdair J, Edwards, Chris D, Clark, Anggie C and Ellis 1997 Estimating Leaf Area Index of Mangrove from Satellite Data. *J.Aqu. Bot.* **58** 11—19

[3] Jensen R R 2000 *Measurement, Comparison, and Use of Remotely Derived Leaf Area Index Predictors* (Ann Arbor: University of Florida) chapter 3 pp 16—38

[4] Breda N N J J 2003 Ground Based Measurement of Leaf Area Index: A Review of Methods, Instruments and Current Controversies *J.of Exp.Bot.* **54** 2403—17

[5] Huete A, Didan K, Miura T, Rodriguez E P, Gao X and Ferreira L G 2002 Overview of the Radiometric and Biophysical Performance of the MODIS Vegetation Indices *J. Rem. Sen. Of Env.* **83** 195—213

[6] Weiss M, Baret F, Smith G J and Jonckheere 2003 Methods for In Situ Leaf Area Index Measurement, Part II. *J. Agr. and Fors. Met.* **121** 17—53

Mangrove Mapping Using SPOT 6 at East Lombok Indonesia

Aninda W Rudiastuti[1*], **Doddy M Yuwono**[1] **and Sri Hartini**[1]

[1]Geospatial Information Agency – Indonesia

aninda.wisaksanti@big.go.id, anindarudiastuti@gmail.com

Abstract. Accumulated biomass estimation in mangrove forest is important for assessing its productivity and sustainability. It is also important on revealing the potential amount of carbon that can be emitted in the form of carbon dioxide when deforestation took place. Mangrove as a part of the coastal ecosystem is related to Blue Carbon, especially in controlling climate change. This is due to its ability in utilizing carbon dioxide for photosynthesis and store it in stock biomass and sediment. This study addressed to estimate the mangrove extent at East Lombok as a part of One Map Mangrove Bali – Nusa Tenggara, and also trying to estimate its above ground biomass (AGB). Those are indispensable, especially for eastern part of Lombok which has the greatest mangroves area and mostly in good condition such as Gili Lawang and Gili Sulat. Forest biomass can be estimated through combination of field measurement and remote sensing - GIS methods. SPOT 6 and another high resolution-imagery were used as primary remotely sensed data. On screen digitation using RGB-NIR bands and Normalized Difference Vegetation Index (NDVI) image transformation are applied for the initial canopy density classification. For field measurement at 2016, plot sample sized 10 m x 10 m were used for mangrove parameter inventory. Non-destructive biomass estimation applied in this research which is applicable for mangrove ecosystems, using allometric equations. AGB estimation model also applied in mangrove area at two small islands. Bruguiera sp. and Rhizopora sp. dominated in East Lombok. While, Sonneratia sp., and Avicennia sp dominantly appeared in seaward zone of mangrove zonation. The spatial analysis resulted mangrove area in East Lombok is 1759.5778 ha. The estimation results of above ground biomass (trees and saplings) using 12 sample plots in East Lombok is 12,667.316 kg, while there are 881,677 kg of AGB of mangrove in Gili Lawang and Gili Sulat based on the AGBs model applied.

Keywords: Mangrove, Above Ground Biomass, SPOT 6

1. Backgrounds

Mangrove as one of the essential ecosystems on coastal are facing many threats[1], [2]. High degradation level, due to its utilization as a source of timber and land cover change becoming the primary cause[3], [4]. In line with that, for sustainable development and coastal spatial planning purpose, information of mangrove extent becoming very essential. In spatial term, Mangrove monitoring can be done using a combination of remote sensing and field surveys. Remotely sensed data still becoming one of the best sources that can reveal mangrove forest condition. Its advantages such as synoptic coverage, provide information for the inaccessible area, historical and near real-time data has brought lots of benefit in short time data processing[5]–[7]. Thus, it is compelling to support land management, spatial planning, rehabilitation, and monitoring of coastal area [7]–[10]. In term of sustainable coastal management, mangrove biomass characteristics are necessary [11]. Biomass estimation has become one crucial step to lead another action such as deforestation deterrence, agroforestry, and forest management [12].

Lombok as one of tourism destination in Indonesia has mangrove as its natural resources that should be preserved. Gili Lawang and Gili Sulat as marine conservation area becoming potency for East Lombok. Some research of mangrove in East Lombok has been done. Sidik and Kusuma [13] have carried study using Formosat to classify mangrove density in Gili Lawang and Gili Sulat. Budhiman et al. [14] have estimated the rate of mangrove forest degradation in Lombok Island using Landsat- TM. Utilizing SPOT 6 imageries, this study aims to map the mangrove extent and its canopy density distribution, and also estimate the aboveground biomass (AGB) map using remote sensing data. Further, mangrove AGB estimation model was produced using vegetation index [11]. Several studies show the estimation of mangrove AGB using remote sensing data such as high-resolution imagery (Worldview-2, GeoEye-1)[15], [16], medium resolution ASTER elevation data[16], and L-band ALOS PALSAR[17]. Along with Anaya *et al.*[18] that implemented enhanced vegetation index (EVI) to estimate aboveground biomass in grassland, primary forest, and secondary forest. This paper explored the possibility of estimate aboveground biomass in a densely vegetated mangrove area using normalized difference vegetation index (NDVI) computed from SPOT 6 imagery. This study was initiated as a step to enrich the mangrove mapping information in Indonesia. Also, field data used in this study were part of Bali – Nusa Tenggara' sampling points for mangrove mapping in 2016.

2. Methodology

Mangrove mapping using remote sensing technology assisted with ground truth or mangrove validation become the major step in this study. Methods of image interpretation were varied depending on the image datasets condition. Visual and digital interpretation are the two primary methods that applied for interpreting the SPOT 6 images.

Following the national technique guidance: Head of Indonesia Geospatial Information Agency Regulation No. 3 2014 about Collecting and Processing of Geospatial Data of Mangrove. Image interpretation of mangrove was varied due to image condition. Mostly, visual interpretation using both of true color (RGB) and false color composite (Green – Red – NIR bands) were occupied to determine the mangrove edges. Canopy density determination was based on NDVI results. Using image enhancement with true and false color composite, made canopy density classes can be easily differentiated[19]. In order to estimate Above Ground Biomass in the mangrove area, allometric equation as cited in national guidance were used [12], [20]. Further, in this paper, AGB also estimated using vegetation index (NDVI) [11]. As part of One Map Mangrove Bali – Nusa Tenggara, several notes that should be noticed in this paper as for the amount and distribution of sampling plots. Brief and clear workflow of the mangrove mapping in East Lombok were captured in **Figure 1**.

2.1. Study Area

As mention in introduction subtitle, this paper utilizes data from Bali – Nusa Tenggara Mangrove Mapping Project that held in 2016 (**Figure 2**). As for initial work of AGB estimation using NDVI, East Lombok chose as a study area. Administratively, Lombok Island divided into Mataram City, Lombok Tengah, Lombok Timur, Lombok Utara, and Lombok Barat. **Figure 3** shows the study area of mangrove mapping, with focused in two areas of East Lombok (Red boxes), and the distribution of survey plots.

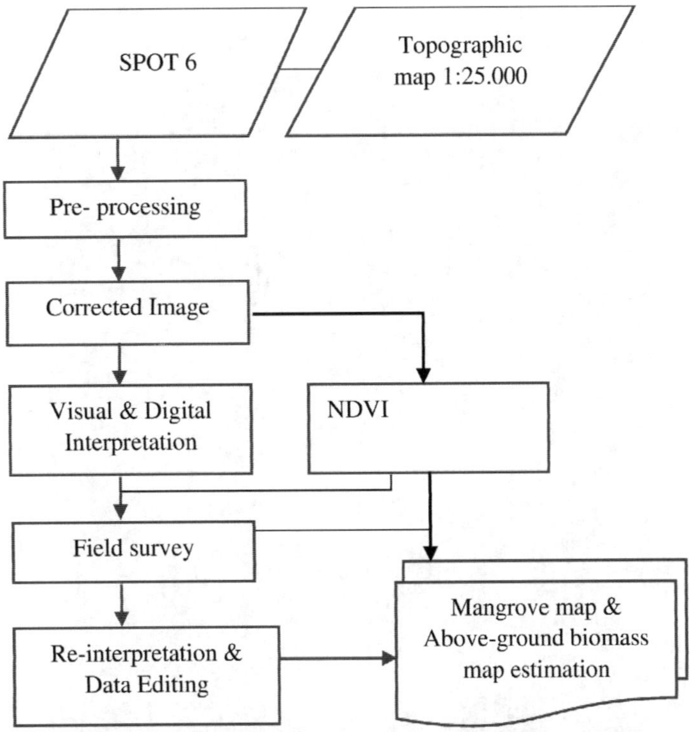

Figure 1. Flowchart of Mangrove Mapping

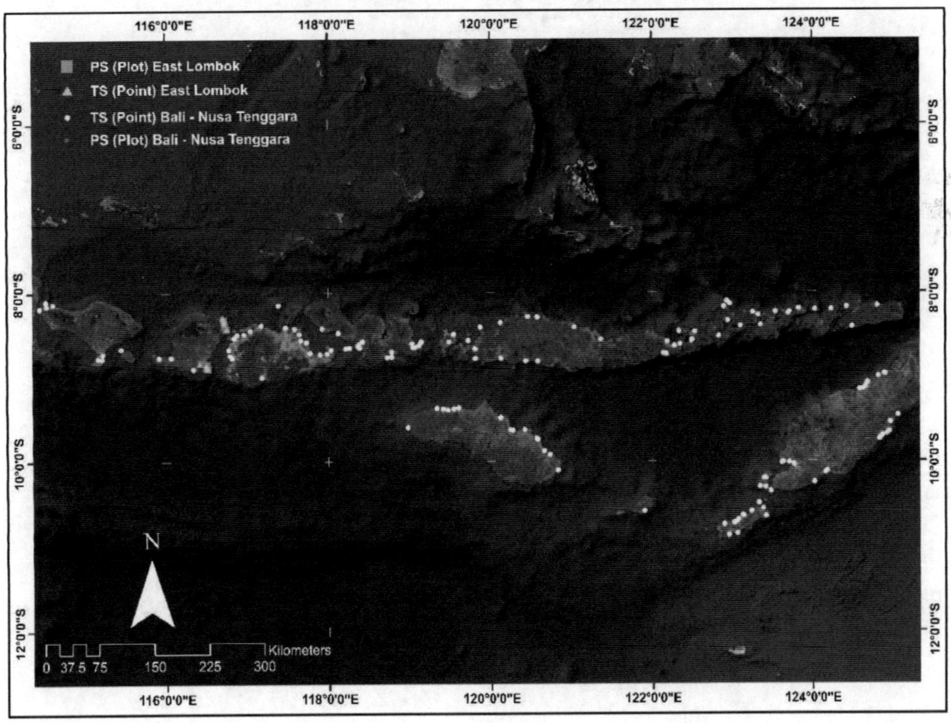

Figure 2. Area of Bali – Nusa Tenggara Mangrove Mapping Project

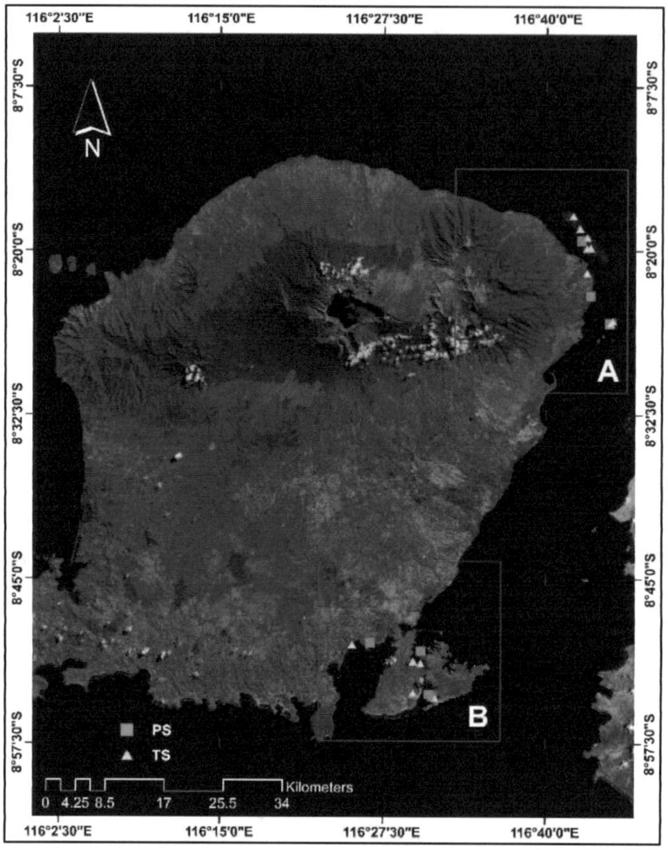

Figure 3. Study area (A: North; B: South)

2.2. *Materials and Methods*

Base map used for image geo-rectification was Indonesia topographic map scale of 1:25.000. Satellite images used in this study are SPOT 6 imageries acquired varied from February 2015 to March 2016, with 6 m of multispectral and 1.5 m panchromatic bands spatial resolution. Those imageries are used for NDVI transformation and also visual interpretation.

2.3. *Vegetation Index*

Classification of canopy density resulted from NDVI image transformation referred to the national standard for Mangrove Surveying and Mapping (SNI 7717: 2011) scale of 1:25.000. The canopy density classes would be: Very dense (5), dense (4), moderate (3), rare (2), and very rare (1). Many vegetation index has been used for vegetation analysis[7], [11], and NDVI can be used for mangrove canopy density mapping [8], [19]. NDVI yielded better correlation between vegetation index and vegetation presentation using ALOS/AVNIR-2 data [21].

NDVI image transformation:

$$NDVI = \frac{Band\ (NIR) - Band\ (R)}{Band\ (NIR) + Band\ (R)} \quad (1)$$

Where:

NIR	: Near Infrared
R	: (visible) Red
Range of NDVI	: -1 to 1

2.4. *Field survey*

The field survey is carried to validate the sample points and polygons as the result of image interpretation. First thing in field survey is validation form of existing mangrove which needed to verify the results of image interpretation. This step is essential to clarify the information derived from the image with cloud existence or other disturbance. Next, inventory of mangrove stands parameters such as trees density, canopy density, value index, and mangrove species (dominance). For 180° view of mangrove canopy, Fisheye lens is used. The percentage of canopies open derived from the photo captured by fisheye lens then it can be calculated to find the percentage of mangrove canopy cover. Secondary data also needed to fit up.

Since this study is part of Bali - Nusa Tenggara Mangrove Mapping Project, survey plots arrangement (number and distribution) are based on the extent of mangrove area along Bali - Nusa Tenggara which divided into 189 map index (NLP). The number of sampling points throughout Bali - Nusa Tenggara was calculated for accuracy assessment. Technically, validation and mangrove parameters measurement carried on a plot sampling (PS) sized 10 x 10 m2. And fewer parameters acquired in point sampling (TS) sized 5 x 5 m2, only canopy density, and species dominance. On average, up to 2 PS distributed in each map index. For Bali - Nusa Tenggara, it resulted in 285 PS and 150 TS (**Table 1**) laying along the coastal, and only 12 PS plotted in in East Lombok, Gili Sulat and Gili Lawang which occupied for AGB estimation based on NDVI (**Figure 3**).

Table 1. Mapping index and number of samples

Areas/ Province	Σ Index	Number of PS	Number of TS
Bali	189 NLP	25	8
Nusa Tenggara Barat (NTB)		89	47
Nusa Tenggara Timur (NTT)		181	104
TOTAL		295	159

Of course, in mapping, accuracy assessment must be done to test the confidence level of the product. Accuracy assessment is required because of another land cover around the mangrove area on the image interpretation results, thus allowing for a less precise or mixing border between mangrove and non-mangrove. Also, interpreting the mangrove area of the image will not always result in accuracy, this may be due to the date of acquisition of the old image irrelevant to the actual land cover conditions. Thus, it is necessary to do the field validation. The assessment of the image interpretation accuracy was calculated using the matrix accuracy developed by Short in 1982:

$$Overall\ Accuracy = \left(\left(\sum_{i=1}^{r} X_{ii}\right)/N\right) x\ 100\% \quad (2)$$

where:
$X_{ii} = diagonal\ matrix$
$N \ \ = sample$

2.5. *Biomass Estimation*

One of biomass estimation method without extraction is sampling. By measuring several parameters such as height (H), diameter at breast height (DBH), diameter (D), and distinguishing between tree and saplings, allometric equations can be applied to extrapolate the above ground biomass [12]. Different allometric equations applied based on range of DBH value. Mangroves with DBH≥ 10 cm, general allometric equation built by [20] and ρ value by Simpson (1996) cited in the national technical guidance of mangrove mapping was applied. Another general allometric equation used for mangroves with DBH

< 10 cm [12]. Beside using allometric equation, above-ground biomass estimation model also implemented using NDVI result [11], but only in Gili Sulat and Gili Lawang which has almost homogeny land cover. The AGBs estimation model was built by field measurements in Segara Anakan, Cilacap and yielded R^2>80%, and claimed that the higher the NDVI will resulted biomass increasing in logistic correlation.
The AGBs estimation model[11] :

$$\frac{(30.97183 - (56.898258 * \mathrm{NDVI}))}{(1 + (3.952476 * \mathrm{NDVI}) - (9.567301 * (\mathrm{NDVI}^\wedge 2))} \quad (3)$$

3. Results and Discussion

The strands of Bali – Nusa Tenggara islands often referred as the "Sunda Kecil". Consisting of three provinces, namely Bali, West Nusa Tenggara, and East Nusa Tenggara, those have different mangrove variations. Bali - Nusa Tenggara Mangrove mapping project was made to support One Map Mangrove Indonesia in 2016.

In Bali, mangroves are found in large areas, especially in Perancak, Bali Barat National Park (Menjangan Island), and Benoa. *Rhizophora* (*R. apiculata* and *R. mucronata*) becoming the dominant genera of mangrove. By the profile, the area with the most mangrove species is Buleleng Regency. *Avicennia* sp., *Sonneratia* sp., *Rhizophora* sp., *Bruguiera* sp., and *Ceriops* sp. inhabited the coastal, sequence from seaward to landward zone. The most extensive mangrove area found at Badung Regency, it is 574.38 ha, while mangrove area based on canopy cover classification in whole Bali province is contained in Table 2. The canopy cover class which is dominant in Bali is "dense" class.
Some of the new species found in West Bali National Park (TNBB) include *Rhizophora lamarckii* (a mixture of *Rhizophora apiculata* species with *Rhizophora stylosa*), *Acanthus ebracteatus, Cerbera odollam, Dolichandrone spathacea, Xylocarpus moluccensis, Xylocarpus granatum, Aegiceras floridum, Heritiera littoralis,* and *Ceriops australis*. Associated with the substrate, the result of grain size substrate test in TNBB area shows that the dominance of silt and clay with percentage is 88.18%. The substrate type is suitable as the habitat of *Rhizophora* sp.

Mangrove observed in the province of NTB spreading on the main island of Lombok and Sumbawa, and also in many small islands "Gili" around Lombok Island. Meanwhile, the mangrove area located on Sumbawa Island is Alas Moyo, Saleh Bay, Seteluk-Taliwang, Lunyuk-Labangka, Teluk Sanggar, Cempi Bay, Waworada Bay, Bima Bay, and Sape Bay. Based on the interpretation result using SPOT 6 image, the mangrove area in NTB province is about 10,667.19 ha, with canopy density class "moderate" having the most extensive area (**Table 2**). In the vegetation phase of trees, the dominant species are *Sonneratia alba*, while in the sapling and seedling phase *R. apiculata* dominated. The grain size test results from the five substrate sample locations of mangrove scattered in NTB showed mostly mangrove substrate is by silt and clay (90%).

Table 2. Mangrove area at Bali – Nusa Tenggara

No	Region	Canopy Density Class	Area (ha)
1	Bali	Very Dense	417,17
		Dense	822,39
		Moderate	472,27
		Rare	197,42
		Very Rare	108,60
2	NTB	Very Dense	155,42
		Dense	3.078,83
		Moderate	3.756,67
		Rare	2.320,14
		Very Rare	1.356,12
3	NTT	Very Dense	473,91
		Dense	5.580,00

No	Region	Canopy Density Class	Area (ha)
		Moderate	9.290,21
		Rare	4.930,07
		Very Rare	1.873,19

Aegialitis annulata found as an endemic species in the province of East Nusa Tenggara, especially in East Sumba, with 1.5 - 3 meter (bonsai) trees height. Mangrove area with extent 22,147.39 ha found throughout the NTT region, where the species of *Sonneratia alba* predominate in the tree-growing phase. In the region of Sabu Raijua and Rote Ndao many *Lumnitzera sp* found, which is a type of mangrove that lies somewhat away from the coastline and is only flooded at the highest tide that occurs just a few days a month. This area is characterized by substrate conditions that tend to be dry and dense. Grain size test of nine substrate samples spread over NTT showed that three of them were dominated by muddy sand with a composition of 3.24% gravel; 58.91% sand; and 37.85% silt and clay, and the remaining six were dominated by mud and clay with a composition of 0.77% gravel; 10.59% sand; and 88.64% silt and clay. By grain size, the substrate composition has not been able to characterize the zonation of a particular species mangrove directly. The composition of the substrate is the general composition of mangrove habitat for all species. In fact, almost all types of mangroves can grow on both types of substrate dominance. Substrate characteristics will be typical when combined with the tides where the substrate slope difference will control the high tidal and puddle frequency at a location which will automatically also control the salinity level at that location.

Mangrove mapping of Bali - Nusa Tenggara region yielded > 85% for the accuracy of above 85% is obtained for mangrove land cover interpretation and canopy cover classification based on NDVI (**Tabel 3**). Some of the technical things that can affect the achievement of such accuracy are, first, the spatial resolution of SPOT 6 where the distinction of mangrove and non-mangrove can be made quite clear. According to Tobler (1987) in Head of Indonesia Geospatial Information Agency Regulation No. 3 2014, the determination of the equivalent image resolution for a certain scale mapping is half the number of map-scale denominators divided by 1000. Since SPOT 6 image has 6 meters for spatial resolution (multispectral), it meets the accuracy rules of using image resolutions on a 1: 25,000 scale mapping. Second, the image used is relatively up-to-date. Date of image acquisition time varied on February 2015 - March 2016 while the field survey was carried on August 2016. Third, the location of PS in the field survey. As has been done, the determination of plot samples (PS) taken in the middle of a polygon interpreted as a mangrove where NDVI values tend to be homogeneous [19]. Four, the delineation process refers to the many SPOT 6 imageries available. It is intended to cover each other the weakness of the image used. Where in one of the cloud-covered imagery, the information can be obtained from another image with the same location and the adjacent time. The process is carried back and forth between the main image and the supporting image. For canopy density, the result of NDVI is tracked back with the green band [19]. Five, Coastal condition of Bali - Nusa Tenggara supports the interpretation of mangrove and non-mangrove areas. Mangroves grow at specific locations, especially the bay. Also, the coastal steepness reliefs in the Nusa Tenggara area and land use conditions in coastal regions consisting of ponds, tourist lands, villas, or garden plots, also bolster the interpretation step.

3.1 Mangrove of East Lombok

Geographically, the mangrove habitat in NTB lays on delta system with an ecosystem width of more than 1 km inland. Mangrove meets the coastal space like the former ponds. Different condition on small islands near Lombok Island, Gili Lawang and Gili Sulat that covered almost 80% with mangrove. Gili Lawang and Gili Sulat are appointed as marine conservation area, and mangrove grows well at those place. Figure 3 shows mangrove mapping results in East Lombok. In the mainland, mangrove grew in the delta, spread like patches in specific coastal. Mangrove's extents based on canopy density slicing are listed in (**Table 4**), where the largest categorized at dense class (4) with 756,0154ha of mangrove area. Using Formosat image, Sidik and Kusuma [13] state that "Dense class" was also found as the largest area (4,409,024 m^2) based on canopy density using NDVI in Gili Sulat and Gili Lawang, East Lombok.

Escalation of mangrove extent had been noticed, where in 2009, only 1,492.56ha mangrove area of East Lombok were recorded [8].

Table 3. Overall accuracy (%) for land cover and canopy density interpretation

	Land cover	Canopy Density
Bali Nusa Tenggara Barat (NTB) Nusa Tenggara Timur (NTT)	89,33%	85,26%

	Canopy Density	Very Rare (Survey)	Rare (Survey)	Moderate (Survey)	Dense (Survey)	Very Dense (Survey)		Total	Comm.	PA (%)
Interpretation	Very Rare	25	2	2	1			30	16.67	83.33
	Rare	4	43	3	3			53	18.87	81.13
	Moderate	1	2	64	10	1		78	17.95	82.05
	Dense			6	79	2		87	9.20	90.80
	Very Dense			1	4	32		37	13.51	86.49
							243			
	Total	30	47	76	97	35		285		
	Omission	16.67	8.51	15.79	18.56	8.57				
	UA (%)	83.33	91.49	84.21	81.44	91.43				85.26

	Land cover	A (Survey)	B (Survey)	C (Survey)	D (Survey)	E (Survey)	F (Survey)	G (Survey)	H (Survey)		**Total**	Comm.	PA (%)
Interpretation	A	105	4		1	1	2	1			114	7.89	92.11
	B		3	1							4	25.00	75.00
	C			7				1			8	12.50	87.50
	D				4				1		5	20.00	80.00
	E					4					4	0.00	100.00
	F										-	-	-
	G	1		1	1			8			11	27.27	72.73
	H					1			3		4	25.00	75.00
										134			
	Total	106	7	9	6	6	2	10	4		150		
	Omission	0.94	57.14	22.22	33.33	33.33	100.00	20.00	25.00				
	UA (%)	99.06	42.86	77.78	66.67	66.67	0.00	80.00	75.00				89.33

*A = Mangrove; B = Forest; C = Plantation; D =Field; E = Settlement; F = Open field; G = Shrubs; H = Ponds

Table 4. Mangrove Area (ha)

No	Code	Class	Area (ha)
1	5	Very Dense	76.4251
2	4	Dense	756.0154
3	3	Moderate	466.9573
4	2	Rare	199.9951
5	1	Very Rare	237.3041
		Total	1736.6970

East Lombok Regency is one of the regencies with the most mangrove species, and the mangrove crossing profile is initiated by Sonneratia sp. followed by *Avicennia* sp., *Rhizophora* sp., and *Bruguiera* sp. In mapping process, the combination of visual interpretation and utilizing vegetation index (**Figure 4**) resulted in the slight difference of mangrove class area.

Canopy density mapping using SPOT 6 with spatial resolution 6 meter provides detail information in terms of canopy separation of individual vegetation. Compare to 30 meter of Sentinel image [22] which has low correlation to canopy density, finer spatial resolution (4 – 8 m) delivers mangrove spatial structure from double tree crown to community [23]. However, another benefit of using multispectral images is spectral resolution. Canopy density is also can be identified by optimizing spectral characteristic of the image. NDVI is chosen to enhance the vegetation parameters and it is perfect indices to canopy density mapping in different environmental settings [24]. Most of NDVI applications are to provide 'greenness' information of vegetation which is closely related to canopy density.

3.2 Above Ground Biomass Estimation

This study utilizes field data from One Map Mangrove Bali - Nusa Tenggara, which aims to achieve mapping accuracy. In East Lombok, there are 12 sample plots (sized 10 m x 10 m) applying vegetation analysis, including DBH measurement. The mangrove vegetation found in 12 sample plots consisted of *Sonneratia alba, Rhizophora mucronata, Rhizophora stylosa, Rhizophora apiculata, Bruguiera gymnorrhiza*, and *Ceriops tagal*. All these species are the primary mangrove plants commonly found in the mangrove area. In 6 sample plots, there were only saplings data to estimate the biomass (**Table 5**). Allometric equation applied to estimate the above ground biomass in 12 plots and resulted in 12667.3158 kgs of mangrove above ground biomass. However, it had noted that using published allometric standard equation could yield significant error in estimating biomass, due to variation of location and species.

From **Table 5**, it can be inferred that the amount of mangrove found in a plot affected the amount of biomass. In PS 41, only nine trees and one sapling of *Rhizophora mucronata* recorded, but allometric equation gives a surprising number of aboveground biomass (5531.751kgs in 100 m2) compared with other plot samples. DBH of mangrove trees inside PS 41 is ranged between 20.701 cm – 28.025 cm and are the largest value of DBH than others. The high-density condition does not mean it has higher biomass than low density because it is determined by the observable size of the existing tree in the field. Thus, an observation plot with a small number of big trees can have larger biomass than a plot with a lot of trees. In general, the largest tree biomass is obtained in trees with the largest diameter as well. According to Kusmana et al.[25], the magnitude of biomass determined by diameter, plant height, wood density, and soil fertility. Moreover, this will affect the amount of carbon that can be absorbed even increases. This is due to biomass is closely related to the process of photosynthesis, where biomass will increase if plants absorb CO2 from the air and convert it into organic compounds from photosynthesis process and resulted in plants growth.

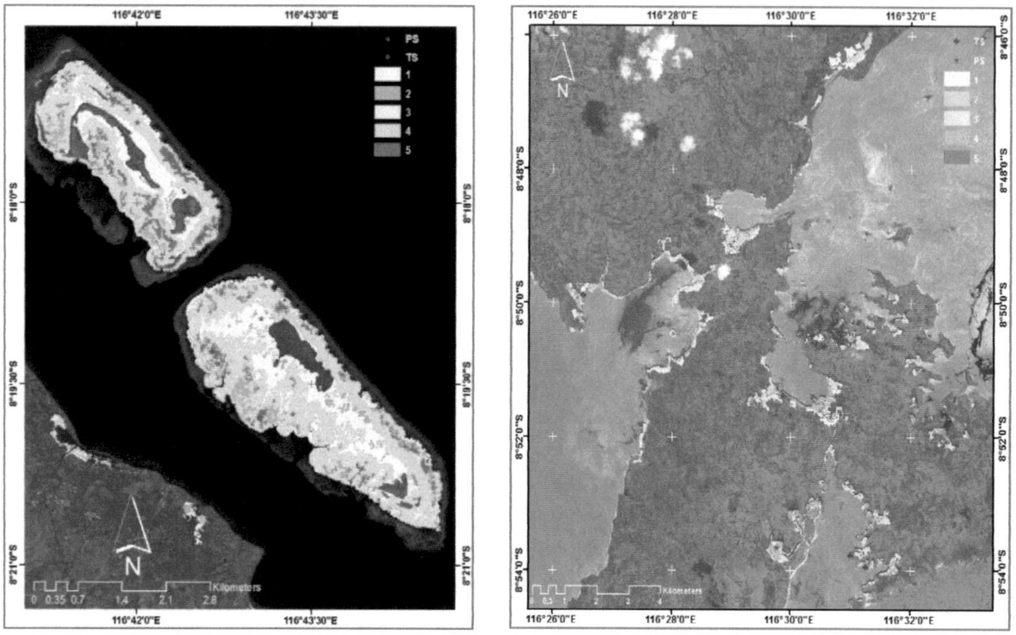

Figure 4. Mangrove area at East Lombok (left: northern part, right: southern part).

Table 5. Above ground biomass (ton) from 12 sample plots at East Lombok

ID	Phase	Genera/ Species	Diameter (cm)	Quantity	AGB (kg)	Total AGB (kg/100m²)	Total AGB (kg/m²)	AGBs Estimation Model (kg/m²)
PS 32	Tree	*Sonneratia alba*	10.032 - 17.197	10	1041.915	1059.741	10.597	0.491
	Sapling	*Sonneratia alba*	2.87 – 5.41	5	17.827			
PS 33	Tree					24.041	0.2404	-
	Sapling	*Rhizophora apiculata*	3.3439 – 6.051	11	24.041			
PS 34	Tree					38.587	0.386	-
	Sapling	*Rhizophora apiculata*	2.5478 – 7.3248	10	38.587			
PS 35	Tree	*Rhizophora mucronata; Sonneratia alba*	11.306 – 16.561	5	673.936	747.631	7.476	-
	Sapling	*Rhizophora mucronata; Sonneratia alba*	5.7325 – 8.5987	10	73.695			
PS 36	Tree					21.514	0.215	0.033
	Sapling	*Ceriops tagal;* *Rhizophora apiculata*	2.2293 – 5.0955	11	21.514			
PS 37	Tree					24.792	0.248	

ID	Phase	Genera/ Species	Diameter (cm)	Quantity	AGB (kg)	Total AGB (kg/100m²)	Total AGB (kg/m²)	AGBs Estimation Model (kg/m²)
	Sapling	*Rhizophora mucronata*	3.1847 – 6.394	8	24.792			
PS 38	Tree					47.900	0.47955	0.027
	Sapling	*Rhizophora mucronata*	2.5478 – 7.3248	9	47.900			
PS 40	Tree					73.564	0.736	0.0097
	Sapling	*Rhizophora mucronata*	5.414 – 7.0064	10	73.564			
PS 41*	Tree	*Rhizophora mucronata*	20.701 – 28.025	9	5527.126	5531.751	55.318	0.0833
	Sapling	*Rhizophora mucronata*	3.8217	1	4.625			
PS 42	Tree	*Bruguiera gymnorrhiza;* *Rhizophora apiculata*	12.102 - 23.5668	12	2107.773	2121.208	21.212	0.0279
	Sapling	*Bruguiera gymnorrhiza*	4.1401 – 6.6978	3	13.435			
PS 43	Tree	*Rhizopora stylosa;* *Rhizophora apiculata*	11.146 – 20.382	10	2089.801	2092.781	20.928	1.625
	Sapling	*Rhizophora stylosa*	2.8662 – 3.8217	3	2.981			
PS 44	Tree	*Bruguiera gymnorrhiza*	11.465 – 17.197	7	789.966	883.809	8.838	0.281
	Sapling	*Bruguiera gymnorrhiza*	3.8217 – 8.9172	6	93.839			
	TOTAL			140	12667.316			

Biomass approach from vegetation is the result of rationing green index using red band and near-infrared (NIR) band. Since red and NIR band have different characteristic when interacting with leaf due to chlorophyll absorption. The red band is weakly reflected. The vegetation index values are in the range of -1 - 1, where the negative value or below 0 correlated with no vegetation. The NDVI results in Gili Lawang and Gili Sulat are shown in **Figure 5**. Manually divided into five classes based on the data distribution form, the canopy grade range based on NDVI on the island area are:

Table 6. NDVIndex reclassification

NDVI	Canopy Density
< 0.001	Non vegetation
0.0014 - 0.2745	Very Rare
0.2745 - 0.3578	Rare
0.3578 - 0.4448	Moderate
0.4448 - 0.5323	Dense
0.5323 - 0.6326	Very Dense

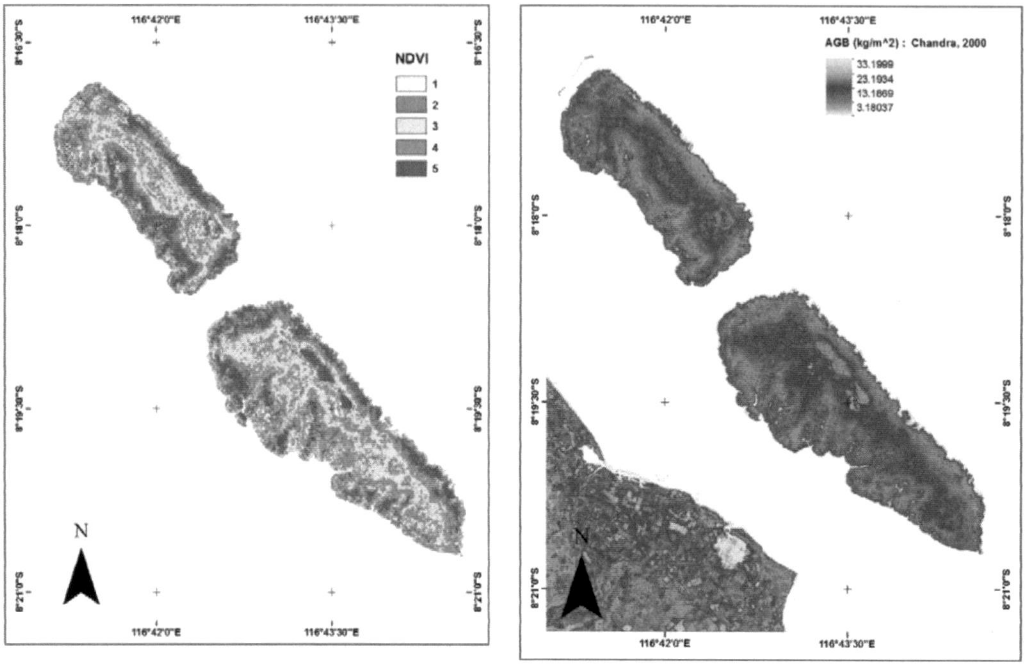

Figure 5. The NDVI result (left) and AGBs estimation using model (right) for Gili Sulat and Gili Lawang, East Lombok

The AGB estimation model implementation results are attached in **Table 5**. There are only 12 results that depicted from the models due to cloud cover on the imagery used. Based on **Table 6**, NDV index found in five PS at Gili Sulat and Gili Lawang are ranged between moderate until very dense category (0.393621 – 0.578368). On the purpose of revealing the spatial information of above-ground biomass in a large mangrove area such as Gili Lawang and Gili Sulat, AGBs Estimation Model built by Chandra was implemented. This step is a preliminary trial of an existing model that require NDVI data. The reason of model [11] implementation instead of own build model are:

- Little amount of field data, particularly on Gili Sulat and Gili Lawang
- The size of plot sample in field measurement did not meet the extent required in producing training area with SPOT 6 imagery (spatial resolution 6m x 6m)
- To estimate aboveground biomass for mangrove area in Gili Sulat and Gili Lawang based on the model that built in the tropical area, particularly southern part of Indonesia and had similar genera variation for mangrove. Further, to avoid extrapolating AGB content from the field into the mangrove area.

Above ground biomass estimation using model derived from the medium resolution of optical image data adopted from Budi [11] found to be low accurate. The high difference of spatial image resolution used between model [11] and this study can be the main reason for the difference of resulted AGB as well. And, less biophysical parameters used to predict AGB also caused inaccurate prediction. A previous study [16] mapped AGB using important ecological attributes such as elevation as a predictor in the model. The regression equation offered biomass simulation and revealed pattern of organic matter that differs due to latitude [26]. Nevertheless, there are some limitations in the model. The model was built based on SPOT XS image with 20 m x 20 m spatial resolution, and it was not the best model to explain the relationship between biomass and vegetation index. The best relation had shown by estimating biomass using Infrared Index between NIR and MIR (Krieger et al. in [11]). The unavailability of MIR band in this research became one of data shortcoming. Employing the AGBs estimation model using NDVI in **Figure 5** yielded 0.00198 kg/m2 of aboveground biomass for all mangroves area at Gili Lawang and Gili Sulat, similar to 881,677 kg of biomass in the whole region.

However, the AGB estimation between using allometric equation and model did not show the same value. The result from allometric equation seemed higher than those predicted from AGB model. It could not conclude that the model does not accurately yield above-ground biomass information since there was no sufficient data for validation.

4. Conclusions

Remote sensing data has been widely used to map biophysical characteristics of mangrove. This study demonstrate that SPOT 6 can be used to extract mangrove canopy density and above ground biomass information. Optical remote sensing data has proven effectively to map mangrove canopy density [24]. Likewise, although it is limited in field data points, this research provide evidence of SPOT 6 data application for mangrove canopy density mapping (accuracy> 85%). And, using this advantage, prediction of above ground biomass can be done as well

NDVI provides better greenness information, however this parameter is not good predictor for AGB mapping. Based on PS 40 point, showed that high NDVI value and canopy density on field survey did not actually represent high AGB. That was different from other PS points which tend to have high AGB with high NDVI value. Mangrove biophysical information extraction using remote sensing data processing needs to be developed. Integration using other geospatial data can be powerful to increase the accuracy of resulted AGB estimation.

5. Suggestions

Due to some of feebleness found in this paper, we suggest to build a survey design that can accommodate both of mapping accuracy and ecological purpose. It is important to digging more information of mangrove or another mapping object while generating spatial accuracy. For AGB estimation, another model based on vegetation index could be implemented to see the comparison.

References

[1] C Giri, E Ochieng, L L Tieszen, Z Zhu, A Singh, T Loveland, J Masek and N Duke 2011 Status and distribution of mangrove forests of the world using earth observation satellite data *Glob. Ecol. Biogeogr*, vol. **20**, no. 1, pp. 154–159
[2] FAO The world's mangroves 1980-2005 *FAO For. Pap.,* vol. **153**, p. 89, 2007.
[3] B Nurkin 1994 Degradation of mangrove forests in South Sulawesi, Indonesia *Hydrobiologia,* vol. **285**, no. 1–3, pp. 271–276
[4] B A Polidoro, K E Carpenter, L Collins, N C Duke, A M Ellison, J C Ellison, E J Farnsworth, E S Fernando, K Kathiresan, N E Koedam, S R Livingstone, T Miyagi, G E Moore, V N Nam, J E Ong, J H Primavera, S G Salmo, J C Sanciangco, S Sukardjo, Y Wang and J W H Yong 2010 The loss of species: Mangrove extinction risk and geographic areas of global concern *PLoS One*, vol. **5**, no. 4
[5] C Kuenzer, A Bluemel, S Gebhardt, T V Quoc and S Dech 2011 Remote sensing of mangrove ecosystems: *A review," Remote Sensing,* vol. **3**, no. 5. pp. 878–928
[6] R A A L M Agris and R A B Arreto 2010 Mapping and assessment of protection of mangrove habitats in Brazil Panam. *J. Aquat. Sci.*, vol. **5**, no. Ong 1995, pp. 546–556
[7] S P Sari and D Rosalina 2016 Mapping and monitoring of mangrove density changes on tin mining area. *Procedia Environ. Sci.*, vol. 33, pp. 436–442
[8] S Hartini, G B Saputro, M Yulianto and Suprajaka 2010 Assessing the used of remotely sensed data for mapping mangroves Indonesia, in 10th WSEAS/IASME International Conference on Electric Power Systems, High Voltages, Electric Machines, POWER'10, 6th WSEAS International Conference on Remote Sensing, *REMOTE'10*, 2010, pp. 210–215.
[9] J Long, D Napton, C Giri and J Graesser 2014 A Mapping and Monitoring Assessment of the Philippines' Mangrove Forests from 1990 to 2010. *J. Coast. Res.*, vol. **294**, no. 2, pp. 260–271
[10] A Aslan, A F Rahman, M W Warren and S M Robeson 2016 Mapping spatial distribution and biomass of coastal wetland vegetation in Indonesian Papua by combining active and passive

remotely sensed data. *Remote Sens. Environ.,* vol. **183**, no. September, pp. 65–81
[11] C Budi 2000 Model Penduga Biomassa dan Indeks Luas Daun menggunakan Data Landsat TM dan SPOT Multispektral (XS) di Hutan Mangrove (Studi Kasus Segara Anakan, Cilacap)," Institut Pertanian Bogor
[12] D Sutaryo 2009 Penghitungan Biomassa: Sebuah pengantar untuk studi karbon dan perdagangan karbon. Bogor: Wetlands International Indonesia Programme
[13] F Sidik and D W Kusuma 2006 Formosat Imagery Formangrove Density Identification In Gili Sulat-Gili Lawangmangrove Forest, East Lombok, Bali
[14] S Budhiman, R Dewanti, C Kusmana and N Puspaningsih 2001 Kerusakan Hutan Mangrove Di Pulau Lombok Menggunakan Data Landsat-TM Dan Sistem Informasi Geografis (SIG)," *Warta LAPAN,* Jakarta, p. 12
[15] O Mutanga, E Adam and M A Cho 2012 High density biomass estimation for wetland vegetation using worldview-2 imagery and random forest regression algorithm *Int. J. Appl. Earth Obs. Geoinf.*, vol. **18**, no. 1, pp. 399–406
[16] N R A Jachowski, M S Y Quak, D A Friess, D Duangnamon, E L Webb and A D Ziegler 2013 Mangrove biomass estimation in Southwest Thailand using machine learning *Appl. Geogr.*, vol. **45**, pp. 311–321
[17] O Hamdan, H Khali Aziz and I Mohd Hasmadi 2014 L-band ALOS PALSAR for biomass estimation of Matang Mangroves, Malaysia *Remote Sens. Environ.*, vol. **155**, pp. 69–78
[18] J A Anaya, E Chuvieco and A Palacios-Orueta 2009 Aboveground biomass assessment in Colombia: A remote sensing approach *For. Ecol. Manage.*, vol. **257**, no. 4, pp. 1237–1246
[19] A W Rudiastuti, D M Yuwono, Niendyawati, G H Pramono and B. D. Rahmanto 2016 Overview of National Thematic Data Integration (An Experience on One Map Mangrove Sulawesi) *IOP Conf. Ser. Earth Environ. Sci.*, vol. 47, no. 1, p. 12012
[20] A Komiyama, J Eong and S Poungparn 2008 Allometry , biomass , and productivity of mangrove forests : *A review," Aquat. Bot.*, vol. 89, pp. 128–137
[21] A R As-syakur and I W S Adnyana 2009 Analisis Indeks Vegetasi Menggunakan Citra Alos/Avnir-2 Dan Sistem Informasi Geografi (Sig). *J. Bumi Lestari*, vol. **9**, pp. 1–11
[22] M Wachid, R Hapsara, R Cahyo, G Wahyu, A Syarif, D Umarhadi, N Fitriani, D Ramadhanningrum and W Widyatmanti 2017 Mangrove canopy density analysis using Sentinel- 2A imagery satellite data Mangrove canopy density analysis using Sentinel-2A imagery satellite data *IOP Conf. Ser. Earth Environ. Sci.*, vol. 70
[23] M Kamal, S Phinn and K Johansen 2014 Characterizing the spatial structure of Mangrove features for optimizing image-based mangrove mapping *Remote Sens.,* vol. **6**, no. 2, pp. 984–1006
[24] D D Torio 2007 Modelling Canopy Density Variations from Remotely Sensed Data: Implication on Monitoring Floristic and Macro-benthic Properties of Mangrove Ecosystems Dante International Institute for Geo-Information Science and Earth Observation
[25] C Kusmana, S Sabiham, K Abe and H. WATANABE 1992 An Estimation of Above Ground Tree Biomass of a Mangrove Forest in East Sumatra, Indonesia. *Tropics,* vol. **1**, no. 4, pp. 243–257
[26] P Saenger and S C Snedaker 1993 Pantropical Trends in Mangrove Aboveground Biomass and Annual Litterfall *Oecologia,* vol. **96,** no. 3, pp. 293–299

Acknowledgement
We thank to these following institution for giving support in this paper: Pusat Pemetaan dan Integrasi Tematik - Badan Informasi Geospasial, and Lembaga Penerbangan dan Antariksa (LAPAN).

Sea Level Rise in The Java Sea Based on Altimetry Satellites Data Over 1993-2015

Ratu Almira Kismawardhani[1*], Anindya Wirastriya[1,2], Dessy Berlianty[3]

[1]Department of Oceanography, Faculty of Fisheries and Marine Science, Diponegoro University, Semarang, Indonesia
[2]Center for Coastal Disaster Mitigation and Rehabilitation Studies, Diponegoro University, Semarang Indonesia
[3]Institute for Marine Research and Observation (BROK), Ministry for Marine Affairs and Fisheries, Bali, Indonesia

ratualmirakismawardhani@gmail.com

Abstract. At any location, sea level changes over time due to many wells understood factors (ocean mass-induced, thermal expansion, etc.). By varied time series data, this research is purposing to get the rising sea level in the Java Sea related to thermal expansion and tide. Sea Level Anomaly (SLA) Altimetry data is supported by SST (Sea Surface Temperature) and SSH (Sea Surface Height) of tide measurement. SST is indicated relating to thermal expansion by surface distribution observed of SLA and SST. Both data trends are calculated by linear regression per grid of Java Sea area. Tidal components are determined by using FFT method to support long period data. The result showed the rising sea level increased up to 0,0039 m/year and followed by the increasing of SST for about 0,027° C/year over 23 years. Sea level rises and related positively to the rises SST produced by 0,56 paired statistic correlation. Temporally both tend to has the similarity in a yearly pattern on long-term period while short period shows the spatial relation. The lowest sea level falling trend occurred in 1997 and the highest rising trend in 2010. It is supported by El Nino and La Nina events. Based on the results, rising and falling rate of sea level in the Java Sea are highly related to the events through ONI (Oceanic Niño Index) value.

Keyword: SLA, SST, Java Sea.

1. Introduction

Sea level rise is the basis, of all regional or global factors, determination of climate change due to global warming. This is because all global climate change factors may have a direct effect on sea level change. The influence of atmospheric pressure due to rising global temperatures triggers a change in sea level position. In a stable state (stationary condition) the sea level can drop or rise up to 1 cm for every 1 mbar atmospheric pressure increased [1]. Changes in sea level rise can also be due to thermal expansion of inputs or heat loss in water bodies due to mass transfer of freshwater between the oceans and continents [2]. The distribution of water masses to the oceans is not always constant but constantly changing constantly and is strongly influenced by seasonal variations. The value of sea-surface changes on the coast of Australia having a change due to climate and regional climate reduction processes. The value is changed from 1.4 ± 0.3 mm per year (1966 to 2009) and 4.5 ± 1.3 mm per year (1993 to 2009) to 1.6 ± 0.2 mm / year (1966 to 2009) and 2.7 ± 0.6 mm / year (1966 to 2009). Changes are generated by reducing the ENSO-related signal factor. After the correction of GIA and atmospheric pressure is done, the re-change occurs to be of 2.1 ± 0.2 mm / year and 3.1 ± 0.6 mm / year [3]. This value is close to the global sea level rise in the last 45 years.

Observations of sea level rise can change in time scale (temporal) and space (spatial). Temporarily sea level rise has the difference due to time series data used varies and even less than 18.6 years [4]. Sea level observation of sea level position requires a long observation time of 18.6 years. Many studies proved that changes occur due to the variation of latitude on earth. Spatially, the range of global sea level rise or decrease ranges from - 10 mm to 10 mm per year. Based on observations from 1992 to present on a global scale, global sea level rise has increased by +3.28 mm annually [5].

Indonesia which located on 6° S - 11° S and 94° E-145° E have a high potential for the impact of global sea level rise. Indonesia's territory is in the range of 2.5 mm to 6 mm per year. Sea level rise in Indonesia until 2012 to reach the value of 5.84 mm per year [6]. This value is almost twice times higher than the average global sea level rise. This certainly has a big effect on the small islands in Indonesia. There are five physical effects of sea-level changes in coastal areas (ie. coastal erosion and headlands) increased flooding and hurricane damage, lowland pools, salt intrusions through groundwater aquifer systems, and an increased water position.

Thus in this paper, we focus on describing the increasing value of sea-level rise and variability around Java Sea supported by its connection to the variability of SST and tidal records.

2. Data and Method

2.1. Satellite Altimetry Data

The altimetry data set used for this study is re-processing daily sea level anomalies (SLA) Level 4 maps composed yearly with the spatial resolution of 0.25°×0.25° grids provided by Copernicus Marine Environment Monitoring Service (http://marine.copernicus.eu/). This data contained multi-mission data result of altimetry satellites since January 1993 to December 2015. It is standardized and validated by all standard geophysical and environmental corrections, including the ionospheric correction, dry and wet tropospheric corrections, solid Earth, and ocean tides, ocean tide loading, pole tide, electromagnetic bias, inverted barometer corrections, and instrumental corrections [7].

Generally, SLA data will be compared with SST and tide gauge data records. SLA and SST have analyzed spatial and temporal analysis and data for such long and short period time series. Trend value is obtained from linear trend calculation. In accordance with tide records, SLA data is computed using the daily data at the closest grid points to the tide gauge location.

2.2. Sea Surface Temperature Data

Specifically, SLA altimetry combined with sea surface temperature (SST) with the spatial resolution of 0.25°×0.25° grids and accessed from National Oceanic & Atmospheric Administration (https://www.esrl.noaa.gov/). Both SLA and SST of Java Sea compiled into yearly calculation with. Those data were also composited towards El Nino and La Nina events, determined by using Oceanic Niño Index (ONI) provided by http://www.cpc.ncep.noaa.gov/. ONI index is the anomalies of SST in the Niño 3.4 region (5°N-5°S, 120°-170°W), based on the basis period of 1993-2015. The threshold for determining El Niño and La Niña period is $\pm$0.5°C.

2.3. Tide Gauge Data

To compare to satellites altimetry data, long-term tide gauge records of six stations are chosen to investigating the present study of sea level rise. There are station Sunda Kelapa, Cirebon, Jepara, Surabaya, Tuban, and Semarang with hourly tide data from 2011 to 2015. Most data are contained some time series with several missing observations. This observation data which distributed by BIG (Geospatial Information Agencies) is separated by its validity to preserve a good data. This study

IOP Conf. Series: Earth and Environmental Science **165** (2018) 012006 doi:10.1088/1755-1315/165/1/012006

implied FFT calculation to gain the tide components. Thus, this data will be used to determine the tidal type of area observation which is led to amplitude value on each station.

3. Result and Discussion

3.1. Long Period Analysis SLA in The Java Sea

The long period spatial analysis of SLA in the Java Sea is presented in fig. 1. The trend map of SLA in the Java Sea was calculated from 1993 to 2015. The trend map shows significant spatial variability (fig.1). In some region, such as region A of the Java Sea, sea level rise up to twice faster than the global mean, while the sea level near the region (112° E, 3° S) exhibited drop. The trend map displays the Java Sea as the region with the fast anomaly. It is indicated that SLA rising trend dominating most of region-observed.

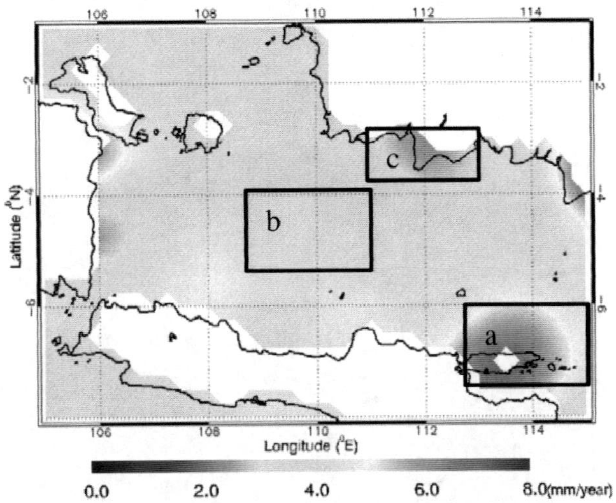

Figure 1. Spatial Variability of SST in the Java Sea. Region-observed is focused on (a) highest rising rate, (b) middle rising rate, and (c) lowest rising rate.

The trend of SLA in the Java Sea (fig.2) shows that there is increasing value since the year 1993 approximately 0.0032 m per year. In 2010, sea level rise in the Java Sea reached the fastest trend of Java Sea for about 0.0985 m while the slowest is occurred in 1994 for about -0.039 m. This interval extends for 16 years spanned the time from the lowest until the highest trend. It is showed the interval in between is about 0.14 m in 22 years. Inter-annual data for sea level trend in the Java Sea reaches its highest rising point in 1997 to 1998 and get the most falling year in 1997 to 1998. Trend value in 1997 rose from -0.035 m to 0.035 m. It means that there is increasing trend value for about 0.07 m or 7 mm on period 1997 to 1998 and decreasing value for about 0.069 m or 6,9 mm on period 1996 to 1997. Both show an extreme change of sea level trend occurred in period 1996 until 1998.

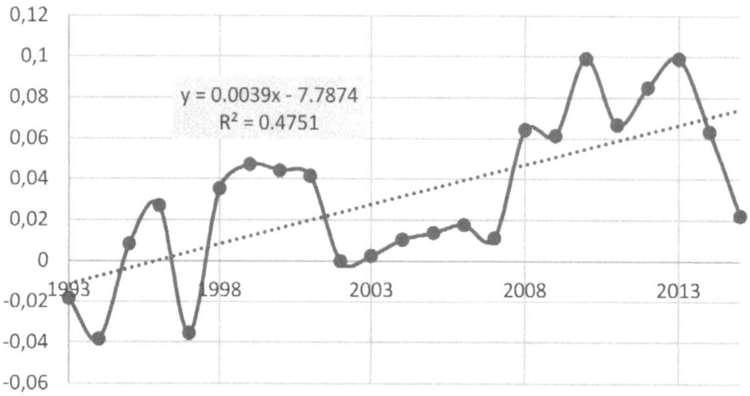

Figure 2.Temporal Variability of Java Sea

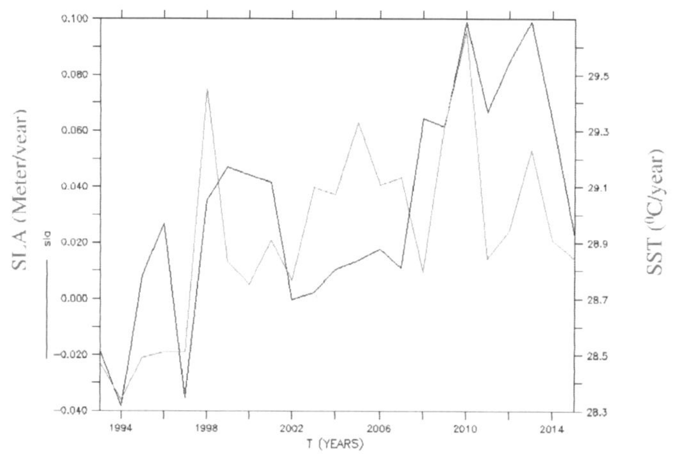

Figure 3.SLA and SST Temporal Variability of Java Sea

In the comparison with SST during the time of observation, the value trend of SST shows a positive trend. The increasing value for about 0,027° C/year. Temporally, the variation between period reaches the highest on 1998 with value of 29,48° C and 2010 for about 29,69° C. The lowest is valued on 1994 by 28,52° C. Those values closely showing the relation between SST toward SLA which SLA also fell to the lowest value in 1994 (-0,038 m/year) and reached the highest in 2010 (0,09895).

The anomaly is caused by the increase of sea level in the Pacific Ocean and giving more stress on the surface, there is occurred mass-induced through the Pacific Ocean to keep the balance in the ocean [8]. This phenomenon supports the Java Sea to surpassing the ocean mass to fill the empty chamber of Pacific Ocean. There are two factors of sea level anomaly such as steric variation caused by thermal expansion in the ocean and mass-induced between the land and ocean [9]. The location of Java Sea which is located between island probably causing mass-induced of land and ocean or vice versa.

In accordance with SST distribution in the Java Sea, ONI value (Table 1) showed the relation of El Nino and La Nina events towards the anomaly of sea level. During 2009 to 2011, the La Nina and El Nino event support the increasing of SLA up to the highest value as it showed below.

Table 1. Oceanic Nino Index *(ONI) on Region Latitude 50^0 N-50^0 S and longitude 120^0 -170^0 W from 2009 to 2011. (*) displayed the event of El Nino and (**) displayed the event of La Nina.*

Year	DJF	JFM	FMA	MAM	AMJ	MJJ	JJA	JAS	ASO	SON	OND	NDJ
2009	**-0,7**	**-0,6**	**-0,4**	**-0,1**	0,2	0,4	0,5*	0,5	0,6	0,9	1,1	1,3
2010	1.3	1.2	0.9	0.5	0.0	-0.4	-0.9**	-1.2	-1.4	-1.5	-1.4	-1.4
2011	-1.3	-1.0	-0.7	-0.5	-0.4	-0.3	-0.3	-0.6	-0.8	-0.9	-1.0	-0.9

3.2. Short Period Analysis of SLA in The Java Sea

To investigate the short period analysis of SLA to the variability of SST in the Java Sea, we composed the SLA in spanned time up to five years and compared with also five years data of SST as shown in Fig. 4. Generally, SST has a strong correlation with the SLA. But SST displays contrary influence toward the rising of SLA spatially in the first period of observation. It is showed from spatial visualization that given in figure 4. The image of the high rising area of SLA displayed red while SST on the same area showed blue. SLA image has the rising level that gradually decreasing from land to open ocean. The correlation is also shown the value of correlation between region a, b, and c as displayed in figure 1. Region a has the lowest correlation value for about 0,488 meanwhile region b has the value of 0,547 and the highest correlation is region c with the value of 0,572. It is showed that the faster rates of SLA caused the lower correlation with the distribution of SST. Thus, it showed complex factors causing the variability of highest SLA in a region.

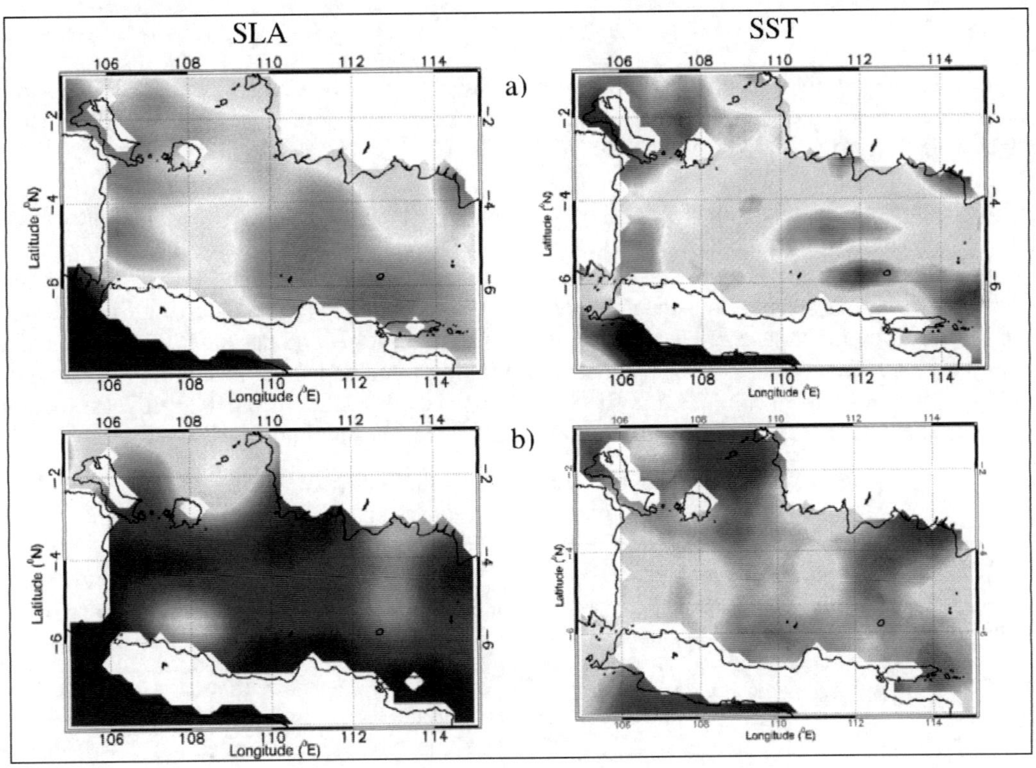

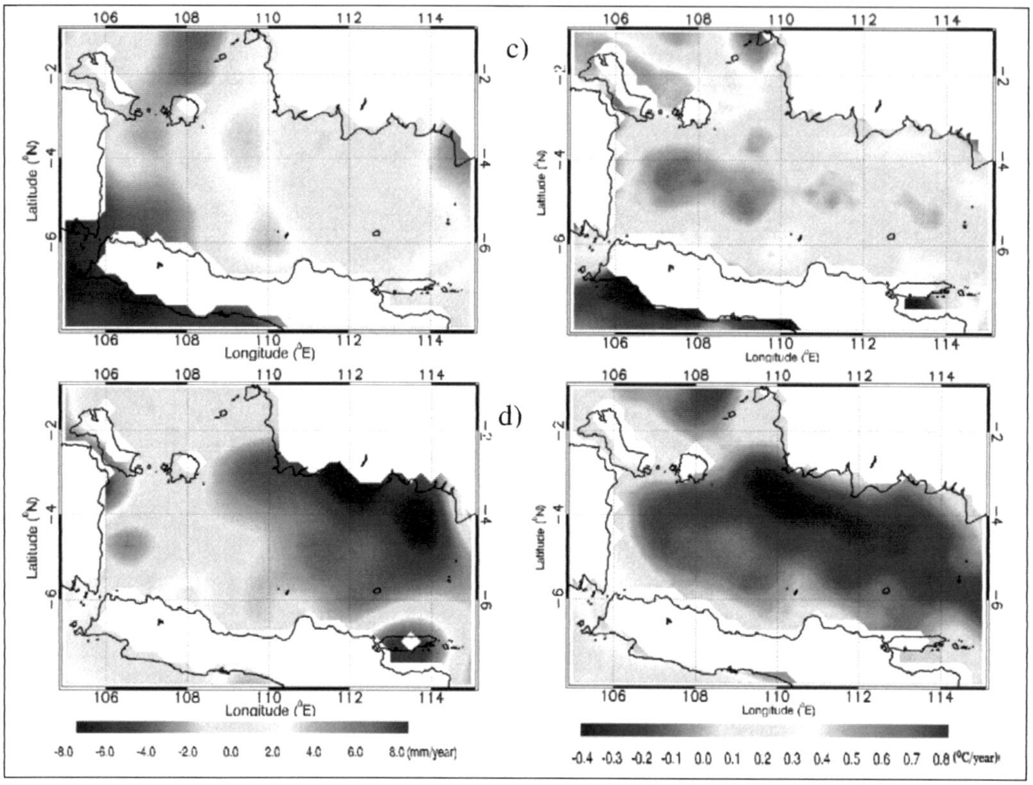

Figure 4. The comparison of the spatial trend between SLA towards SST in the Java Sea on short time series. The variability consists of several short period such as (a) 1993- 1997, (b) 1998 – 2002, (c) 2003 – 2007, and (d) 2008 – 2015.

For investigating thermal expansion, SST parameter is chosen yet displayed the main movement of thermal expansion in the Java Sea. Due to lack in-situ data, spatial observation of thermal expansion is not quite displayed. Both data showed the higher rate of the anomaly since the first time span and increasing over time. The first time span observation rates highly due to the occurring El Nino and the fourth time span by La Nina. Spatial observation showed that the second decade (2004-2015) the Java Sea has a higher rising rate in both of parameter and falling trend happened for the first decade (1993-2003). As it is displayed in figure 3, the first decade reaches the value of 0,01025 m/year. In another case, the value of the second decade (during 2004 to 2015) reaches up to 0.051091 m/year.

Table 2. The comparison between satellite altimetry and tide gauge data at the same observation time span

Tide Gauge Station	Location		Time Span	Distance (km)	Trends (m/yr)	
	Longitude	Latitude			Tide Gauge	Altimetry
Sunda Kelapa	106°48'34.21" E	6° 7'30.77" S	June 2012 - December 2015	3,29	-0,000052 ± 0,00065	-0,00025 ± 0,00014
Semarang	110°25'25.49" E	6°56'30.77" S	September 2011 – October 2015	2,63	-0,00093 ± 0,0011	-0,0013 ± 0,00066
Jepara	110°38'55.04" E	6°35'30.46" S	April 2011 – June 2015	7,24	-0,00064 ± 0,47	-0,0005 ± 0,0017

Tide Gauge Station	Location		Time Span	Distance (km)	Trends (m/yr)	
	Longitude	Latitude			Tide Gauge	Altimetry
Cirebon	108°35'4.50" E	6°44'1.91" S	December 2013 – July 2015	2,52	-0,0033 ± 0,00040	-0,00028 ± 0,003
Tuban	111°56'47.78" E	6°45'49.38" S	April 2011 – December 2015	2,73	-0,0052 ± 0,0000059	-0,00065 ± 0,00044
Surabaya	112°44'26.01" E	7°12'0.23" S	August 2011 – December 2015	35,02	-0,00066 ± 0,00071	-0,001 ± 0,00059

Comparing with tide gauge data, SLA showed less related to tide gauge trend spanned in five years observation. Each station has different time span based on the records. That is shown in Table 1 which every station stated different time observation. It is due to lack of the data records and also we only separated the valid value of tide gauge data. Table 1 showed a quite significant different value of observation between the two parameters. It is left for future study to have longer time span and more accurate tide data. In another case, rising and falling pattern between tide gauge and altimetry data showed more correlation by its time span. The longer data used the better correlation will be made.

4. Conclusion

Sea level rise in the Java Sea has been investigated by using long-term observation of altimetry data (1993-2015). The conclusions are as follows:

1. The Java Sea shows increasing the value of sea level trend for about 0,032 m/year during 23 years observation. It is nearly 0,09895 m/year with the highest rising anomaly on 2010. Meanwhile, the lowest falling trend reaches on 1994 for about -0,039 m/year.
2. The first decade of observation (during 1993 to 2003) results in the lower trend of sea level anomaly by the value of 0,01025 m/year. In another case, the value of the second decade (during 2004 to 2015) reaches up to 0.051091 m/year which is indicated that rising trend happen mostly in the second decade.
3. Sea level anomaly trend increases over time and in line with the increasing of SST. Particularly, during extreme El Nino (1997-1998) and La Nina (2009-2010) events. In another case, SLA is correlated with tide gauge data yet particularly with the longest data such as station of Surabaya and Sunda Kelapa.
4. The lowest and highest SST anomaly throughout the year of observation occurs exactly the same as SLA by the value of 28,34^o C in 1994 and 29,66^o C in 2010 with the average value of 28,93^o C. SLA and SST is closely correlated by the correlation value of 0,56. Yet, shows the mass-induced from thermal expansion as the main factor of SLA anomaly trend in the Java Sea.

References

[1] Lisitzin E 1974 Sea Level Changes Elsevier Scientific Publishing Company Netherlands 295
[2] Willis Josh K 2010 Global Sea Level Rise Recent Progress and Challenges for The Decade to Come Oceanography **23 (4)** 26-35
[3] White Neil J 2014 Australian Sea Levels—Trends Regional Variability and Influencing Factors Earth-Science Review **136** 155-174
[4] Ali M, D K Miharja and S Hadi 1994 *Pasang Surut Laut* (Bandung: Institut Teknologi Bandung)
[5] AVISO 2015 Through Indonesia https://www.aviso.altimetry.fr/en /news/idm/2015/apr-2015-through-indonesia.html
[6] Nababan B 2015 Dinamika Anomali Paras Laut Perairan Indonesia. *Journal Ilmu Kelautan and Teknologi Kelautan Tropis* **7 (1)** 259-272

[7] Geruo A, Wahr J and Zhong S 2013 Computations of the viscoelastic response of a 3-D compressible Earth to surface loading: an application to glacial isostatic adjustment in Antarctica and Canada. Geophys *J. Int.* **192(2)** 557
[8] Trujillo Alan P and Harold V Thurman 2011 Essentials of Oceanography Ed 10 Prentice Hall United Stated 577
[9] Wei Feng and Zhong Min 2015 Global Sea Level Variations from Altimetri GRACE and Argo Data over 2005-2014 Geodesy and Geodynamics **6 (4)** 274-279

Detection of SST Wake in Belitung Island Induced by The Monsoon Wind

Hariyadi[1,2], Anindya Wirasatriya[1,2*], Jarot Marwoto[1,2], Nur Taufiq-Spj[3] and Destina Verika Currie[1]

[1]Department of Oceanography, Faculty of Fisheries and Marine Science, Diponegoro University, Semarang, Indonesia
[2]Center for Coastal Disaster Mitigation and Rehabilitation Studies, Diponegoro University, Semarang Indonesia
[3]Department of Marine Science, Faculty of Fisheries and Marine Science, Diponegoro University, Semarang, Indonesia

aninosi@yahoo.co.id

Abstract. The Java Sea is one of the monsoon wind path blowing over Indonesia Seas. Belitung Island is located in the center of the monsoon path in the Java Sea. Although the topography Belitung Island relatively flat, the existence of mountainous with 200-300m height in the center of the Island may form the island wake during the strong wind event. Island wake phenomena in Belitung Island was investigated by using MODIS SST Lv3. From 2003 to 2016 period of observation, 245 cases of SST wake were found with the distribution for each season are 39, 35, 135 and 36 cases for NW monsoon, Transition I, SE monsoon and Transition II, respectively. The direction of SST wake depends on the direction of the surface wind. During NW monsoon and Transition I, the direction of SST wake is southeastwardly while during SE monsoon and Transition II is northwestwardly. The high Chl-a concentration outside the SST wake patch indicates that the occurrence of SST wake in the seas of Belitung Island may influence biological productivity in that area.

Keyword: Belitung Island, monsoon, SST wake, topographic effect, blocked wind

1. Introduction

In geophysics "island wakes" is a definition used to refer to atmospheric effects induced by the island mountains' (atmospheric wakes), as well as ocean effects induced by the island's bathymetry (ocean wakes). Ocean wakes, can be grouped into two main categories (i) wakes induced by atmospheric phenomena (e.g. wind induced wakes), and (ii) wakes induced by the oceanic phenomena (e.g. ocean current wakes). Both atmospheric and oceanic wakes have been the subject of intense study by the scientific community. The most robust evidence of island wake phenomenon is found in the Hawaiian Islands which has been reported by many studies [e.g., 1,2,3,4].

Island wake in the Hawaiian Islands is generated by the existence of easterly trade wind that persistently blowing throughout the year. The trade wind is blocked by the islands creating wind shadow in the lee of the islands. Using satellite data, [1] detected a wind wake trailing westward behind the Hawaiian Islands for 3000 kilometers, a length many times greater than observed anywhere else on Earth. This wind wake drives an eastward ocean current that draws warm water from the Asian coast 8000 kilometers away, leaving marked changes in surface and subsurface ocean temperature. Moreover, [4] revealed that the distributions of high SST, low wind and high concentration of cloud liquid water (CLW) in the lee of Hawaiian Islands are caused by the Hawaiian Lee Counter Current (HLCC) and that this current is driven by the wind-curls induced by the orographic effect of the islands. It is also shown that wind changes around the Hawaiian Islands can further affect the speed of the North Equatorial Current (NEC) and SST over the current, and intra-annual variability in CLW to the west of the islands is governed, not only by SST but also by wind speed. Thus, island wake in the Hawaiian Islands has an important role in regulating the global climate through ruling the speed of NEC which is a part of the north Pacific Gyre.

Lying on the confluence of the Eurasian Plate, the Indo-Australian Plate, and the Pacific Plate is the Indonesian Archipelago. The Indonesian region is also known as the "Maritime Continent" [5]. Because of its unique position, the Indonesian maritime continent experiences seasonal (i.e., monsoons) to inter-annual (i.e., El Nino-Southern Oscillation, ENSO) climate variations. There is two monsoon seasons every year, i.e. southeast (SE) monsoon and northwest (NW) monsoon. The SE monsoon (dry season) is associated with easterlies from Australia that carry warm and dry air over the region. On the other hand, the NW monsoon (rainy season) is associated with westerlies from the Eurasian continent that carry warm and moist air to the Indonesian region [6]. The pathways of the monsoon wind across Indonesian Seas is illustrated in Fig. 1. As an archipelagic state, Indonesia has 16.056 islands, spread over 5.180.053 km^2 area (http://www.big.go.id). This large number of the island may interact with the monsoon wind creating typical island wakes phenomena which may differ from the Hawaiian Islands due to the change of monsoon wind direction in every half year which makes the island wake is not persistence throughout the year.

Belitung Island is the second largest island after Bangka Island in Bangka Belitung Province. Besides the two mainlands, this province possesses a number of small islands like Sellu Island, Mendanau Island, Nangka Island ring, Memperak, Buku Limau, Sekunyit and Long Island ring. Belitung Island lies on the strategic location in trading and economic position in within trading and economic route in South East Asia region, Asia, and Indonesia. Belitung Island is located at the one of the monsoon path i.e., Karimata Strait with an areal size of 4.800 km^2. Although the topography of Belitung Island is relatively flat with mainly less than 100 m height, in the center of the Island, as shown by the red arrow, mountainous area exists with approximately 200-300 m height (Fig.1). This high topography is potential for blocking the monsoon wind for creating island wake in its lee. Since monsoon wind changes its direction seasonally, the island wake may also occur not only on one side of the island which makes the island wake in Belitung Island is different with it in the Hawaiian Island. Thus, Belitung Island is the potential area for island wake study in Indonesia. In this study, we identified the high SST appearances in the lee of Belitung Island referred as the SST wake phenomena influenced by the bidirectional monsoon by using infrared based SSTs.

2. Data and Method

Since SST wake is the air-sea-land interactions, the topography is the important parameter. We used a 10m resolution of Digital Elevation Model and the Global 30 arc sec elevation (GTOPO30) for topography. For island wake identification, we use Sea Surface Temperature (SST) from Moderate Resolution Imaging (MODIS) Level 3 onboard Aqua and Terra satellites with 4 km of spatial resolution [7] Terra passes over the equator at approximately 10:30 am and 10:30 pm each day, while Aqua satellite is at approximately 1:30 pm and 1:30 am. Thus, we can get 2 images of SST during daytime which is important for identifying SST wake in the Belitung Island. Moreover, because this SST data is generated based on an infrared sensor, it has the advantage to observe SST only during clear sky condition which is favorable for the generation of SST wake.

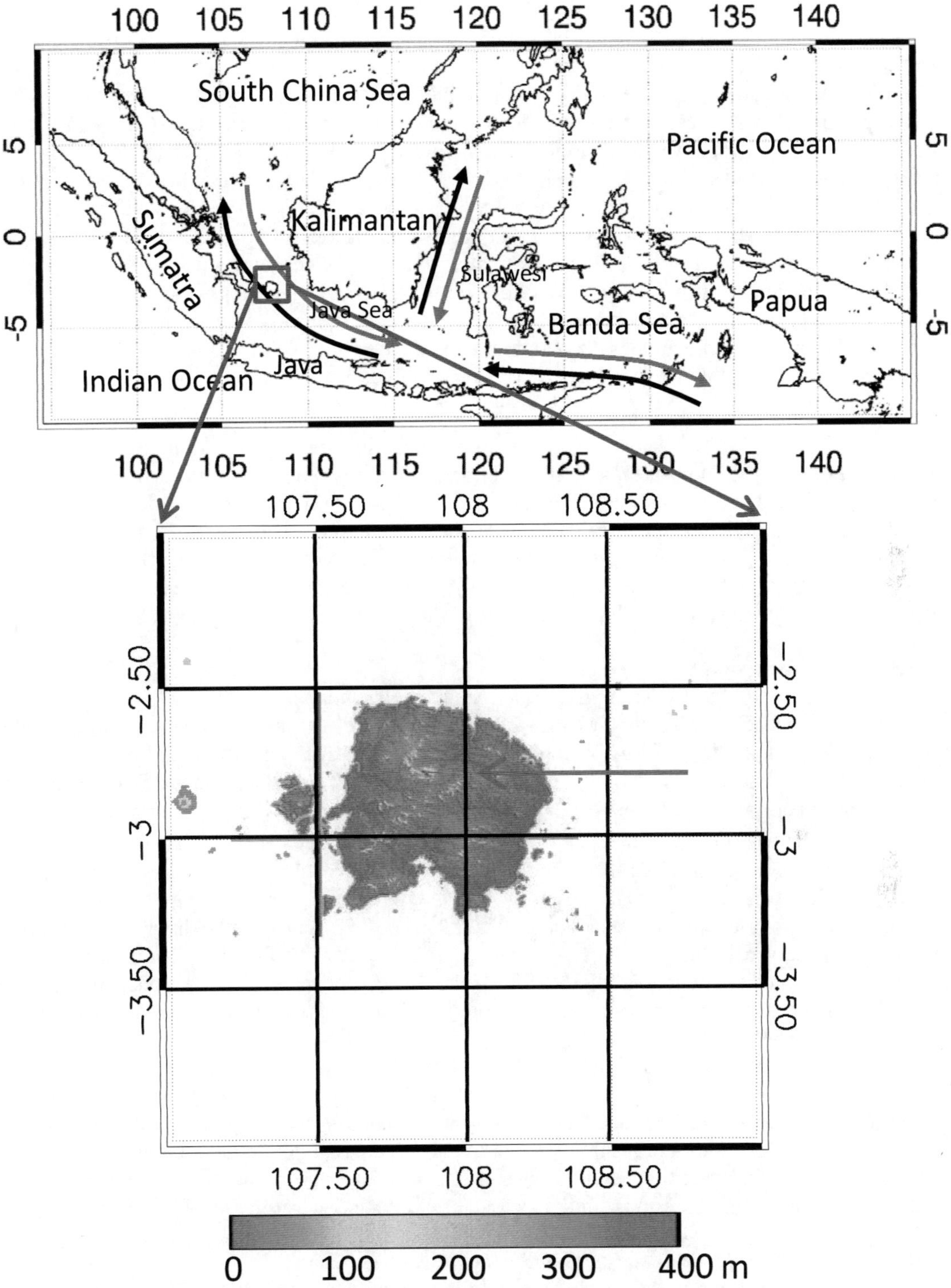

Figure 1. Monsoon wind path of Indonesian Archipelago (Upper figure) and the Topographic map of Belitung Island (Lower figure). Black arrows denote the paths of southeast monsoon season (Jun, July, and August) and gray arrows denote the paths of Northwest monsoon season (December, January February).

For investigating the mechanism of the SST wake, we used The Cross-Calibrated Multi-Platform (CCMP) gridded surface vector winds [8]. This product is produced using satellite, moored buoy, and model wind data, and as such, are considered to be a Level-3 ocean vector wind analysis product. The new CCMP V2.0 dataset now available from Remote Sensing Systems (RSS). The V2 CCMP processing now combines Version-7 RSS radiometer wind speeds, QuikSCAT and ASCAT scatterometer wind vectors, moored buoy wind data, and ERA-Interim model wind fields using a Variational Analysis Method (VAM) to produce four maps daily of 0.25 degree gridded vector winds. For diurnal SST data, we used deltaSST_v7.1.3 produced by Tohoku University, Japan with grid interval of 0.1°. The method to estimate diurnal SST amplitude is explained in [9] following the parametric model described in [10]. For investigating the oceanic response on island wake, surface chlorophyll-a distribution was investigated. We used daily MODIS Chlorophyll-a Lv3, with spatial resolution 4 km [11]. All parameters derived from remote sensing and reanalysis data were composted into monthly climatology to obtain seasonal pattern following [12]. First, all parameters were calculated into monthly means and we derived monthly climatologies by using this formula:

$$\bar{X}(x,y) = \frac{1}{n}\sum_{i=1}^{n} xi\,(x,y,t) \qquad (1)$$

where $\bar{X}(x,y)$ is monthly mean value or a monthly climatology value at position (x,y), $xi(x,y,t)$ is i^{th} value of the data at (x,y) position and time t. Furthermore, n is number of data in 1 month and the number of monthly data in 1 period of climatology (i.e., from 2003 to 2016 = 14 data) for monthly calculation and monthly climatology calculation respectively. Pixel xi is excluded in the calculation if it is a hollow pixel. The investigation of island wake was conducted by displaying all semi-daily data of SSTs for 14 years observation (i.e., 2003-2016). Then, island wake phenomena were detected manually by eye. The robust cases were chosen for further analysis, investigating the oceanic response.

3. Result and Discussion

3.1. Seasonal variation of SST in the seas around the Belitung Island

The seasonal variation of SST in the seas around the Belitung Island is shown by the monthly climatology of SSTs in Fig. 2. The NW monsoon, the transition I, SE monsoon and transition II are represented by December, January, February; March, April, May; Jun, July, August; and September, October, November, respectively. In NW and SE monsoon, SST tends to be low ranges about 27°C to 29°C. Lowest SST was detected in the north-western part of the study area in January. In April and November, SST increased to more than 29°C. The peak of SST cooling (warming) occurred in NW and SE monsoon (Transition I and II) season. Thus, semi-annual variation characterizes the SST variation in the seas around the Belitung Island. This result is consistent with the distribution of SST in the Java Sea observed by the previous studies [e.g., 13, 14]. The SST wake phenomena were not detected by using climatological analysis. The short occurrence period of this phenomenon was dissembled by the averaging process in the climatological analysis. Thus, daily data observation is needed to capture the SST wake phenomena.

The physical parameters underlying the SST climatology distribution in the seas around the Belitung Island is surface wind speed. Fig. 3 shows the monthly climatology of surface wind. The minimum SST occurs during NW and SE monsoon was related to the maximum wind speed during that season. Conversely, during transition season, the SST increased due to the minimum wind speed. Thus, the stronger wind speed induced stronger mechanical mixing and heat release which reduced SST. This wind-driven mechanism is similar as found in Maluku Sea as found by [12, 15].

3.2. Detection of SST wake in the seas of Belitung Island

The detection of SST wake in the seas of the Belitung Island was conducted by investigating semi-daily MODIS SST onboard Aqua and Terra satellite visually from 2003 to 2016. Thus, about 18.000 scenes of SST was investigated to obtain SST wake.

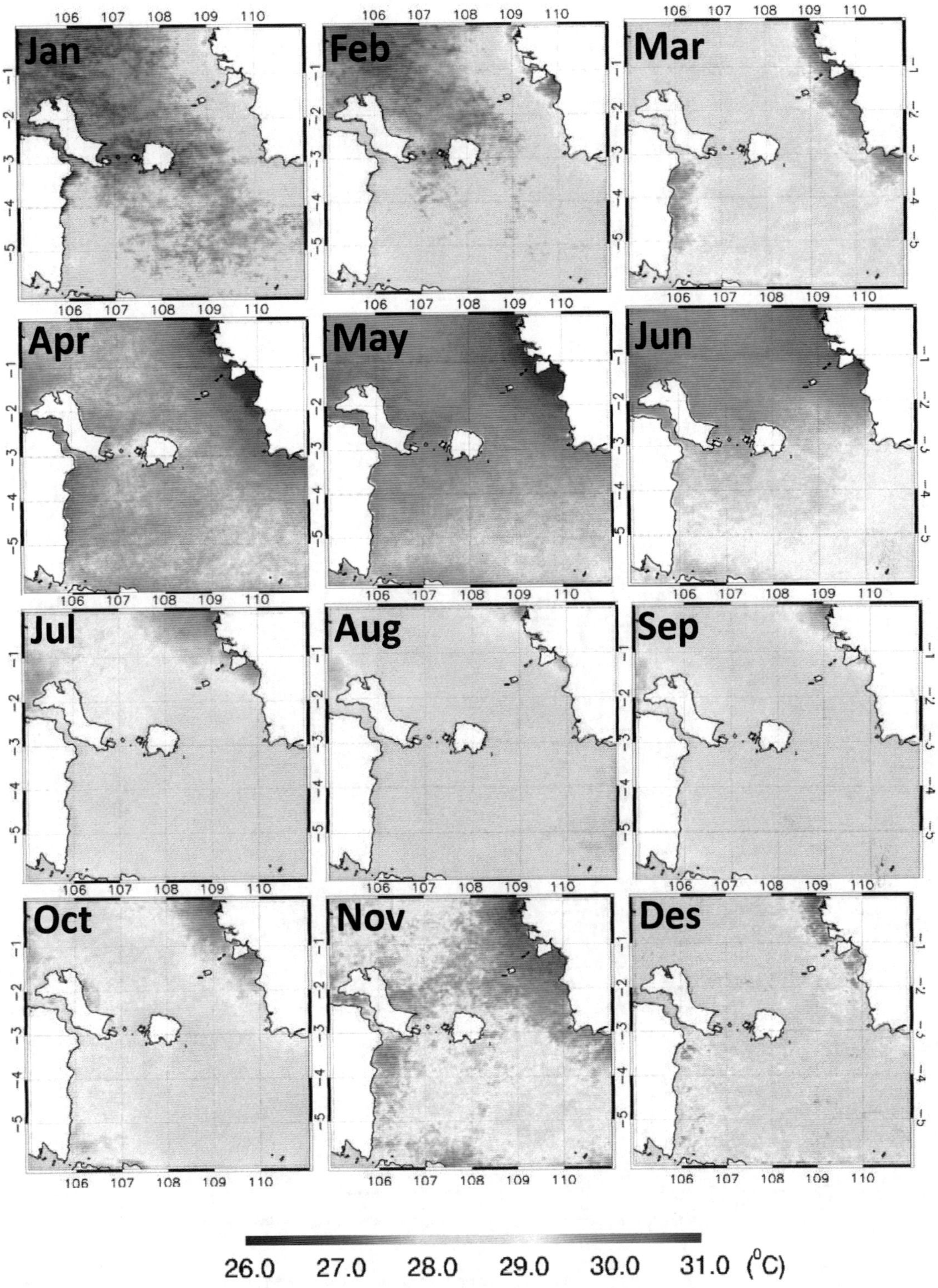

Figure 2. Monthly climatology of SST in the seas around Belitung Island calculated from semi-daily MODIS SST Lv3 from 2003 to 2016

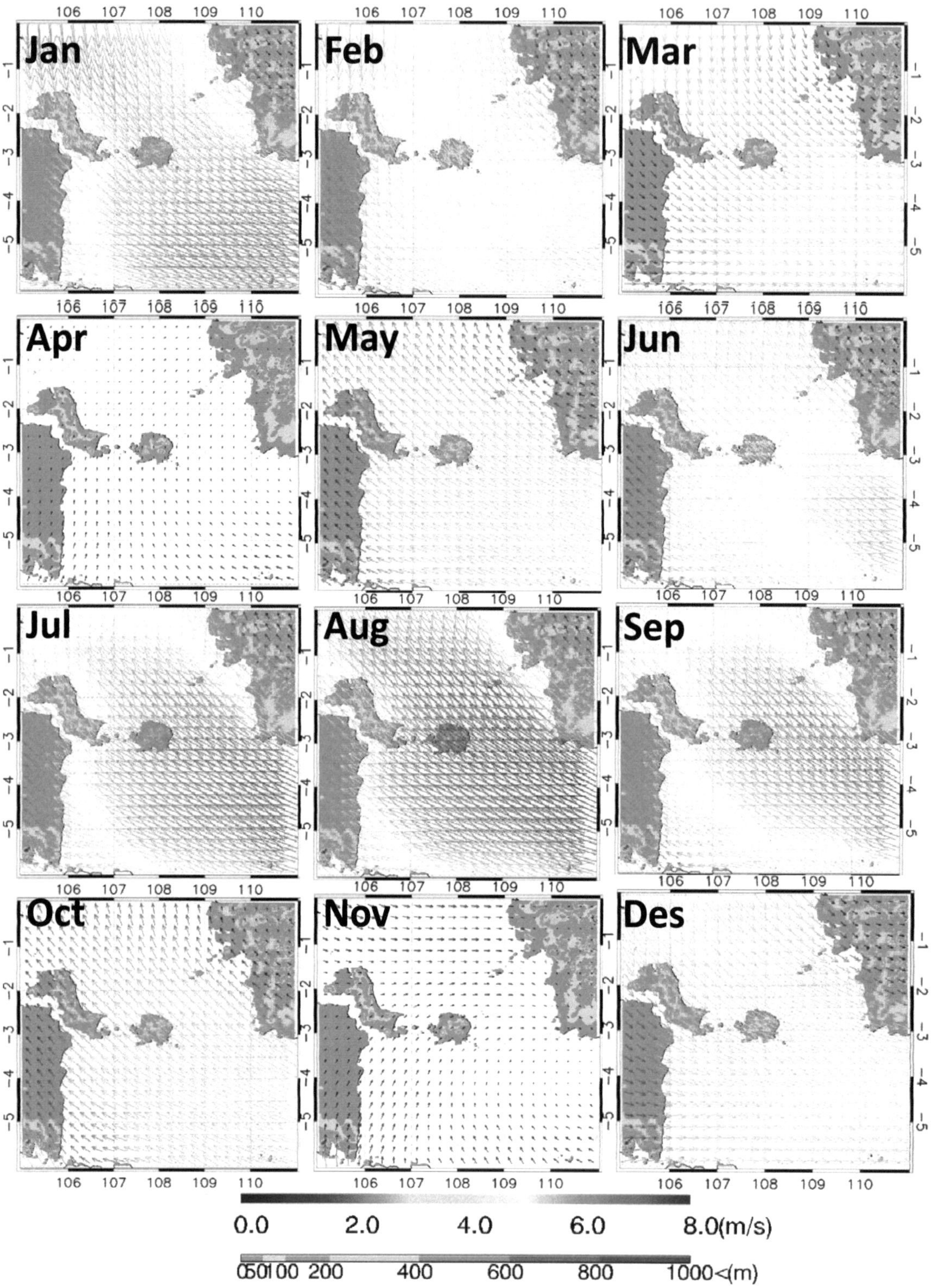

Figure 3. Monthly climatology of surface wind around Belitung Island calculated from semi-daily CCMP wind product from 2003 to 2016

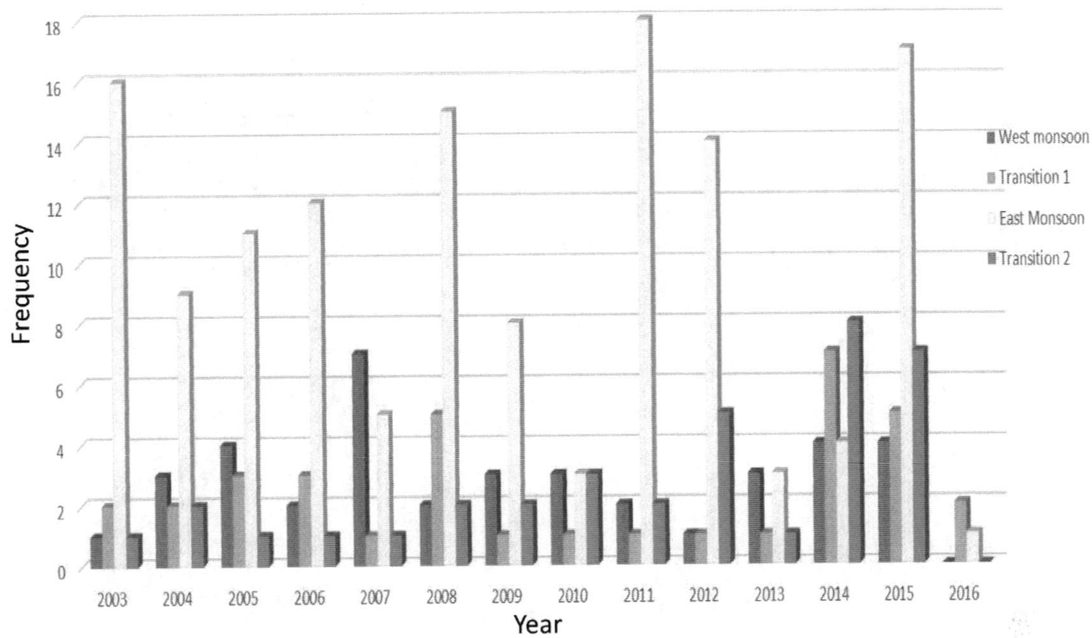

Figure 4. Number of detected SST wake in the seas around Belitung Island from 2003 to 2016

Because MODIS SST product is collected by an infrared sensor, it has the advantage that SST is observed only during clear sky condition which is favorable for the generation of SST wake. The result is shown in the histogram in Fig. 4. We found 245 SST wake cases from 2003 to 2016 period. Each season distributions were 39, 35, 135 and 36 cases for NW monsoon, Transition I, SE monsoon and Transition II, respectively. The highest occurrence of SST wake during SE monsoon may be related to the dry season which occurs during this period. During the dry season, less cloud was formed caused the high solar radiation frequently occurred during this season. The high solar radiation was needed for the generation of high SST as well as SST wake. Conversely, NW monsoon is a rainy season. Thus, the cloud formation increased during this season, reduced the solar radiation penetrating to the sea surface. During transition season, wind direction intermittently changed which SST wake was difficult to be generated. The detection of SST wakes indicates that the high topography in Belitung Island was able to block the monsoon wind, created SST wake in its lee.

The examples of SST wake cases are shown in Fig. 5,6,7 and 8 for NW monsoon, Transition I, SE monsoon and Transition II, respectively. The SST wake phenomena are denoted by the high SST patchiness (i.e., more than 29°C) extended from the lee of the island toward the sea. The length of SST wake was more than 100 km and the longest one was observed during the Transition I for more than 200 km. Thus, comparing to the SST wake in Hawaiian Island as found by [1,3,4], SST wake in Belitung Island is much shorter which is caused by the much lower topography and wind speed.

During the NW monsoon and Transition I (SE monsoon and Transition II) season, the direction of SST wake was southeastwardly (northwestwardly). This direction was related to the surface wind direction which wes southeastwardly (northeastwardly) during the NW monsoon and Transition I (SE monsoon and Transition II) season. Moreover, SST wake was located in the lee of the Island where the wind speed was low due to the existence of mountainous. However, the wind wake phenomena are not clearly seen in the result. This may be caused by the spatial resolution of wind data is not sufficient to observe the braking effect of the wind in the lee of the island. For more detail analysis of the wind effect on the generation of SST wake, the much higher spatial resolution of wind data is needed. This is left for future study.

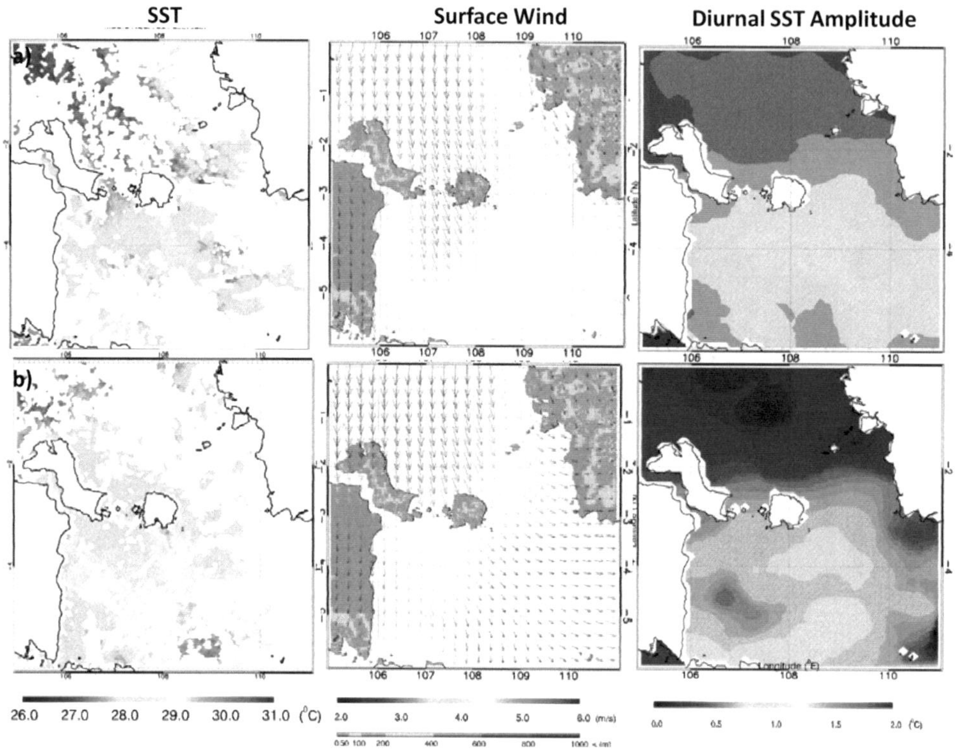

Figure 5. Example cases of SST wake during NW monsoon on (a) 07 Feb. 2007 and (b) 22 Jan. 2004.

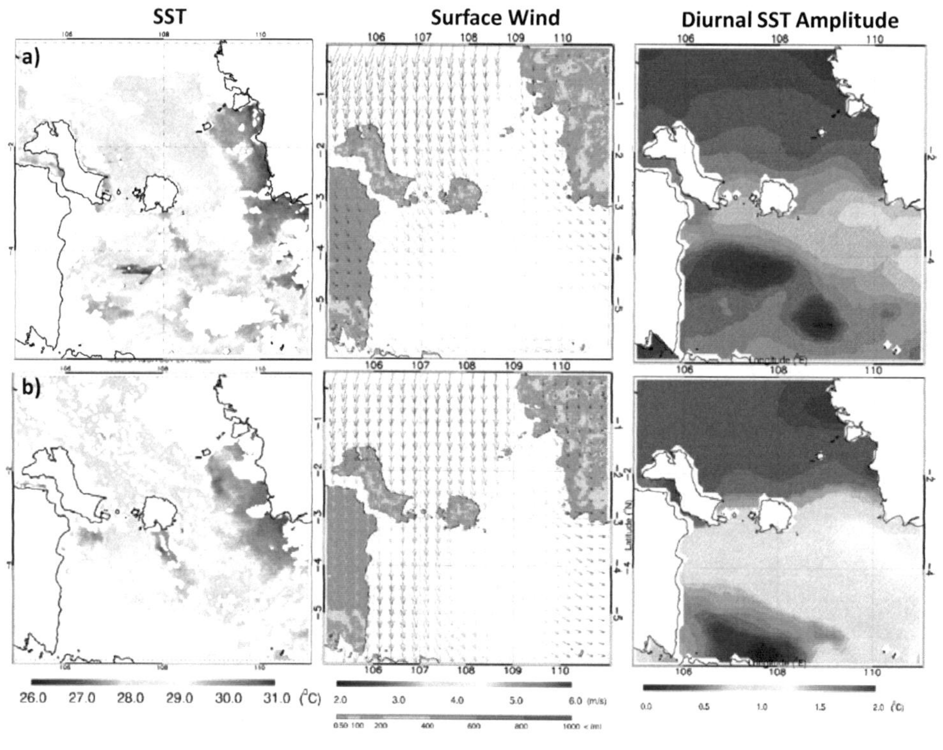

Figure 6. Example cases of SST wake during Transition I on (a) 14 March 2005 and (b) 23 March 2014.

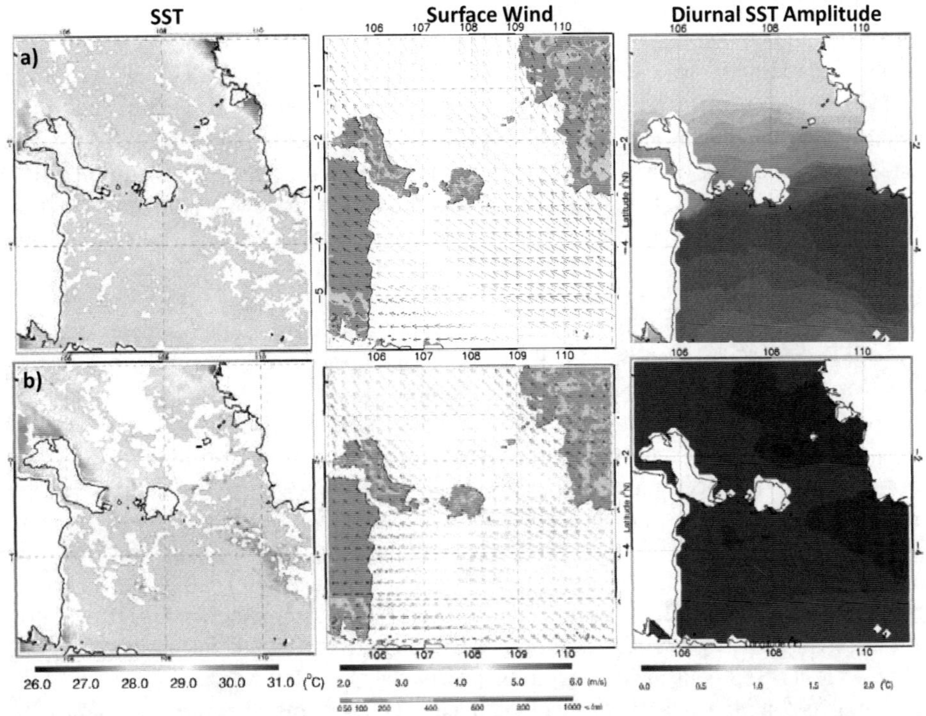

Figure 7. Example cases of SST wake during SE monsoon on (a) 01 Aug. 2008 and (b) 13 Aug. 2003.

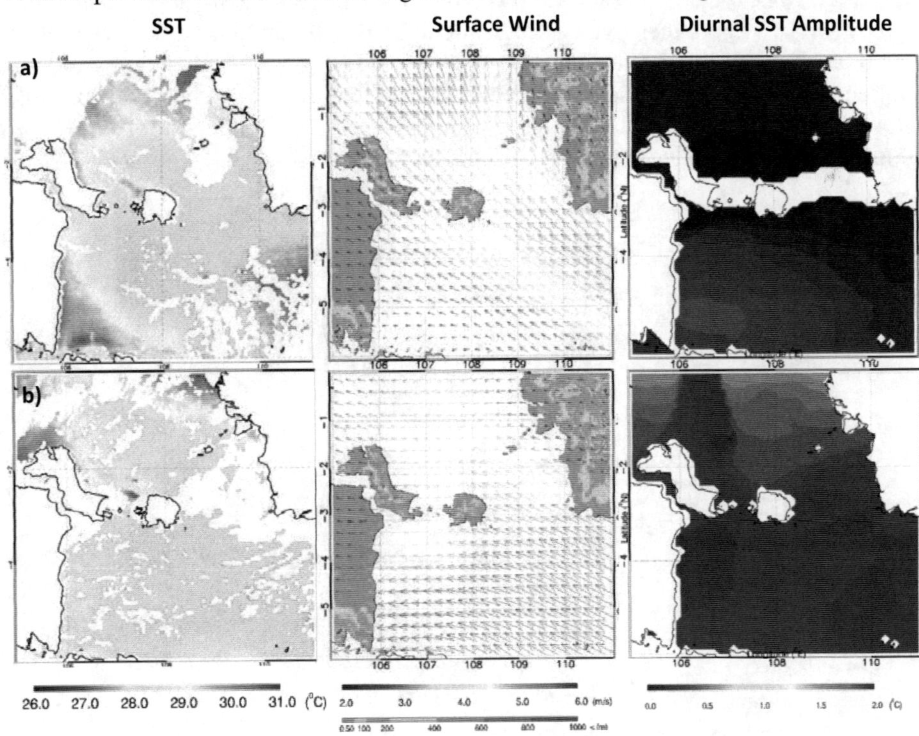

Figure 8. Example cases of SST wake during Transition II on (a) 09 Sep. 2015 and (b) 19 Sep. 2014

The diurnal SST amplitude data also demonstrate the same tendency with the surface wind. High diurnal SST amplitude means that the SST difference between the maximum and minimum SST within a day is high. High diurnal SST amplitude is caused by high solar radiation and low wind speed. Thus, lee of the island is potential for generating high DSST amplitude. Fig. 5-8 show that although the SST wake was located at the high DSST area of more than 0.5°C, the area of SST wake was not well coincided with the area of high DSST amplitude. The area of high DSST amplitude coincides with the low wind speed area. This is caused by the calculation of DSST amplitude data used partly the surface wind data as CCMP wind.

3.3. Effect of SST wake on the biological productivity

To investigate the effect of SST wake on the biological productivity of the seas around the Belitung Island, we plotted the Chl-a distribution during the SST wake event. An example of Transition I season is shown in Fig. 9a. It is clearly seen that the low Chl-a concentration coincided with the SST wake patch on 7 February 2007. Outside SST wake area, Chl-a concentration was higher. The high Chl-a concentration may be generated due to the occurrence of eddies around the SST wake patch. This phenomenon also has been reported in other areas with the more robust signal. Island wakes are also related to the concept of ''island mass effect.'' which was first described by [16] as an increase in ''biological productivity,'' that is, chlorophyll concentration, on the east side of Oahu Island in the Hawaiian Archipelago. Subsequently, the effect of island bathymetry on ambient oceanography was generalized to include island effects around entire islands [17] and island mass effects now include differences in phytoplankton, zooplankton and fish biomass due to the formation of island-induced wakes, eddies, fronts, filaments and upwelling processes [see also 18-22]. Thus, the blocky wind induces the high potential vorticity at the lateral island, triggering the eddy formation and finally transporting the nutrient from the deeper layer to the sea surface which increases Chl-a. However, the occurrence of SST wake was not always followed by the increase of biological productivity. Fig. 9b shows that on 10 March 2006, there was no indication of Chl-a bloom during the event of SST wake. This may be related to the different response of blocky wind by the high topography on inducing the eddy generation in the seas around the island. Nevertheless, this indication should be examined further in the Belitung Island by using a much higher resolution of surface wind data for calculating the wind curl and Ekman pumping velocity. This is left for future study.

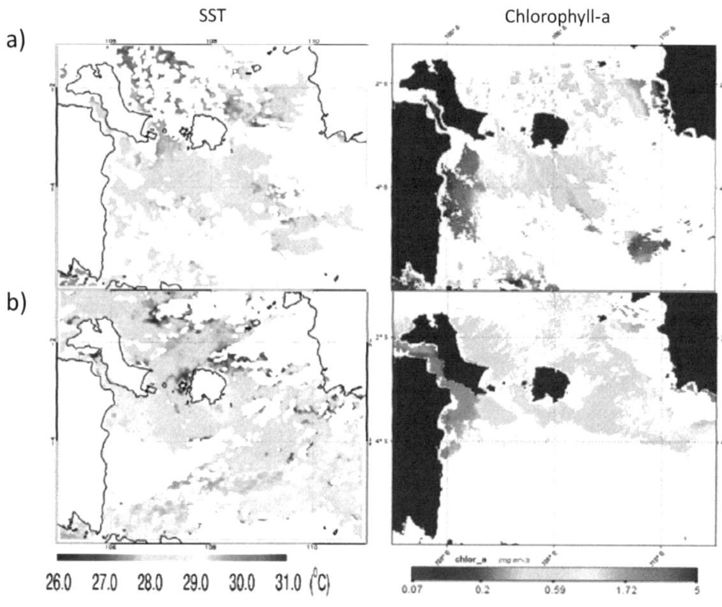

Figure 9. SST and Chl-a concentration on (a) 07 February 2007 and (b) 10 March 2006

IOP Conf. Series: Earth and Environmental Science **165** (2018) 012007 doi:10.1088/1755-1315/165/1/012007

4. Summary

Island wake phenomena in Belitung Island has been investigated by using MODIS SST. The result can be summarized as follows :

a. Although the topography Belitung Island relatively flat, the existence of mountainous with 200-300m height in the center of the Island may form the island wake during the strong wind event.
b. The peak of SST cooling (warming) occurs in NW and SE monsoon (Transition I and II) season following the wind speed distribution. The higher wind speed the lower SST occur in the seas around Belitung Island.
c. From 2003 to 2016 period of observation, 245 SST wake cases were found with the distribution for each season are 39, 35, 135 and 36 cases for NW monsoon, Transition I, SE monsoon and Transition II, respectively
d. The direction of SST wake depends on the direction of the surface wind. During NW monsoon and Transition I, the direction of SST wake is southeastwardly while during SE monsoon and Transition II is northwestwardly.
e. The high Chl-a concentration outside the SST wake patch indicates that the occurrence of SST wake in the seas of Belitung Island may influence biological productivity in that area.

References

[1] Xie SP, Liu WT, Liu Q and Nonaka M 2001 Far-Reaching Effects of the Hawaiian Islands on the Pacific Ocean-Atmosphere System *Sci.* **292** 2057 DOI: 10.1126/science.1059781
[2] Chavanne C, Flament P, Lumpkin R, Dousset B and Bentamy A 2002 Scatterometer observations of wind variations induced by oceanic islands: Implications for wind-driven ocean circulation *Can. J. Rem. Sens.* **28(3)** 466
[3] Hafner J and Xie SP, 2003 Far-field simulation of the Hawaiian Wake: Sea surface temperature and orographic effects *J. Atmos. Sci.* **60** 3021
[4] Sakamoto TT, Sumi A, Emori S, Nishimura T, Hasumi H, Suzuki T and Kimoto M 2004 Far-reaching effects of the Hawaiian Islands in the CCSR//NIES//FRCGC high-resolution climate model *Geophys. Res. Let.* **31** L17212 **DOI**:10.1029/2004GL020907
[5] Qu T, Du Y, Strachan J, Meyers G and Slingo J 2005 Sea Surface Temperature and its Variability in the Indonesian Region *Oceanography* **18(4)** 50 http://dx.doi.org/10.5670/oceanog.2005.05.
[6] Setiawan RY and Habibi A 2010 SST cooling in the Indonesian Seas *Indo. J. Mar. Sci.* **15(1)** 42
[7] O'Reilly J, Maritorena S, Mitchell BG, Siegel D, Carder KL, Garver S, Kahru M and McClain C 1998 Ocean color chlorophyll algorithms for SeaWiFS *J. Geophys. Res.***103** 24937
[8] Atlas R, Hoffman RS, Ardizzone J, Leidner SM, Jusem JC, Smith DK and Gombos D 2011 A cross-calibrated, multiplatform ocean surface wind velocity product for meteorological and oceanographic applications *Bull. Amer. Meteor. Soc.* **92** 157 DOI: 10.1175/2010BAMS2946.2
[9] Hosoda K 2013 Empirical method of diurnal correction for estimating sea surface temperature at dawn and noon *J. Oceanogr.* **69** 631 doi:10.1007/s10872-013-0194-4
[10] Kawai Y and Kawamura H 2002 Evaluation of the Diurnal Warming of Sea Surface Temperature Using Satellite-Derived Marine Meteorological Data *J. Oceanogr.* **58** 805
[11] Brown OB and Minnett PJ 2009 MODIS infrared sea surface temperature algorithm theoretical basis document http://modis.gsfc.nasa. gov/data/atbd/atbd_mod25.pdf
[12] Wirasatriya A, Setiawan RY and Subardjo P The effect of ENSO on the variability of chlorophyll-a and sea surface temperature in the Maluku Sea 2017 *IEEE J. Sel. Topics Appl. Earth Observ. Remote Sens. (JSTARS)* **10(12)** 5513 **DOI:** 10.1109/JSTARS.2017.2745207
[13] Hendiarti N et al 2005 Seasonal variation of pelagic fish catch around Java *Oceanography* **18(4)** 112
[14] Susanto RD, Moore II T and Marra J 2006 Ocean color variability in the Indonesian Seas during the SeaWiFS Era *Geochem. Geophys. Geosyst.* **7(5)** 1 doi:10.1029/2005GC001009.
[15] Setiawan RY and Habibi A 2011 Satellite Detection of Summer Chlorophyll-a Bloom in the Gulf of Tomini *IEEE J. Sel. Topics Appl. Earth Observ. Remote Sens. (JSTARS)*, **4(4)** 944

[16] Doty L and Oguri M 1956 The island mass effect *J. Cons. Cons. Perman. Int. Explor. Mer* **22** 33
[17] Hamner W and Hauri I 1981 Effects of island mass: Water flow and plankton pattern around a reef in the Great Barrier Reef lagoon, Australia *Limnol. Oceanogr.* **26(6)** 1084
[18] Aristegui J et al 1994 Island-induced eddies in the Canary Islands *Deep-Sea Res., Part I* **41** 1509
[19] Aristegui J et al 1997 The influence of island-generated eddies on chlorophyll distribution: A study of mesoscale variation around Gran Canaria *Deep Sea Res Part I* **44** 71
[20] Hernandez-Leon S 1991 Accumulation of mesozooplankton in a wake area as a causative mechanism of the island-mass effect *Mar. Biol.* **109** 141
[21] Hernandez-Leon S 1988 Gradients of mesozooplankton biomass and ETS activity in the wind shear area as evidence of an island mass effect in the Canary Island waters *J. Plankton Res.* **10(6)** 1141
[22] Rissik D et al 1997 Enhanced particle abundance in the lee of an isolated reef in the south Coral Sea: The role of flow disturbance *J. Plankton Res.* **19(9)** 1347

Acknowledgments
We thank the Faculty of Fisheries and Marine Science, Diponegoro University for funding this research. NASA Goddard Space Flight Center, Ocean Ecology Laboratory, Ocean Biology Processing Group; (2014): Sea-viewing Wide Field-of-view Sensor (SeaWiFS) Ocean Color Data, NASA OB.DAAC. http://doi.org/10.5067/ORBVIEW-2/SEAWIFS_OC.2014.0. Accessed on 2017/06/01. CCMP Version-2.0 vector wind analyses are produced by Remote Sensing Systems. Data are available at www.remss.com. DeltaSST_v7.1.3 belongs to Satellite Oceanography Laboratory, Tohoku University.

IOP Conf. Series: Earth and Environmental Science **165** (2018) 012008 doi:10.1088/1755-1315/165/1/012008

The Influence of Indian Ocean Dipole (IOD) on The Variability of Sea Surface Temperature and Precipitation in Sumatera Island

Muhammad Hafidz Ibnu Khaldun[1], Anindya Wirasatriya[1,2*], Agus Anugroho Dwi Suryo[1] and Kunarso[1]

[1]Oceanography Departement, Faculty of Fisheries and Marine Science, Diponegoro University. Prof. H. Sudarto, SH Street, Semarang. 50275 INA
[2]Center for Coastal Disaster Mitigation and Rehabilitation Studies, Diponegoro University, Semarang. Prof. H. Sudarto, SH Street, Semarang. 50275 INA

aninosi@yahoo.com

Abstract. Sumatera Island is located adjacent to the ocean and directly influenced by the dynamic of Indian Ocean. Indian Ocean Dipole (IOD) as one of the global climate factor has a big impact towards this dynamic. As it aims to get the influence of IOD towards Sumatera water area, high-resolution remote sensing data is needed. Those are Sea Surface Temperature (SST), precipitation, and Dipole Mode Index (DMI). DMI is calculated by the differentiation of SST anomaly. IOD condition is emphasized by using precipitation data. Area observed is divided into three areas which are in series West Sumatera water, East Sumatera water, and Bukit Barisan highland. During IOD negative, the precipitation values were estimated 5,43; 4,63; 4,5 mm/hr sequentially. Meanwhile, during IOD positive, those values were 2,74; 3,47; 3,11 mm/hr sequentially. It is shown that IOD has the impact of the precipitation values in Sumatera waters. By showing the higher value of IOD negative and the lower value on IOD positive. West Sumatera water is observed as the most influencing area by the dynamic of IOD. Topography is also influencing towards the precipitation and it is shown by its values weather on IOD negative or positive.

Keywords: SST, Precipitation, IOD, Sumatra Waters

1. Introduction

Indonesia is geographically located between two Continents and two Oceans and traversed by the equator which causes Indonesia to be affected by global climate change. Its strategic location makes Indonesia one of the centers of the world's climate control system so that the phenomenon of climate change occurring in Indonesia and globally will be interconnected [1]. One of the global climate variability that influences Indonesia is Indian Ocean Dipole (IOD) [2,3,4].

A more detailed assessment of global climate variability on the effect of sea surface temperature and precipitation is important. Based on the results of various studies that have been done show that there is a fairly clear connection between the phenomenon of sea surface temperature with the incidence of rain in an area [5,6]. This explains that the variability of precipitation is strongly influenced by changes in sea surface temperature and this is associated with changes in sea surface temperature anomalies either spatially or temporally [e.g., 7,8].

Anomalous sea surface temperature in the Indian Ocean Dipole Mode (IOD) greatly affects the intensity of precipitation in Indonesia. At the time of dipole fashion in the Indian Ocean, generally, precipitation in Indonesia has decreased. This is due to the positive anomalies of the southeast wind as the dipole mode progresses so that the southeast wind that blows in the area gets stronger. The strength of the southeastern winds that contain only a small amount of water causes more difficult precipitation to rain [9]. When IOD positive precipitation intensity will decrease and when IOD negative precipitation intensity will increase [e.g., 3,10,11]. Therefore, there is a need for further study on the study of the variability of sea surface temperature and precipitation and correlation with IOD climate variability in Sumatera Island. An understanding of the effect of Indian Ocean Dipole climate variability on SPL variability and precipitation and its physical processes on the Island and Waters of Sumatra can help to determine the mitigation measures of the disaster caused by the increase and decrease in precipitation.

2. Data and method

Precipitation and sea surface temperature data are obtained from daily data of Tropical Precipitation Measuring Mission (TRMM) satellite and daily data of Administration-Advanced Very High-Resolution Radiometer (AVHRR) and Advanced Microwave Scanning Radiometer on the Earth Observing System (AMSR-E) resolution $0.25^{o} \times 0.25^{o}$ with observation period from 1998 to 2016. These data are developed and distributed by the National Oceanic and Atmospheric Administration (NOAA). Data of precipitation and sea surface temperature were analyzed by compiling the data into monthly data until becoming monthly climatology data by using equation as follows [12]:

$$\overline{X}\,(x,y) = \frac{1}{n}\sum_{i=1}^{n} xi(x,y,t) \tag{1}$$

Where, $\overline{X}\,(x,y)$ is monthly or monthly climatology mean value at position (x,y), $xi(x,y,t)$ its the value of the data at (x,y) position and time t. Furthermore, n is the number of data in 1 month and the number of monthly data in 1 period of climatology (i.e., from 1998 to 2016 = 18 data) for monthly calculation and monthly climatology calculation, respectively. If xi is a hollow pixel, that pixel is excluded in the calculation.

The Indian Ocean Dipole climate variability data was obtained using Dipole Mode Index (DMI). DMI is a value derived from differences in sea surface temperature anomalies in the West Indian Ocean Hindia (10°N-10°S and 50°E-70°E) and East Indian Ocean (10°S-0°U and 90°E-110°E) [13]. The value of DMI is obtained by the weight equation:

$$DMI = West - East \tag{2}$$

Where DMI is IOD Index, $West$ is the mean sea surface temperature value in the western area of the Indian Ocean (10°N-10°S and 50°E-70°E). Furthermore, $East$ is the mean sea surface temperature value in the southeastern area of the Indian Ocean (10 ^{o}S-0°N and 90°E -110°E). DMI data used has a range of time range from 1998 to 2016. Anomaly values obtained are used to classify Dipole Mode Index (DMI) data in 2 (two) phenomena namely Negative DMI and DMI Positive with threshold value $\pm$ 0.4.

The phenomenon of Indian Ocean dipole always happens every year as shown in Figure 1. Data processing of sea surface temperature from January 1 1998, to December 31 2016, obtained results that show the strongest positive IOD phenomenon occurred in October 2006 with an anomaly index value of 1.211. The strongest negative IOD phenomenon occurred in July 2016 with an anomaly index value of -1.043.

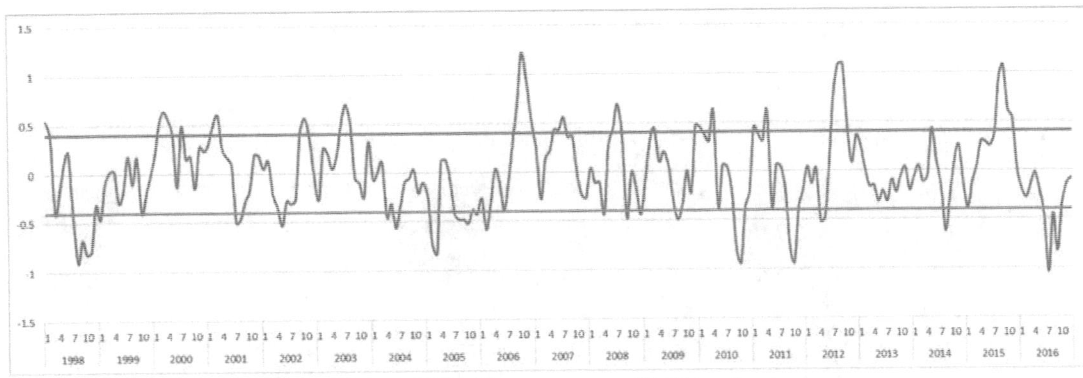

Figure 1. Anomalous Index of Sea Surface Temperature (DMI)

3. Result and Discussion

The distribution of sea surface temperature distribution pattern and monthly climatology precipitation in Figures 2 and 3 shows the difference of sea surface temperature and precipitation value in Sumatera waters when influenced by climate variability. Figures 2 and 3 show lower sea-surface temperature values when IOD positive and increases when IOD negative which indicates that IOD phenomena affect the variability of sea surface temperature and precipitation in Sumatera island. Sea surface temperature values in area 2 and area 3 are obtained from sea surface temperature values close to the area. The value can be interpreted as the value of sea-surface temperature that has an effect on area 2 and area 3.

The distribution of sea surface temperature distribution and monthly climatology precipitation in Figure 3 shows an interesting pattern of distribution on the value of the pattern of precipitation distribution in December whose territory is located in South East Sumatra waters which tend to be higher than the East of Sumatera Island. Therefore, the authors divide the study concentration area into 3 areas. The determination of this area is based on the value of precipitation contour in December climatology which has the highest precipitation value compared to other months which can be seen in Fig. 4.

During IOD negative, the sea-surface temperature values were estimated 27,78; 27,96; 27.85°C sequentially. Meanwhile, during IOD positive, those values were 27,63; 27,87; 27,71°C sequentially. And during IOD negative, the precipitation values were estimated 5,43; 4,63; 4,5 mm/hr sequentially. Meanwhile, during IOD positive, those values were 2,74; 3,47; 3,11 mm/hr sequentially. The area I has a relatively higher precipitation value than the other two areas with the highest intensity of 5.43 mm/hr. This is caused by the increase of sea surface temperature which is more dominant than other areas reaching 27.78°C so that the intake becomes more leverage. This is also due to the difference in pressure that occurs in the eastern Indian Ocean area is higher than the western Indian Ocean causing the mass of water to move towards Sumatera [e.g., 3,9,10,14,15,16].

Area III has the lowest precipitation value than in the other two areas, ie 4-6 mm/hr. At the highest precipitation area III has an intensity of 5.02 (mm/hr) and when the lowest intensity is only 0.68 (mm/hr). This is because the temperature of the sea surface in the eastern waters is lower when compared to the value of sea surface temperature watered west. In addition, the eastern part of the island of Sumatera has a low topography so that the condensation process is not running optimally. On a spatial scale, precipitation variability is strongly influenced by topography [e.g., 17,18,19,20].

Area II precipitation value is relatively higher than the area III that is 4.63 mm/hr. This is due to the difference in altitude along the western coast of Sumatra that there is a plateau of Barisan. This resulted in the condensation process resulting in an increase in the value of precipitation. Based on the results in Fig. 4 it can be seen that variability of sea surface temperature and precipitation is influenced by IOD phenomenon. At the time of negative IOD, precipitation intensity becomes higher than normal while in positive IOD precipitation intensity will be lower. When IOD positive precipitation intensity will decrease and when IOD negative precipitation intensity will increase [10].

(a) IOD Positive (b) Monthly Climatologies (c) IOD Negative

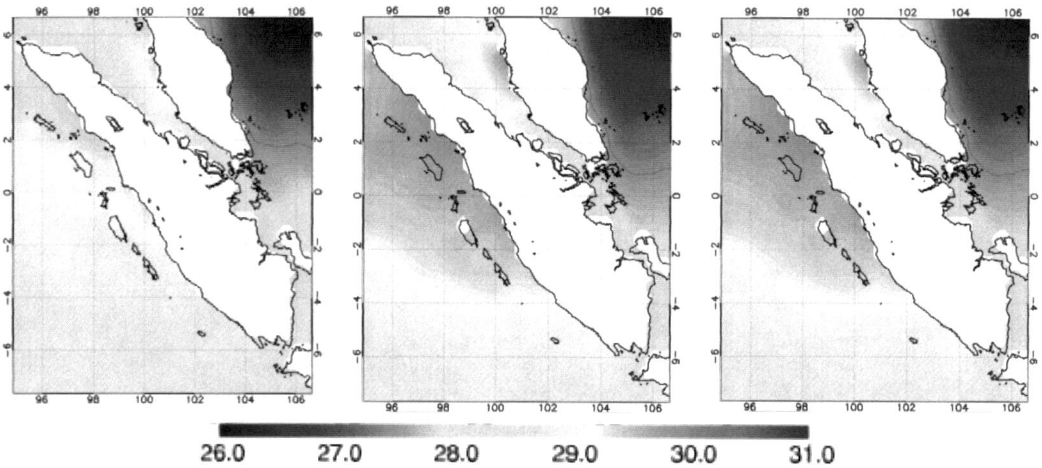

Figure 2. The pattern of distribution of sea surface temperature in December

(a) IOD Positive (b) Monthly Climatologies (c) IOD Negative

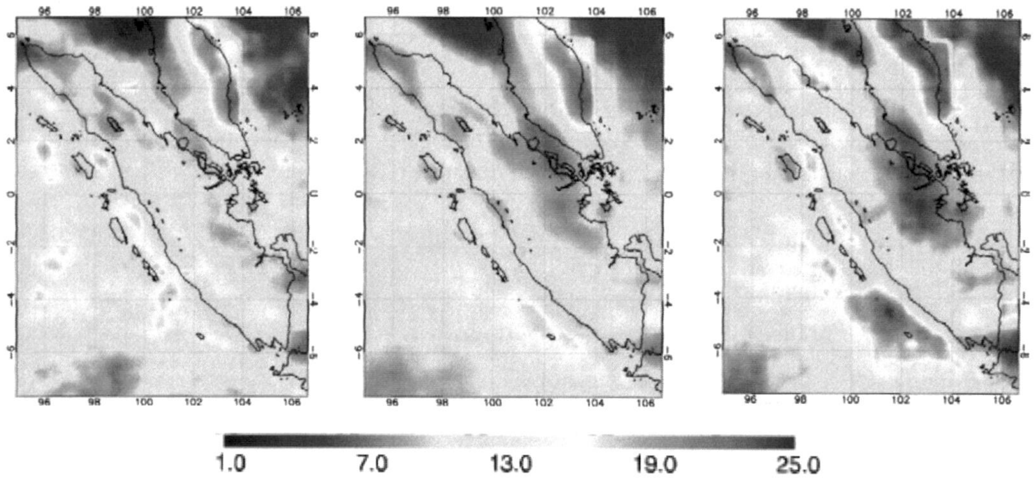

Figure 3. The Pattern of Distribution of Precipitation in December

During the IOD phenomenon, Negative sea surface temperature values and precipitation intensity will be higher when compared to when IOD is positive. This is due to the differences in the pressure of the East Indian Ocean and the West Indian Ocean resulting in the mass transfer of air containing moisture. At IOD Negative sea surface temperatures in the East Indian Ocean will be higher than in the West Indian Ocean, resulting in an increase in precipitation in the West waters of Sumatra. In contrast to positive IOD sea surface temperatures in the West Indian Ocean will be higher than the East Indian Ocean so that there is a decrease in precipitation in West waters of Sumatra [e.g., 3,9,10,14,15,16].

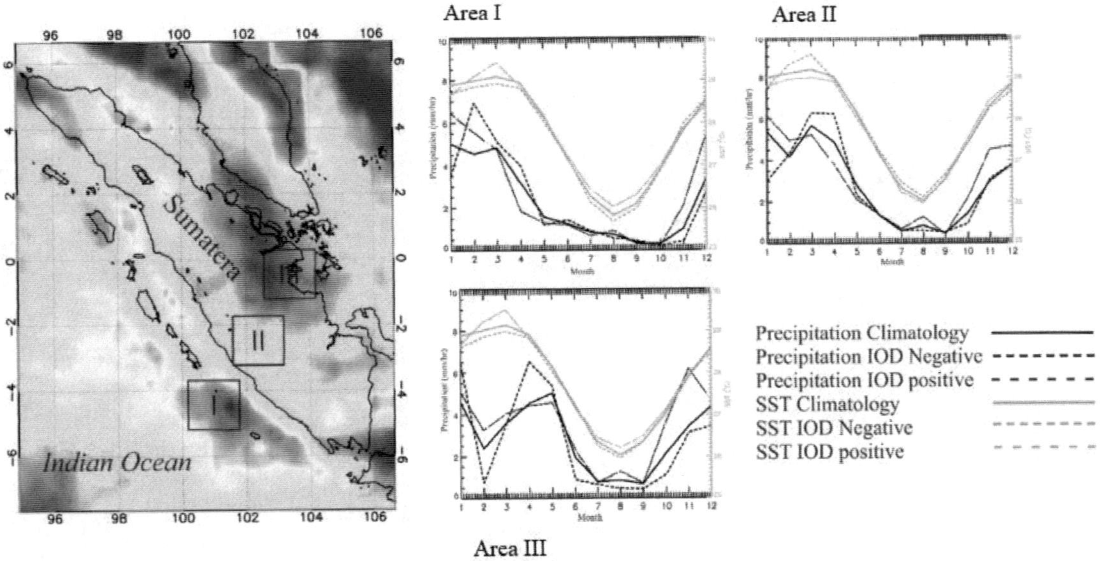

Figure 4. Precipitation distribution pattern in December and Graph of SPL and Precipitation relationship

4. Conclusion

IOD phenomenon affects the variability of sea surface temperature and precipitation that have been studied using remote sensing data of TRMM and AVHRR-AMSRE satellite with resolution 0.25° × 0.25°. During the phenomenon of IOD (IOD Negative) occurs, in general, an increase in sea surface temperature intensity of 0.05°C and precipitation of 0.3 mm/hr. This is due to the differences in the pressure of the East Indian Ocean and the West Indian Ocean resulting in the mass transfer of air containing moisture. Increased precipitation also occurs in areas with relatively high topography caused by more optimum condensation.

References

[1] Robertson A, Moron V, Qian J, Chang C P, Tangang F, Aldrian E, Koh T Y and Juneng L 2011 *The Maritime Continent Monsoon, in The Global Monsoon System: Research and Forecast, 2nd Ed.* Eds. CP Chang, Y Ding, NC Lau, RH Johnson, B Wang and T Yasunari, (World Scientific Publication Company) Vol 5. 608 p
[2] Webster P J, A Moore, J Loschnigg and M Leban 1999 *Nature* **401** 356-360
[3] Saji N H, B N Goswami, P N Vina yachandran and T Yamagata 1999 *Nature* **401** 360-363
[4] Fadholi A 2013 *J. Ilmu Lingkungan* **2(1)** 43-50
[5] Estiningtyas, W Ramadhani F and Aldrian E 2007 *J. Agromet Indonesia* **21(2)** 46–60
[6] Lindzen R S and S Nigam 1987 *J.Atmos. Sci.* **44** 2418–2436
[7] Shailendra Rai and A C Pandey 2008 *Atmosphere-Ocean* **46 (3)** 361-376 doi:10.3137/ao.460305
[8] Arif Suryantoro and Bambang Siswanto 2008 *J.Sains Dirgantara* **6(1)** 1-21
[9] Mulyana E 2002 *J. Jurnal Sains & Teknologi Modifikasi Cuaca*. **3(1)** 39-43
[10] Kunarso, Hadi S, Ningsih N S and Baskoro M S 2012 *J. Ilmu Kelautan*. **17(2)** 87-89
[11] Karumuri Ashok, Zhaoyong Guan and Toshio Yamagata 2001 *Geophys. Res. Lett.* **28(23)** 4499- 4502
[12] Wirsatriya A, Setiawan R Y and Subardjo P 2017 *IEEE J. Sel. Topics Appl. Earth Observ. Remote Sens. (JSTARS)*, **10(12)** 5513 - 5518 doi: 10.1109/JSTARS.2017.2745207
[13] Han Soo Lee 2015 *Water* **7** 1751-1768 doi:10.3390/w7041751
[14] Aurelie P Harou, Robert F Lajoie, Dominic R Kniveton and Michael R Frogley 2006 *Int. J.*

Climatol, **26** 45–54 doi: 10.1002/joc.1239
[15] Clark C, Webster P and Cole J 2003 *J. Climate* **16** 548–554
[16] Richard Washington and Anthony Preston 2006 *J.Geophysical Research,* **111** 1 – 15 doi:10.1029/2005JD006724
[17] Tongkukut S 2011 *J. Ilmiah Sains.* **11(1)** 102-108
[18] Alison M Anders, Gerard H Roe, David R Montgomery and Bernard Hallet 2008 *Geology* **36(6)** 479 – 482 doi: 10.1130/G24821A.1
[19] Ntwali D, Ogwang B A and Ongoma V 2016 *Atmospheric and Climate Sciences* **6** 145-157 http://dx.doi.org/10.4236/acs.2016.62013
[20] Alan Basist, Gerard D Bell and Vernon Meentemeyer 1994 *J. Climate* **7** 1305 – 1315

TRMM PR Observed Spatial Patterns of The Convective-Stratiform Rainfall Over Indonesia and Their Response to ENSO

Abd. Rahman As-syakur [1*]

[1]Marine Science Department, Faculty of Marine and Fisheries, Udayana University, Bukit Jimbaran, Bali 80361, Indonesia

assyakur@unud.ac.id

Abstract. This study investigates the variability of convective and stratiform rainfall over Indonesia from 5 years (2004–2008) data of Tropical Rainfall Measuring Mission (TRMM) Precipitation Radar (PR). The remote sensing data was employed to analyze the seasonal means and seasonal differences between convective and stratiform rainfall values. In addition, the El Niño Southern Oscillation (ENSO) analysis impact on two rainfall types was also done with exemplification in 2006 and 2007 event. Two datasets were used namely the probability and the rainfall accumulation from both rainfall types. The results showed that the probability of Indonesian stratiform rainfall type was higher than convective rainfall type. In contrast, convective rainfall type more contributed to rainfall values. Spatial patterns of convective-stratiform rainfall types have different spreads. Generally, convective rainfall supplied more rainfall over the land, whereas stratiform rainfall supplied more rainfall over the sea, except in December-January-February (DJF) season. The ENSO impact was clearly seen during September-October-November (SON) season from both rainfall types over the south. Moreover, the stratiform rainfall type was more affected by ENSO during June-July-August (JJA) season.

Keywords: Rainfall, Convective, Stratiform, TRMM PR, ENSO, Indonesia

1. Introduction

Understanding space-time variations of rainfall is an important topic in climate research, in which modern, high quality, global-scale rainfall observations are essential [1]. The tropical Indonesian archipelago is a central atmospheric heat source characterized by huge quantities of rainfall, which are important contributions to understanding the world's climate system [2]. Rainfall information is especially important in the tropics because such a large fraction of the planet's rain falls within low latitudes and because there are large variations therein that are related to climatic events [e.g. El Nino–Southern Oscillation (ENSO) and Indian Ocean Dipole (IOD)] [3,4,5,6]. Many studies have shown that Indonesia rainfall tends to decrease (increase) during warm (cold) ENSO events [e.g. 7,8,9]. Changes in rainfall can be attributed to the anomalous sea surface temperature (SST) and the associated anomalous Walker circulation over the tropical Pacific [10]. Annual and interannual climate variability in Indonesia is somewhat unusual, as it is not homogenous over the whole region and the coherence of rainfall patterns varies seasonally [11,12,13]. That's way Indonesian rainfall information is necessary for the study of climatology, flood predictions, irrigation plans, and water resource problems.

Tropical precipitation system is characterized by extensive stratiform cloud directly associated with deep convective [14]. The separation between convective and stratiform is important for climatological studies because convective and stratiform precipitation is characterized by different types of weather events. Stratiform precipitation falls from nimbostratus clouds, while convective precipitation falls from active cumulus and cumulonimbus clouds. These cloud types may occur separately or entangled with each other in the same cloud complex. Convective areas are associated with heavy rainfall rates of shorter duration than stratiform areas. Although stratiform rain is associated with generally lower rain rates, it covers large areas contributing a significant portion of the rainfall region [15]. The convective and stratiform rain components are the outcome of distinct cloud dynamical processes that influence the atmospheric circulation [16]. The convective and stratiform modes of rainfall are especially important in the Tropics, where precipitating clouds often organize into mesoscale systems containing the two distinctive environments [17, 18].

Indonesia, covered mostly by the ocean, is the world's largest archipelago. Therefore, there are several problems in studying and simulating rainfall of the region for an appropriate land–sea representation and a complex topographical distribution [19]. Rainfall displays high space-time variability that requires frequent observations for adequate representation [15]. Rain gauge observations yield relatively accurate point measurements of precipitation. Thus, meteorological satellites observation may be the best solution for adequate temporal and spatial coverage of rainfall. The use of remote sensing which has better spatial and temporal resolution data in investigates the variability of convective and stratiform rainfall over Indonesia, thus, offers an exciting opportunity.

The Precipitation Radar (PR) onboard the TRMM satellite, which was developed by the Japan Aerospace Exploration Agency (JAXA, previously known as the National Space Development Agency, or NASDA) in cooperation with Communications Research Laboratory [20], is the first spaceborne precipitation observation radar designed to measure the vertical structure of tropospheric precipitation and three-dimensional rain structures over the tropics and subtropics [21,22]. The main components of the TRMM PR retrieval algorithm convert the intensity of backscattered energy from hydrometeors within a defined vertical interval [20,23]. The PR satellite enables capture of the three-dimensional rainfall structure over the ocean and land of tropical oceans and continents [20]. The PR operates at the frequency of 13.8 GHz and 250 m vertical resolution, allowing penetration of even the heaviest obscuring cloud layers and detection of underlying precipitation. PR is an effective platform for investigating shallow precipitation [24], able to detect rainfall intensities as low as 0.7 mm/h [21]. The algorithm of TRMM PR to separating convective, stratiform, or other rainfall types is based on the horizontal and vertical radar echostructure in each pixel. Several factors affect the accuracy of PR rain rate estimation: radar calibration, light rains sensitivity (minimum detectable signal), attenuation correction (moderate-to-high rain rates), drop size distributions variations (reflectivity-rain rate (Z-R) and attenuation-reflectivity (k-Z) relationships), horizontal/vertical gradients in Z (partial beam filling), and surface clutter rejection [25]. Over the years, several groups have been validated PR data with ground data (e.g. 22,26,27,28,29).

An especially useful aspect of the TRMM PR is its ability to discern convective rain from stratiform [30]. The TRMM PR observations can detect the vertical structure of precipitation which is important in convective-stratiform separation and thus provide the best means with which to classify rain as convective or stratiform [31]. Convective and stratiform rainfall are characterized by distinctly different processes and have distinctly different latent heating profiles. Convective systems are generally identified with intermittently strong vertical velocities (> 1 m/s in magnitude), high rainfall rates (> 3 mm/h), and small areas (1-10's of km), intense, horizontally inhomogeneous radar echo. Conversely, stratiform systems are characterized by small vertical velocities (< 1 m/s), low rain rates (< 5 mm/h), and widespread areas (10-100's of km), horizontally homogeneous radar echo [16]. The algorithm [32] to separation convective-stratiform from TRMM PR 2A23 product is based on Steiner et al.'s [33] classification method for ground radars.

Convective and stratiform rainfall distribution information and the structure of precipitation systems from large areas of Indonesia are important. The spatial and temporal variations of convective and stratiform rainfall across Indonesia not yet clear break down, therefore now, is a great opportunity to conduct this research. This study examines the variation of the convective and stratiform rainfall determined by TRMM 3A25 products across Indonesian, showing the capability of these products to contribute to an analysis of climatic-scale rainfall in Indonesia.

2. Data and Method

The research was conducted in the archipelago of Indonesia composed of 17,508 islands of various sizes. Spatial data covered 8°00' N to 12°00' S and 93°00' E to 142°00' E. Figure 1 indicates the distribution of Indonesian topography. Indonesia is located between two continents and oceans with a population reaching 237,641,326 people in 2010. Sumatra, Kalimantan, Jawa, Sulawesi and Papua are five major islands with diverse topographical distributions. Several important mountains in Indonesia are Jayawijaya in Papua, Bukit Barisan in Sumatra, Kendeng in Jawa, Fenema and Gorontalo in Sulawesi, and Muller in Kalimantan.

Figure 1. The Study area includes the Indonesian topography

Monthly Rainfall data from 2004 to 2008, measured and collected by TRMM PR 3A25 version 6 (V6) satellite data, were employed to observe Indonesian convective-stratiform rainfall variability. The PR 3A25 is a space-time average of PR products [20] containing monthly 0.5°× 0.5° gridded rainfall estimates and structure statistics. The monthly rain accumulation for convective and stratiform of 3A25 (mm month^{-1}) is obtained by multiplying the unconditioned rain rate with the number of hours in a month. In fact, all 3A25 statistics are conditioned on the existence of one or more variables. To convert to an unconditioned mean rain rate the quantity is first multiplied by the probability of rain, given by the ratio of the number of rain counts to the total number of observations for the month [25].

The PR, operating at 13.8 GHz, is a 128-element active phased array that allows fast and sophisticated cross-track scanning over a swath width of 215 km with a cross-range spatial resolution of about 4.3 km [21]. After boosting the satellite from 350 to 402.5 km on August 22, 2001, the horizontal resolution increased to nearly 5 km and swath width increased to 245 km. The PR is subject to attenuation from cloud water, rain, and partially melted hydrometeors [22]. The major roles of the PR in TRMM are to provide 3-D rainfall distribution information, to achieve quantitative rainfall estimation over land and ocean, and to utilize data to improve the TRMM Microwave Imager (TMI) estimations of rainfall [21]. The TRMM PR data processing and analysis algorithms were developed by the International TRMM PR team [20], while the PR 3A25 algorithm was developed by Meneghini et al. [25]. The algorithm 3A25 gives various space-time statistics of conditional rain characteristics over a month from the TRMM level 1C21, 2A21, 2A23 and 2A25 products [33,34,35). There are four types of statistics calculated: probabilities of occurrence, mean standard deviations, histograms, and correlation coefficients [25,33]. The statistics analyzed are conditioned to quantities such as the presence of rain, stratiform rain, or the presence of a bright-band.

Several statistical scores were used to determine the Indonesian convective-stratiform rainfall characteristics. Rainfall probability means, rainfall accumulation means, and value of rainfall mean differences, were used in the analysis. El Niño and La Niña events on 2006/07 and 2007/08, respectively were selected to illustrate spatial patterns of Indonesian convective-stratiform rainfall anomalously. Shuanglin and Qin [36] and Takahashi et al. [37] classified 2006/07 and 2007/08 as El Niño events and 2006/07 as a La Niña event. JJA and SON were chosen as the main season on the seasonal rainfall anomaly analysis because the ENSO usually have a peak in JJA and SON [3,9]. The percentage of rainfall anomaly (%) is given by the ratio of the values of rain to the rainfall average of observations for the month.

Seasonal analyses were conducted in this study. The monthly analysis compared data from the same months of annual observation. The seasonal analysis is conducted based on the monsoon activity, described by Aldrian et al. [12], over the entire observation period; in this analysis, the seasons were classified as December-January-February (DJF), March-April-May (MAM), June-July-August (JJA), and September-October-November (SON). DJF represents the peak of northwest Australia–Asia monsoon, and JJA represents the peak of the southeast Australia–Asia monsoon, and MAM and SON represent monsoon transitions. Likewise, long-term analysis observed all-time series data over the entire observation period [38]. The analysis is done using the coordinate as the reference identity and conducted in each pixel. Data extracted from the TMPA in each pixel to generate point-by-point data. The point has the information about coordinate, month, year, and rainfall values. Then, the data are sorted in accordance with the purposes of analysis. After obtaining the values, the point data is converted into a raster data format which has the same spatial resolution with the original data ($0.5° \times 0.5°$).

3. Results

The probability of rain is defined as the percentage of the total rain pixel numbers with total pixel numbers within each 0.5x0.5-degree box (pixel for 3A25). The original horizontal resolution of PR data is nearly 5 km. In addition, the TRMM PR records locations once approximately every 3.6 days in the tropical region because of the non-sun-synchronous satellite orbit. Thus over the tropical areas has the numbers of the total pixel in each 0.5x0.5-degree box ranged among the 988 to 1759. Given by the explanation, Figure 2 and 3 illustrates the distributions of seasonally averaged of convective and stratiform rainfall probability over Indonesia during 2004–2008. Generally, the probability of stratiform rain values is higher compared with the probability of convective rain values, but with similar north-south seasonal changes. Seen from the land-sea spatial pattern, most of the convective rainfall type occurs over the land, in contrast with stratiform rainfall type but except during DJF season.

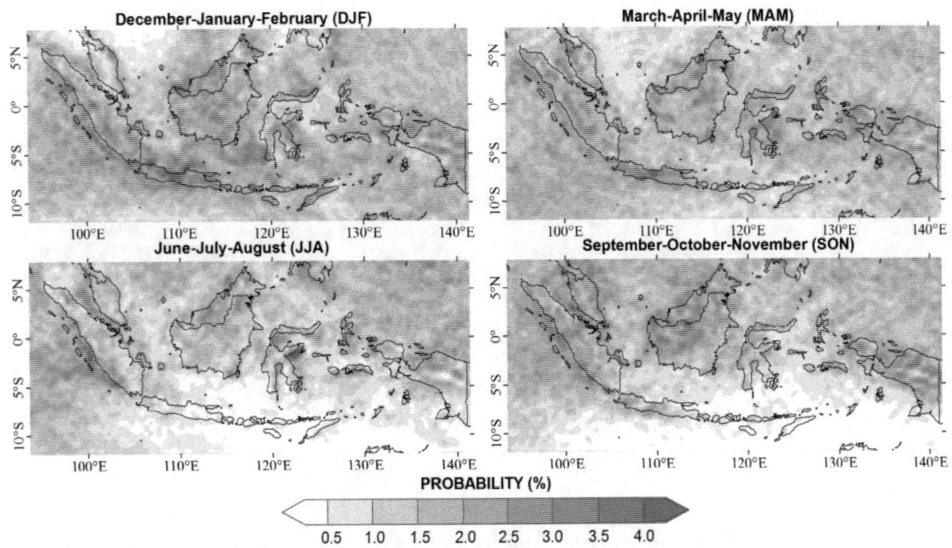

Figure 2. Spatial distributions of convective rainfall probability based on seasonal averages from January 2004 to December 2008.

Monthly rainfall accumulation is defined as the multiplying of the rainfall probability with a mean rain rate of the total rain pixel and the number of hours in a month within each 0.5x0.5-degree box. Figure 4 and 5 displays the spatial pattern of seasonal averaged of monthly convective and stratiform rainfall accumulation over Indonesia during 2004–2008. Low probability of convective rainfall does not impact the rainfall accumulation values because convective areas are associated with heavy rainfall rates than stratiform areas. Most high convective rainfall is occurring over the land, except over JJA season that happens outside the land such as in the west coast of Sumatra. An interesting phenomenon is seen from stratiform rainfall pattern, where the high rainfall values occur at the top of mountains such as in Papua and Kalimantan Islands.

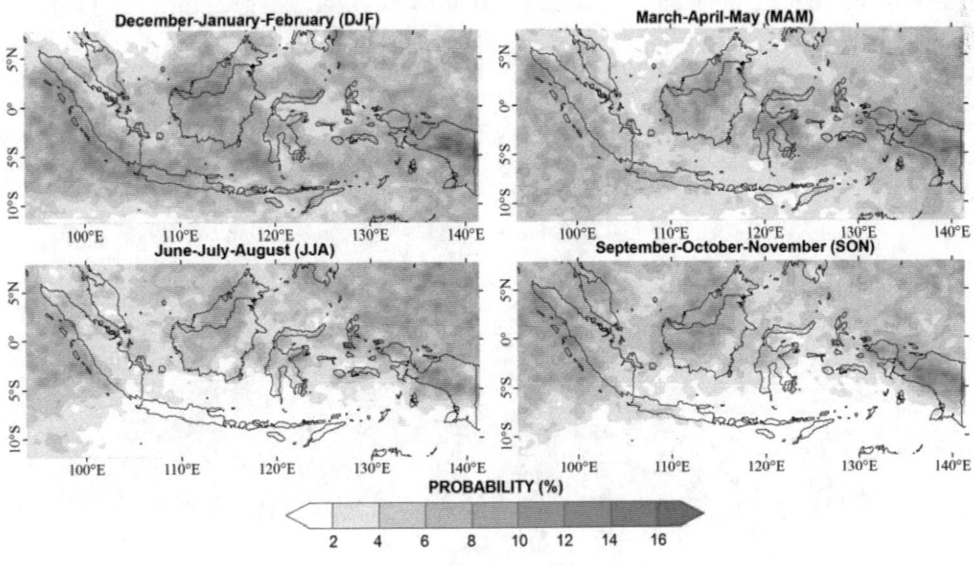

Figure 3. Same as Figure 2, but for stratiform rainfall type probability.

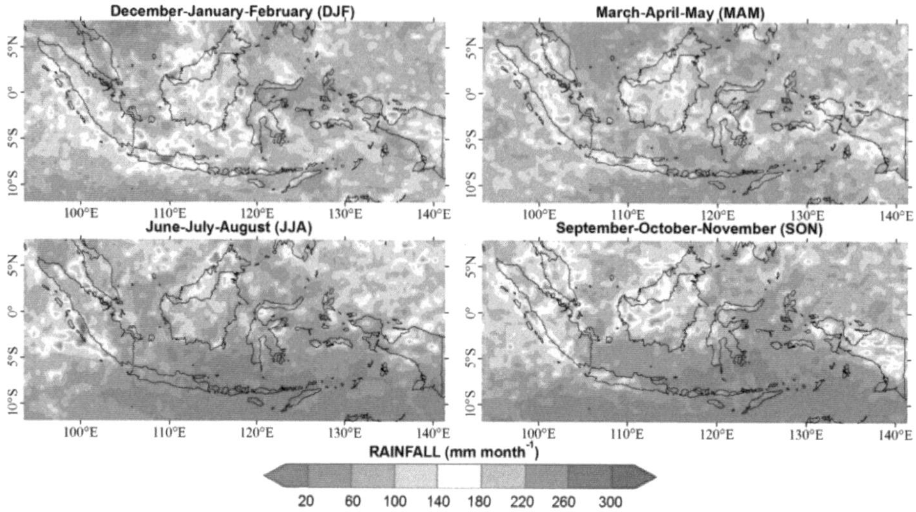

Figure 4. Seasonal mean distributions of convective rainfall accumulation based on monthly composites from January 2004 to December 2008.

The seasonal differences between convective and stratiform rainfall values are displays in Figure 6. Here the white-green colors show areas of more stratiform rainfall and the white-blue colors show areas of more convective rainfall clearly seen from that figure the values of convective rainfall are higher compared with stratiform rainfall values, especially over the land except during DJF season. During DJF, most of the convective rain occurred in the south, whereas the stratiform rain arises over the north. In addition, during JJA and SON, most of the stratiform rain happened the areas of poor rain.

Figure 7 and 8 shown spatial distributions of rainfall anomaly JJA and SON El Nino 2006 and JJA and SON La Nina 2007, respectively. Generally, stratiform rainfall type is dominantly influenced by ENSO compare with convective rainfall type. However, convective rainfall type is seen impacted by ENSO during El Nino 2006 with the same location as stratiform. The figure shows the areas of stratiform rainfall anomaly is wider than convective rainfall, where the most of stratiform rainfall anomaly is occurring over the south and influence rainfall in dry season over Indonesia.

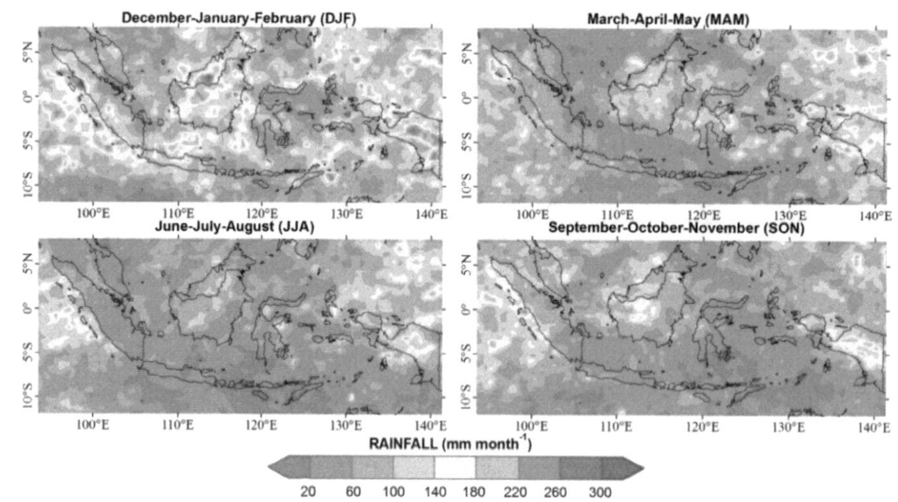

Figure 5. Same as Figure 4, but for rainfall accumulation of stratiform rainfall type.

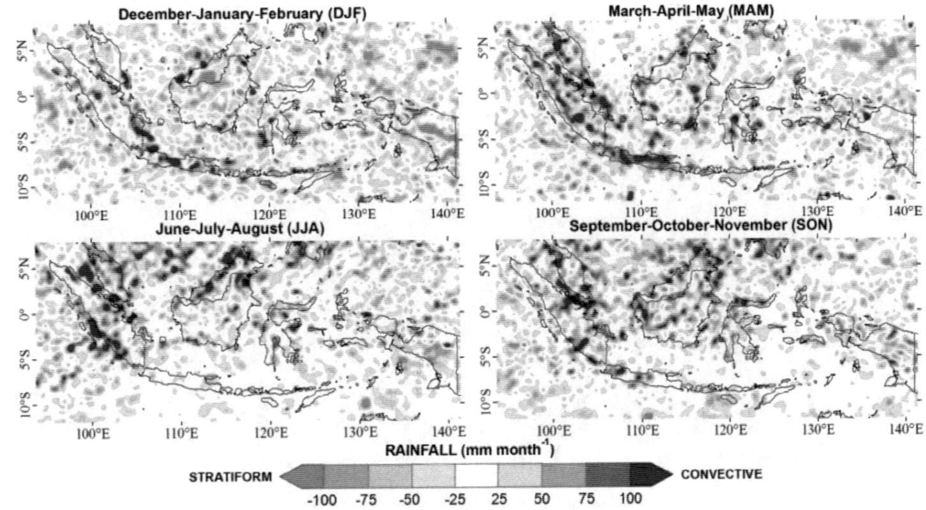

Figure 6. Seasonal analysis of rainfall differences. Blue color indicated high rainfall during convective fall, while green color indicated for stratiform fall.

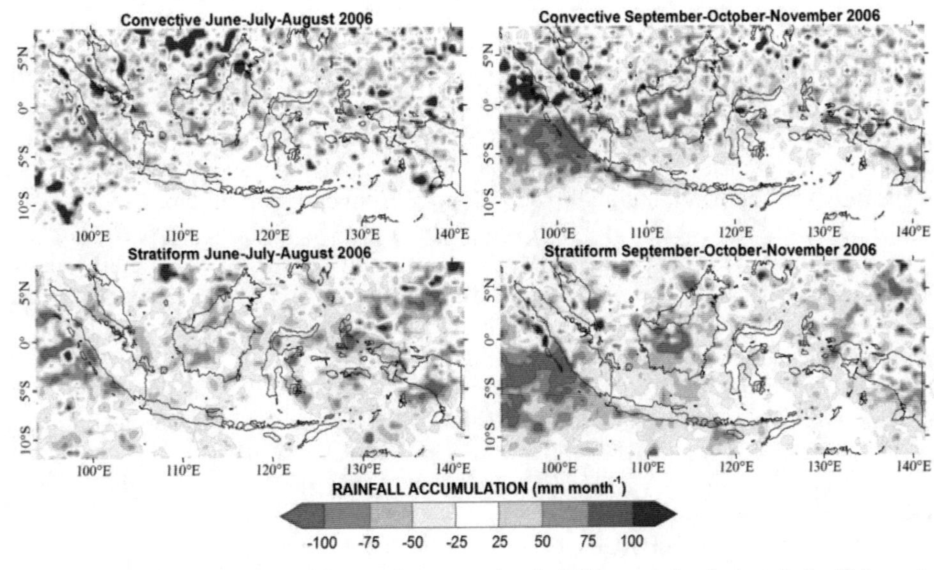

Figure 7. Spatial distributions of seasonal rainfall anomaly for periods JJA and SON 2006. The top is for convective rainfall type and bottom for stratiform rainfall type

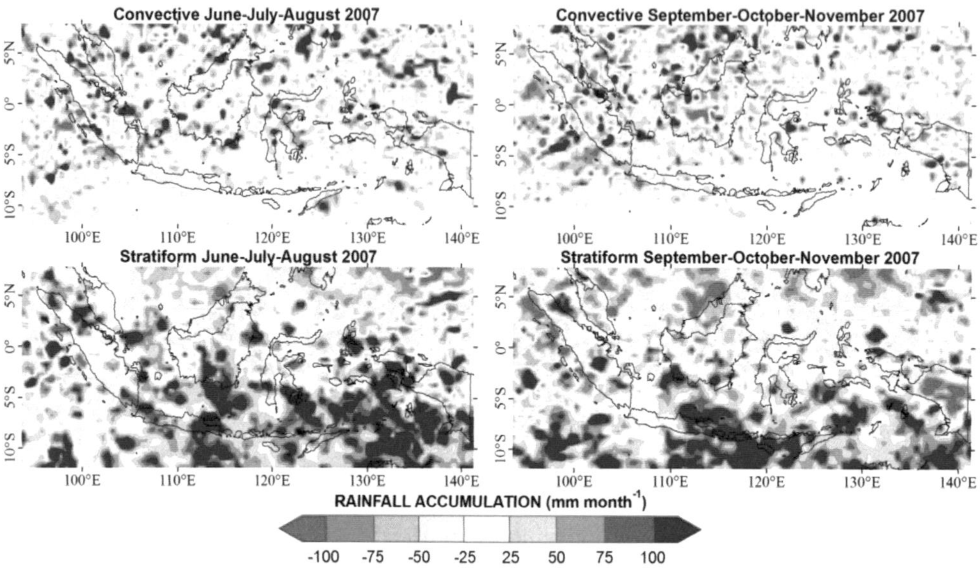

Figure 8. Same as Figure 7, but for rainfall anomaly for periods JJA and SON 2007

4. Conclusions

An investigation of Indonesian convective and stratiform rainfalls variability using satellite data from TRMM PR 3A25 over 5 years (from January 2004 to December 2008) is presented here. Monthly, seasonal and long-term time series analyses were conducted by applying several statistical methods. The seasonal mean difference in rainfall accumulation from convective and stratiform was analyzed to determine the most impact on total rainfall over Indonesia from both rainfall types.

The results show the probability of stratiform rainfall type is higher compared with convective rainfall type. In contrast, convective rainfall type more contributes to total rainfall values. Spatial patterns of convective-stratiform rainfall types have a different spread. Analysis of seasonal mean difference in rainfall accumulation from both rainfall types showed convective rainfall supply more rain values over the land, whereas stratiform rainfall supply more rain over the sea except in DJF season. In another analysis, the ENSO impact is clearly seen during SON season from both rainfall types over the south. Moreover, during JJA season the ENSO just have an impact on stratiform rainfall type.

References

[1] Yang S and Smith E A 2008 Convective-stratiform precipitation variability at seasonal scale from 8 yr of TRMM observations: Implications for multiple modes of diurnal variability *J. Clim.* **21** 4087-4114.

[2] Ramage C S 1971 *Monsoon Meteorology International Geophysics Series 15* (New York USA Academic Press).

[3] As-syakur A R, Adnyana I W S, Mahendra M S, Arthana I W, Merit I N, Kasa I W, Ekayanti N W, Nuarsa I W and Sunarta I N 2014 Observation of spatial patterns on the rainfall response to ENSO and IOD over Indonesia using TRMM Multisatellite Precipitation Analysis (TMPA) *Int. J. Climatol.* **34** 3825-3839.

[4] Adler R F, Huffman G J, Bolvin D T, Curtis S and Nelkin E J 2000 Tropical rainfall distributions determined using TRMM combined with other satellite and rain gauge information *J. Appl. Meteor.* **39** 2007-2023.

[5] Ropelewski C F and Halpert M S 1987 Global and regional scale precipitation patterns associated with the El Niño/Southern Oscillation *Mon. Weather Rev.* **115** 1606-1626.

[6] Saji N H, Goswami B N, Vinayachandran P N and Yamagata T 1999 A dipole mode in the tropical Indian Ocean *Nature* **401** 360-363.

[7] McBride J L, Haylock M R and Nicholls N 2003 Relationships between the Maritime Continent heat source and the El Niño-Southern Oscillation phenomenon *J. Clim.* **16** 2905–2914.

[8] Slezak M 2014 World is unprepared for major El Nino later this year *New Sci.* **222**, 8-9.

[9] As-syakur A R, Osawa T, Miura F, Nuarsa I W, Ekayanti N W, Dharma I G B S, Adnyana I W S, Arthana I W and Tanaka T 2016 Maritime Continent rainfall variability during the TRMM era: the role of monsoon, topography and El Niño Modoki *Dyn. Atmos. Oceans.* **75** 58-77.

[10] Julian P R and Chervin R M 1978 A study of the Southern Oscillation and Walker Circulation phenomenon *Mon. Weather Rev.* **106** 1433-1451.

[11] Haylock M and McBride J 2001 Spatial coherence and predictability of Indonesian wet season rainfall *J. Clim.* **14** 3882-3887.

[12] Aldrian E, Gates L D and Widodo F H 2007 Seasonal variability of Indonesian rainfall in ECHAM4 simulations and in the reanalyses: The role of ENSO *Theor. Appl. Climatol.* **87** 41-59.

[13] As-syakur A R, Tanaka T, Osawa T and Mahendra M S 2013 Indonesian rainfall variability observation using TRMM multi-satellite data *Int. J. Remote Sens.* **34** 7723-7738.

[14] Houze R A 1989 Observed structure of mesoscale convective systems and implications for large-scale heating *Quart. J. Roy. Meteor. Soc.* **115** 425-461.

[15] Feidas H and Giannakos A 2012 Classifying convective and stratiform rain using multispectral infrared Meteosat Second Generation satellite data *Theor. Appl. Climatol.* **108** 613-630.

[16] Houze R A 1997 Stratiform precipitation in regions of convection: A meteorological paradox? *Bull. Am. Meteor. Soc.* **78** 2179-2196.

[17] Goldenbergl S B, Houze Jr R A and Churchill D D 1990 Convective and stratiform components of a winter monsoon cloud cluster determined from geosynchronous infrared satellite data *J. Meteor Soc. Japan, Ser. II* **68** 37-63.

[18] Tokay A, Short D A, Williams C R, Ecklund W L and Gage K S 1999 Tropical rainfall associated with convective and stratiform clouds: Intercomparison of disdrometer and profiler measurements *J. Appl. Meteor.* **38** 302-320.

[19] Qian J H 2008 Why precipitation is mostly concentrated over islands in the Maritime Continent *J. Atmos. Sci.* **65** 1428–1441.

[20] Iguchi T, Kozu T, Meneghini R, Awaka J and Okamoto K I 2000 Rain-profiling algorithm for the TRMM precipitation radar *J. Appl. Meteor.* **39** 2038-2052.

[21] Kawanishi T, Kuroiwa H, Kojima M, Oikawa K, Kozu T, Kumagai H, Okamoto K, Okumura M, Natatsuka H and Nishikawa K 2000 TRMM precipitation radar *Adv. Space Res.* **25** 969-972.

[22] Liao L and Meneghini R 2009 Validation of TRMM precipitation radar through comparison of its multiyear measurements with ground-based radar *J. Appl. Meteor. Clim.* **48** 804-817.

[23] Iguchi T, Kozu T, Kwiatkowski J, Meneghini R, Awaka J and Okamoto K I 2009 Uncertainties in the rain profiling algorithm for the TRMM precipitation radar *J. Meteor. Soc. Japan.* **87A** 1–30.

[24] Short D A and Nakamura K 2000 TRMM radar observations of shallow precipitation over the tropical oceans *J. Clim.* **13** 4107-4124.

[25] Meneghini R, Jones J A, Iguchi T, Okamoto K and Kwiatkowski J 2001 Statistical methods of estimating average rainfall over large space–timescales using data from the TRMM precipitation radar *J. Appl. Meteor.* **40** 568-585.

[26] Robinson M, Kulie M S, Silberstein D S, Marks D A, Wolff D B, Amatai E, Ferrier B S, Fisher B L and Wang J 2000 Evolving improvements to TRMM ground validation rainfall estimates *Phys. Chem. Earth, Part B, Hydrol., Oceans Atmos.* **25** 971-976.

[27] Franchito S H, Rao V B, Vasques A C, Santo C M and Conforte J C 2009 Validation of TRMM precipitation radar monthly rainfall estimates over Brazil *J. Geophys. Res. Atmos.* **114** D02105.
[28] Prakash S, Mitra A K, Momin I M, Rajagopal E N and Basu S 2014 Agreement between monthly land rainfall estimates from TRMM-PR and gauge-based observations over South Asia *Remote Sens. Lett.* **5** 558-567.
[29] Prasetia R, As-syakur A R and Osawa T 2013 Validation of TRMM Precipitation Radar satellite data over Indonesian region *Theor. Appl. Climatol.* **112** 575-587.
[30] Okamoto K I and Kozu T 1993 TRMM precipitation radar algorithm *Proc. IEEE Int. Geosci. Remote Sens. Symp.* (Tokyo-Japan) 426-428.
[31] Schumacher C and Houze Jr R A 2003 Stratiform rain in the tropics as seen by the TRMM precipitation radar *J. Clim.* **16** 1739-1756.
[32] Awaka J, Iguchi T, Kumagai H and Okamoto K I 1997 Rain type classification algorithm for TRMM precipitation radar *Proc. IEEE Int. Geosci. Remote Sens. Symp.* (Singapore, Singapore) 1633-1635.
[33] Okamoto K, Meneghini R, Iguchi T, Awaka J, Takahashi N, Shimizu S, Kwiatkowski J and Stout J 2004 TRMM Precipitation Radar Version 6 Algorithms *Proc. 2nd TRMM Int. Sci. Conf.* (Nara, Japan) 1-4.
[34] Nakazawa T and Rajendran K 2008 10-year climatology of tropical rainfall characteristics from TRMM Data *Proc. 28th Conf. Hurricanes Trop. Meteo.* (Orlando, Florida, USA) 1-4.
[35] Shimizu S, Oki R and Igarashi T 2001 Ground validation of radar reflectivity and rain rate retrieved by the TRMM precipitation radar *Adv. Space Res.* **28** 143-148.
[36] Shuanglin L, and Qin W 2012 A new approach for classifying two types of El Niño events *Atmos. Oceanic Sci. Lett.* **5** 414–419.
[37] Takahashi H, Su H, Jiang J H, Luo Z J, Xie S P and Hafner J 2013 Tropical water vapor variations during the 2006–2007 and 2009–2010 El Niños: Satellite observation and GFDL AM2. 1 simulation *J. Geophys. Res. Atmos.* **118** 8910–8920.
[38] As-syakur A R, Tanaka T, Prasetia R, Swardika I K and Kasa I W 2011 Comparison of TRMM multisatellite precipitation analysis (TMPA) products and daily-monthly gauge data over Bali *Int. J. Remote Sens.* **32** 8969-8982.

Acknowledgments
I gratefully acknowledge data received from the National Aeronautics and Space Administration (NASA) and the Japan Aerospace Exploration Agency (JAXA) for the TRMM PR 3A25 v6 data.

IOP Conf. Series: Earth and Environmental Science **165** (2018) 012010 doi:10.1088/1755-1315/165/1/012010

MOLA Model for Optimization of Nickel Mining Management in North Morowali District

Y Suwarno[1*] **and I Nahib**[1]

[1]Researcher at Geospatial Information Agency

yatinsuwarno@yahoo.com

Abstract. Geology of North Morowali Regency is dominated by ultrabasic rock compounds, most of which have weathered to form rich nickel. The Local Government has issued 52 letters of Nickel Mining Business Licence with a total area of 177,249.20 Ha. Issues for local government are: there are indications of overlapping of land Licences, areas for other limited uses, and many abandoned mine land. Local governments should have clear rules in the provision of Mining Business Licence in order not to cause spatial conflicts. Decisions must be supported by a system that can help local governments to minimize the occurrence of problems. The purpose of this research is to select mine land allocation by using MOLA (Multi Objective Land Allocation) Model. The method used is multicriteria analysis by weighting and scoring in Geographic Information System. The model results show scenario 2 with the proposed 80,000 ha mine area is the best choice, and recommended to the local government. The allocation of mining area is 69, 72%, so there is still 30, 28% for other used. In this scenario, mining area with Exploration Mining Business Licence status decreased 49.60%, while with Production Operation Mining Business Licence status decreased 24.41%.

1. Introduction

North Morowali Regency is an expansion of Morowali Regency, Central Sulawesi Province, officially endorsed on 12 April 2012. The Regency with the capital of Kolonedale has a land area of 10,004.28 km2 or 1,000,428 ha [1]. Geologically the area of North Morowali Regency is dominated by ultramafic rock compound, consisting of serpentinite metamorphic rock, peridotite igneous rocks, gabbro, and dunit partially covered by marine sediments ie coral limestones, clastic limestone. and claystone. Most of the ultrabasic rocks on the surface have been weathered to form laterite soils with a widespread and rich spreading of nickel minerals [11]. The process of forming the lateritic nickel precipitate begins with the process of weathering ultramafic rocks such as peridotite, serpentinite and dunite with mineral content of olivine, pyroxene, magnesium silicate and iron, making it very susceptible to weathering processes because these minerals are unstable [10].

The large potential of nickel causes many mining companies to apply for mining licences. As of July 31, 2013, the North Morowali Regency Government has issued 52 Mining Business License (MBL) for Nickel Minerals. A total of 52 MBLs, 28 MBLs with an area of 135,991 Ha are exploration status, while 24 MBLs of 41,258.20 Ha are Production Operations status [3]. Mining Business License Exploration status is a Licence given by local government to mining companies to conduct exploration (research). If considered a prospect, the government may issue a Production Operation status, which allows the company to mine and produce its mining products. The long time of Exploration status that has not been changed into Production Operation status is due to the research but not the prospect or even the research has not been done at all.

The obligation to build a smelter contained in Law number 4/2009 on Minerals and Coal, among them regulating the export of unprofable minerals and prohibits mining companies in Indonesia to export raw materials from January 2014 [7]. The government will give sanctions on halting contracts of work for mining companies in Indonesia that do not meet the obligation to build smelters by the end of 2014, according to the Ministry of Energy and Mineral Resources. Local governments must have clear rules in granting mining Licences so as not to cause spatial conflicts. Potential conflicts can arise against spatial patterns, forest areas, and existing land use. A system is needed that can help local governments to make decisions in minimizing the occurrence of spatial conflicts.

Geographic Information Systems have the capacity to integrate information from multiple sources into spatial contexts and are highly suited to support decision-making procedures. Geographic Information Systems can act as a tool to help decision makers evaluate alternatives, visualize options and explore specific alternatives. Ultimately, it is the decision maker, who determines the criteria, factors, constraints, individual weights and decision rules [9]. MOLA (Multi Objective Land Allocation) is a multi-criteria and multi object model for determining land allocation in a Geographic Information System [4]. This model is widely used for the determination of land allocation, such as nature reserves, tourist areas, but there is no utilization of this model for the selection of mine land allocation.

The objective of this research is to select the optimal allocation of mining land in the management of nickel mine in North Morowali Regency based on several scenarios using MOLA Model.

2. Materials and method

2.1. Research Location

The research location in North Morowali Regency of Central Sulawesi Province is shown by Figure 1. North Morowali Regency is bordered by: Tolo Bay in the east, Morowali district in the south, Poso district in the western, Tojo Una-una and Banggai district in the north.

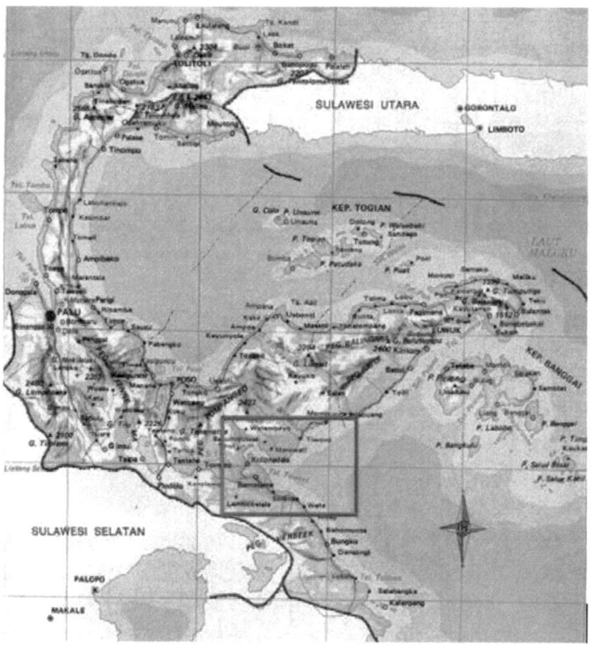

Figure 1. Location of North Morowali Regency

IOP Conf. Series: Earth and Environmental Science **165** (2018) 012010 doi:10.1088/1755-1315/165/1/012010

2.2. Data Used

The material used in this study consists of several spatial data (maps), which consists of:

a. Map of Mining Permit until 2013, presented on the map scale 1: 250,000;
b. Map of the latest Forest Areas of Central Sulawesi Province in accordance with SK.635 / Menhut-II / 2013, presented on the map scale of 1: 250,000;
c. Land Coverage / Land Use Map scale1: 50.000, results of Landsat satellite image recording 2015.
d. Spatial Planing Map, which is spatial pattern North Morowali Regency 2014 of scale 1: 50.000.

2.3. MOLA Model

The method used is multicriteria analysis by weighting and scoring by using MOLA (Multi Objective Land Allocation) model in Geographic Information System. MOLA is a multi-criteria model and multi object to determine land allocation in a Geographic Information System [4]. The basic theory of MOLA is to solve multi-purpose land allocation problems for many cases [2]

2.4 Method of Analysis

The use of this model, the criteria used are to avoid possible conflicts that may occur in any object. The analysis in the MOLA model is basically done by weighting and clarifying the parameters used. Weighting is done to differentiate the influence of each parameter on the MOLA model that will be generated. Weighting is given in units of percent (%), high-impact parameters are given a high weight, and vice versa. Scoring is done to determine the rank of each class on each parameter. Scoring is made relatively from small to large. The class with the highest score means the most Licenceted, and vice versa.

2.5 Weighting and Scoring

Weighting is a decision-making technique in a process that involves multiple factors simultaneously by weighting each of these factors. In this study, weighting is done subjectively with the determination based on certain consideration. The determination of the weight is subjectively based on the understanding of the process. Each factor has a different role indicated by the weight difference between these factors. Total weighting is 100%, and will be distributed to each parameter based on the factors that influence it. The greater the factors that affect the greater the weight will be given, and vice versa. The designated area in the Spatial Map is assumed to have the highest influence, so it is given 40% weight, then the second is the Forest Area Map with the weight of 35%, and the least the Land Use Map with 25% weight.

In the Spatial Map, the Mine Zone is given the highest score of 7, because it is in the space pattern intended for mining, it means that the area is most feasible to be maintained. In contrast, the residential area is considered to be the most vulnerable if evicted for mining, so it is given the lowest score that is 1. Forest Area, Other Use Areas (most likely, thus gave the highest weight 5. The Protected Area / Natural Protected Area are given the lowest weight that is 1. The existing land use, shrub is given the highest weight 5. Water body (lake, river) is the most unlikely area to be freed for mining area and given weight 1. Weighting of each parameter and scoring each class is presented in **Table 1**.

Table 1. Weights and Scores of each parameter

A	Allocation Area (weight = 40%)	Score
1	Mine Area	7
2	Fishery Area	6
3	Plantation Area	5
4	Dray land Farming Area	4
5	Limited Production Forest Area	3
6	Permanent Production Forest Area	2

7	Settlement Area	1
B	Forest Area (weight = 35%)	Score
1	Other Use Areas	5
2	Limited Production Forest	3
3	Production Forest	2
4	Protected Forest	1
5	Nature Reserve Area/ Natural Protected Area	1
C	Land Use (weight = 25%)	Score
1	Shrubs	5
2	Agriculture	4
3	Plantation	3
4	Settlement	3
5	Forest	2
6	Water Body	1

2.6 Research Flow chart

Flow chart of research as presented in **Figure 2**. Spatial data used are Forest Area Map, Map of Mining Permit, Spatial Map, and Land Cover Map. All of the spatial data is overlaid to get the Analysis Area Map, then by weighting and scoring, processed using MOLA Model. Some scenarios are based on land needs assumptions, i.e. to get priority weight, in the end will be obtained the direction of map of land allocation in each scenario.

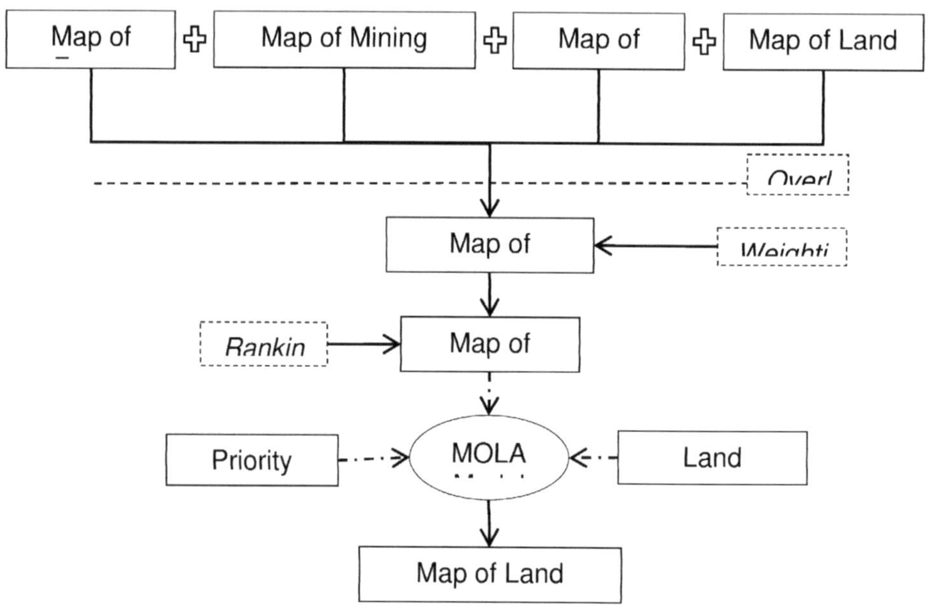

Figure 2. Flow chart of Research

3. Results and Discussion

3.1. Scenario 1 (Mine Land Allocation 100,000 Ha)

The assumption in this scenario 1, the allocation of land for nickel mine area of 100,000 ha or about 87.15% of the area of North Morowali Regency (114,743.62 Ha). The assumption in scenario 1 is too broad, since it leaves only 12.85% of land for other uses. Nevertheless, this scenario was created to give companies an opportunity already licensed by the local government to improve its management.

Based on the model results, it is known that nickel mine area with exploration status must be maintained at 67,445.70 Ha, with the company's name and its sub-Regency location as shown in **Table 2**. Compared to the previous status of 135,991 Ha, that is the reduction area of 68,545.30 Ha (50.40%). This reduction is quite significant, and is expected to address the issue of abandoned mine land, so that it can be converted for other uses.

Nickel mining area with Production Operation status to be maintained in this scenario is 32,678.96 Ha (**Table 3**). Compared with the status of the previous status of Production Operation that is 41.258,20 Ha, there is a reduction of 8,579.24 Ha (20.79%). This reduction is especially expected for mining sites that stop production because they can not afford to build their own smelters. The results of spatial model for scenario 1, both for Exploration status and Operation Production status are presented in **Figure 3**.

Table 2. Status of Exploration for Land Requirement 100,000 Ha.

Num.	Mining Company	Sub District	Area (Ha)
1	Anugrah Selaras Sejati	Petasia	2.451,5
2	Arnins Corporation	Bungku Utara	5.783,3
3	Bangun Bumi Indah	Mamosalato	2.794,3
4	Bumi Indah Sultra	Lembo	5.408,0
5	Global Samudera Atlantik	Petasia	85,8
6	Graha Sumber Mining Indonesia	Petasia	45,0
7	Lampion Alaskah Sekawan	Soyo Jaya	74,9
8	Latanindo Mining	Petasia, Lembo	5.171,4
9	Mahkota Mega Lestari	Soyo Jaya	688,8
10	Pringgondani Berseri	Mamosalato	14.158,9
11	Pringgondani Berseri	Mamosalato	234,6
12	Samudera Megah Abadi	Momosalato	909,7
13	Senamas Energindo Mining	Bungku Utara	14.417,4
14	Sharon Sindo Sejahtera	Petasia	86,7
15	Sinosteel Indonesia Mining	Petasia dan Lembo	1.264,2
16	Surya Borneo	Petasia	10.057,6
17	Sumberjati Pratama Selatan	Soyo Jaya	3.561,71
18	Surya Borneo	Petasia	49,5
19	Tiga Samudra Perkasa	Mamosalato	202,34
		Total	67.445,7

Table 3. Status of Operation Production for Land Requirement 100,000 Ha.

Num.	Mining Company	Sub District	Area (Ha)
1	Bangun Bumi Indah	Petasia	149,7
2	Bukit Makmur Istindo Nikeltama	Petasia Timur	4.729,9
3	Cahaya Murni Sejahtera	Petasia	3.531,9
4	Cipta Hutama Maranti	Soyo Jaya	2.852,7
5	Cocoman	Petasia	5.424,3
6	Gema Ripah Pratama	Soyo Jaya	148,4

Num.	Mining Company	Sub District	Area (Ha)
7	Genba Multi Mineral	Petasia	220,0
8	Ghanesa Wana Utama	Mamosalato	215,5
9	Halmahera International Resources	Petasia	418,9
10	Hffmen International	Petasia	338,4
11	Integra Service Nusantara	Soyo Jaya	154,6
12	Integra Technology Nusantara	Petasia	990,6
13	Itamarta Nusantara	Petasia	3.703,6
14	Karya Agung Permai	Petasia	29,3
15	Keinz Ventuta	Petasia Timur	533,5
16	Mulia Pacific Resources	Petasia	1.173,1
17	Persada Agung Inti Lestari	Soyo Jaya	143,5
18	Rezky Utama	Petasia	130,2
19	Sarana Mineralindo Perkasa	Lembo dan Petasia Timur	106,89
20	Sumber Permata Selaras	Petasia	55,8
21	Sumber Permata Selaras	Petasia Timur	4.559,3
22	Sumber Swarna Pratama	Petasia dan Soyo Jaya	256,9
23	Tiga Utama	Mamosalato	2.641,5
24	Trinusa Dharma Utama	Petasia	170,7
		Total	32.679,0

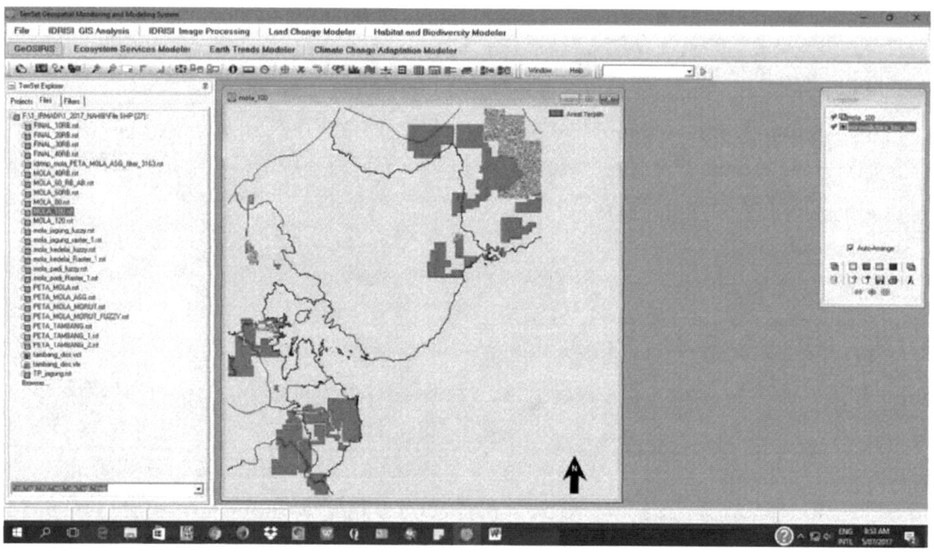

Figure 3. Results of MOLA Model for scenario 1 (mine area of 100,000 ha)

3.2. Scenario 2 (Mine Land Allocation 80,000 Ha)

The assumption in scenario 2 is that the land allocation for nickel mine is 80,000 ha or about 69,72% from Morowali Regency (114,743,62 Ha). The assumption in scenario 2 is more realistic, i.e. the comparison between land for mine allocation (± 70%) and land for other uses (± 30%). The result of the model is known that the area of nickel mining area with the status of Exploration to be maintained is 48,414.42 Ha, with the names of the companies and the location of the Sub District as shown in **Table**

4. Compared with the status of the original mining permit which is 135,991 Ha, that is the reduction area of 87,576.58 Ha (64.40%). The nickel mine area with the status of Operating Production must be maintained for 31,186.99 Ha (**Table 5**). Compared to the status of Mining Permit originally covering 41,258.20 Ha, there was a reduction of 10,071.21 Ha (24.41%). The results of the spatial model for scenario 2, both for the Exploration status and the Production Operation status are presented in **Figure 4.**

Table 4. Status of Exploration for Land Requirement 80,000 Ha

Num.	Mining Company	Sub District	Area (Ha)
1	Anugrah Selaras Sejati	Petasia	19,0
2	Arnins Corporation	Bungku Utara	5.097,3
3	Bangun Bumi Indah	Mamosalato	1,8
4	Bumi Indah Sultra	Lembo	85,9
5	Global Samudera Atlantik	Petasia	44,5
6	Graha Sumber Mining Indonesia	Petasia	8.726,2
7	Lampion Alaskah Sekawan	Soyo Jaya	196,0
8	Latanindo Mining	Petasia, Lembo	243,2
9	Mahkota Mega Lestari	Soyo Jaya	5.250,6
10	Pringgondani Berseri	Mamosalato	13.920,0
11	Samudera Megah Abadi	Momosalato	688,3
12	Senamas Energindo Mining	Bungku Utara	4.869,0
13	Sharon Sindo Sejahtera	Petasia	5.783,3
14	Sinosteel Indonesia Mining	Petasia dan Lembo	713,0
15	Sumberjati Pratama Selatan	Soyo Jaya	884,2
16	Surya Borneo	Petasia	1,6
17	Tiga Samudra Perkasa	Mamosalato	1.890,6
		Total	48.414,4

Table 5. Status of Operation Production for Land Requirement 80,000 Ha.

Num.	Mining Company	Sub District	Area (Ha)
1	Bangun Bumi Indah	Petasia	149,82
2	Bukit Makmur Istindo Nikeltama	Petasia Timur	4.729,8
3	Cahaya Murni Sejahtera	Petasia	3.531,9
4	Cipta Hutama Maranti	Soyo Jaya	2.851,4
5	Cocoman	Petasia	4.673,4
6	Gema Ripah Pratama	Soyo Jaya	148,4
7	Genba Multi Mineral	Petasia	220,2
8	Ghanesa Wana Utama	Mamosalato	215,5
9	Halmahera International Resources	Petasia	310,9
10	Hffmen International	Petasia	338,4
11	Integra Service Nusantara	Soyo Jaya	154,5

Num.	Mining Company	Sub District	Area (Ha)
12	Integra Technology Nusantara	Petasia	988,5
13	Itamarta Nusantara	Petasia	3.703,6
14	Karya Agung Permai	Petasia	26,4
15	Keinz Ventuta	Petasia Timur	533,5
16	Mulia Pacific Resources	Petasia	1.173,1
17	Persada Agung Inti Lestari	Soyo Jaya	139,8
18	Rezky Utama	Petasia,	130,2
19	Sarana Mineralindo Perkasa	Lembo dan Timur	106,9
20	Sumber Permata Selaras I	Petasia	55,8
21	Sumber Permata Selaras	Petasia Timur	4.558,9
22	Sumber Swarna Pratama	Petasia dan Soyo Jaya	208,42
23	Tiga Utama	Mamosalato	2.068,2
24	Trinusa Dharma Utama	Petasia	169,6
		Total	31.187,0

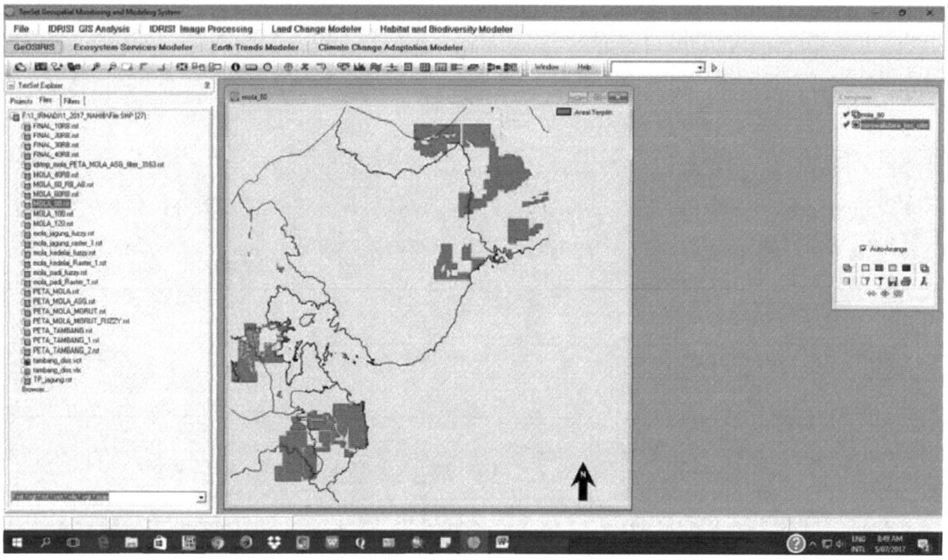

Figure 4. Results of MOLA Model for scenario 2 (mine area of 80,000 ha)

3.3. Scenario 3 (Mine Land Allocation 60,000 Ha)
The MOLA model in scenario 3, the mine area that wants to be maintained is 60,000 Ha or 52.29% of the area of Morowali Regency (114,743.62 Ha). This scenario is designed to provide the widest possible opportunity for abandoned mine land to be converted into productive land. Many mining sites have been licensed since 2009 but are still abandoned.

The result of model in scenario 3 is obtained nickel mining area with status of Exploration Mining Permit is 37,645,51 Ha (**Table 6**). Compared with the status of the original Mining Permit (135,991 Ha), there was a reduction in the area of 98,345.49 Ha (72.32%). Many mining areas already have Licences with Mining Business status of Exploration but are still abandoned. The cause is suspected, the licence owner has never done any research at all because it does not have sufficient capital and or expertise, or has been done exploration but its potential is considered low

Mining land with the status of Operating Production that must be maintained in scenario 3 is 21,621.46 Ha (**Table 7**). Compared with the status of Mining Operations previously 41.258,20 Ha, there is a reduction of 19,636.74 Ha (47.59%). Mining land with abandoned Production Operating status is generally after 2012, which is suspected as the company is unable to build smelters for mining companies that export, as required by the government. The spatial model for scenario 3, both for the Exploration status and the Production Operation status is presented in **Figure 5**.

Table 6. Status of Exploration for Land Requirement 60,000 Ha

Num.	Mining Company	Sub District	Area (Ha)
1	Arnins Corporation	Bungku Utara	5.688,4
2	Bangun Bumi Indah	Mamosalato	1,6
3	Bumi Indah Sultra	Lembo	3.447,3
4	Global Samudera Atlantik	Petasia	1,2
5	Graha Sumber Mining Indonesia	Petasia	15,7
6	Lampion Alaskah Sekawan	Soyo Jaya	0,11
7	Latanindo Mining	Petasia, Lembo	4.948,7
8	Mahkota Mega Lestari	Soyo Jaya	0,3
9	Pringgondani Berseri	Mamosalato	8.642,3
10	Samudera Megah Abadi	Momosalato	826,2
11	Senamas Energindo Mining	Bungku Utara	4.769,5
12	Sharon Sindo Sejahtera	Petasia	0,1
13	Sinosteel Indonesia Mining	Petasia dan Lembo	191,8
14	Sumberjati Pratama Selatan	Soyo Jaya	1.837,3
15	Surya Borneo	Petasia	7.080,2
16	Tiga Samudra Perkasa	Mamosalato	194,8
		Total	37.645,5

Table 7. Status of Operation Production for Land Requirement 60,000 Ha

Num.	Mining Company	Sub District	Area (Ha)
1	Bangun Bumi Indah	Petasia	149,9
2	Bukit Makmur Istindo Nikeltama	Petasia Timur	3.373,4
3	Cahaya Murni Sejahtera l	Petasia	2.082,3
4	Cipta Hutama Maranti	Soyo Jaya	2.851,2
5	Cocoman	Petasia	1.947,6
6	Gema Ripah Pratama	Soyo Jaya	148,2
7	Genba Multi Mineral	Petasia	219,6
8	Ghanesa Wana Utama	Mamosalato	215,5
9	Halmahera International Resources	Petasia	31,5
10	Hffmen International	Petasia	338,4
11	Integra Service Nusantara	Soyo Jaya	154,6
12	Integra Technology Nusantara	Petasia	520,4
13	Itamarta Nusantara	Petasia	3.666,9

Num.	Mining Company	Sub District	Area (Ha)
14	Karya Agung Permai	Petasia	19,9
15	Keinz Ventuta	Petasia Timur	450,2
16	Mulia Pacific Resources	Petasia	243,9
17	Persada Agung Inti Lestari	Soyo Jaya	121,4
18	Rezky Utama	Petasia	130,1
19	Sarana Mineralindo Perkasa	Lembo, Petasia Timur	106,9
20	Sumber Permata Selaras	Petasia	55,8
21	Sumber Permata Selaras	Petasia Timur	3.895,0
22	Sumber Swarna Pratama	Petasia, Soyo Jaya	72,7
23	Tiga Utama	Mamosalato	659,3
24	Trinusa Dharma Utama	Petasia	167,4
		Total	21.621,5

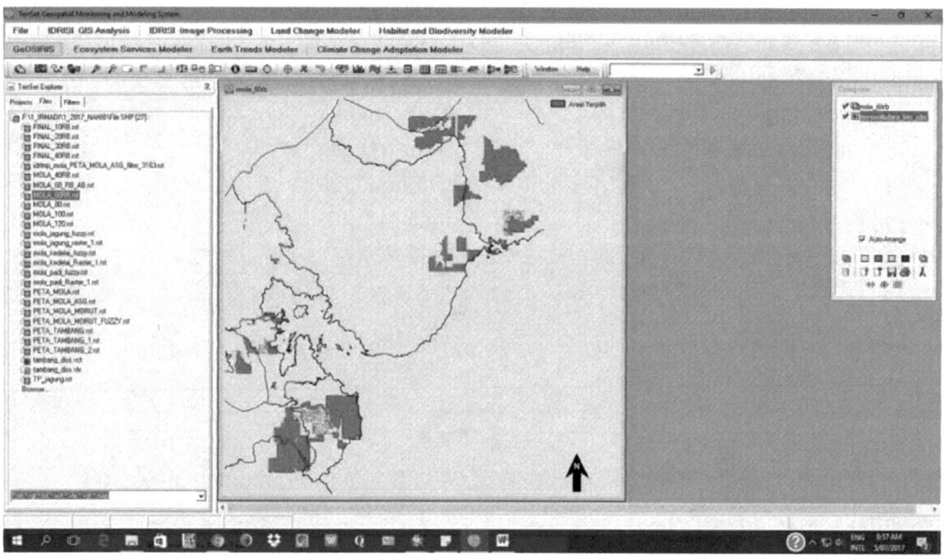

Figure 5. Results of MOLA Model for scenario 3 (mine area of 60,000 ha)

3.4. Discussion

The scenario studied is 3 options, scenario 2 is the best choice, that is by maintaining 69,72% from North Morowali Regency area for mine land allocation, so there is still another 30,28% for other allocation of use.

In scenario 1 this is by maintaining the mining area with Exploration status of 67,445.70 Ha, or a reduction of 68,545.30 Ha (49.60%). The reduction of mining land will not have a significant impact on mining companies that have exploration. Exploration status that has not changed into Production Operation status due to 2 things. Firstly, the company holding Exploration does not conduct exploration activities at all because it does not have the financial capability and / or expertise, so the land becomes abandoned. Second, the company holding Exploration Business Mining Licence has conducted exploration activities, but it is suspected that the mine land is not prospective or not economically valuable.

For mining area under the status of Operating Production, the area maintained is 31.186.99 Ha, or there is a reduction of 10,071.21 Ha (24, 41%) The reduction of mining area with the status of Operation Production is still possible, since there are many companies which stopped the production operation because it does not have a smelter yet. In the Government Regulation of the Republic of Indonesia Number 1 (2014), the Contract Holder is required to refine the domestic mining products (Article 112C.1), and shall perform the processing and refining of domestic mining products (Article 112C.2) sales to a certain amount of abroad (Article 112C.3). The comparison of the MOLA model results for the 3 scenarios is presented in **Table 8**.

Table 8. Comparison of 3 MOLA model scenarios

Area	**scenario**		
	1	**2**	**3**
Regency area (Ha)	114.743,6	114.743,6	114.743,6
Mine land allocation (Ha)	100.000	80.000	60.000
Mine land allocation (%)	87,2	69,7	52,3
Previous Exploration status (Ha)	135.991	135.991	135.991
Exploration status (Ha)	67,445.7	48.414	37,645.5
Reduction (Ha)	68.545,3	87.577	98.345,5
Reduction (%)	50,4	64	72,3
Previous Production Operation status (Ha)	41.258,2	41.258,2	41.258,2
Production Operation status (Ha)	32,679.0	31,187.0	21,621.5
Reduction (Ha)	8.579,2	10.071,2	19.636,7
Reduction (%)	20,8	24,4	47,6
Non-Mining land use (%)	± 17	± 70	± 48
Robustness	bad	good	bad

4. Conclusion

The MOLA model is good enough for mine land allocation, especially for mineral-rich areas. The MOLA model provides a solution, how much land should be allocated, where location, what kind of land, consideration of licensing status, and its relation to spatial planning. With this model can be determined how much area and where mining land will be allocated based on local government policy. Selection of mine land allocation is based on the rank and weight of each parameter used.

In this research, there are 3 scenarios, with proposed different mining land area, that is 100.000 Ha, 80.000 Ha, and 60.000 Ha. Scenario 2 with a proposed 80,000 ha mine is considered the best option, since it only allocates 69.72% of the total area of North Morowali Regency, while the rest (30.28%) can be utilized for other purposes. In this scenario 2, the mining area with Exploration status was reduced by 49.60%, while the status of Production Operation decreased by 24.41%.

References

[1] Bappeda Kabupaten Morowali Utara 2016 *Kabupaten Morowali Utara Dalam Angka* p 276
[2] Carver S J 2007 Integrating Multi-Criteria Evaluation with Geographical Information Systems *International Journal of Geographical Information Systems* Vol **5** pp 321-339
[3] Dinas Pertambangan dan Energi Provinsi Sulawesi Tengah 2016 Data Ijin Usaha Pertambangan di Wilayah Kabupaten Morowali, Provinsi Sulawesi Tengah
[4] Hajehforooshnia S. Soffianian Mahiny A S andFakheran S 2011 Multi objective land allocation (MOLA) for Zoning Ghamishloo Wildlife Sanctuary in Iran *Journal for Nature Conservation* Vol **19** pp 254-262
[5] Peraturan Menteri Energi dan Sumber Daya Minearal RI 2015 Nomor 43 tentang Tata Cara Evaluasi Penerbitan Izin Usaha Pertambangan Mineral dan Batubara. *Berita Negara Republik Indonesia Nomor 2014 Tahun 2015*
[6] Peraturan Pemerintah Republik Indonesia Nomor 1 Tahun 2014 tentang Perubahan Kedua atas Peraturan Pemerintah Nomor 23 Tahun 2010 tentang Pelaksanaan Kegiatan Usaha

Pertambangan Mineral dan Batubara. *Lembaran Negara Republik Indonesia Nomor 5489 Tahun 2014*
[7] RI (Republik Indonesia) 2009 Pertambangan Mineral dan Batubara. *Lembaran Negara Republik Indonesia Nomor 4959 Tahun 2009*
[8] RI (Republik Indonesia) 2015 Surat Keputusan Menteri Lingkungan Hidup dan Kehutanan RI No. 2382/Menhut-VI/BRP-UK/2015 tentang Peta Arahan Pemanfaatan Hutan Produksi untuk Usaha Pemanfaatan Hutan Provinsi Sulawesi Tengah
[9] Sante-Riveira I Boullo´n-Magan M Crecente-Maseda R anad Miranda-Barros D 2008 Algorithm Based on Simulated Annealing for Land-use Allocation *Computers & Geosciences* Vol **34** pp 259-268
[10] Ningsih S A 2012 Eksplorasi Awal Nikel Laterit di Desa Lamontoli dan Lalemo, Kecamatan Bungku Selatan, Kabupaten Morowali, Propinsi Sulawesi Tengah *Jurnal Ilmiah MTG* Vol **5**, No 2
[11] Mulyaningsih S 2013 Indikasi Gunung Api Purba di Daerah Morowali Sulawesi Tengah *Prosiding Seminar Nasional Kebumian Ke-6 Teknik Geologi Universitas Gadjah Mada* 11-12 Desember 2013

Acknowledgements
The authors wish to acknowledge the Head of the Promotion and Cooperation Research Centre and Head of Research at the Geospatial Information Agency, who have facilitated everything so that the research and writing of this scientific paper can be accomplished.

IOP Conf. Series: Earth and Environmental Science **165** (2018) 012011 doi:10.1088/1755-1315/165/1/012011

Spatial Analysis of Environmental Critically due to Increased Temperature in The Built Up Area With Remote Sensing

Bandi Sasmito[1*] **and Andri Suprayogi**[1]

[1]Geodetic Engineering Department, Faculty of Engineering, Diponegoro University, Semarang Central Java, Indonesia

bandy.geo96@gmail.com

Abstract. This study aims to detect the environmental crises phenomena as an effect temperature increases in the built area by remote sensing technology. This study is taking place in Surakarta City, Central Java, Indonesia. The methods are used in this research is to spatially compare the distribution of surface temperature that detected using Land Surface Temperature (LST) algorithm with the distribution of built-up area that detected using Built-Up Index (BU) algorithm. Spatial data are used is multi-temporal satellite imagery of Landsat. The result of this research, the surface temperature is detected an increase in Surakarta City every year. The spread of high temperatures is increasingly and widespread followed by the widespread of Built-Area. The central area of Surakarta City is identified as the most critical environmental area based on the temperature and built the region. The northern border area of the city can also be detected with increased Build-Area distribution and surface temperature distribution.

1. Introduction

A city is a place that functions for the central of peoples activities. The city is the central of government, economics, tourism, and entertain. That's way city become large and many of development. The development of various fields in urban areas is very rapid than other areas. The magnitude of opportunities to improve the quality of live has made the urban areas become populous. Peoples have been coming from suburb-area and the villager who try their luck in the big city.

This study takes Surakarta City as a study area. Surakarta is also known Solo City. Surakarta City is in Central Java, Indonesia. Surakarta City is growing rapidly which is included in the secondary city in Indonesia. Development of infrastructure is growing with the Hotel, Mall, and others. The development of the city of Surakarta this day can be quite high. The city of Solo also has problems, namely slums, unlicensed dwellings and basic infrastructure needs; regional disparities; limited land; and management capacity [1].

Solo or also called Surakarta is a city in Central Java which has an area of 46 km2 is bordered by Karanganyar and Boyolali regencies in the north, Karanganyar and Sukoharjo regencies in the west and east also Sukoharjo regency in the south. In the eastern part of Solo City shows the area of the Bengawan-Solo river. Solo also has a population of 510,077 in 2014. The development of development in this city has increased every year which can be seen from the incessant development of high buildings in recent years is located in the central of Surakarta starting from the construction of malls, and the addition of hotels and apartments. The existence of this event then conducted research on the effect of increasing buildings in the city of Surakarta with temperature changes.

The construction of a central building activity increases every year caused some densely built area resulting in reducing vegetation area. The surfaces of concrete and asphalt replaced the growing variety of vegetation. Air temperature will increase drastically in the built area when compared with areas that are still dominated by vegetation. The nature of the materials used in the construction of urban structures leads to heat emissivity. The heat wastes generated by factories, air conditioners (AC) and motor vehicle fumes contribute also to temperature increase [2–5]. Negative impacts cause environmental criticality that causes increased energy consumption, increased air pollution and greenhouse gases, affect human health and comfort, and disturbed water quality. [6].

Detecting the critically of environment by measuring the density of the built area and temporarily temperature changes will be difficult and consuming much time if done by field measurement. So in this study offers a fast, effective and efficient method of utilizing Remote sensing technology. The general conditions that make this technique more interesting than conventional methods such as data obtained in the form of digital data, the measurement includes a more distributed area compared with conventional measurement only certain points (samples). Remote sensing has advantages such as data obtained in the form of digital data; spatial based, and analysis can cover a large area [7,8]. The method used in detecting the density of the built area is Built-up Index (BU) algorithm, and detecting temperature is Land Surface Temperature (LST) algorithm. Critically of Environment will analyze from BU and LST results in multi-temporal with spatial data.

The result of this research is the spatial analysis of environmental criticism index in Surakarta City and its surroundings by measuring the accumulation of heat energy as the impact of built area density over time and presented in the form of Thematic Map.

2. Method

2.1. Data and Method

This research was conducted in Surakarta City Central Java Indonesia. The data used is Landsat Image. Imagery is downloaded from the USGS (United States Geological Survey) site at http://glovis.usgs.gov/ [9]. The image used in this study is an archive on the path: 120, and row: 65 in which the recording coverage contained Semarang City. The image selected time series 3 years ie on 6 August 2008, August 20, 2013, June 14, 2015, and 11 May 2017. Terms of image selection include cloud coverage less than 20% and in the same month each year. Terms in the adjacent month are used to minimize seasonal differences during image recording. The list of imagery used is shown in Table 1.

Tabel 1. List of Landsat Imagery used in this Study

Scene Id	Date of acquisition	Output format	Data Type
LT51190652008219BKT00	August 6th 2008	GEOTIFF	L1T
LC81190652013232LGN00	August 20th 2013	GEOTIFF	L1T
LC81190652015174LGN00	June 23th 2015	GEOTIFF	L1T

Based on the flow chart of the processing method in Figure 1, the first step is to detect Land Surface Temperature (LST). LST is the measurement of how hot (temperature) of the ground surface. Temperature is measured from surface soil emissions due to solar radiation. Soil surface heat emissions are captured by satellite image sensors in the range of thermal infra-red spectra. The sensor converts heat emission into a brightness difference on an imagery band (band). Landsat 8 is one of the images that have a thermal infrared channels, the channel is TIRS (Thermal Infrared Red Sensor) on channel number 10 and 11. In this research LST is made from Landsat Image 8 data of TIRS sensor recording from image data list in Table 1. Before the Landsat image is formed into LST, the conversion of the original Digital Number (DN) image becomes Spectral Radiance. Step conversion of Landsat 8 DN TIRS to Spectral Radiance Value using Equation (1). [9]

$$L\lambda = MLQcal + AL \tag{1}$$

Lλ : Value of Spectral Radiance,
ML : Radiometric rescaling grup on radians mult band
Qcal : Digital Number TIRS Band,
A_L : Radiometric rescaling grup on radians add band

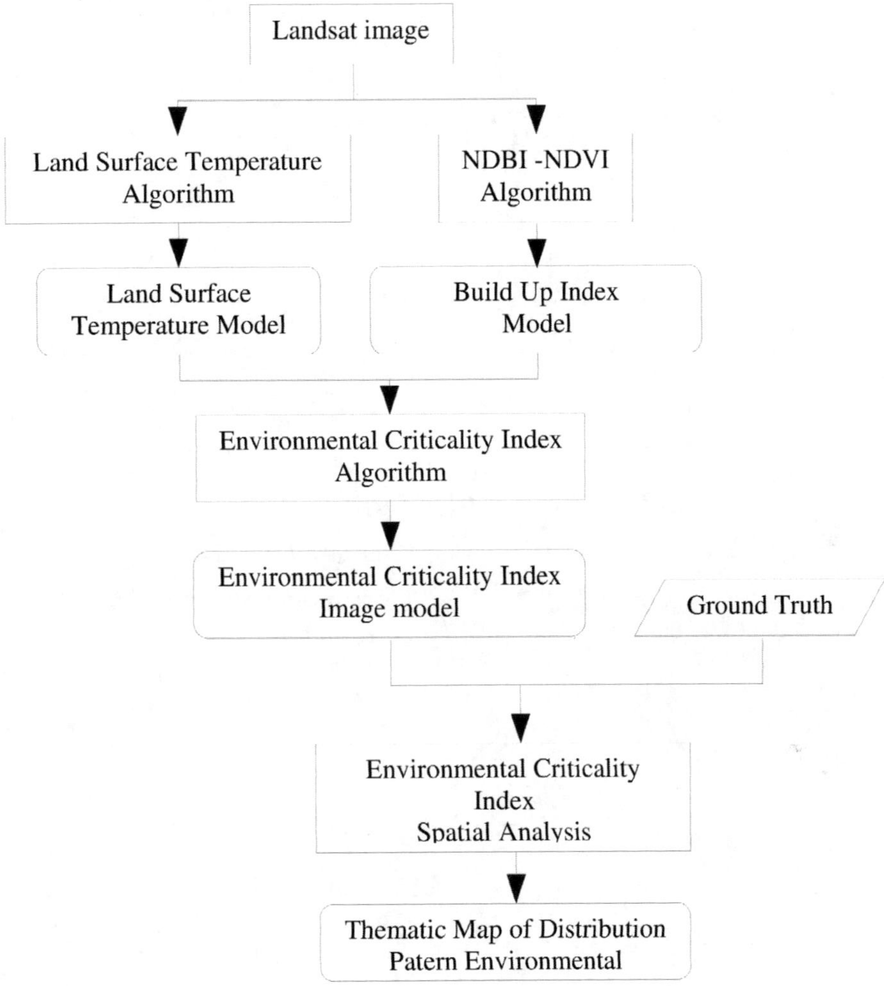

Figure 1. Flow chart of processing method

Next is the radiance values obtained from Equation (1) are converted to LST temperature values, the temperature values having a Kelvin unit. LST is obtained by applying the Planck algorithm as shown in Equation (2) [9–11].

$$Ts = \frac{K2}{\ln(\frac{K1}{L\lambda}+1)} \tag{2}$$

Ts : LST in Kelvin unit,
K_1 K_2 : Calibration Constanta of Spectral Radiance
L_λ : Value of Spectral Radiance

The LST temperature value in the Kelvin (K) unit is converted to degrees Celsius (C) using Equation (3). The reason used is Celsius because it has a better range for clarity in image interpretation [2]. Figure 2 shows the distribution of surface temperature from three different years in the study area.

$$T(C) = T(K) - 273.15 \tag{3}$$

T(C) : LST in Celcius Unit,
T(K) : LST in Kelvin, and value of 273.15 is the constant to convert Kelvin into Celcius

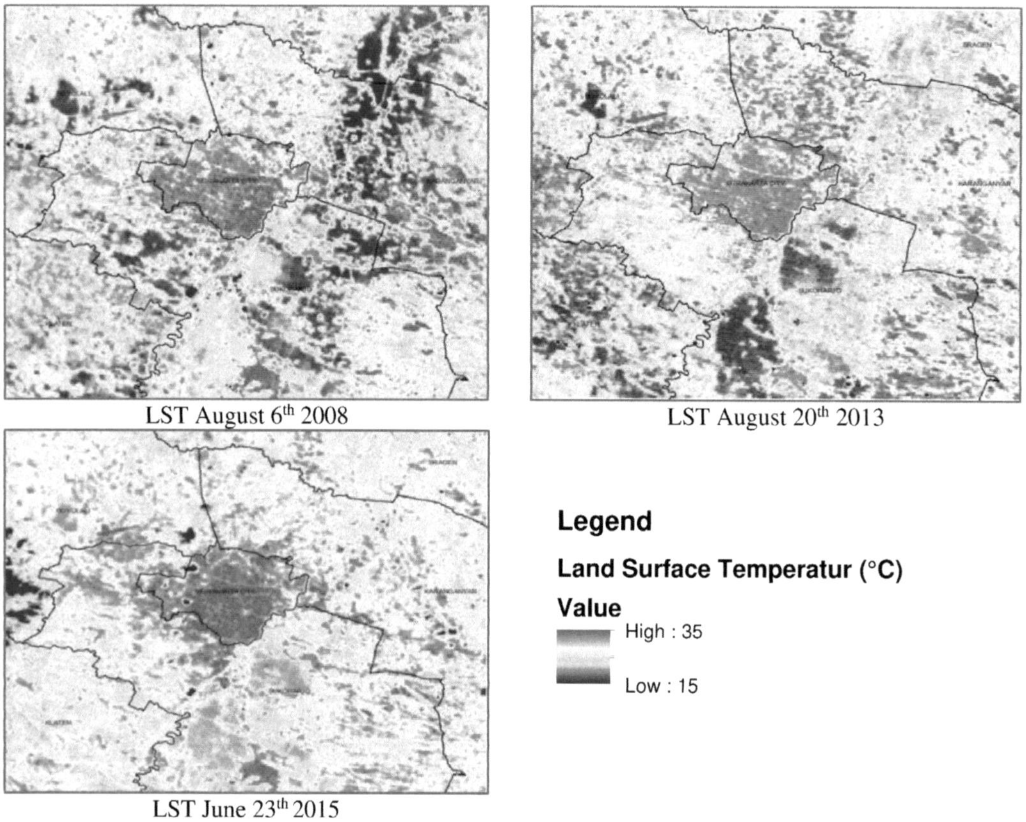

Figure 2. Land Surface Temperature (LST) Multi Temporal of Surakarta City.

Next is to build a Build-Up Area (BU) index which is derived from the differential of the Normalized Difference Build-up Index (NDBI) and Normalized Difference Vegetation Index (NDVI). Prior to the original NDVI processing, the Landsat original image corrected the Top of Atmospheric (TOA) radiometric, this correction was made to improve the visual quality of the image and simultaneously improve the pixel values that did not match the actual reflectance or spectral value of the object. Band 4 (visible RED) and band 5 (NIR) are given a correction with Equation (4) [9].

$$\rho\lambda = \frac{M_p\, Q_{cal} + A_\rho}{Sin(\theta_{SE})} \tag{4}$$

Pλ : Corrected pixel value Top Of Atmosphere (TOA) reflectance,
Mρ : Band-specific rescaling factor multiplication of metadata,
Aρ : Band-specific rescaling additive factor from metadata,
Qcal : Digital Number (DN) Band, and
θSE : sun angle during image recording.

According to zha et. all. [12] uses the NDVI analogy to develop a built-in area index called the Normalized Difference Built-up Index (NDBI) used to calculate the Built-up area. The NDBI index will focus on highlighting urban areas or built areas where there is usually a higher reflection in the Short-Wave Infrared (SWIR) area when compared to Near-Infrared (NIR) areas. Near Infrared is very sensitive to detect vegetation, whereas the reflectance for open land and built land is very low. Short Wave Infrared (SWIR) can reflect moisture content on various land uses. Therefore, NDBI utilizes both bands. NDBI spectral range values range from 0.1 to 0.3. Equation (5) describes the method of calculating NDVI.

$$NDBI = \frac{(\rho SWIR - \rho NIR)}{(\rho SWIR + \rho NIR)} \tag{5}$$

ρNIR : Value of reflectance SWIR,
ρRED : Value of reflectance NIR

NDVI is well known and widely used. NDVI was first revealed by Rouse in 1973 [13,14]. NDVI is an index that describes the greenness of a plant. The vegetation index is a mathematical combination between a red band and NIR (Near-Infrared Radiation) band. Vegetation appears dark in the visible but visible in the near infrared wavelength range [2]. NDVI is calculated on a per-pixel basis of the normalization difference between the red-spectral band of sight and near-infrared/NIR in the image.

In calculating NDVI, the red band represents the visible spectrum compared to the infrared (NIR) spectrum. Equation (6) describes the method of calculating NDVI.

$$NDVI = \frac{(\rho NIR - \rho RED)}{(\rho NIR + \rho RED)} \tag{6}$$

ρNIR : Value of reflectance NIR,
ρRED : Value of reflectance Visible-Red.

Build-Up (BU) Index is built from the difference between NDBI and NDVI. The wake area will stand out if the reflectance of the vegetation is reduced. Equation (7) describes the method of calculating Built-Up (BU) Index. Figure 3 describes the Build-Up Index of three 3 different times. [12]

$$BU_{area} = NDBI - NDVI \tag{7}$$

NDBI : Normalized Difference Built-up Index,
NDVI : Normalized Difference Vegetation Index

From LST and BU data, the ECI (Environmental Criticality Index) is identified. Senanayake et al (2013) established the ECI by comparing the LST value with the availability of vegetation cover (NDVI algorithm). The spatial layers of LST and NDVI before being compared first equate their spectral values. Spectral values are equated by spreading the pixel value histogram equalization to 1-255 to increase the clarity and contrast of the spatial layer to avoid the limitations of the ECI index value due to the value 0 on the pixels of the NDVI algorithm. The ECI equation is described in equation (8) [2].

$$\mathrm{ECI}_{(\mathrm{LST-VEG})} = \frac{\mathrm{LST}_{(\mathrm{Streched\ 1-255})}}{\mathrm{NDVI}_{(\mathrm{Streched\ 1-255})}} \tag{8}$$

ECI : Environmental Critically Index,
LST (*Streched* 1-255) : LST stretched to 1-255,
NDVI (*Streched* 1-255) : NDVI stretched to 1-255.

IOP Conf. Series: Earth and Environmental Science **165** (2018) 012011 doi:10.1088/1755-1315/165/1/012011

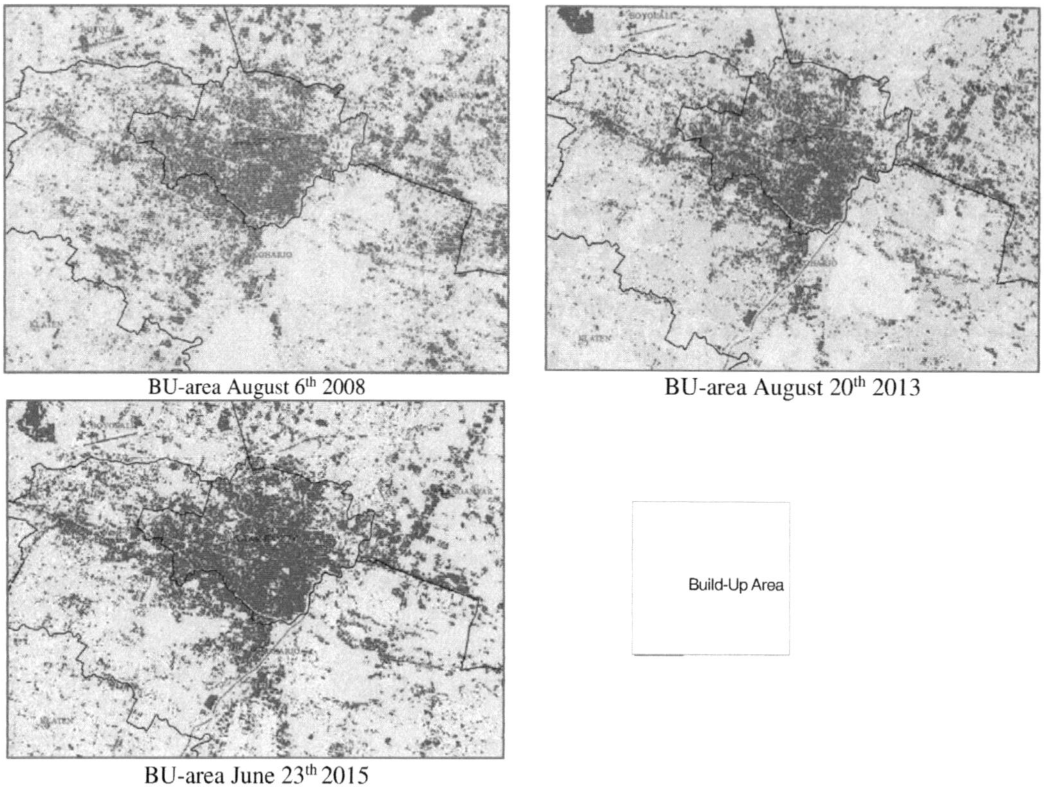

Figure 3. Multi Temporal of Built Up Index of Surakarta City.

While in this research will use the Senanayake's analogy [2], the difference is using Built-Up Index. From Figure 2 and Figure 3 we can assume LST and BU distribution is directly proportional or it can be said that LST distribution follows BU distribution so that ECI algorithm will be adjusted, ECI will be directly proportional to Built-Up (BU) Index. For that reason the ECI equation will be modified will not dividing LST by BU but multiplying LST and BU. The ECI equation with the BU index is described in equation (9).

$$\mathrm{ECI}_{(LST-BU)} = \mathrm{LST} * \mathrm{BU} \tag{9}$$

ECI : Environmental Critically Index,
LST : Value of LST
BU : Value of BU

The results of the ECI algorithm with BU Index are shown in Figure 4. The red colour indicates a higher vulnerability index which means the wake area will result in increased temperature.

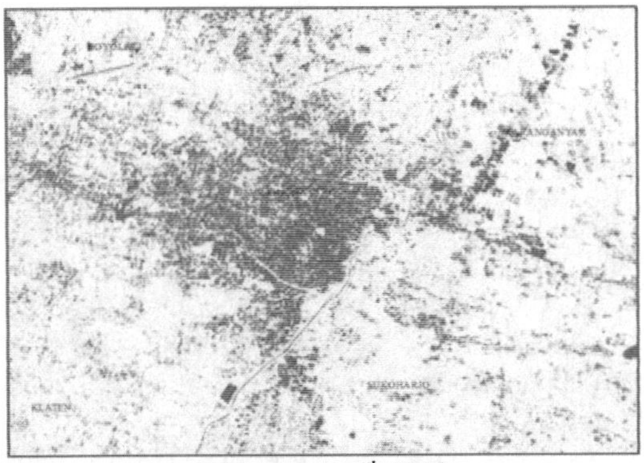

ECI August 6th 2008

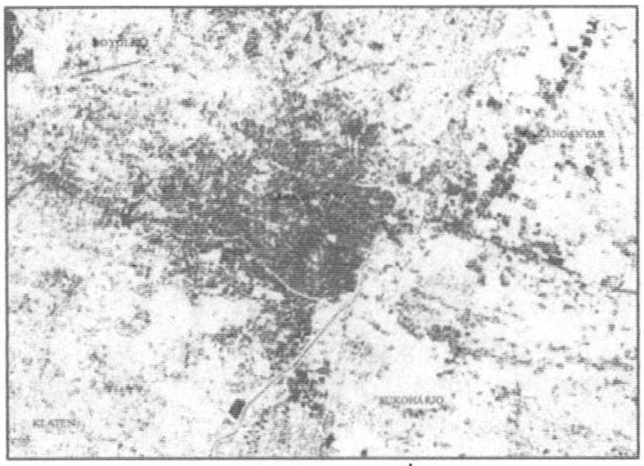

ECI area August 20th 2013

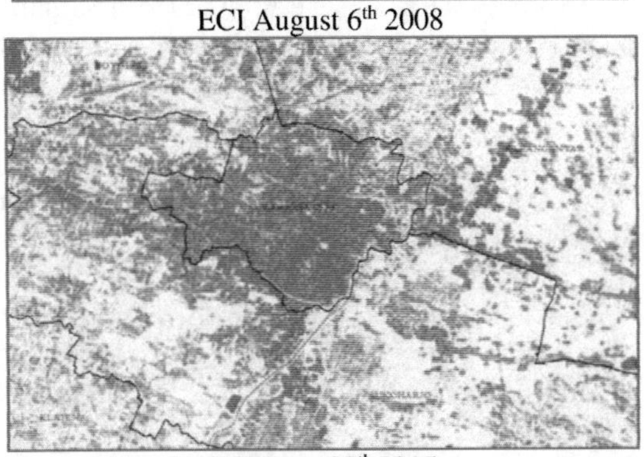

ECI June 23th 2015

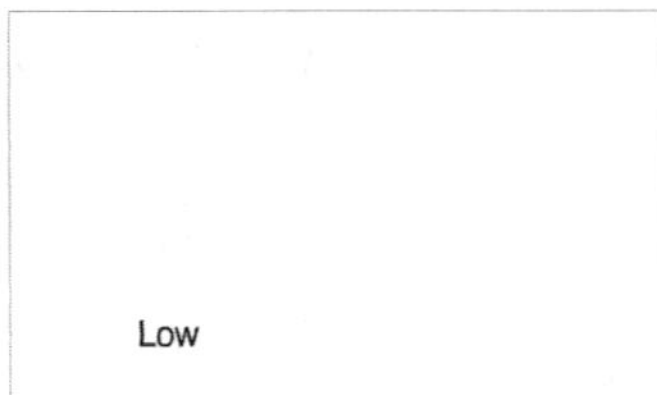

Figure 4. Multi Temporal of Environmental Critically Index (ECI) of Surakarta City.

2.2. Validation

The validation in the field is based on the temperature classes. The Classes has been made previously on the Land Surface Temperature map and evenly distributed the sample point. Temperature sampling is done one by one at each point of the plan by using two temperature measuring devices such as the infrared thermometer. The devices look like in Figure 5.

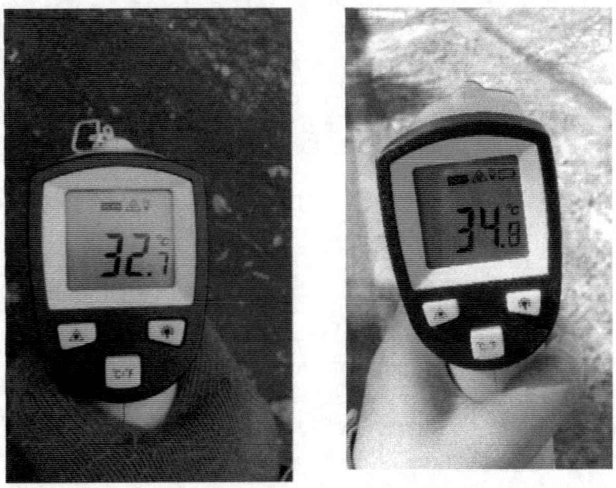

Figure 5. Infrared Thermometer using in temperature validation

Validation is done following the Landsat 8 tracks. The satellite passes above the equator at ± 10.00 GMT +7. The validation survey shall be conducted within 2 hours from 09.00 - 11.00 hours. Each of the two devices is measured at the planned sample point then recorded according to the sample location point. Meanwhile, position measurement used handheld Global Positioning System tool.

3. Result

The result of surface distribution analysis (LST) as shown in Figure 2 shows that the area of Surakarta City has significant temperature difference to the surrounding area. Surface temperature distribution (LST) according to previous studies indicates higher temperatures are present in densely built areas [15,16]. The surface of the concrete in buildings more absorbs heat energy than to reflect it so as to make the temperature rise in the area around the building [2,17].

The distribution of Build-Up Area shown in Figure 3 shows that from year to year the area built in Surakarta City is increasingly widespread and penetrated into other administrative areas around the border. In accordance with previous research, the distribution of vegetation density is inversely related to the rise in surface temperature (LST) [2,17].

The diminished vegetation and rising temperatures affect the quality of the environment. Environmental quality is assessed by the ratio result index of LST and NDVI, a ratio called the Environmental Criticality Index (ECI) [2]. ECI results Figure 4 shows the distribution of high critical classes in the central city area and spread to areas adjacent to other districts. The high environmental criterion resulted from the ECI algorithm was identified as a hot island (UHI) [18–20]. To clarify the critical area due to UHI is done the overlay with a map that has a larger scale. ECI maps are only used in the next "High" class in overlays with maps from OpenStreet Map, shown in Figure 5 and Figure 6.

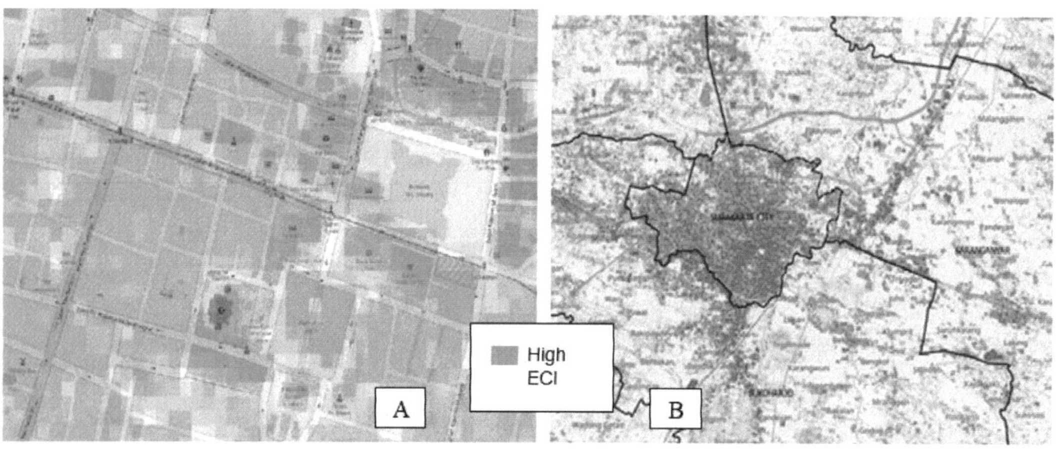

Figure 6. A) Environmental Critically Index (ECI) in central of Surakarta City. (B) Environmental Critically Index (ECI) border of Surakarta City.

The Map of Environmental Criticality Index in Figure 4, it can be seen that a region with high criticality is indicated by the phenomenon of the hot island (UHI) [18–20]. The environmental critique of the "High" class is overlaid with OpenStreetMap to show the most critical areas of Surakarta. The central area with surrounding infrastructure is identified as having high critically and severe heat islands (UHI), the spatial distribution is shown in Figure 6 (A).

In Figure 6(B) describe the border area of Surakarta also seen to be critical. The expansion of the city to the border is accompanied by the level of environmental critically. In this figure, the city expansion seems more dominant towards the south.

The relationship of LST temperature processing results to the temperature of the field validation can be known by using a simple linear regression test. This test yields an equation that can infer a relationship or not from both aspects. The following Table 2 is a field survey compared to LST.

Table 2. Temperature from Survey with LST

No	UTM Zone 49S Coordinate		Temperature (°C)	
	X	Y	Survey	LST
1	477905.490	9166403.547	32.0	33.4
2	477998.662	9166362.841	33.6	33.2
3	478043.480	9166465.451	34.1	35.7
4	478909.025	9165730.158	33.2	34.3
5	479028.505	9165779.679	34.9	35.7
6	479221.143	9165603.463	35.7	36.2
7	479892.589	9165365.957	34.7	35.1
8	480336.516	9164629.528	35.4	36.3
9	480497.558	9164609.828	35.6	37.2
10	480594.454	9164114.969	33.9	35.0
11	480839.515	9163877.027	37.3	38.5
12	480832.097	9164372.690	34.5	34.4
13	482338.924	9164048.572	37.3	38.4
14	482046.337	9164138.369	37.6	37.1
15	483655.254	9164623.883	32.7	35.3
16	483799.359	9164707.864	32.8	33.8
17	483885.418	9164342.505	35.0	35.7
18	482479.313	9163930.975	33.2	34.6
19	477278.418	9164732.458	39.7	39.8
20	482630.488	9162059.253	37.3	36.8
21	481125.322	9162808.846	37.0	33.8
22	481149.654	9162724.214	34.3	34.2
23	481020.767	9162638.395	38.8	39.2
24	480358.768	9162850.002	36.5	36.6
25	480226.407	9161695.869	35.7	35.3
26	480293.992	9161481.291	36.4	36.3

The relationship of LST temperature processing results to the temperature of the field validation can be known by using a simple linear regression test. This test yields an equation that can infer a relationship or not from both aspects. Here is the result of simple linear regression of LST temperature processing relationship and Surakarta field temperature.

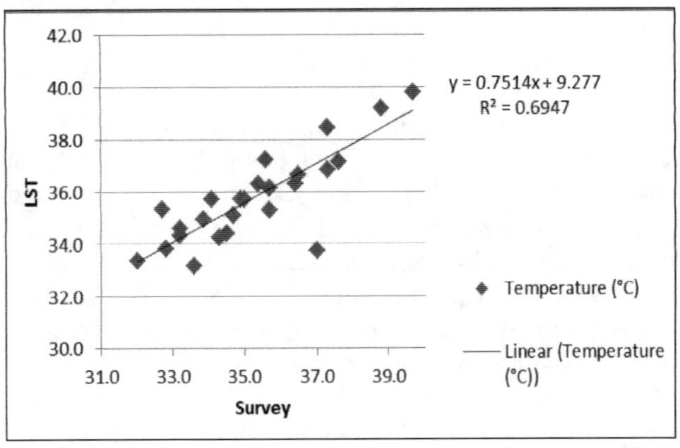

Figure 7. Linier regression of LST vers survey of field temperature

Figure 7, describes the relationship between temperature processing of LST and field temperature using infrared thermometer tool which regression equation is y = 0.7514x + 9.277. The positive sign on the equation states that there is a direct correlation between LST and the field temperature with the coefficient of determination of $R^2 = 0.6947$ or 69.47%.

4. Conclusion

Surface temperature in Surakarta City increases every year and the spread of high-temperature class is increasingly widespread in line with the extent of Built-Area distribution in every year. The phenomenon of the rise in temperature indicates a direct relationship with the number of Built-Up Areas, in accordance with previous studies by [2,15,16,18–20].. The environmental criterion due to temperature is directly proportional to the vegetation density.

The central area of Surakarta City is identified as the most critical environmental area based on the temperature and built the region. The development of the city towards the south can also be detected by increasing the distribution of the built area and the surface temperature.

The environmental critically index model with LST and Built-Up Index algorithms can be utilized and adopted in other urban. The results of the environmental index criterion assessment can be recommended as one of the complementary data in urban development and planning.

References

[1] Adhi I S 2016 Development of the city the city of Solo became a reference in urban development planning in Indonesia *Solopos.com* 2016 Available from: http://www.solopos.com/2016/05/25/pembangunan-kota-kota-solo-jadi-rujukanperencanaan-dan-rancangan-ruang-perkotaan-722909

[2] Senanayake I P, Welivitiya WDDP and Nadeeka PM 2013 Urban Climate Remote sensing based analysis of urban heat islands with vegetation cover in Colombo city Sri Lanka using Landsat-7 ETM + data *Urban Climate* (Elsevier B V*)* 2013 **5**:19–35 Available from: http://dx.doi.org/10.1016/j.uclim.2013.07.004

[3] Fazeli R, Ruth M and Davidsdottir B 2016 Urban Climate Temperature response functions for residential energy demand – A review of models Urban *(Elsevier BV)* 2016 **15**:45–59 Available from: http://dx.doi.org/10.1016/j.uclim.2016.01.001

[4] Li X and Norford L K 2016 Urban Climate Evaluation of cool roof and vegetations in mitigating urban heat island in a tropical city Singapore *Urban Climate* (Elsevier B V) 2016 **16**:59–74 Available from: http://dx.doi.org/10.1016/j.uclim.2015.12.002

[5] Isima K, Chan A, Aekbal S, Chel M, Ooi G and Yaasiin M 2016 Urban Climate Numerical study on the urbanisation of Putrajaya and its interaction with the local climate over a decade *Urban Climate* (Elsevier B V) 2016 **16**:1–24

[6] EPA 2008 Reducing Urban Heat Islands: *Compendium of Strategies Urban Heat Island Basics* 2008 **1–22**.

[7] Loveland T R and Irons J R 2016 Remote Sensing of Environment Landsat 8: The plans, the reality, and the legacy *Remote Sensing Environment* (Elsevier B V) 2016 **m185**:1–6 Available from: http://dx.doi.org/10.1016/j.rse.2016.07.033

[8] Roy D P, Wulder M A, Loveland T R, Woodcock C E, Allen R G and Anderson M C 2014 Remote Sensing of Environment Landsat-8: Science and product vision for terrestrial global change research *Remote Sensing Environment* (Elsevier B V) 2014 **145**:154–72 Available from: http://dx.doi.org/10.1016/j.rse.2014.02.001

[9] USGS 2015 *Landsat 8 (L8) data users handbook* Vol. 8 USGS 2015 1993-1993 p

[10] Li Y, Zhang H and Kainz W 2012 International Journal of Applied Earth Observation and Geoinformation Monitoring patterns of urban heat islands of the fast-growing Shanghai metropolis, China : Using time-series of Landsat TM / ETM + data *International Journal of Applied Earth Observation and Geoinformation* (Elsevier B V) 2012 **19:**127–38 http://dx.doi.org/10.1016/j.jag.2012.05.001

[11] Shahmohamadi P, Che-Ani A I, Abdullah N A G, Tahir M M, Maulud K N A and Mohd-Nor M F I 2010 The link between urbanization and climatic factors: A concept on formation of urban heat island *WSEAS Transactions On Environment And Development* 2010 **6** (11):754–68.

[12] Zha Y 2003 Use of normalized difference built-up index in automatically mapping urban areas from TM imagery *Internanional Journal Remote Sensing* 2003 **24**:583–94 Available from: https://is.muni.cz/el/1431/podzim2012/Z8114/um/35399132/35460312/ndbi.pdf

[13] Gonsamo A and Pellikka P 2012 The sensitivity based estimation of leaf area index from spectral vegetation indices *ISPRS Journal of Photogrammetry and Remote Sensing* 2012 **70** **(15)** :25

[14] Julien Y, Sobrino J A, Mattar C, Noz N and Oria G S 2011 Temporal analysis of normalized difference vegetation index (NDVI) and land surface temperature (LST) parameters to detect changes in the Iberian land cover between 1981 and 2001 *International Journal Remote Sensing* 2011 **32** (7):2057–68.

[15] Sharma R and Joshi P K 2016 Mapping Environmental Impacts of Rapid Urbanization in the National Capital Region of India using Remote Sensing Inputs *Urban Climate* 2016 **15**:70–82

[16] Lauwaet D, De Ridder K, Saeed S, Brisson E, Chatterjee F, Van Lipzig N P M, Maiheu B and Hooyberghs H 2016 Assessing the current and future urban heat island of Brussels *Urban Climate* 2016 **15** pp: 1-1:1–15.

[17] Comarazamy D E, González J E, Luvall J C, Rickman D L and Mulero P J A 2010 Land-atmospheric interaction study in the coastal tropical city of San Juan Puerto Rico *Earth Interaction* 2010 **14** (16)

[18] Sarrat C, Lemonsu A, Masson V and Guedalia D 2006 Impact of urban heat island on regional atmospheric pollution *Atmospheric Environment* 2006 **40** (10):1743–58

[19] Kikegawa Y, Genchi Y, Kondo H and Hanaki K 2006 Impacts of city-block-scale countermeasures against urban heat-island phenomena upon a building's energy-consumption for air-conditioning *Applied Energy* 2006 **83** (6):649–68

[20] Liu L and Zhang Y 2011 Urban heat island analysis using the landsat TM data and ASTER Data: A case study in Hong Kong *Remote Sensing* 2011 **3** (7):1535–52.

Acknowledgments

The authors are grateful to the support of the Laboratory of the Department of Geodesy Engineering - Faculty of Engineering, Diponegoro University, and USGS and NASA in Landsat data collection.

IOP Conf. Series: Earth and Environmental Science **165** (2018) 012012 doi:10.1088/1755-1315/165/1/012012

Application of Remote Sensing and GIS for Malaria Disease Susceptibility Area Mapping in Padang Cermin Sub-District, District of Pesawaran, Lampung Province

Adhina Azdah Hanifati[1], Anggi Permata[1], Dian Mustofa[1], Dita Eko Wulandari[1*], Intan Dewi Ratnasari[1], Nabila Anggita Ekafitri[1], Yusron Hidayat Ridho[1] and Prima Widayani[1]

[1]Cartography and Remote Sensing, Departement of Science Information Geography, Faculty of Geography, Gadjah Mada University, Yogyakarta 55281, Indonesia

dita.eko.w@mail.ugm.ac.id

Abstract. Malaria is epidemiologically a specific local infectious disease and can be studied spatially by taking into account the characteristics of malaria vector habitat. Padang Cermin Sub-District, District of Pesawaran, Lampung Province is one of the endemic areas of malaria which is geographically a coastal area dominated by water-flooded land use. That area supports the breeding and living place of *Anopheles sp* mosquitoes. The objectives of this study were mapping the malaria disease susceptibility area in Padang Cermin sub-district by using Landsat 8 imagery, identified the influence of physical environmental factors on the spread of malaria disease, and analyzed the spread of malaria disease based on environmental factors. The research method used scoring and overlay spatial analysis of spatial parameters supporting the breeding of malaria vector. Some parameters that can be extracted through Landsat 8 imagery for identification of malaria susceptibility, i.e. land use, soil texture, vegetation density with overall mapping accuracy of vegetation density, land use, and soil texture are 83.2%, 88.7%, and 83.7%, while other parameters resulted from spatial analysis of non-remote sensing data, i.e. temperature, rainfall, slope, altitude, and distance to river. These parameters are environmental factors that influence the spread of malaria disease. The results of the identification of malaria susceptibility showed that the areas with high levels of susceptibility are Sidodadi, Sukajaya Lempasing, Gebang, Padang Cermin, Hanau Berak, Way Urang, Tambangan, Hanura, Banjaran and Sanggi Village. This study showed that areas which are located on the coast are prone to malaria disease.

1. Introduction

Malaria is a disease caused by the Plasmodium parasite which is transmitted by the female *Anopheles sp.* mosquito [7]. Malaria is a life-threatening health problem that almost happened in every Indonesia's regions. Malaria disease problem is not only experienced by Indonesian, but this endemic disease can cause death and also threaten communities in various parts of the world, particularly in tropical and subtropical countries. According to World Health Organization data (2016), nearly half of the world's population is at risk of malaria. In 2015, there were roughly 212 million malaria cases and an estimated 429000 malaria deaths.

Lampung Province is one of the malaria-endemic regions in Indonesia. In 2015, there were nearly 26722 malaria cases and 2 sufferers died. The percentage of malaria incidence confirmed by Health Department of the Lampung Province during 2010 to 2015 tended to increase. Lampung Province had Annual Parasite Incidence (API) index or indicator to measure malaria incidence as much as 0.41; 0.49; and 0.51; per 1000 population during 2013 to 2015. Lampung Province overall was categorized as a low level of malaria endemicity because of API Lampung Province index was still below the national average rate in 2010, i.e 1,96. However, some regions in Lampung Province showed a high level of malaria endemicity, such as Padang Cermin and Punduh Pidada sub-district, District of Pesawaran.

Malaria is epidemiologically a specific local infectious disease and can be studied spatially by taking into account the characteristics of malaria vector habitat. Remote sensing and Geographic Information System (GIS) can be utilized in planning and mapping the spread of malaria disease quickly and efficiently. The mapping results can be used to cope with and reduce the spread of malaria disease. This study is also related to supporting Indonesia's Health Minister Program to eliminate malaria in the decentralization era. Global Malaria Programme (GMP) stated that malaria is a disease requires continuous observations, monitoring and evaluation, and appropriate formulation of policies and strategies.

Padang Cermin sub-district is one of the endemic areas of malaria which is geographically a coastal area dominated by water-flooded land use supports the breeding and living place of Anopheles sp. mosquito. Physical environmental factors, such as land use, vegetation density, soil texture temperature, rainfall, slope, altitude, and distance to the river may affect Anopheles sp. breeding. This study was aimed to map the malaria disease susceptibility area in Padang Cermin sub-district by using Landsat 8 imagery, identified the influence of physical environment factors on the spread of malaria disease, and analyzed the spread of malaria disease based on environmental factors.

2. Data and Methodology

2.1. Study Area and Data

This research was conducted in Padang Cermin Sub-District, District of Pesawaran, Lampung Province (Figure 1). This study used some secondary data are as follows: Indonesia Topographic Map scale 50,000; Landsat 8 OLI (June, 21st 2016); Geological Map of Tanjung Karang Sheet; boundary administration data of Pesawaran District; average daily temperature data (2014-2016) of Pesawaran District; average annual rainfall (2014-2016) data of Pesawaran District; and the amount cases of malaria disease data of Padang Cermin Sub-District 2016 from Health Department of the Lampung Province.

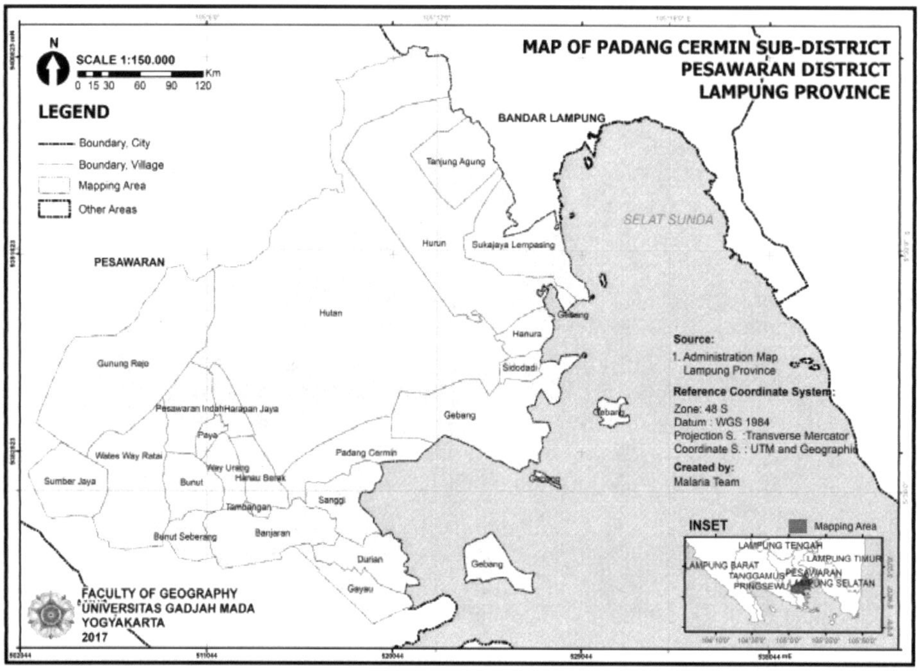

Figure 1. Research Location, Map of Padang Cermin Sub-District

2.2. Methodology

This study used quantitative spatial approach, and was divided into several stages (Figure 2), namely:

2.2.1. Data Pre-Processing

1) Radiometric Correction

Radiometric correction is performed to improve the visual quality of the image as well as correct the errors of pixel values that do not match the actual object reflectance. These errors can be caused by atmospheric disturbances during image recording or sensor error. The radiometric correction process in Landsat 8 OLI imagery can be performed according to the process issued by USGS (2013).

2) Geometric Correction

The process of image quality improvement in order to produce ready-to-use imagery for specific applications not only include radiometric correction but also need geometric correction. Geometric is related to the geographical position or distribution (geo-referenced data), either the position in the coordinate system or the information contained therein. Remote sensing imagery's geometric must have errors, such errors can arise because of the movement of satellites, earth rotation, the movement of sensor scanning mirror, and also earth curvature [4].

2.2.2. Data Processing

1) Vegetation Density Map

The vegetation density map was derived from the value of vegetation density pixel using Normalized Difference Vegetation Index (NDVI) transformation which is utilized near-infrared and red wavelengths on Landsat 8 OLI. The results of the index value were classed into three classes, i.e low, moderate, and high then further field checking was performed on each class (Table 1). The use of vegetation density in this research was related to the best breeding and living place of Anopheles sp. mosquito. The more dense vegetation will increase the convenience of a mosquito to grow and breed, whereas the low of vegetation density will reduce the convenience of a mosquito to grow, it is because the direct contact habitat with sunlight can reduce the humidity of mosquito's habitat. Classification and the score of vegetation density classes are as follow:

Table 1. Classification and the Score of Vegetation Density Classes

Vegetation Density	Score
Low (<30%)	1
High (>60%)	2
Average (30-60%)	3

Source : Eiten (1968) in Ridwan, 2012

2) Land Use Map
Land use map was created from Landsat 8 OLI by visual interpretation using ArcGIS software to limit a homogeneous object that distinguishes with other appearance. Land use classification used I Made Sandy (1977) land use classification scheme on 1 : 100.000 scale with some modifications (Table 2). Land use related to the main habitat of the community as well as the habitat for breeding and survival of the mosquito *Anopheles sp.* Classification and the score of land use classes are as follow:

Table 2. Classification and the Score of Land Use Classes

Land Use	Score
River, mangrove, paddy fields, shrimp ponds	3
Bushes, crops, vacant land, forest	2
Settlement, garden	1

Source: Dulbahri (1992) with modifications in Sulistiarini (2011)

3) Soil Texture Map
Soil texture information was derived from the integration of geology map with landform map. The soil-formation is assumed from parent rock underneath or not the result of transformation nor translocation. The making of soil texture tentative map was performed by visual interpretation of rock types. Soil texture tentative map is used as the basis for sampling field to measure the accuracy of soil texture. The soil has physical properties in terms of the size of the soil particles. The more smooth soil texture causes the difficulty of soil to be passed by the water so the land will easily flood. Classification and the score of soil texture classes are as follow:

Table 3. Classification and the Score of Soil Texture Classes

Criteria	Soil Texture	Score
Fine – fairly smooth	Clay, dusty clay, sandy loam, loam, loamy clay, dusty clay loam	3
Average	Dusty, dusty loam, loam	2
A bit rough - rough	Sandy loam, loamy sand, sand	1

Source: Achmad, 2003 with modifications

4) Rainfall Map
Rainfall map was created from secondary data processing of annual mean rainfall data (mm/year) of rainfall stations around the research area. The annual mean rainfall data was processed using interpolation Inverse Distance Weighting (IDW) to produce an isohyet map, i.e. the lines connect the points that have the same rainfall intensity. The higher intensity of rainfall will increase the relative humidity that supports the survival of *Anopheles sp.* mosquito, which resulted in an increase the mosquitoes density. Classification and the score of rainfall classes are as follow:

Table 4. Classification and the Score of Rainfall Classes

Criteria	Rainfall (mm/year)	Score
Low	<1000	1
Average	1000-2000	2
High	>2000	3

Source: Achmad, 2003 with modifications

5) Altitude Map
Altitude is related to the air temperature of an area that affects the breeding of malaria vector. The higher altitude of an area then the air temperature will tend to be lower. It will cause malaria vector inactive to bite and effect to the lower transmission level of malaria disease. Malaria disease vector lives in an ideal area that is not too high, i.e. between 100-500 above the sea level. The altitude map was created by interpolation contour data using topo to raster method. Classification and the score of altitude classes are as follow:

Table 5. Classification and the Score of Altitude Classes

Altitude (m)	Score
< 150	3
2.1. – 700	2
>700	1

Source: Idung, R. Fiolenta, M. et al. (2006) with modifications

6) Distance to River Map
A river is one of the ideal locations for malaria vector breeding place. The making of distance to river map is based on multiple buffer spatial analysis. The closest area to the river causes the risk of malaria disease will increase. Especially, if the river has a steady flow or tends to stagnate, it will make the risk of malaria disease higher because the larvae of Anopheles sp. prefers a steady flow to breathe through the spiracles. Classification and score of distance to river classes are as follow:

Table 6. Classification and the Score of Distance to River Classes

Distance (meter from river)	Score
<50	3
50- 300	2
>300	1

Source: Sunaryo, 2009 in Ridwan, 2012

7) Slope Map
The slope map is derived from Digital Elevation Model (DEM) which is produced from contour interpolation to slope information on percent units. Slope factor is related to the availability of puddles as the breeding place of malaria vector. On steep slopes, water tends to flow as surface runoff gravitationally to the lower area, whereas on the gentle slope has a high potential for the formation of puddles and may increase the breeding of malaria vector. Classification and score of slope classes are as follow:

Table 7. Classification and Score of Slope Classes

Slope	Score
0-8%	3
8-15%	2
15-25%, 25-45 %	1

Source: Bruce-Chwatt (1985) with modifications

8) Temperature map
The temperature map is created by converting the elevation data (DEM) into the value of temperature using Braak law. It used an assumption that temperature will reduce 0,61°C every 1000 meter increase on the areas with elevation under 2000 meter above sea level. Then, the temperature also reduces 0,52°C every 100 meter increase in the area with elevation above 2000 meter [9]. Classification and score of temperature classes are as follow:

Table 8. Classification and score air temperature factor

Temperature (°C/Year)	Score
<25	1
>27	2
25-27	3

Source: Bruce-Chwatt (1985) with modifications

2.2.3. Field Checking. Field checking was conducted to match the result of parameters interpretation and the extent of malaria susceptibility, as well as records of changes in the field based on samples that have been established in the previous stage. The checked field parameters were vegetation density, land use, and soil texture, while other parameters were extracted from secondary data so field checking was not needed. The sampling technique was used purposive random sampling with the total samples to be field tested were at least 30 samples. Purposive random sampling is a sampling of random population members by considering land units, reliefs, and accessibility levels. Each sample represents a specific land unit, consist of land use, vegetation density and soil texture parameter. This sampling method is based on the presence of unequal (heterogeneous) population members.

2.2.4. Interpretation Accuracy Measurement. The accuracy measurement was performed using a confusion matrix that matches the interpreted pixels by the pixel test sample, in which one test sample can consist of many pixels. Accuracy measurement is expressed in the form of producer's accuracy, user's accuracy, and overall accuracy. Accuracy measurement can be performed by multivariate or Kappa index [3] using formula as follows:

$$K = \frac{N \sum_{i=1}^{r} x_{ii} - \sum_{i=1}^{r} (x_{i+} \quad X \quad x_{+i})}{N^2 \; - \; \sum_{i=1}^{r} (x_{i+} \quad X \quad x_{+i})} \qquad (1)$$

2.2.5. Reinterpretation. Reinterpretation was performed after the field checking and accuracy measurement of the tentative map. Reinterpretation is intended to justify misinterpretation based on field checking results in order to analyze using data that has been in accordance with actual conditions in the field. The reinterpreted map in this research is field checked parameter map, i.e vegetation density, land use, and soil texture map. The results of reinterpretation parameters map then can be used to perform further analysis of malaria susceptibility.

2.2.6. Spatial Analysis of The Influence of Each Parameter on Malaria Susceptibility. The influence of physical environment factors to the spread of malaria disease is analyzed by using quantitative analysis method, that is correlation analysis to know the relationship between two variables, the direction of variable and variable relation. Based on the use of physical environmental parameters that have been performed by field checking and reinterpretation, then malaria-prone areas map was generated by using overlays method. The class of malaria susceptibility is divided into three classes, i.e high, moderate and low susceptibility. The malaria susceptibility map is then validated to malaria incidence data and also conducted a descriptive qualitative analysis of the spread of malaria susceptibility.

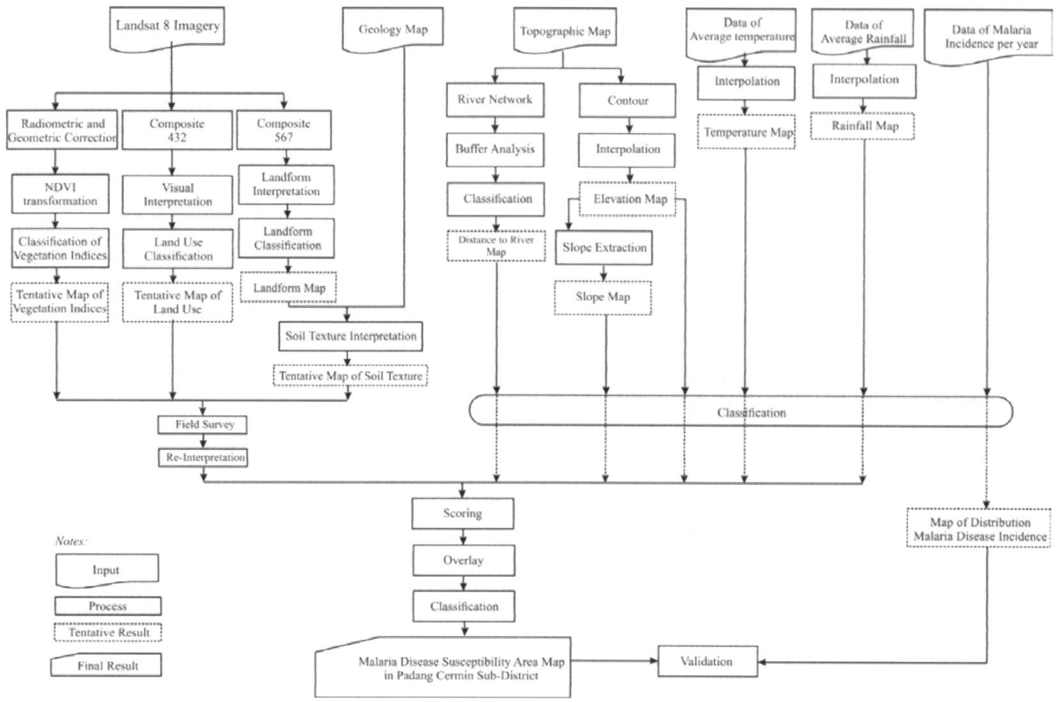

Figure 2. Malaria Disease Susceptibility Area Mapping Flowchart

3. Result

3.1. The Linkages between Anopheles Sp. and Field Conditions

The presence of mosquito species is influenced by environmental conditions, there are physical, chemical, and biological environments. Research shows Anopheles sundaicus sp. is the dominant Anopheles species in Pesawaran District. In addition, there are also other species influenced by diverse ecological circumstances. An. Sundaicus primarily liked the place of breeding with the brackish water such as lagoons, swamps, ponds, and embankment. Embankment land use is commonly found in Pesawaran District, especially in Padang Cermin Sub-District. Some of the existing embankment in the area are still active, but in some places, abandoned embankment are found no longer maintained. The interaction between soil on the abandoned embankment has increased the potential of Anopheles sundaicus mosquito breeding in the area. In addition, another inundated land use became one of the breeding places of Anopheles mosquito is mangrove.

Padang Cermin sub-district is located along the coast. Coastal areas have relatively higher humidity caused by large water evaporation. Mosquitoes generally liked the area with high humidity that has the optimal conditions for mosquito breeding. The faster mosquito breeding will increase the possibility to contact with humans, so the risk of transmission of malaria disease will be greater. Padang Cermin sub-district is also a region with high level of vegetation existence. The existence of vegetation can affect the life of mosquitoes, such as a breeding, shelter and foraging place, as well as a resting place for adult mosquito. In optimal conditions areas for mosquito breeding, the higher vegetation density level will increase the convenience level of mosquito growth. This condition is found in high-density mangrove vegetation.

3.2 Analysis of the Distribution of Malaria Hazard in Padang Cermin sub-district

Based on the ecological unit malaria disease susceptibility area map in Padang Cermin sub-district (Figure 3) shows most of the region is malaria-prone areas, especially in areas close to the coast. The areas close the coast are dominated by a high level of susceptibility due to the environmental factors of land use, namely, embankment is the preferred land use by Anopheles sp. mosquito. Not all parts of coastal areas have the high level of susceptibility, there are also some coastal areas that do not have the high level of susceptibility. This is due to the influence of other ecological factors that actually reduce the probability of malaria susceptibility.

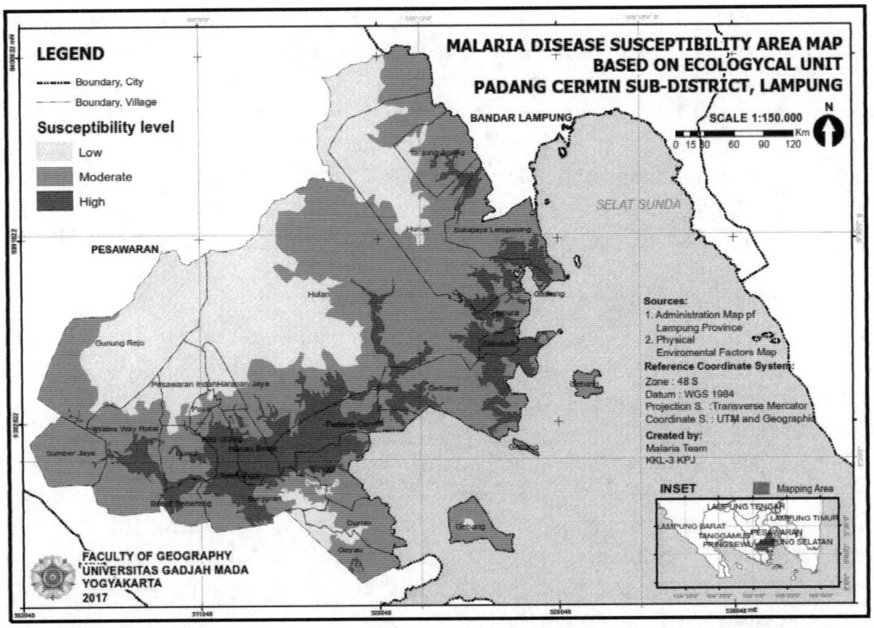

Figure 3. Map of the distribution of malaria vulnerability based on ecological unit

The high level of malaria susceptibility in coastal areas shows a link between malaria susceptibility and the spread of Anopheles sp. mosquito habitat, especially An. sundaicus which is the most dominant mosquito species in Padang Cermin sub-district. The areas in Northwest of Padang Cermin sub-district have the low level of susceptibility due to malaria susceptibility parameters in the region less support the existence of Anopheles sp. mosquito breeding habitat. The low level of susceptibility areas is found around the Forest area, Ratai mount, Pesawaran mount, and Betung mount. It caused by the ecological conditions do not suit with the characteristic of Anopheles sp. mosquito habitat: have a fairly steep slope and the altitude more than 700 meters. These characteristics may inhibit the development of Anopheles sp. Steep slopes can cause the high level of land erosion, so water tends to flow as surface runoff gravitationally to the lower area and puddles as Anopheles sp. breeding place are not formed. Meanwhile, the altitude more than 700 m closely related to the temperature that can affect the biological condition of Anopheles sp. Malaria susceptibility mapping result shows that areas close to coastal areas have higher levels of susceptibility, but these results should be compared with malaria incidence data to be more valid and appropriate.

3.3 Interview Results and their Relation to Malaria Incidence and Occurrence

Based on interviews with 10 scattered randomly interviewees in all samples of Padang Cermin sub-district, showed that the level of effectiveness of counseling or the introduction of a danger of malaria disease is 70%. The government has not been doing maximal training or counseling about malaria disease. This was marked by 30% of people said that they have never followed malaria education and lack of fogging routine. The government also has not provided the maximum malaria vaccination yet. Some interviewees said, there were a net allotment and medical assistance to deal with malaria but the aid has not evenly distributed yet and there was no clear information.

The dominant livelihoods of the Padang Cermin community are farmers and traders. The dominant community is domiciled on the edge of the rice field, has a cattle pen near to the house and does not have mosquito larva-eating fish. It may potentially increase the spread of malaria vector. In terms of daily habits, the community has been aware to use mosquito nets and anti-mosquito as preventive efforts of malaria disease. The most prevalent malaria incidents are found in Sukajaya Lempasing Village. There are several villages prone to malaria disease but has no occurrence of malaria disease. It can be influenced by mosquitoes fly patterns or non-sedentary human activity. Way Urang, Hanau Berak, Bunut Seberang and Tambangan are villages that do not have incident data.

Based on interviews results in Hanura Village, there are some areas that have abandoned embankment. The embankment and slum settlements are one of the land use that triggers the breeding of Anopheles sp. mosquito, causes many malaria cases occur. Rice cropping system that does not alternate with Palawija provide opportunities to Anopheles sp. mosquito to breed around the rice field.

3.4 Malaria Susceptibility and Incidence Map

In this research, malaria susceptibility map was conducted validation with malaria incidence map by the year 2016 (Figure 4). Malaria incidence map in 2016 was selected by adjusting the remote sensing imagery time of acquisition to know the relationship between physical conditions in the field with malaria incidence level. Based on the data, it was known that the malaria incidence in a period of 1 (one) year (Annual Parasite Incidence) in 2016 was the highest in Sukajaya Lempasing village with the number of occurrences as many as 913 cases. The high amount of malaria incidence most likely occurs due to physical environmental factors, and an individual factor of sufferers. Physical environmental factors were examined, namely, land use, soil texture, and vegetation density. The individual factors influence malaria incidence, namely gender, ages, type of work, and community behavior to consciously against malaria disease.

Based on malaria susceptibility map, administrationally can be aware that areas have the high level of susceptibility namely, Sidodadi, Sukajaya Lempasing, Gebang, Padang Cermin, Hanau Berak, Way Urang, Tambangan, Hanura, Banjaran, and Sanggi Village (Figure 5). The spread of malaria susceptibility on administrative units based on the extent dominant susceptibility area percentage. Areas with the high level of susceptibility do not always have a high incidence of malaria, as well as in areas with low to moderate level of susceptibility. It was caused by the individual and environmental factors. The existence of the settlements and the activities of the community in the area with high level of susceptibility will greatly affect the presence or absence of malaria incidence. While other factors that affect malaria incidence is the type of work, community behavior to consciously against malaria, population mobility, and the Anopheles sp. mosquito transmission.

Based on malaria incidence data, malaria sufferers are dominated by men aged over 17 years, because those age and gender group have a high mobility, especially related to the demands of the type of work. The dominant type of work that exists is embankment farmers who demand to work in embankment area as Anopheles sp. breeding place.

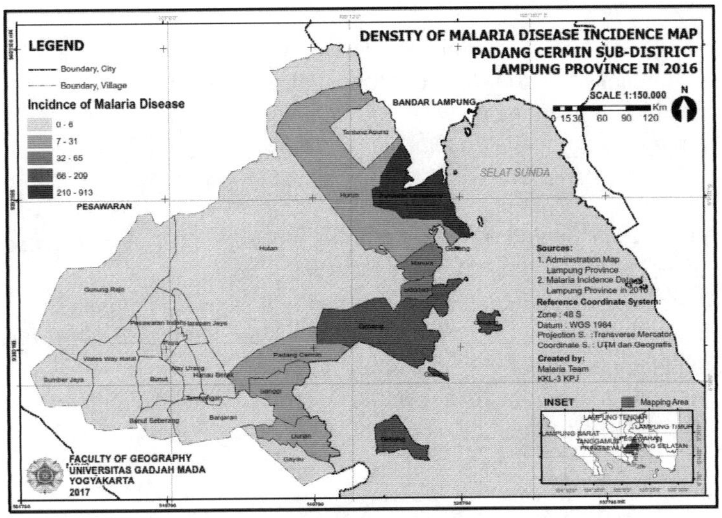

Figure 4. Map of Malaria Disease Incidence

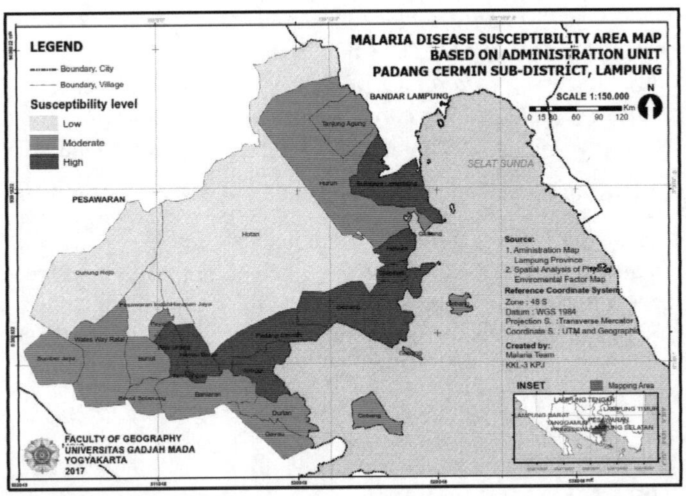

Figure 5. Map of the distribution of malaria vulnerability based on administrative unit

4. Conclusions

1) Physical environmental factors that affect the high level of malaria susceptibility in Padang Cermin sub-district are embankment and mangrove land use, moderate to high level of vegetation density, low altitude with the optimum temperature between 25-27°C, fine soil type, and the gentle slope that affect the presence of puddles as the malaria vector habitat.
2) The malaria susceptibility mapping result in Padang Cermin sub-district shows high-level susceptibility areas are found in Hanau Berak, Way Urang, Bunut Sebarang, Sanggih, Tambangan, Padang Cermin, Hanura, Sidodadi, Gebang and Sukajaya Lempasing village, while low-level susceptibility areas are found in Pesawaran Indah, Harapan Jaya, Hutan, and Gunung Rejo village.
3) The area of malaria susceptibility distribution in Padang Cermin sub-district belongs to high categorized spatially distributed in the coastal area with the influence of mangrove and embankment land use, low altitude and optimal temperature, while areas with the low level of malaria susceptibility are influenced by high altitude, low temperature, and steep slope.

References

[1] Achmad H, Mardihusodo S J, Sutanto, Hartono and Kusnanto H 2003 Estimasi Tingkat Intensitas Penularan Malaria dengan Dukungan Penginderaan Jauh (Studi Kasus di Daerah Endemis Malaria Pegunungan Menoreh Wilayah Perbatasan Provinsi Jawa Tengah dan Daerah Istimewa Yogyakarta) *Jurnal Ekologi Kesehatan* **2** 157-164

[2] Bruce Chwatt L J *1985 Essential Malariology: 2nd Ed*ition (New York: John Wiley and Sons) p 52

[3] Congalton R G and K Green 2009 *Assessing the Accuracy of Remotely Sensed Data: Principles and Practices: Second Edition* (USA: CRC Press, Inc.)

[4] Danoedoro Projo 2012 *Pengantar Penginderaan Jauh Digital* (Yogyakarta: Penerbit Andi)

[5] Dinas Kesehatan Provinsi Lampung 2015 *Profil Kesehatan Provinsi Lampung* (Lampung: Dinas Kesehatan Lampung)

[6] Idung R Fiolenta 2006 *Penyusunan Model Spasial Prediksi Lingkungan Sebaran Malaria (Anopheles sp)* (Bogor: Institut Pertanian Bogor)

[7] Levis Joel T and Gus M Garmel 2009 *Clinical Emergency Medicine Casebook* (United Kingdom: Cambridge University Press)

[8] Ridwan Muhammad 2012 *Kajian Aplikasi Penginderaan Jauh dan Sistem Informasi Geografis untuk Estimasi Wilayah Rawan Penyakit Malaria di Kabupaten Purworejo, Provinsi Jawa Tengah* (Postgraduate Program: Universitas Gadjah Mada)

[9] Rokhis MK 1998 *Pewilayahan Tanaman Mangga dan Jambu Mente di Sulawesi Tenggara* (Undergraduate Program: Institut Pertanian Bogor)

[10] Sulistiarini Yani 2011 *Pemanfaatan Citra ALOS AVNIR 2 dan Sistem Informasi Geografis untuk Menentukan Tingkat Kerawanan Wilayah Terhadap Malaria secara Spasio-temporal di Kabupaten Purworejo* (Undergraduate Program: Universitas Gadjah Mada)

Acknowledgments

This research is a part of the Cartography and Remote Sensing 3rd fieldwork. The authors deeply thank the Health Department of the Lampung Province, Lampung Province Government, all the Departement of Science Information Geography lecturers for supporting our research. Also, sincere gratitude is extended to the anonymous reviewers for corrections and insightful comments.

IOP Conf. Series: Earth and Environmental Science **165** (2018) 012013 doi:10.1088/1755-1315/165/1/012013

Peatland Non Use Value Survey In Siak Regency

R Wisnu Ali Martono[1*], Meuthia Djoharin Puspo[1] and Hari Prayogi[1]

[1]Pustekwil BPPT, Gedung Geostech, Pupiptek, Serpong, TangerangSelatan, Banten

wisnu.ali.martono@gmail.com

Abstract. Complementing the peatland depth survey using Ground Penetrating Radar di Siak Regency conducted by Pustekwil (Pusat Teknologi Sumberdaya Wilayah) BPPT, a Non-use Value survey was conducted in May 2017. Peatland is a natural resource whose value has long been underestimated. In order to correctly assign for peatlands' value, a different approach, namely the Total Economic Value, must be used. The Total Economic Value (TEV) method comprises of Use Value and Non-Use Value. Whereas the Use Value component may use secondary data (and market price data), the Non-Use Value data must be gathered directly from the respondents using Contingent Valuation Method. Data gathered during a recent survey in Siak Regency, using stratified random sampling method, shows a Bequest Value of IDR/ha/year 1,413,647.95 (WTP based) dan IDR/ha/year 29,025,305.57 (WTA based).The same survey shows Existence Value of IDR/ha/year IDR 82,725 (WTP based) and of IDR/ha/year 9,472,281.78 (WTA based). While the gathered values seem small compared to Direct Use Value (of converting peatland for palm oil plantation), they need to be collected as part of the peatland's Total Economic Value calculation.

Keywords: Siak peatland, WTP, WTA, total economic value, non-use value, existence value, bequest value

1. Introduction

Peat is a heterogeneous mixture of more or less decomposed plant (humus) material that has accumulated in a water-saturated environment and in the absence of oxygen. Its structure ranges from more or less decomposed plant remains to a fine amorphic, colloidal mass. The warmer the climate, the quicker the plant material will decompose. The rate of accumulating plant material is greatest in areas where the temperature is high enough for plant growth but too low for the vigorous microbial activity that breaks down the plant material. Such conditions are found more frequently in the northern hemisphere.

Indonesia has approximately 15 to 20 million hectares of tropical peatlands: the fourth-largest peatland area in the world (BRG, 2016). According to [1] Indonesia's has 20,6 million ha peatlands area (10.8% of the country's area), of which 7,2 million hectares (35%) is in Sumatera. The Indonesian peatlands represent 80% of SE Asian tropical peatlands and 50% of the world's tropical peatlands [2]. Economically, an undisturbed peatland often considered as unproductive natural resources. Therefore, there have been pressures to use peatlands for commercial plantations purposes.

In Indonesia, the peat swamp forests vanishing at a critical rate (2.6%/year in Sumatra and Kalimantan) between 2007 and 2015 [3]. Growing demand for arable land especially for oil palm, attractiveness in relation to their flat topography and non-active agricultural use have all led to intense conversion over past decades.In response to global market, oil palm has become one of the most economically attractive crops to cultivate in tropical regions. Indonesia is the leading global producer of crude palm oil with a production rate growing exponentially over time (Murdiyarso et al. 2010).The contribution of oil palm expansion to peatland deforestation also tends to follow an exponential pattern [3], even though oil palm development is not the sole driver of peat swamp forest disappearance.

Notwithstanding with first general impression that peatland is an unproductive natural resource, Wetlands/peatlands is a natural resource with multiple functions. They offer many services, namely in water resources protection, flood control, seawater intrusion defense, supporting biodiversity, climate control (through its ability to act as the carbon sink), and many more. Owing to this multi products nature of peatlands, peatlands' value must not be judged in the conventional way, i.e. the products directly produced and sold at the market. A Total Economic Value approach must be used.

According to Plottu et al (2007), Total economic value (TEV) is a concept in cost-benefit analysis that refers to the value derived by people from a natural resource, a man-made heritage resource or an infrastructure system, compared to not having it. It appears in environmental economics as an aggregation of the (main function based) values provided by a given ecosystem.
The value of an ecosystem can be distinguished as:
Use Value – Can be split into Direct and Indirect use values:
Direct use value: Obtained through a removable product in nature (i.e. timber, fish, water).
Indirect use value: Obtained through a non-removable product in nature (i.e. sunset, waterfall).
Non-use value – Values for existence of the natural resource. For example, knowing that tigers are in the wild, even though you may never see them.

Option value: Placed on the potential future ability to use a resource even though it is not currently used and the likelihood of future use is very low. This reflects the willingness to preserve an option for potential future use.Bequest value or existence value: Placed on a resource that will never be used by current individuals, derived from the value of satisfaction from preserving a natural environment or a historic environment (i.e., natural heritage or cultural heritage) for future generations. Both Existence and Bequest Values are concerned with abstract benefits, which there is no actual market. Therefore, market prices cannot be used to measure the values respondents put on peatland. To alleviate this valuation problem, a contingent valuation method (CVM) is used for the survey.

The contingent valuation method involves directly asking people, in a survey, how much they would be willing to pay for specific environmental services. In some cases, people are asked for the amount of compensation they would be willing to accept to give up specific environmental services. It is called "contingent" valuation, because people are asked to state their willingness to pay (WTP), contingent on a specific hypothetical scenario and description of the environmental service.

Other than WTP, in a CVM survey respondents are often asked their willingness to accept a (hypothetical) financial compensation, in order to forgo for specific environmental services. Theoretically, both values (WTP and WTA) should not differ greatly. However, when asked to state their WTP and later their WTA for a specific hypothetical case, respondents often give very much different values. The WTA is almost always higher than WTP.

2. Objective

The survey's objective is to evaluate the peat non-use values perceived by the Siak regency's inhabitants. The non-use values are very important parameters to correctly evaluate natural resources value, together with the 'traditional' use value. Owing to the necessity to ask many respondents to collect this perceived values, non-use value survey is rarely conducted. This survey was part of combined studies in Siak regency's peatland (including the peatland's depth study using Ground Penetrating Radar and water level study using drone). Collecting the peatland's non-use value, combined with previously collected use value of Siak Regency peatland, will provide a better calculation of the peatland's total economic value. Other than that, as non-use values survey requires intensive human resources, it is seldom conducted. Therefore the resulted non-use values may be used in another peatland total economic value study.

3. Methodology

The method used for this survey is Stratified proportional random sampling. A Stratified random sampling is a method of sampling that involves the division of a population into smaller groups known as strata. In stratified random sampling, the strata are formed based on members' shared attributes or characteristics. A random sample from each stratum is taken in a number proportional to the stratum's size when compared to the population. These subsets of the strata are then pooled to form a random sampling. The strata in this survey is the chosen districts. Out of 14 districts in Siak regency, seven districts have been chosen as the target area. Firstly, the districts are ranked according to their peatlands areas. Since this is a Non-Use value survey, the lesser the peatland area in a district, the higher the rank. Secondly, the districts are re-ranked according to their distance and accessibility from Siak city. Thirdly, and lastly, the districts are chosen until the required respondent numbers are met. To calculate the respondent requirement, a formula is used, namely:

$$n = N/(1 + Ne2) \quad (1)$$

in which:
n = the required respondents, representing a household
N = populations (households)
e = margin of error (%)

Having chosen the districts to be surveyed and the required respondents in those districts, the conducted survey follow the following algorithm. While the main survey objective is to gather Non-Use Values of peatland (namely the respondents' WTP and WTA values concerning with peatland preservation), when the engaged respondent(s) is later known cultivating the peatland, Use Value data will be gathered from them, instead. This algorithm is used in order to quickly switch the survey questions, from the Non-use Value questionnaire to Use-value questionnaire. See attached survey questionnaires.

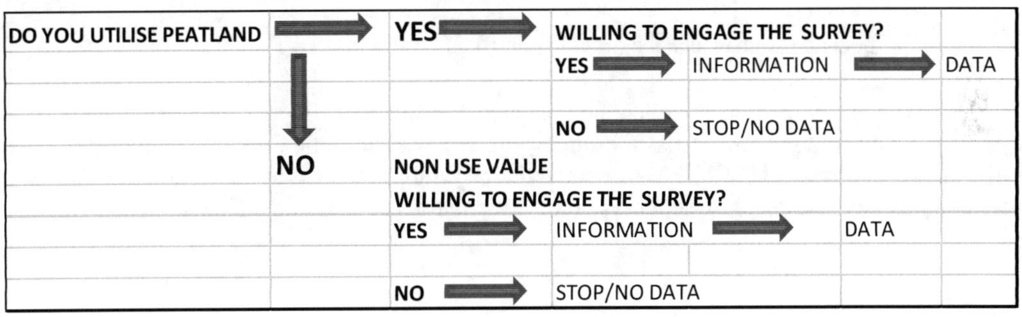

Figure 1. Survey Algorithm

4. Result

4.1 Peatlands In Siak Regency

Geographically, the Siak regency lies between 10 16' 30" — 00 20' 49" North Latitude and 100 54' 21" 102° 10' 59" East Longitude. BujurTimur. Physically, the regency has a coastal area which shares borders with other ASEAN countries (Malaysia and Singapore). The Siak regency's physical area mostly consists of lowland (in eastern region) and highland on the western side of the regency. The Siak regency's soil structure generally consists of the yellow-red podsolic in rock and alluvial form, as well as organosol soil and gley humus in wetland and wet soil s. This area has a tropical climate with 25° -- 32° Celsius air temperature, and high humidity and rainfall.

Out of 869.72 hectares Siak regency area, 53,22percents (459.193 hectares) is peatland, which is spread over 12 districts. Of this peatland area, 34,76 percent is wetland (302.291 Ha, with deep peat deposit), while the rest (18,47 percent or 160.615 Ha) is shallow peat wetland.

Of those peat wetlands, mostly located on forest area, forest concession and plantation companies. Outside of those areas, the deep peat wetland lies in Koto Gasip, Dayun, Mempura and SungaiApit districts. There are three small peat domes and four big peat domes. The small peat domes are Sungai Mandau, Buatan and Marempan domes.

The big peat domes are the Siak Kecil dome (which lies in Siak and in Bunga Raya districts), the Zamrud dome (lies in Siak and Pelelawan regencies), the Bukit Batu dome (lies in Siak and Bengkalis regencies) and the Kandisdome (lies is Siak, Rokan Hulu and Bengkalis regencies). Of the utilized peatlands, the largest is used for industrial woods plantation (HTI, HutanTanamanIndustri) in Zamrud dome (46.920 Ha) and in Sungai Mandau dome (29.487 Ha), planted with acacia for pulp industry. The deep dome utilization has resulted in peatland subsidence, decreasing peatland biodiversity, as well as introducing new pests for the flora. The following table show peatland distribution among the Siak Regency's districts.

Table 1 Peatland Areas in 7 Districts (in Siak Regency)

Peatland Depth (metres)	**1-2**	**2-4**	**>4**
Siak District	44.76		221.65
Mempura Dist	134.13	82.52	135.33
Dayun Dist	30.95	59.27	911.45
Kerinci Kanan Dist		1.03	16.65
Bunga Raya Dist	43.84		58.57
Lubuk Dalam Dist			7.04
Koto Gasib Dist	48.78	26.25	99.01
Total 7 districts (ha)	302.45	169.07	1,449.70
Total			192,123.00

Source: Peta LahanGambut Indonesia, BBSDLP, 2014

From the original 14 districts Siak peatland map (from which the above table is calculated), it shows that while the Sungai Apit district has the largest area of peatland, the Siak and Bunga Raya districts have the big peat domes. This 192,123 ha peatland area figure will be used throughout the analysis. The following map shows peatland distribution in the Siak districts.

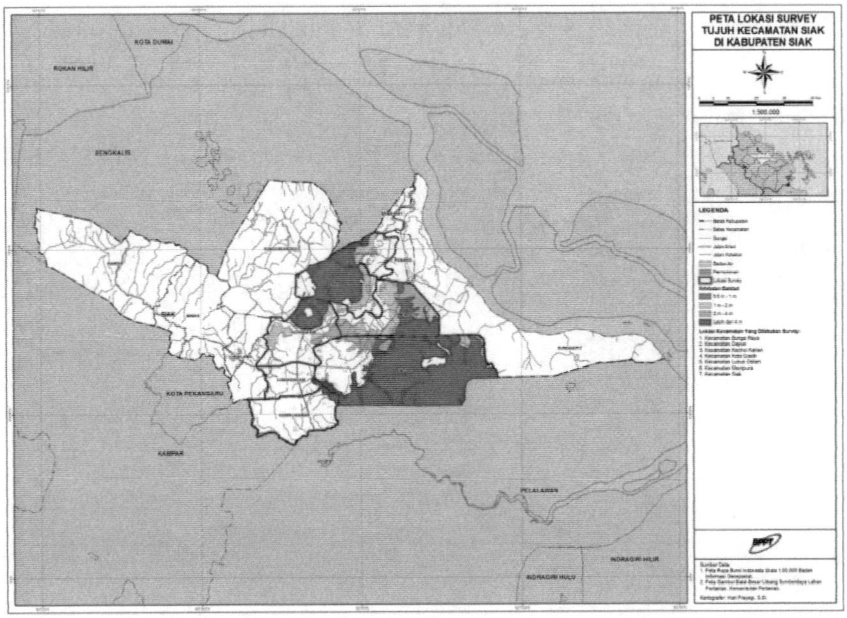

Figure 2. Peatland Thickness of Siak District

From the above map can be noted that soil or peat located at Siak has two Subclasses of the thickness of the peat moss. Peat with a thickness of 1 meter to 2 meters and peat with a thickness of more than 4 meters. For the distribution of peatland in peat thickness with Siak over 4 meters is located in the northern part, central part, and the southern part. For peat with a thickness of 1 meter to 2 meters is located in the central part with a little part towards the East area. The extensive peat with a thickness of 1 meter to 2 meters in sub-district of Siak around 44,762 km^2. while the area of peat with a thickness of more than 4 meters in sub-district of Siak covering an area of 221,647 km^2. The red circle appears on the image map above is showing the area of the site survey team in finding the respondents. Other districts' peatland map is attached at the end of this paper.

4.2 Number of Respondents Requirement

For the Non-use Value Survey purpose, out of 14 Siak regency's districts, 7 districts have been chosen as the survey areas. The selection is based on the methodology used (see Ch III). The following table summarises the required data and the resulting sample requirement for each selected district.

There area 116,950 households in the Siak Regency. The chosen margin of error is 2,5%. Entering the data into the formula (see Methodology in Ch III) resulting in 1,578 households respondents requirement. As result of limitation (budget, time, manpower), it was decided to take only 25% of the required samples, namely 400 respondents. Therefore, each respondent representing 265 populations.

Table 2. Number of Respondents In Siak Regency's Selected Districts

Districts	Area (km^2)	Population	Pop Density	Households	Sample
8. TUALANG	343.60	126,442	367.99	30,548.00	115
2. KANDIS	1493.65	82,349	55.13	19,711.00	74
10. DAYUN	232.24	32,591	140.33	8,371.00	32
1. MINAS	346.35	31,939	92.22	7,647.00	29
4. SUNGAI APIT	1346.33	30,929	22.97	7,808.00	29

Districts	Area (km^2)	Population	Pop Density	Households	Sample
3. SIAK	894.17	27,548	30.81	6,940.00	26
11. BUNGA RAYA	151.00	26,192	173.46	6,979.00	26
6. KERINCI KANAN	128.66	25,827	200.74	6,737.00	25
9. KOTO GASIB	704.70	23,488	33.33	5,963.00	23
7. LUBUK DALAM	155.09	20,800	134.12	5,206.00	20
12. MEMPURA	437.45	16,323	37.31	4,187.00	400
13. SABAK AUH	73.38	12,789	174.28	3,231.00	
5. SUNGAI MANDAU	1705.00	7,445	4.37	1,885.00	
14. PUSAKO	544.47	6,668	12.25	1,737.00	
Total	8556.09	471,330		116,950.00	

4.3 Survey Results

The survey conducted in 8 days and has met the required 400 respondents. Not all the gathered data are good for analysis purposes. There are incomplete data, too low or too high WTP and/or WTA values. For data analysis purpose, the complete data must also be cleaned from outliers. Outliers are data values which are too low or too high compared to data's mean.

Having been cleaned and smoothed (outliers deleted), the following steps are taken to get the Non-Use Values result.

- Step 1, calculate the average Existence/Bequest WTP/WTA values from original data
- Step 2, multiply with 7 districts' population (172,769.00) to get 7 districts' Existence/Bequest WTP/WTA values
- Step 3, multiply with 12 to get total annual figures
- Step 4, divide with 7 districts peatland area to get IDR/ha/year value. This step is required as the Use Value components (of TEV, Total Economic Value) also using IDR/ha/year unit.

The resulting Step 1 – Step 4 is shown in the following tables.

Table 3. Existence Value Results (WTP)

Total WTP as surveyed (IDR/month)	8,186,000
Average WTP as Surveyed (Rp/mo/person)	91,977.53
Rata-rata WTP Survei (IDR/thn/orang)	98,232,004
Total WTP in 7 Districts (mi IDR/annum)	15,893
WTP in 7 Kec (IDR/ha/annum)	82,725

Table 4. Existence Value Results (WTA)

Total WTA as surveyed (IDR/month)	78,987,667.00
Average WTA as Surveyed (Rp/mo/person)	877,640.74
Rata-rata WTA Survei (IDR/thn/orang)	10,531,688.93
Total WTA in 7 Districts (mi IDR/annum)	1,819,833.72
WTA in 7 Kec (IDR/ha/annum)	9,472,281.78

The above tables show that when asked to state how much the respondent will ask for a compensation, in return of his/her willingness to accept the peatland disappearance, the answer is almost 10 times (877,640/91,977) as much as their willingness to pay to prevent the peatland disappearance.

Table 5. Bequest Value Results (WTP)

Total WTP as surveyed (IDR/month)	14,931,666.67
Average WTP as Surveyed (Rp/mo/person)	130,979.53
Rata-rata WTP Survei (IDR/thn/orang)	1,571,754.39
Total WTP in 7 Districts (mi IDR/annum)	271,592.87
WTP in 7 Kec (IDR/ha/annum)	1,413,647.95

Table 6. Bequest Value Results (WTA)

Total WTA as surveyed (IDR/month)	306,580,000.00
Average WTA as Surveyed (Rp/mo/person)	2,689,298.25
Rata-rata WTA Survei (IDR/thn/orang)	32,271,578.95
Total WTA in 7 Districts (mi IDR/annum)	5,576,399.76
WTA in 7 Kec (IDR/ha/annum)	29,025,305.57

Similar with Existence Value data, when asked to state how much the respondent will ask for a compensation, in return of his/her willingness to accept the peatland disappearance, the answer is more than 20 times as much (2,689,289/130,979) as their willingness to pay to prevent the peatland disappearance.

5. Conclusions

When compared with Direct Use of converting peatland for palm oil plantation (which can produce 19,64-25,53 tonnes of fresh fruit bunch/ha/year, [4] the survey result shows lower values of (WTP) Existence and BequestValues in peatland. However, when the respondents are asked the compensation values (WTA), the resulting Non-Use Value are much higher than the Direct Use Value of palm oil plantation in peatland. Notwithstanding with this mixed result, a Non-Use value survey has to be conducted in peatland conversion, to catch as much as value as possible.

References

[1] Wahyunto et al 2003 Peta Luas SebaranLahanGambut Dan KandunganKarbon Di Pulau Sumatera 1999-2002, *Wetlands International*

[2] Hooijer A, Page S, Canadell J G, Silvius M, Kwadijk J, Wösten, H and Jauhiainen J 2010 Current and future CO2 emissions from drained peatlands in Southeast Asia.*Biogeosciences* **7**, 1505–1514, in Managing Peatland Fire Risk in Central Kalimantan, Indonesia, World Resources Report Case Study.

[3] Miettinen J, Shi C and Liew S C 2016 Land cover distribution in the peatlands of Peninsular Malaysia, Sumatra and Borneo in 2015 with changes since1990. *Global Ecology and Conservation* **6**:67–78, in Managing peatlands in Indonesia Challenges and opportunities for local and global communities, Kristell Hergoualc'h et al, 2016.

[4] Wahyunto et al 2013 Prospect of Peatland Utilization For Oil Palm Plantation In Indonesia, *Perspektif* Vol. **12** No. 1/Juni 2013. Hlm 11-22)

Attachments

Survey Questionaires

1 Existence Value

Q1: How much will you pay per month in order to preserve the peatland in your district from being converted to oil palm plantations?

Q2: How much will you agree to be compensated per month to allow the peatland in your district to be converted to oil palm plantations ?

2. Bequest Value

Q1: How much will you pay per month in order to preserve the peatland in your district from being converted to oil palm plantations, so that your future generations can see the peatland?

Q2: How much will you agree to be compensated per month to allow the peatland in your district to be converted to oil palm plantations, such that your future generations may not be able to see the peatland?

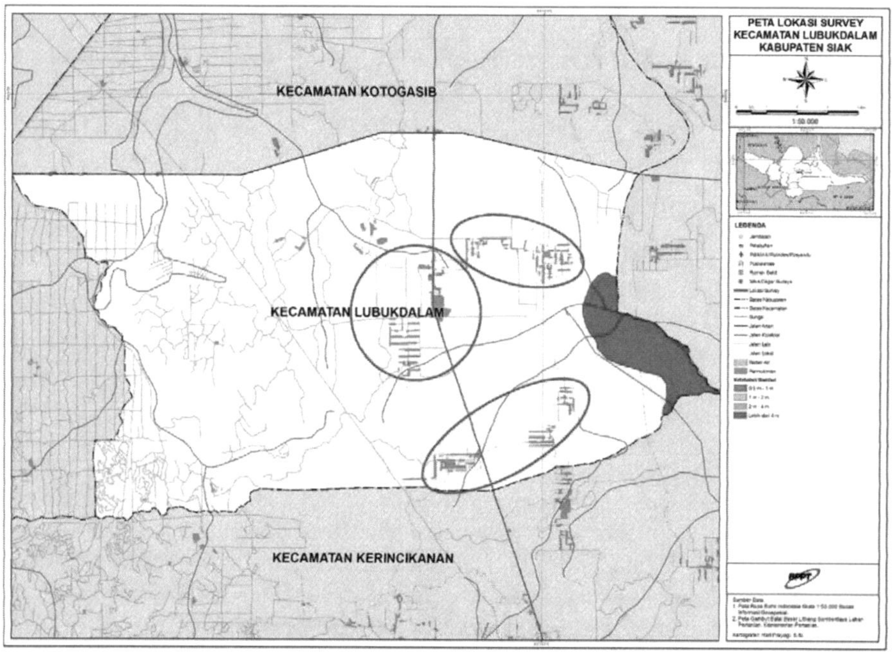

Figure 3. Peatland Thickness Distribution and Survey Locations Map of Lubuk Dalam District

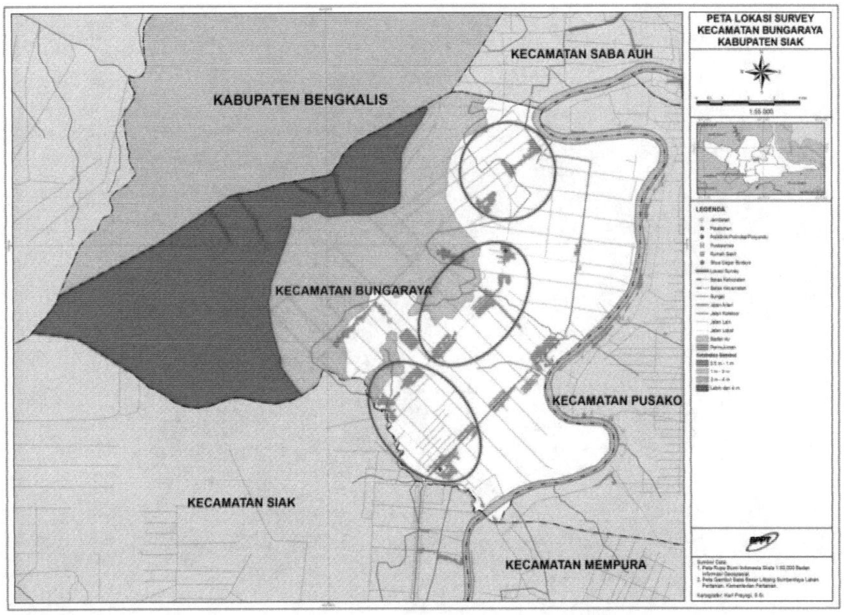

Figure 4 . Peatland Thickness Distribution and Survey Locations Map of Bunga Raya District

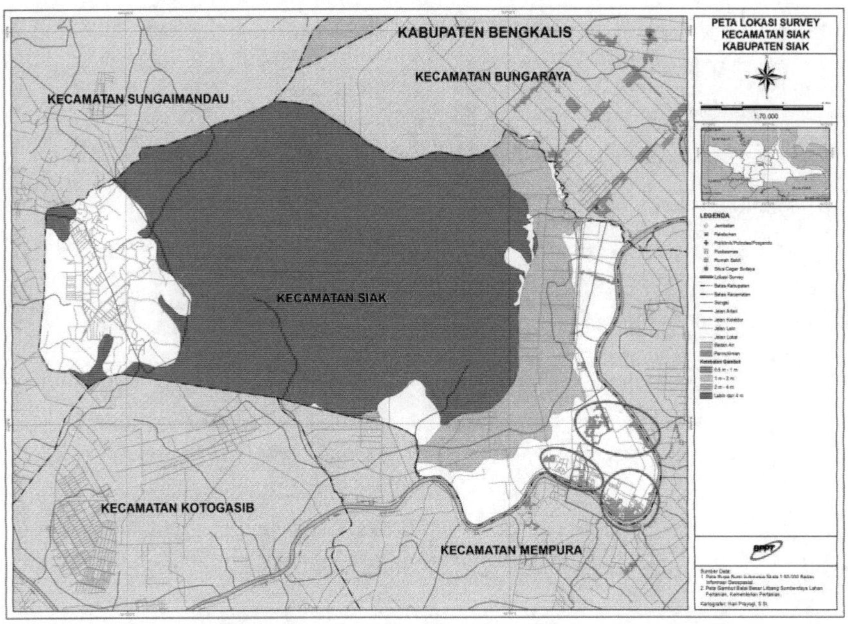

Figure 5 . Peatland Thickness and Survey Locations Map of Siak District

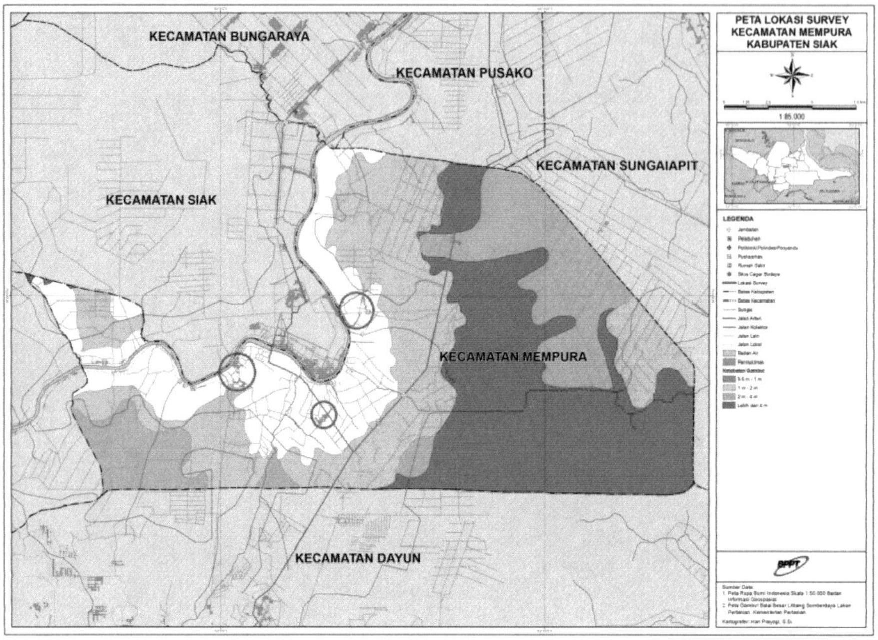

Figure 6 . Peatland Thickness Distribution and Survey Locations Map of Mempura District

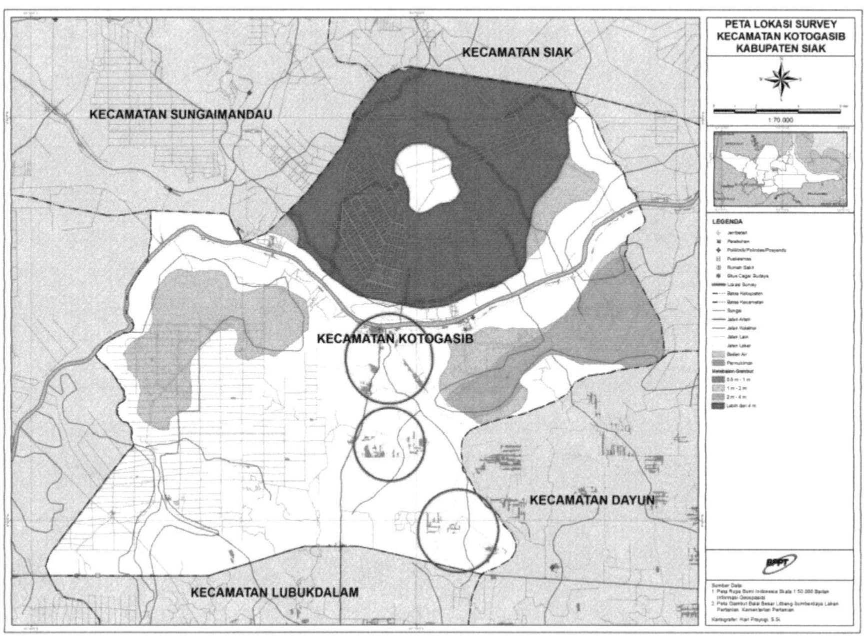

Figure 7 . Peatland Thickness Distribution and Survey Locations Map of Koto Gasib District

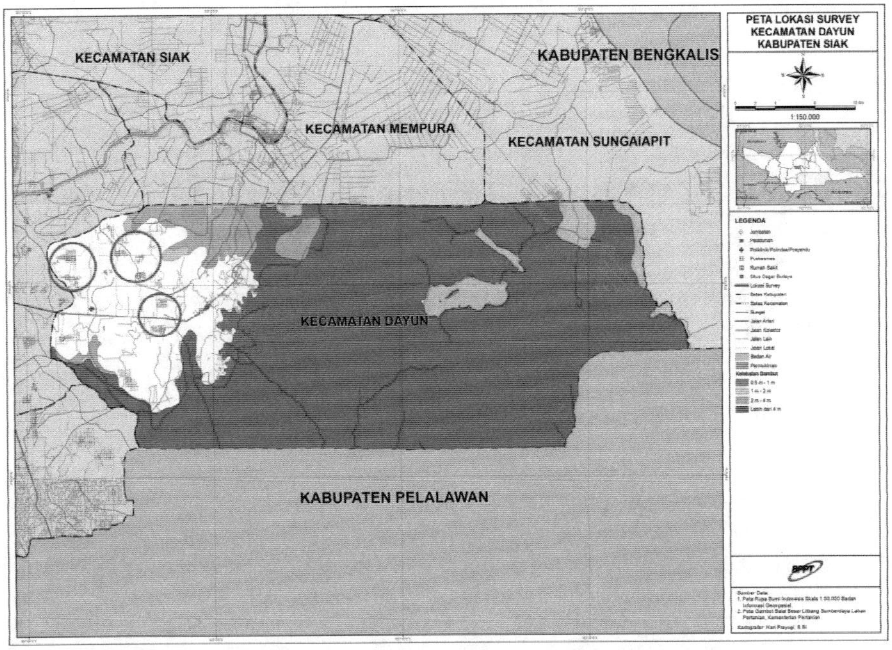

Figure 8 . Peatland Thickness Distribution and Survey Locations Map of Dayun District

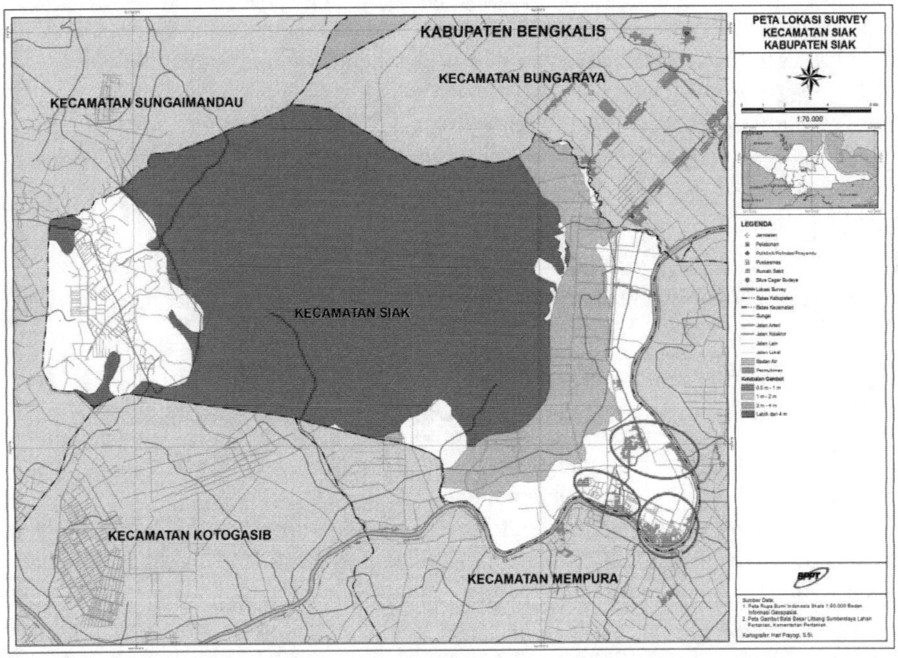

Figure 9 . Peatland Thickness Distribution and Survey Locations Map of Siak District

Identification of Tropical Peatland Using ALOS 2 Palsar

F Marpaung[1*], S Putiamini[1], D Fernando[1], E Avianti[1], H Priyadi[1] and A Darmawan[1]

[1]Agency for the assessment and application of technology (BPPT), Jakarta, 10340, Indonesia

fiolenta.marpaung@bppt.go.id

Abstract. In Indonesia, a huge amount of peat has accumulated under peat swamp forest over millennia. However, a repeatable large-scale fire as a result of land clearing has rapidly devastated these tropical peatlands. These disturbances (e.g. climate change, deforestation, peatland drainage, forest and peatland fires, and land conversion) change the ecosystem respiration and resulting a high carbon emitted to the atmosphere. To restore the peatland ecosystem, Indonesia government makes a special force by using three activities, namely rewetting, replantation, and communal empowerment. These activities need an accurate peatland monitoring. However, accessing the spatial variation of the peatland is quite limited, difficult and laborious. Therefore, we investigated the existence of tropical peatland in Siak, Riau Province, Indonesia. We measured soil moisture at four different types of land cover (namely; forest, oil palm, shrubs, and agriculture) a long with two types of peat conditions. We estimated the Constanta Dielectric (ε) values obtained from a fully polarimetry for ALOS 2/ PALSAR. Results indicated that the ε of oil palm was lower in non-peat area than in peat area, whereas the decrease in ε probably in peat area resulted from the rapid growth of oil palm.

Keywords: Polarimetric SAR, Constanta Dielectric, Soil Moisture, Land Cover Types, Peat

1. Introduction

Peat is soil comprised of partly decayed plant material formed in wetlands. The peatlands are habitats with a highly specialized flora and fauna. The peatlands also play an important role in global climate regulation. It covers cover only 3% of the land, but contain more carbon than the entire forest biomass of the world. The peatland located in tropic area, is widely distributed over 7×10^5 km^2 (about 5.21 % of the total global peatland area). It is mainly located in Indonesia and stores soil carbon up to 57.4 Pg., which accounts for ~ 12% of the total global peatland area [11]. However, these tropical peatland have been rapidly devastated by logging, poor land management practices, such as extensive plantation development, and enormous drainage in the last decades [9]. These activities drained the peatland, and well-preserved carbons were released as greenhouse gas to the atmosphere. The peatlands have also experienced repeatable large-scale fires as a result of land clearing [10] , and lowered groundwater level (GWL) in the late dry season, especially when strong El Niño drought arise [7]. The fires alters peat soil properties, reduces plant cover, plant composition; whereas severe drought can increase tree mortality, and blench huge amount of carbon into the atmosphere. It also produces long lasting smoky; underground blazes that impacts widespread public health. These conditions along with associated environmental and phenological characteristic give different impact on ecosystem respirations, and ultimately change the flux of carbon to the atmosphere. To restore the peatland ecosystem, Indonesia government makes limitation on the extent of plantation development in order to minimize the rate of this peat loss [13]. The government also makes a special force by using three activities, namely rewetting, replantation, and communal empowerment [14;15].

Land monitoring have been widely conducted using surface reflectance, which are derived from satellite data. This information is important for stakeholders and governments, as they are able to detect trends, seasonally and anomalies from a frequent and large-scale observations. However, monitoring and mapping tropical peatland is quite limited. The lack of this information is due to high attributes of cloud and aerosol influences the usability of satellite images. On the other hand, Synthetic Aperture Radar (SAR) with the use of multi-polarized L-band airborne images is highly suitable for detecting coastal wetlands [20]. They indicated that the HV polarization was the best for mapping abrupt boundaries in the coastal zone, the VV was the best for recognizing intertidal area morphology under low spring tide conditions, and the HH was the best for mapping coastal environments covered with forest and scrub vegetation such as mangrove and vegetated dunes. The radar system with a low incidence angle of the beam was also suitable for estimating information on soil moisture [3].

In this paper, radar scattering mechanisms for peatlands are examined in order to determine the existence of tropical peatland. It is known that peat and volumetric water content of a soil are closely related, and the possibility of a peat loss occurring becomes high when the volumetric water content of a soil is low. Thus, we evaluated soil moisture from the field, and estimated its spatial variation from the Advanced Land Observation Satellite (ALOS 2) Phased Array type L-band Synthetic Aperture Radar (PALSAR).

2. Material and Methods

2.1. Site Descriptions

The study sites (1.275 °N, 100.906 °E – 0.347 °N, 102.183 °E) are located at two different conditions of peat in Siak, Riau Province (Figure 1, Table 1). The study area has a total area of 8275 km^2 and is in a tropical rainy climate zone. The annual temperature ranges between 25 °C and 32°C, with the mean annual temperature is 28 °C.

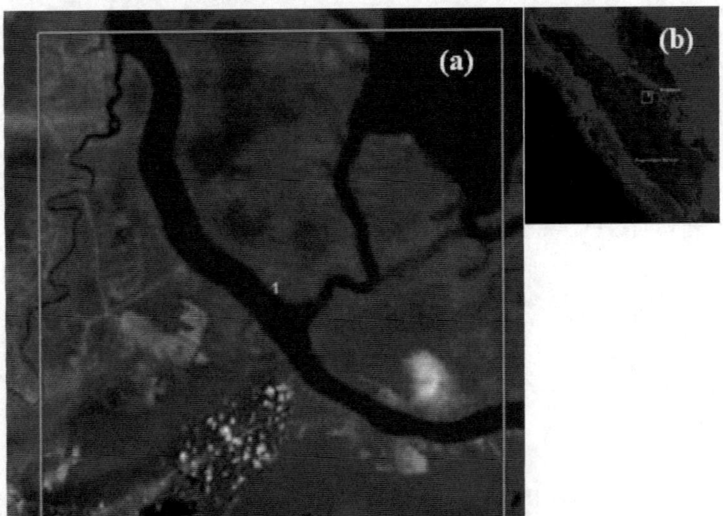

Figure 1. Study site (a) in Sumatra Island, Indonesia (b).

This region has two peaks of precipitation, in October – November and in March to May, which are associated with the southward and northward movement of the inter-tropical convergence zone (ITCZ) [1]. The main land covers in the study site area include forested peatland, oil palm, shrub, ferns plants, agriculture, and open water.

2.2. Data sources and field measurements

For the investigation, an ALOS 2/ PALSAR has been used. PALSAR provides an L-band SAR data and almost monthly temporal resolution for all weather, day and night observations. PALSAR with a dual polarimetry and a full polarimetry are two of the five different science data modes in which PALSAR can operate with the polarization of HH, HV, VV, and VH. The data were retrieved from PASCO Geospatial Group (http://en.alos-pasco.com) with time acquisitions of 11 February 2017 and 25 March, respectively for a dual and a full polarimetry data with the pixel spacing of 6.25 m. Detailed descriptions of ALOS 2/PALSAR are reported in literatures (e.g., http://www.eorc.jaxa.jp/ALOS-2/en/doc/format.htm).

The study areas were inaccessible due to flooding and lack of transportation facilities. Therefore, sampling sites were selected by field survey considered the accessibility, spatial distribution of inventoried peatland data, and area of each class. A total of 5-separated sampling sites were selected. Each sampling site (segment) with 1.5-km quadrants was recognized as discrete areas dominated by the same land cover types. The *in-situ* data were collected during the field campaigns from 4 May to 16 May 2017. The volumetric soil water content (SWC) of the top 5-cm layer and the top 10-cm layer were measured with Soil Moisture Meter Model: PMS-714. To avoid diurnal variation of SWC as the effect of soil temperatures and respirations, we set the time of the measurements in the morning between 07h and 10h local time. The numbers of chosen measurement points were equal to 250. The geographic coordinates were recorded using a Global Positioning GPS receiver with the accuracy of 5 m. Type of peatlands, as well as vegetation cover in each sites have been described during the field surveys (Table 1). In addition, air temperature and precipitation were measured at height of 1.5 m at three locations, namely Station R1 (101.81° E, 1.60° N), Station T1 (102.95° E, 0.86° N) and station T2 (102.80° E, 0.93 N). Hourly means of air temperature and precipitation which were measured at interval of 10 min, were recorded with a data logger (Campbell Scientific Inc., USA)

Table 1. List of site locations and vegetation types

Site Name	Latitude (deg-min-s)	Longitude (deg-min-s)	Elevation (m)	Peat conditions (peat depth*)	Dominant vegetation covers
A1	102°07'46.9"E	1°08'40.9"N	5.834	Disturbed and no/less peat	Oil palm
A2	102°10'07.0"E	1°09'56.2"N	7.104	Disturbed and thick peat (3.46 m)	Shrubs / fern plants, oil palm, and agriculture (pineapples)
D2	102°08'25.3"E	0°58'46.5"N	11.934	Disturbed and thick peat (3.68 m)	Oil palm, Shrubs
E1	102°12'19.9"E	0°55'16.0"N	8.391	Disturbed and thick peat (4.91 m)	Oil palm, Shrubs
J1	102°15'55.3"E	0°40'00.5"N	12.152	Undisturbed and thick peat (> 7.6 m)	Peat swamp forest

* averaged peat depth which measured by coring

2.3. Data Analysis

2.3.1. SAR analysis. Speckles found in SAR data due to random multiplicative noise can be recognized by the random pattern and the noise levels increases with average grey level of specified window area in an image scene. The speckle effects can be minimized using adaptive filter and convolution filters. In this study, we used Lee Refined Filter with a 7 x 7 sliding window owing to high contras areas. Furthermore, the raw Digital Number (DN) values of each pixel in SAR image were transformed to normalized backscattering coefficients (σ°), expressed in dB, according to the following equation :

$$\sigma^{\circ} = 10 * log_{10}(DN^2) + CF \tag{1}$$

where CF is the calibration coefficient for PALSAR standard products, and equal to -83 dB [16]. Prior to speckles filtering, the data had been georefenced using the coordinated provided by the header. All the processes for ALOS 2/ PALSAR image calibration and filter were performed using Sentinel Application Platform (SNAP) (Version 5.0).

2.3.2. Soil Moisture. Near-surface soil moisture can be approximated from PALSAR data by using Dubois Model [5]. The model has two questions that relate to the radar backscatter to sensor and soil parameters; one applicable for H-H polarized data, and the other for V-V polarize data, as shown below:

$$\sigma^{o}_{HH} = 10^{2,75}\frac{\cos\theta^{1,5}}{\sin\theta^{5}}\,10^{0,28\varepsilon\tan\theta}(kh\sin\theta)^{1,4}\lambda^{0,7} \tag{2}$$

$$\sigma^{o}_{VV} = 10^{2,5}\frac{\cos\theta^{3}}{\sin\theta^{3}}\,10^{0,46\varepsilon\tan\theta}(kh\sin\theta)^{1,1}\lambda^{0,7} \tag{3}$$

where σ^{o}_{HH}, and σ^{o}_{VV} are respectively the horizontally emitted-horizontally, and the vertically emitted-vertically received radar backscattering value (dB), θ is the incidence angle (37°), k is the free space wave number given by $k = 2\pi/\lambda$, λ is the free space wavelength (23.5 cm), ε is the real part of the dielectric constant, and h is the RMS height at the surface. As mentioned above, this model has been derived from bare soil measurements, but is also applicable to a moderately dense vegetation cover [5]. The $kh\sin\theta$ can be omitted from the relationship with ε_r by referencing Equations (2) and (3) [6]:

$$\varepsilon_r = \frac{1}{\tan(\theta)(0.0305)}log_{10}\left(10^{2.4}\frac{(\sigma^{VV})^{1.27}}{\sigma^{HH}}\cos(\theta)^{-2.3}\sin(\theta)^{-1.2}\lambda^{-0.19}\right) \tag{3}$$

and the volume soil moisture percentage (mv) were estimated using scattering model of the inter-compared Dubois – Oh [17]:

$$mv = -5.3\,x\,10^{-2} + 2.92\,x\,10^{-2}\varepsilon - 5.5\,x\,10^{-4}\varepsilon^2 + 4.3\,x\,10^{-6}\varepsilon^3 \tag{4}$$

To minimize the influence of roughness and shape of surface to SAR backscattering, we used 3x3 windows moving average of estimated soil moisture.

3. Results and Discussion

3.1. Profile of SAR backscatter and Precipitation in Siak Regency

Figure 2a and Figure 2b show the profile of horizontal-horizontal (HH) polarization and horizontal-vertical (HV) polarization that were derived from a dual mode and a full mode polarization in 5 locations. In general, the backscattering value of a dual mode data was lower than a full mode data. Rainfall or prior the days mainly affected this difference (Figure 2c). During the recording of ALOS 2 (Dual Mode) on 11 February 2017 (DOY 42), precipitation was relatively low with a 6.5 mm-rainfall of 10 February 2017 (DOY 41), whereas the precipitation was relatively high during the recording of the Full-mode data on 25 March 2017 (DOY 84). Precipitation of 13.5 mm and 31.5 mm were occurred on 22 March 2017 (DOY 81), and 24 March 2017 (DOY 83), respectively. In addition, precipitation was low during field measurements. A 2 mm- and a 24.5 mm-rainfall were measured respectively, for 2 May 2017 (DOY 122) and 4 May 2017 (DOY 124). A low difference occurred between HV polarization data, whereas a higher difference occurred on HH polarization. These conditions were suggested due to the strength relationship between HH and soil moisture in the L band [4; 8].

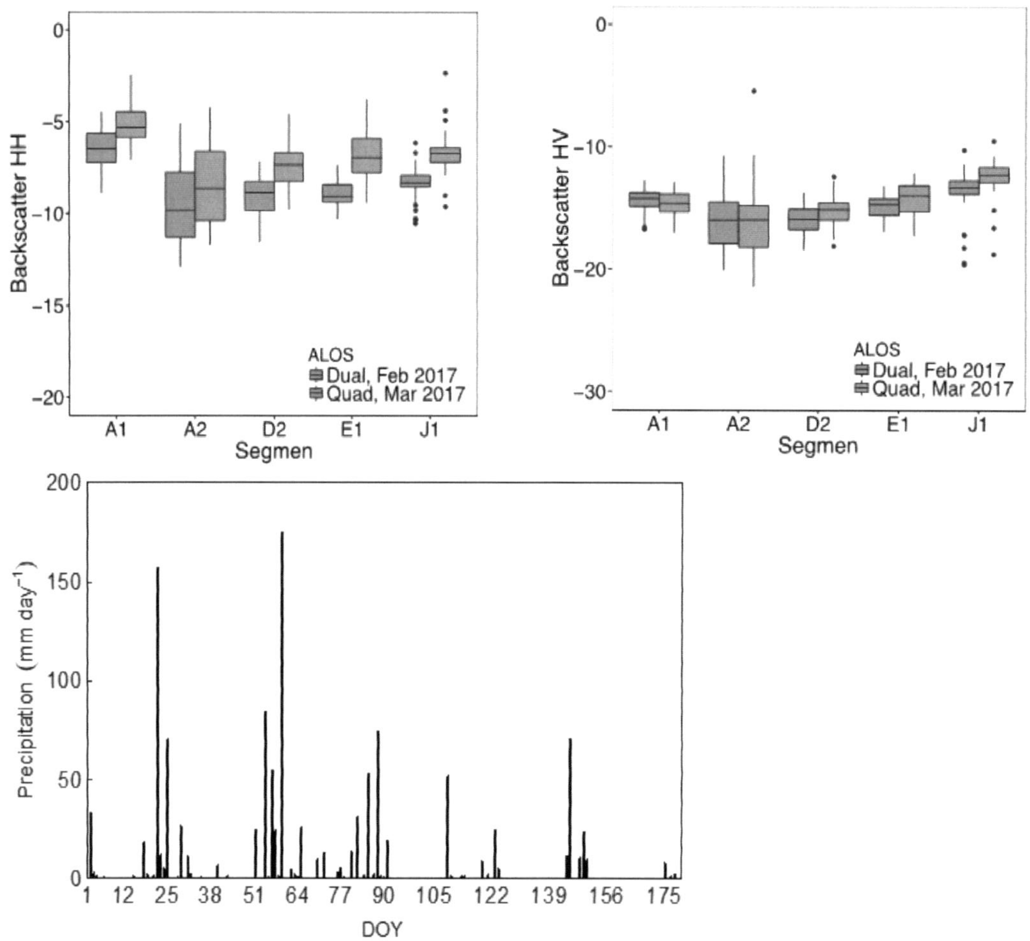

Figure 2. Backscatter values of ALOS 2 Palsar with a HH polarization (a) and a HV polarization from a dual mode data (11 February 2017) and a full mode (25 March 2017) in 5 locations along with precipitation (c) in Siak Regency.

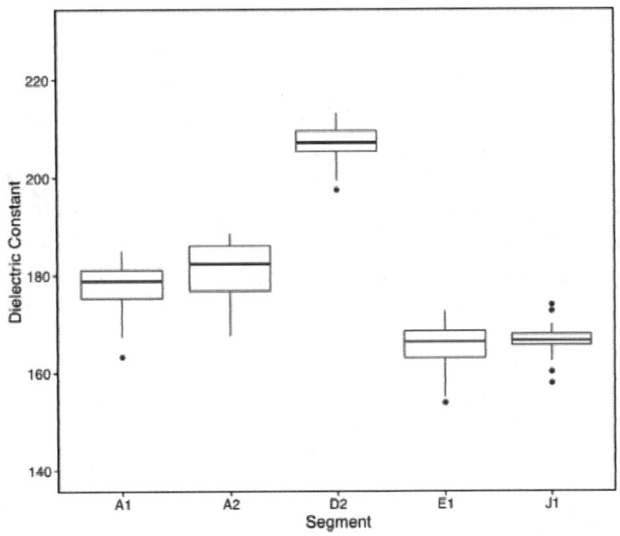

Figure 3. Dielectric constant of ALOS 2 Palsar – Full Mode.

3.2. Dielectric constant in Siak Regency

Although the fundamental shortcoming of the Oh and Dubois models in terms of dielectric constant and soil moisture retrieval described as Eq. (3) is that they are valid only for bare or sparsely vegetated soil surfaces, however, in the implementation, this model are used for many real soil surfaces that are covered to some extent by vegetation. Dielectric constant of A2 has a high variance with a 5.3 deviation, whereas the J1 has a low variance with a 2.7 deviation among the segments (Figure 3). These conditions had been resulted from discrepancy of different backscattering on land cover and the heights in each segment (Table 2). Oil palm, pineapple, rubber, and shrubs cover A2 segment while J1 is in the protected national park with an undisturbed peat swamp forest. Meanwhile, D2 segment with land cover of young oil palm and shrubs on a 3.7 m-peat has a high dielectric constant among the segments. This was affected by different total plant among segments [12]. This was affected by land cover height. Oil palm in D2 was relatively lower than in A1 and E1. The average was about 1 m, 4 m, and 3.5 m respectively, for D2, A1 and E1.

Table 2. Dielectric constant of land cover in each site/segment

Site/ Segment	Land cover*						
	Closed to canal	Oil palm	Oil palm & Shrubs	Pineapple & Shrubs	Rubber & Shrubs	Undisturbed Forest	ε Segment
A1		177.9±4.7					177.9±4.7
A2		179.3±6.1	178.7±3.0	184.9±3.0	174.4±4.0		181.1±5.3
D2		208.6±4.8	206.8±2.7				207.2±3.4
E1	171.7		165.4±4.5				165.6±4.5
J1	169.9					166.4±2.7	166.5±2.7
ε Land over	170.8±1.3	183.4±12.6	185.3±18.8	184.9±3.0	174.4±4.0	166.4±2.7	

* average ± standard Deviation
ε dielectric constant

Table 2 shows the dielectric constant of different land cover in each segment. In general, the dielectric constant of non-peat area was lower than the peat area. The dielectric constants were about 177.9±4.7 and 183.1±3.8 respectively, for the non- and peat area. The non-peat area of A1 with a dominant land cover of oil palm has a 178 – dielectric constant, whereas the peat areas of A2 and D2 have a higher dielectric constant with the ε of 179 and 209, respectively for A2 and D2 segments. The increase in dielectric constant of oil palm in D2 compared to A2 probably resulted from the rapid growth of oil palm under the conditions of moist peat soil. A thick peat contributes to an increasing of soil moisture. In contrast, peat areas of E1 and J1, which their ground mainly covered by thick- leaf litter and -woody debris, have low dielectric constants than in A1. The leaf litter and woody debris contributed to L band backscattering and decreasing the strength of the relationship to soil moisture.

3.3. Soil moisture in Siak Regency

Soil moisture of respective segments was calculated by using the Eq. (4). The estimation was an average of 3x3-moving window data. Figure 4 shows the profile of soil moisture estimations and measurements in in different land cover types, both in non-peat and peat area. Despite of different acquisition date between measurements and ALOS, soil moisture in non-peat area was lower than in peat area. Moreover, it relatively varied by 2.6%, 4.4%, and 3.5% in peat areas respectively, for ALOS, the measurements at 5-m, and 10-m depth, and was higher than in non-peat areas. Derived soil moisture in peat areas from ALOS ranges from 12.4% to 24.6% with mean value 20.1% (±5.4), whereas in non-peat area, it ranges from 12.6% to 17.1% with mean value 15.4% (±1.0%). Soil moisture measured at 5-cm depth in the peat area has a range between 7.3% and 22.1% with mean value 15.4% (±4.6%), and a range of 12.0% - 23.9% with mean value 19.4% (±3.4%) at 10-cm depth, whereas in non-peat at the 5-cm (10-cm) depth, it ranges between 2.3% and 26.5% (7.5% and 22.5%) with the mean value 11.9% (13.9%) and standard deviation 5.5% (5.5%).

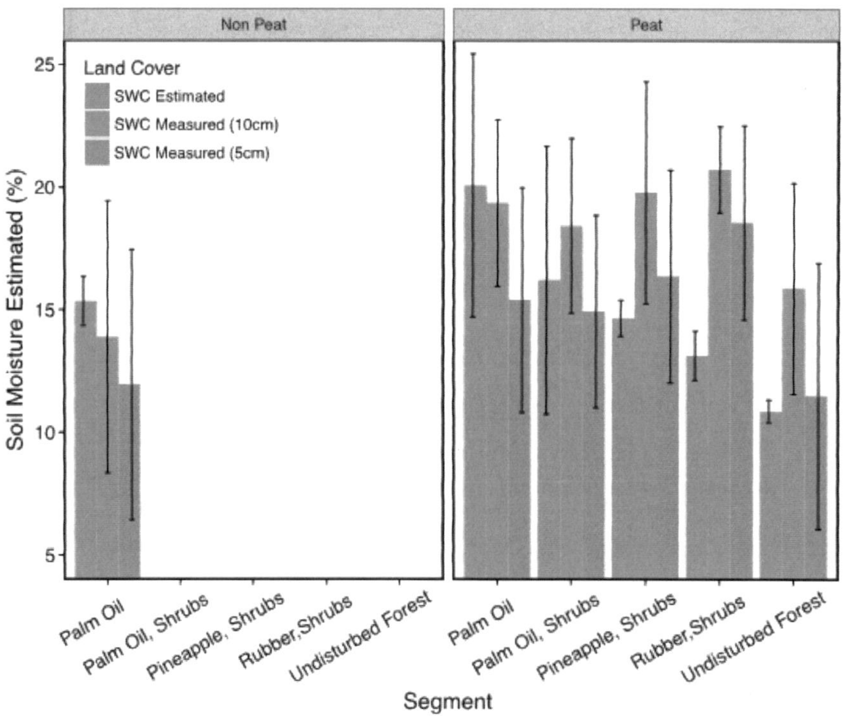

Figure 4. Soil moisture estimations and measurements in different land cover types, both in non-peat area (left panel) and in peat area (right panel) in Siak Regency.

4. Conclusions

In this study, profiles of dielectric constant derived from ALOS 2 Palsar in non-peat area were lower than in peat area. The profiles were about 177.9±4.7 and 183.1±3.8 respectively, for the non- and peat area. It was correlated with peat conditions, land cover, and the variations in precipitation. In a similar land cover, dielectric constant increases as peat depth decreases. However, in undisturbed forest with a thick peat, the profiles decreased. We estimated soil moisture using backscatter inversion model for bare or sparsely vegetated soil surfaces. The model underestimated the soil moisture in sparsely vegetated soil surfaces of pineapples.

References

[1] Aldrian E and Susanto R D 2003 Identification of three dominant rainfall regions within Indonesia and their relationship to sea surface temperature. *Int. J. Climatol.* **23**, 1435–1452. doi:10.1002/joc.950

[2] Argenti F, Lapini A, Bianchi T and Alparone L 2013 A Tutorial on Speckle Reduction in Synthetic Aperture Radar Images. In *IEEE Geoscience and Remote Sensing Magazine*, **vol. 1 no. 3**, pp. 6-35. - ISSN 2168-6831

[3] Dabrowska-Zielinska K, Gruszczynska M, Lewinski S, Hoscilo A and Bojanowski J 2009 Application of remote and in situ information to the management of wetlands in Poland. J. *Environ. Manag.* **90**, 2261–2269.

[4] Della Vecchia A, Ferrazzoli P, Guerriero L, Ninivaggi L, Strozzi T and Wegmueller U 2008 Observing and modeling multifrequency scattering of maize during the whole growth cycle. *IEEE Trans. Geosci. Remote Sens.* **46(11)**, 3709–3718.

[5] Dubois P C, Vanzyl J J and Engman T 1995 Measuring Soil Moisture with Imaging Radars. *IEEE transactions on geoscience and remote sensing*. **33**:916-926.

[6] Hosseini R, Newlands N K, Dean C B and Takemura A 2015 Statistical Modeling of Soil Moisture, Integrating Satellite Remote-Sensing (SAR) and Ground-Based Data. *Remote Sens.*, **7**, 2752-2780; doi:10.3390/rs70302752

[7] Langner A and Siegert F 2009 Spatiotemporal fire occurrence in Borneo over a period of 10 years. *Glob. Chang. Biol.* **15**, 48–62. doi:10.1111/j.1365-2486.2008.01828.x

[8] Mattia F, Satalino G, Pauwels, V R N and Loew A 2009 Soil moisture retrieval through a merging of multi-temporal L-band SAR data and hydrologic modelling. Hydrol. *Earth Syst. Sci.* **13(3),** 343–356.

[9] Miettinen J, Shi C, Liew, S C 2011 Deforestation rates in insular Southeast Asia between 2000 and 2010. *Glob. Chang. Biol.* **17**, 2261–2270. doi:10.1111/j.1365-2486.2011.02398.x

[10] Murdiyarso D and Adiningsih E S 2007 Climate anomalies, Indonesian vegetation fires and terrestrial carbon emissions. *Mitig. Adapt. Strateg. Glob. Chang.* **12**, 101–112. doi:10.1007/s11027-006-9047-4

[11] Page S E, Rieley J O and Banks C J 2011 Global and regional importance of the tropical peatland carbon pool. *Glob. Chang. Biol.* **17**, 798–818. doi:10.1111/j.1365-2486.2010.02279.x

[12] Patel P, Srivastava H S, Panigrahy S and Parihar J S 2006 Comparative evaluation of the sensitivity of multi-polarized multi frequency SAR backscatter to plant density. I*nt. J. Remote Sens.* **27 (2),** 293-305

[13] President of Indonesia 2011 Instruction of the President of the Republic of Indonesia number 10 of 2011 about suspension of granting of new licenses and improvement of governance of natural primary forest and peatland.

[14] President of Indonesia 2014 Government Regulation Number 71 of year 2014 about Protection and Management of Peat Ecosystems.

[15] President of Indonesia 2016 Presidential Regulation Number 1 of year 2016 about Peat Restoration Agency.

[16] Rosenqvist A, Shimada M, Ito N and Watanabe M 2007 ALOS PALSAR Apathfinder mission for global-scale monitoring of the environment. *IEEE Transactionson Geoscience and Remote Sensing*, **45**, 3307–3316.

[17] Rao S Srinivasa, Kumar S Dinesh, Das S N, Nagaraju M S S, Venugopal M V, Rajankar P, Laghate P,Sivaprasad Reddy M, Joshi A K and Sharma J R 2013 Modified Dubois Model for Estimating Soil Moisture with Dual Polarized SAR Data. December 2013. *Journal of the Indian Society of Remote Sensing*. **41(4)**:865–872 DOI 10.1007/s 12524-013-0274-3

[18] Shimada M and Ohtaki T 2010 Generatinglarge-scale high-qualitySAR mosaic datasets: Application to PALSAR data for global monitoring. *IEEE Journal of Selected Topics in Applied Earth Observations and Remote Sensing*, **3**, 637–656.

[19] Shimada M 2011 Model-based polarimetric SAR calibration method using forest and surface-scattering targets. *IEEE Transactions on Geoscience and Remote Sensing,* **49**,1712–1733

[20] Souza-Filho P W M, Paradella W R, Rodrigues S W P, Costa F R, Mura J C and Gonçalves F D 2011 Discrimination of coastal wetland environments in the Amazon region based on multi-polarized L-band airborne Synthetic Aperture Radar imagery. Estuar. *Coast. Shelf Sci.* **95**, 88–98.

[21] Takada M, Mishima Y and Natsume S 2009 Estimation of surface soil properties inpeatland. Consortium of Landscape and Ecological Engineering and Springer 2009. *Landscape Ecol Eng (2009)* **5**:45–58

[22] Thapa R, Itoh T, Shimada M and Watanabe M 2014 Evaluation of ALOS PALSARsensitivity for characterizing natural forest cover in wider tropical areas. *Remote Sensing of Environment* **155** : 32–41

[23] Top G C, Davis J L and Annan A P 1980 Electromagnetic Determination of Soil Water Content: Measurenments in Coaxial Transmission Lines. *Water resources res.* **16**:574-582.

Acknowledgements
The Agency and Application and Assessment of Technology (BPPT) funded the field measurements in Siak, Riau Province, Indonesia. This work was supported by BPPT No. 04 Tahun 2017. We thank to JAXA for providing ALOS-2 data through PI. No. 1248002.

IOP Conf. Series: Earth and Environmental Science **165** (2018) 012015 doi:10.1088/1755-1315/165/1/012015

Spatial Analysis of Water Pollution using Multitemporal Satellite Imagery (Case Study: River Lamong Estuary, Surabaya)

Bangun Muljo Sukojo[1] **and Renita Elizabeth Sianipar**[1]

[1]Department of Geomatics Engineering, Faculty of Civil, Environment, and Earth Engineering Institut Technology of Sepuluh Nopember

bangunms@gmail.com

Abstract. Water is a natural resource that has a very important function for human life and promotes the welfare of humanity itself. Water is also the important component of the environment for the survival and life of humans and other living beings. It's necessary to manage and control water pollution wisely by considering the ecological balance. On the border of the Surabaya and Gresik, there's an industrial center located on the north coast of Surabaya. River Lamong is estimated to have water quality degradation problems due to growing numbers of industries in River Lamong and factory wastes flowing downstream. By looking at the condition, it's suspected that there is pollution in River Lamong. Conditions of water pollution can be indicated through several parameters such as the physical parameter, chemical parameter, and biological parameter. Physical parameters that can be observed include temperature, dissolve residue, and suspended residue (TSS). As for the chemical parameters such as pH, BOD (Biochemical Oxygen Demand), (COD) Chemical Oxygen Demand (COD), and other metal content. The result of this study indicates that there are polluted points based on BOD parameters at 4 points.

Keywords: Water Pollution, River Lamong, Total Suspended Solid

1. Introduction

The existence of industry can improve the welfare of the community and harm the community such as the emergence of environmental pollution due to industrial waste disposal that generally contains heavy metals. There are 93 industries registered with PT Kawasan Industri Gresik and some of its industries are right on the outskirts of River Lamong.

According to PP RI. 82 year 2001 on the Management of Water Quality and Water Control, water pollution is the entry or inclusion of living things, substances, energy, or other components into the water by human activities, so that water quality down to a certain level that causes water can't function in accordance with its designation. Conditions of water pollution in waters can be indicated through several parameters, namely physical, chemical, and biological parameters. Physical parameters that can be observed include temperatures, dissolved residues, and suspended residues. As for the chemical parameters such as pH, BOD, COD, and other metal content. In this research, two approaches are

IOP Conf. Series: Earth and Environmental Science **165** (2018) 012015 doi:10.1088/1755-1315/165/1/012015

conducted directly and indirectly. For indirect approach through remote sensing, 3 image types used are very high-resolution image (Pleiades) and medium (Landsat 8) to obtain the approach value of TSS (Total Suspended Solid) parameter.

Physical parameters that can be observed include temperature, dissolved residue, and suspended residue. As for the chemical parameters such as pH, BOD, COD, and other metal content. In this research, two approaches are conducted directly and indirectly. For indirect approach through remote sensing, two types of images that were used are the very high-resolution image (Pleiades) and medium (Landsat 8) to obtain approximate values of the parameters of TSS (Total Suspended Solid). Total Suspended Solids (TSS) is the dry-weight of particles trapped by a filter. With some combinations of band formula, a value of TSS in a point can be obtained.

2. Research Methodology

2.1. Research Site

The study was conducted in the northern coastal area of Surabaya (Teluk Lamong) located between 7° 11' 0" - 7° 13' 0" LS dan 112° 38' 0" - 112° 43' 0" BT. Here is a picture of the research location:

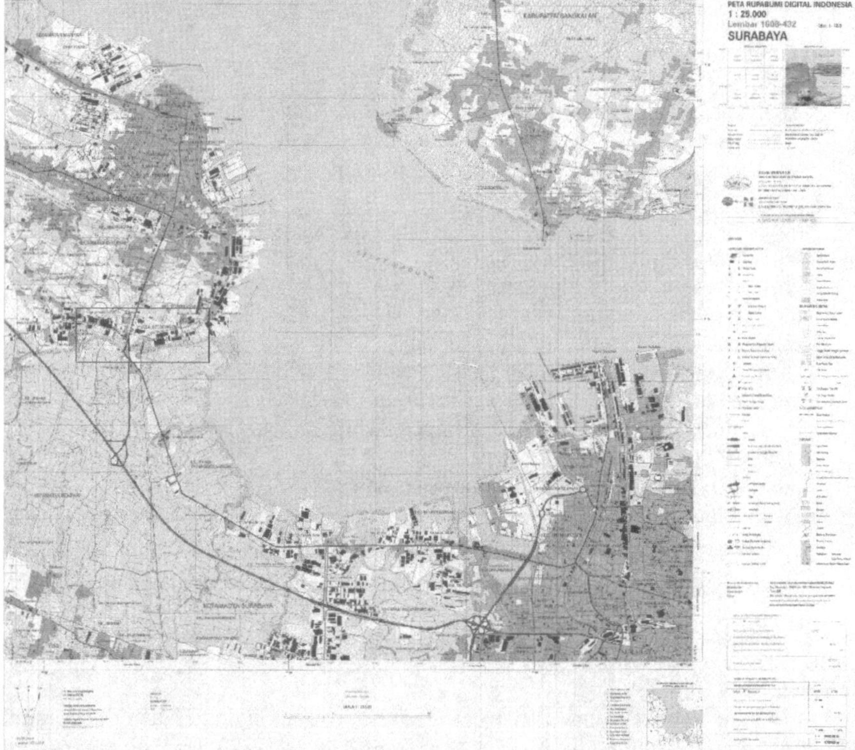

Figure 1. RBI Surabaya Map

2.2. Data and Equipment

Data

1. Pleiades 1A satellite image (Acquisition date: August 28, 2015)
2. Landsat 8 satellite image (Acquisition date: July 4, 2016 and February 13, 2017)
3. RBI Map scale 1 : 25,000
4. GCP and ICP coordinate measurement results
5. Water sample parameters result

Equipment used in this research:

1. Hardware
   a. Notebook
2. Software
   a. Image processing software
   b. Statistical processing software
   c. Layouting processing software
3. Field device:
   a. Geodetic GPS
   b. Navigation GPS

2.3. Data Processing

The data processing can be seen in Fig. 2. The explanation of data processing diagram described as follows:

a. The initial stage includes refining imagery (pansharpening) is a process improvement to the quality of the image so that the image has a sharper display by combining the panchromatic band with multispectral bands to obtain satellite imagery in accordance with a spatial resolution in the panchromatic image. Cutting the image (cropping) cuts the image is based on observed area. The purpose of this image cutting is for research to focus on the study area.

b. Design of Control Point Nets and SoF Calculations. The net design is carried out for GCP measurement positions performed using a Geodetic GPS. Design nets that have been made a calculation of Strength of Figure (SoF) to find out the strength of the net.

c. Geometric correction is a geometric image correction so that the map projection and coordinate system used in accordance with the real world. Geometric Correction (rectification) on the image of the Pleiades 1A to help decision-point Ground Control Point (GCP) in the field using geodetic GPS while the reference to Landsat 8 using RBI map scale of 1: 25,000. In performing geometric correction the result of RMSE should be <1 pixel [2].

d. Radiometric Correction to correct the pixel values to fit the supposedly normally considering atmospheric disturbance factor as the main source of error [3]. Radiometric corrections include radiometric calibration and atmospheric correction. Radiometric calibration converts DN (Digital Number) to Radiance / reflect and then atmospheric correction with 6SV method.

e. Calculation of TSS Algorithm by using Laili algorithm (2015) as follows:

$$\text{TSS}\ (\text{mg} / \text{l}) = 31.42 * ((\text{Log}\ (\text{RRS2}) / (\text{LogRRS4})) - 12.719 \qquad (1)$$

Laili algorithm was used because it had the closest value according to water sampling and produced the largest R (correlation) among other algorithms.

f. Field observations (Ground Truth) : the retrieval of field data in the form of GCP coordinates and water samples parameter. GCP is obtained through several steps, including control point nets designing and measurement using Geodetic GPS. Water samples were taken and processed directly in laboratory on the same day.

g. The correlation test is performed to find out the relation between y variables.

h. Analysis of pollution conditions in River Lamong with samples of parameters.

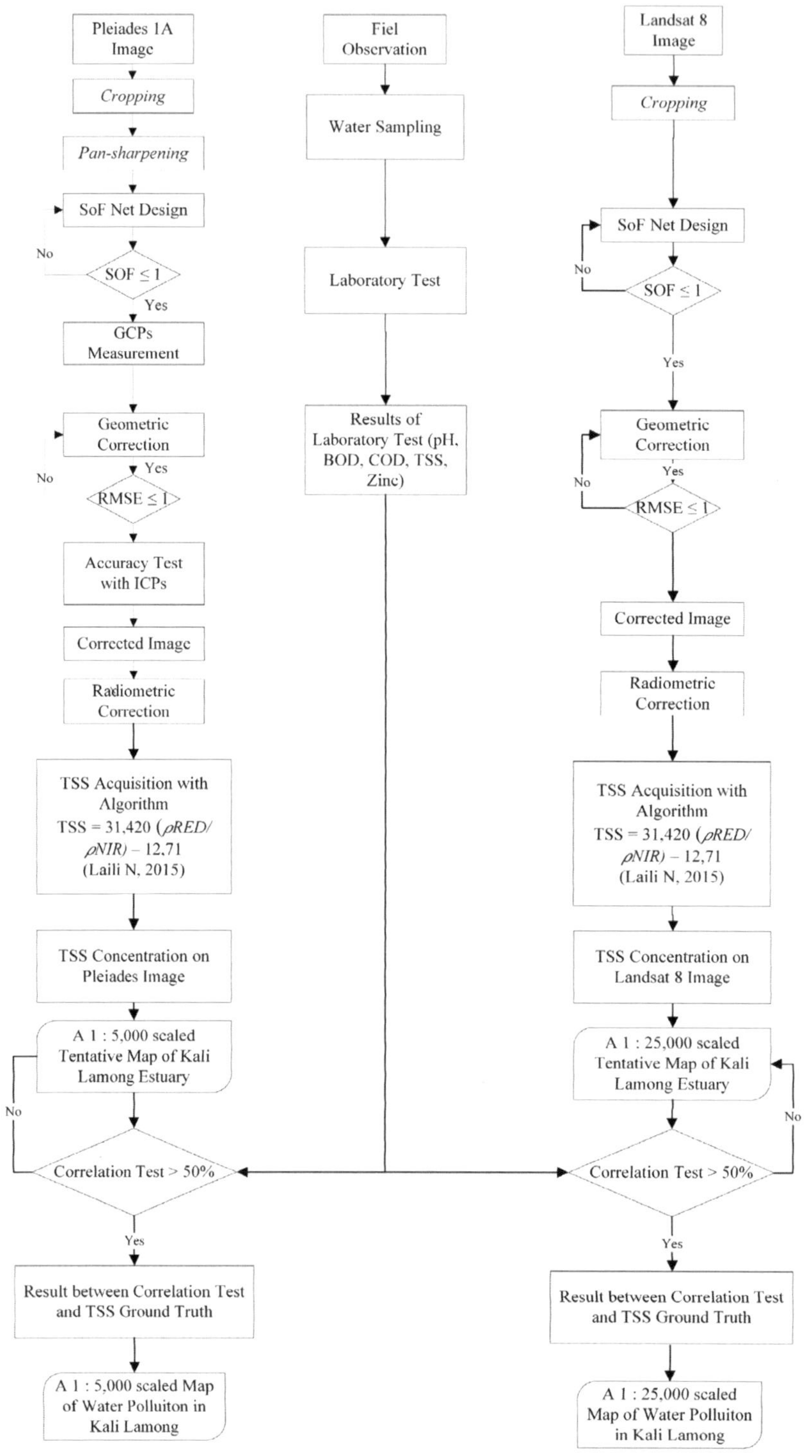

Figure 2. Data Processing Flow Chart

3. Results And Analysis

3.1 Geometric Correction and SoF Calculation

The result of GCP point coordinate has bound to SRGI control point and obtained RMSE image of

IOP Conf. Series: Earth and Environmental Science **165** (2018) 012015 doi:10.1088/1755-1315/165/1/012015

Pleiades 1A at 0.41. Geometric correction of Landsat 8 dated July 4, 2016 and February 13, 2017 using the RBI maps of 1: 25,000 with a 5 point GCP RMSE obtained for 0203 and 0219 so that it has entered the required tolerance of less than 1 pixel [2]. The GCP point has been converted in accordance with SRGI 2013. The calculation point of Strength of Figure (SoF) is 0.29 and 0.36 and has entered the required tolerance of less than 1, so the SoF net design is considered strong [5].

3.2 Water Sampling Result

The followings are the data of the field sample test results:

a. pH

pH is a numeric scale used to specify the acidity or basicity of an aqueous solution.

Table 1. pH parameter

Sample Code	Coordinate		pH	Status
	X (m)	Y (m)		
KK-1	683953.89	9204992.23	7.85	4th Class
KK-2	683933.05	9204995.54	7.80	4th Class
KK-3	683914.86	9204999.17	7.80	4th Class
PGN-1	683545.32	9204579.08	7.80	4th Class
PGN-2	683539.37	9204598.26	7.75	4th Class
PGN-3	683527.46	9204613.48	7.80	4th Class

b. COD

COD or Chemical Oxygen Demand is an indicative measure of the amount of oxygen that can be consumed by reactions in a measured solution.

Table 2. COD Parameter

Sample Code	Coordinate		COD (mg/L O2)	Status
	X (m)	Y (m)		
KK-1	683953.89	9204992.23	23	4th Class
KK-2	683933.05	9204995.54	20	4th Class
KK-3	683914.86	9204999.17	23	4th Class
PGN-1	683545.32	9204579.08	26	4th Class
PGN-2	683539.37	9204598.26	26	4th Class
PGN-3	683527.46	9204613.48	20	4th Class

c. BOD

BOD or Biological Oxygen Demand is the amount of dissolved oxygen needed (i.e. demanded) by aerobic biological organisms to break down organic material present in a given water sample at certain temperature over a specific time period.

Table 3. BOD Parameter

Sample Code	Coordinate		BOD (mg/L O2)	Status
	X (m)	Y (m)		
KK-1	683953.89	9204992.23	11	4th Class
KK-2	683933.05	9204995.54	13	Polluted
KK-3	683914.86	9204999.17	13	Polluted
PGN-1	683545.32	9204579.08	15	Polluted
PGN-2	683539.37	9204598.26	15	Polluted
PGN-3	683527.46	9204613.48	11	4th Class

Based on the result of the sampling test of each parameter, it is found that there are 4 points in Lamong River which are indicated to be polluted according to BOD parameters: Point KK1, KK3, PGN1, and PGN2 with successive value 13, 13, 15, 15 mg / L O2 based on 4th class range in PP No.82 of 2001[7]. Thus it can be concluded that Lamong River is polluted and does not meet the requirements of the IV class qualification because the BOD parameters are out of bounds.

The polluted BOD spot in the PGN code area lies on the southern side of Lamong River, which is directly adjacent to the settlement. The polluted BOD point in the KK code area is directly adjacent to Karang Karang Beach tourism. High levels of BOD are thought to be related to the location of the point closer to human activity. At the point, PGN found household waste, and in KK point there are domestic garbage tourists who visit the beach Karang Kering.

d. TSS

Table 4. TSS Parameter

Sample Code	Coordinate		TSS	Status
	X (m)	Y (m)		
KK-1	683953.89	9204992.23	72.00	4th Class
KK-2	683933.05	9204995.54	26.00	4th Class
KK-3	683914.86	9204999.17	30.00	4th Class
PGN-1	683545.32	9204579.08	84.00	4th Class
PGN-2	683539.37	9204598.26	90.00	4th Class
PGN-3	683527.46	9204613.48	34.00	4th Class

e. Zinc

Table 5. Zinc Parameter

Sample Code	Coordinate		Seng (mg/L Zn)	Status
	X (m)	Y (m)		
KK-1	683953.89	9204992.23	1.88	4th Class
KK-2	683933.05	9204995.54	1.92	4th Class
KK-3	683914.86	9204999.17	1.86	4th Class
PGN-1	683545.32	9204579.08	1.96	4th Class
PGN-2	683539.37	9204598.26	1.89	4th Class
PGN-3	683527.46	9204613.48	2.05	4th Class

3.3Correlation Between TSS Sample Result and Satellite Image

a. Pleiades Image August 2015

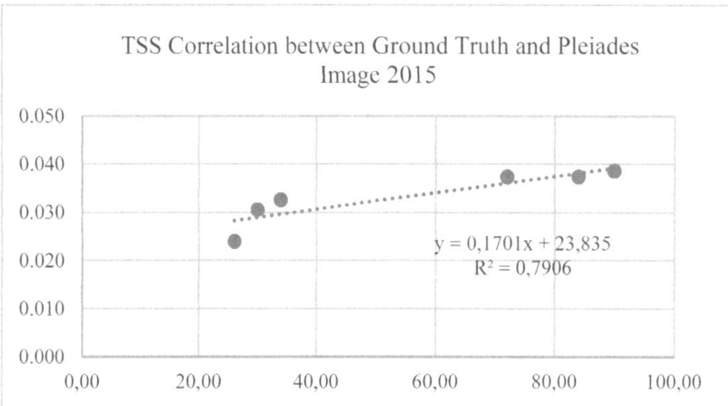

Figure 3. Correlation between TSS Sample and Pleiades Image 2015

IOP Conf. Series: Earth and Environmental Science **165** (2018) 012015 doi:10.1088/1755-1315/165/1/012015

Correlation test between water sampling and TSS Pleiades Image August 2015 using Laili algorithm (2015) has 79,06% relation coefficient value. Based on the correlation value according to Sugiyono (2008) [6], it can be said that TSS of sample test result with TSS Pleiades image has a strong relationship level.

b. Landsat 8 Image July 2016

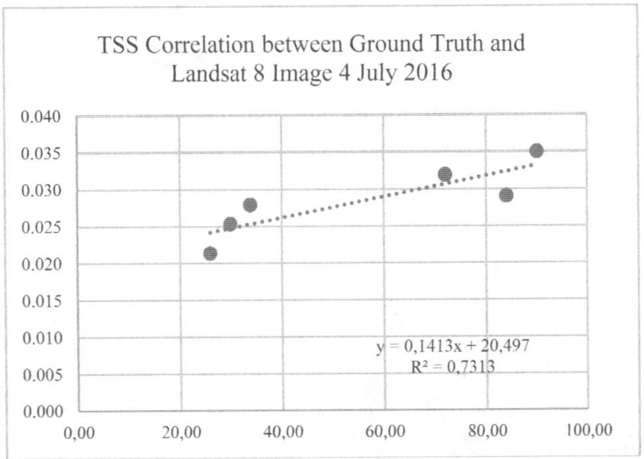

Figure 4. Correlation between TSS Sample and Landsat Image 4 July 2016

Correlation test between water sampling and TSS Landsat Image 4 July 2016 using Laili algorithm (2015) has 68,36% relation coefficient value. Based on the correlation value according to [6], it can be said that TSS of sample test result with TSS Pleiades image has a strong relationship level.

c. Citra Landsat-8 13 February 2017

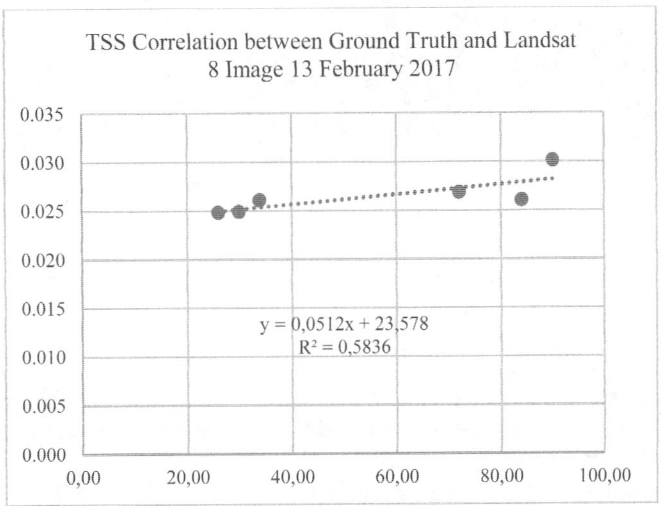

Figure 5. Correlation between TSS Sample and Landsat Image February 13, 2017

IOP Conf. Series: Earth and Environmental Science **165** (2018) 012015 doi:10.1088/1755-1315/165/1/012015

Correlation test between water sampling and TSS Landsat Image 13 February 2017 using Laili algorithm (2015) has 58,36% relation coefficient value. Based on the correlation value according to [6], it can be said that TSS of sample test result with TSS Pleiades image has a medium relationship level.

d. TSS Result

Table 6. TSS Value of Image

Titik	TSS Pleiades	TSS Landsat July 4, 16	TSS Landsat Feb 13, 17
KK1	37.323	31.892	26.822
KK2	23.919	21.318	24.799
KK3	30.525	25.318	24.872
PGN1	37.311	29.015	26.025
PGN2	38.528	35.015	30.119
PGN3	32.569	27.907	26.052

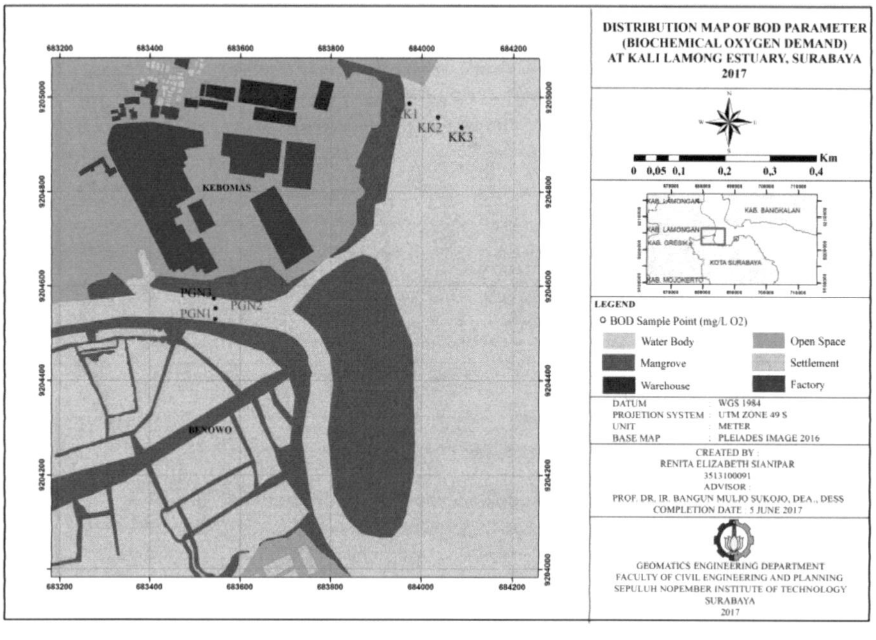

Figure 6. Distribution Map of BOD Parameter

4. Conclusion and Recommendation

4.1 Conclusion

This research above can be concluded as follows:

1. Based on the results of the sampling test of each parameter, there are 4 points in Lamong River which are indicated to be polluted according to BOD parameter. Thus it can be concluded that Lamong River is polluted and does not meet the requirements of class IV qualification based on PP. 82 Year 2001 due to unbound BOD parameter level.
2. The more to the estuary then the BOD level will be lower
3. BOD levels are related to household waste and domestic waste

4.2 Suggestions

1. It is necessary to increase the awareness of people about disposing of garbage where it should

not disturb water quality

2. It is necessary to develop the latest TSS algorithm because the algorithm that researchers use is the old algorithm because the current algorithm does not show good results

References

[1] Danoedoro P 1996 *Pengolahan Citra Digital: Teori dan aplikasinya dalam bidang Penginderaan Jauh* (Yogyakarta: Fakultas Geografi UGM)

[2] Abidin H Z 2002 *Survei dengan GPS.* (Jakarta: Pradnya Paramita)

[3] Soenarmo S H 2009 *Penginderaan Jauh Dan Pengenalan Sistem Informasi Geografi Untuk Bidang Ilmu Kebumian* (Bandung: Institut Teknologi Bandung)

[4] Ministry of Forestry 2003 *Buku Panduan Rehabilitasi Lahan dan Perhutanan Sosial. Penerbit Badan Penelitian dan Pengembangan Kehutanan Departemen Kehutanan* (Jakarta)

[5] Purwadhi S H 2001 *Interpretasi Citra Digital* (Jakarta: Grasindo)

[6] Sugiyono 2013 *Metode Penelitian Kuantitatif, Kualitatif dan R&D* (Bandung: CV Alfabeta)

[7] Government Regulation 2001 *Peraturan Pemerintah No. 82 Tahun 2001 tentang Pengelolaan Kualitas Air dan Pengendalian Pencemaran Air* (Jakarta)

Acknowledgements :

This research was supported by Department of Geomatics Engineering, Faculty of Civil, Environment, and Earth Engineering, Institut Technology of Sepuluh Nopember (ITS), Surabaya Indonesia and Institut of Research and Public Service, Institut Technology of Sepuluh Nopember (ITS), Surabaya Indonesia

IOP Conf. Series: Earth and Environmental Science **165** (2018) 012016 doi:10.1088/1755-1315/165/1/012016

Priority of The Forestry Extension Officer Distribution Based on Geographic Information System In Environment Problem Solutions

Lidya Ernawati[1*] **and Sutrisno Anggoro**[2,3]

[1] Master Program of Environmental Science, School of Postgraduate Studies, Diponegoro University, Jl. Imam Barjo SH Pleburan, Semarang, Indonesia
[2] School of Postgraduate Studies, Diponegoro University, Jl. Imam Barjo SH Pleburan, Semarang, Indonesia
[3] Management of Water Resources Department, Faculty of Fisheries and Marine Science, Diponegoro University, Jl. Prof. Soedharto SH Tembalang, Semarang, Indonesia

Lidya Ernawati@yahoo.com

Abstract. Magelang Regency is one of the regencies in Central Java Province that has forests as much as 32% of its area. The forest area continues to drop along with the many threats to the existence of the forest. Deforestation, critical land and forest fire are a widespread environmental problem in protected areas and cultivation. This will have an impact on the environmental quality that is essential for sustainable living. To minimize the impact, community awareness efforts need to be conducted through extension activities. Therefore, the existence of forestry extension as the forming actors of the environment-caring community becomes very important. The current competence and placement of forestry extension is not ideal compared to environmental issues and community needs. Utilization of Geographic Information System (GIS) through priority distribution modeling can help overcome this.

Keyword : Environmental, Forestry extension, GIS

1. Introduction

Forest is a potential natural resource that should be preserved for its function and its advantages. Forest gives a huge contribution to mankind, either from the ecology side, economy and social side. Forest as the capital of national development, should be managed, preserved, protected and utilized sustainably for the present and the future generation.

In Java, the recent issue happened is the increased amount of critical land, forest fire and also land cover change rate from forest into non-forest land/ deforestation. Statistics from KLHK [1] shows that the wide of critical land in Java on 2013 (critical & very critical category) is about 1.738.800 ha. The wide of forest & land fire in Java on 2015 reaches 11.263,69 ha. Deforestation land either inside or outside forest zone in Java during 2014-2015 is about 20.900 ha [2]. Those three things threatened the sustainability function of the forest.

The forest damage influence on human life. Impact of changes in land cover looks at changes in atmospheric temperature, humidity, clouds, circulation, and precipitation. On a regional scale, changes in land cover caused a decrease in temperature of 1 ° C in summer and -0.8 ° C in winter [3], has resulted in extreme conditions ie floods and droughts become more frequent with greater volume and wide [4]. Those damaged forest condition becomes a tough job in forest development. Ministry of Environment & Forestry commits an accelerated movement in order to decrease and rehabilitate the damaged forest through the Priority Policy on Forestry and National Development Program of United Indonesia Cabinet, one of them are the empowerment of people living close to the forest [5]. Through the empowerment program, people are encouraged to analyse the potential, the condition and forestry issues that needs to be resolved in order to create sustainable forest. The people empowerment program for forestry development is also executed by conducting a forestry extension.

Forestry extention is a knowledge, behaviour and attitude development process of the target community so they will know, want, understand, execute and manage the forestry business to increase their income and their welfare as well as to let the community active in preserving forest and the environment itself. [6]. The role of forestry extention officer is important yet strategic to persuade and coach the community to contribute into sustainable forest management. Sufficient number of forestry extention officer become a consideration for a successful extention.

Magelang Regency is one of the regencies in Central Java Province which suffer from damaged forest and lands. In Mount Merbabu National Park, the forest land closure only have 30% remaining from the whole area of the National Park with deforestation rate about 3% per year [7]. The wide of the critical land in Magelang Regency is about 24.907,8 ha. Forest fire recorded around 17 times and the width of the damaged area around 62,9 ha [8].

From the human resources itself, Magelang Regency only have 13 Forestry extension officer. This condition is far from ideal amount. The following chart shows the number of forestry extension gaps ideal and existing.

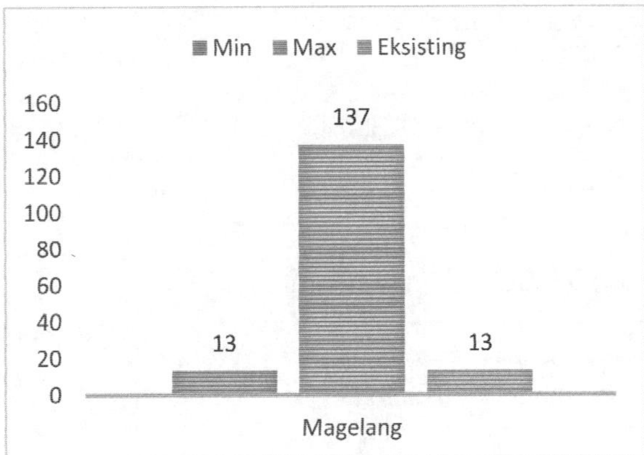

Figure 1. Proportion of Forestry Extension Officers in Magelang Regency

With the limited staff as well as the complicated issues, it is a must to have an information regarding the work area priority as the policy foundation for staff assignation. The purpose of this research also aim to create a mapping model of Priority Of The Forestry Extension Officer Distribution In Environment Problem Solutions in Magelang Regency according to Geographical System Information.

2. Methods

The approach used on this research is descriptive quantitative. The data being used is the critical land data, land cover data during 2000-2015 and hotspot data during 2012-2017. Method used to measure

each variable is Analytical Hierarchy Process (AHP). AHP is designed to collect the perception from person who is related to specific matters through a procedure to achieve a specific preference scale from various scales [9]. The perception as the main input came from the experts who understood about the matters given, know the effect of the issue or even have certain responsibility with certain issues. In this research, there are 3 respondents from structural officials and forestry extention officer. The AHP Method is needed to measure each indicator in order to decide the priority of the work area of the forestry specialist. AHP is necessary for weighting of each indicator in the determination of priority work areas of forestry extension. Converting data into a GIS database realize the results of AHP weighting and determine a score on a thematic map through the formulation of mathematical models.
Model of Priority Of The Forestry Extension Officer Distribution Based on Environment Problems generated by justification through weighting on each independent variable, that is:

a. Critical land
b. Deforestation
c. Hotspot

The result of weighting is the basic data in making priority map of placement of forestry extension workers. The classification is divided into 3 (three) based on the value of the weighted result, namely:

a. Low Priority.
b. Medium Priority.
c. Main priority.

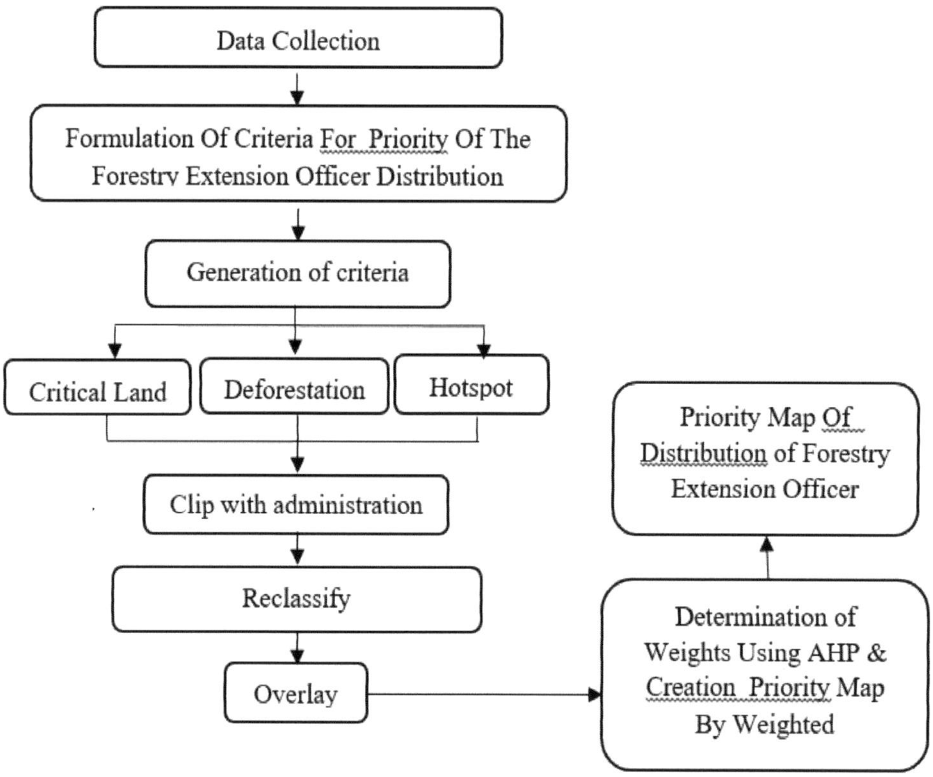

Figure 2. Skematic Representation of Methodology

3. Results & Discusion

The analysis process begins with determining the forest damage indicator, the calculation of indicator weight, and overlay process through GIS to generate priority distribution of forestry extension officer.

IOP Conf. Series: Earth and Environmental Science **165** (2018) 012016 doi:10.1088/1755-1315/165/1/012016

From interviews with experts, including variables of forest destruction are critical land, deforestation and forest fires.

3.1. Critical Land Distribution

Parameter of critical land is based on Permenhut Nomor P.32/Menhut-II/2009 which including land cover, land slope angle, erosion danger rate, productivity and management. Data distribution of critical land in Magelang Regency obtained from the Institute for Watershed Management Serayu Opak Progo. The classification of critical land for each district is presented in the following table:

Tabel 1. Critical Land Distribution per District

No	District	Non critical (ha)	Critical potential (ha)	Moderate (ha)	Critical (ha)	Total (ha)
1	Bandongan	1.481	3.102	1.981	4	6.567
2	Borobudur	2.939	176	1.478	1.092	5.685
3	Candimulyo	1.193	2.904	343		4.440
4	Dukun	442	2.839	2.351	12	5.644
5	Grabag	538	5.873	2.518	10	8.938
6	Kajoran	1.253	2.031	4.792	571	8.648
7	Kaliangkrik		1.394	2.084	1.888	5.365
8	Mertoyudan	4.557	45			4.602
9	Mungkid	3.773	329			4.102
10	Muntilan	2.194	823			3.017
11	Ngluwar	1.223	1.012	50	1	2.287
12	Pakis		4.586	4.732	1.352	10.670
13	Salam	1.348	1.277	218		2.843
14	Salaman	2.809	815	2.316	914	6.854
15	Sawangan	1.530	2.284	2.230	504	6.548
16	Secang	4.126	617	245		4.988
17	Srumbung	1.119	3.648	634	532	5.933
18	Tegalrejo	3.143	1.541	105		4.789
19	Tempuran	2.217	296	42		2.556
20	Windusari	438	2.793	2.526	1.337	7.093
	Total	36.323	38.383	28.645	8.218	111.569

Source: Critical Land Map of Magelang Regency (KLHK, 2017).

Critical lands are included in the class is not critical and potentially critical is 74.615 ha (66.87%). From 21 districts, almost all districts have either critical or potentially critical degraded land. Mertoyudan, Mungkid and Muntilan districts do not have critical land but only potentially critical. These districts are a sub-district in the lowlands with these district is in the lowlands with the land use of settlements and rice fields.

The area of critical land belonging to moderate critical is 28,645 ha (25,67%). Kajoran & Pakis district are the district that have largest area in moderate critical class. Kajoran & Pakis are the districts which located in the upper section of the river, the area with wave topograhy & steep slope. Land moderate critical category is land that has been eroded mild to moderate, including surface erosion and flow erosion. If it is left without any repair effort then in a relatively short time the land will become critical.

IOP Conf. Series: Earth and Environmental Science **165** (2018) 012016 doi:10.1088/1755-1315/165/1/012016

The land area of critical category is 8,218 ha (7.36%). Critical land is a soil that has experienced severe erosion, with a degree of gully erosion. They generally have medium-shallow solum (<60 cm) with a horizon A thickness less than 5 cm, slope 15 to> 30%. This condition illustrates the need for conservation cultivation techniques on lands classified as critical, such as conservation techniques bench terracing, rolls or alley cropping techniques [10].

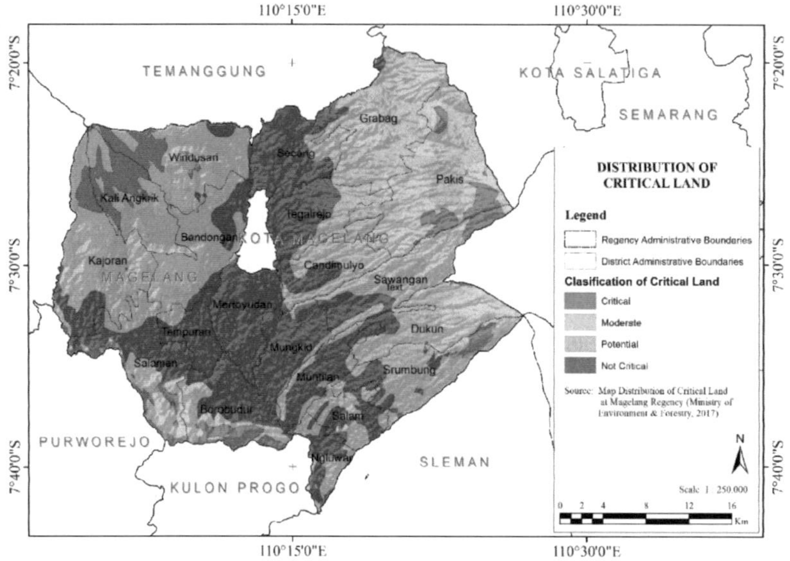

Figure 3. Critical Land Distribution Map

3.2. Deforestation

Deforestation is a condition of forest land cover change to be not forested [11]. Deforestation rate is based from the land cover map (during 2000-2015) which released by BPKH Region XI Java – Madura. Calculations performed on the condition of land cover in the coverage in 2010 was Forest while the coverage of the 2015 has been changed into Areas Non (Non-Forest) is reduced conditions the land cover in coverage in 2010 is not Forested (Non Forests), while the coverage of the 2015 changes to be Forested.

Table 2. Deforestation rate during 2000 - 2015

No	District	Deforestation (Ha)
1	Dukun	49
2	Pakis	97
3	Sawangan	104
4	Srumbung	15
5	Windusari	5
	Total (Ha)	271

Source: The results of the land cover map analysis from 2000 to 2015, Scale 1: 250.000 (BPKH Region XI Yogyakarta, 2017).

Of these results, it turns out deforestation occurs only in 5 districts. The highest deforestation occurred in the district of Sawangan. Sawangan is a district located between the upstream and downstream. in this district, the forest turned into rice fields and vegetable land. Sawangan, Dukun and Pakis known as district of vegetable suppliers for Magelang and surrounding districts.

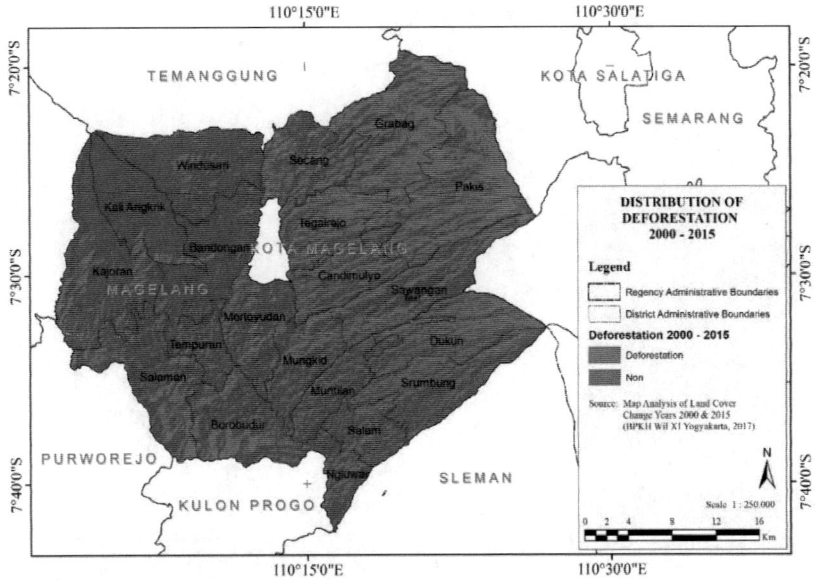

Figure 4. Deforestation Distribution Map from 2000 to 2015

3.3. Hotspot

Hotspot indicated that certain region is vulnerable to fire. The main objective of forestry extention is to inform the community about the danger of forest fire and encourage them to contribute to prevent forest fire. So, by knowing the hotspot distribution, forestry specialist could educate the community about how to deal with forest fire.

Table 3. Hotspot Distribution During 2011-2017

No	District	2011	2012	2013	2014	2015	2017	Total
1	Borobudur				2			2
2	Kajoran	1						1
3	Kaliangkrik	1	1			5		7
4	Mertoyudan	1		2		2		5
5	Pakis					3		3
6	Salaman				1			1
7	Sawangan					1		1
8	Secang		2	1				3
9	Srumbung					7		7
10	Tempuran						1	1
11	Windusari					3		3
	Total	3	3	3	3	21	1	34

Source: MODIS C6 NASA FIRMS Fire Archive, start date: 2010-01-01, end date: 2017-08-15, accessed on https://firms.modaps.eosdis.nasa.gov/download/tmp/DL_FIRE_M6_16167.zip, date August, 18, 2017 (NASA, 2017).16

The research result shows that minimum hotspot rate is found in Magelang regency. In the last 6 years, only 34 hotspots distribution found in 11 districts. Here are 3 districts (Srumbung, Kaliangkrik, Mertoyudan) with the highest score. This means that forest fires need to be wary of these three districts.

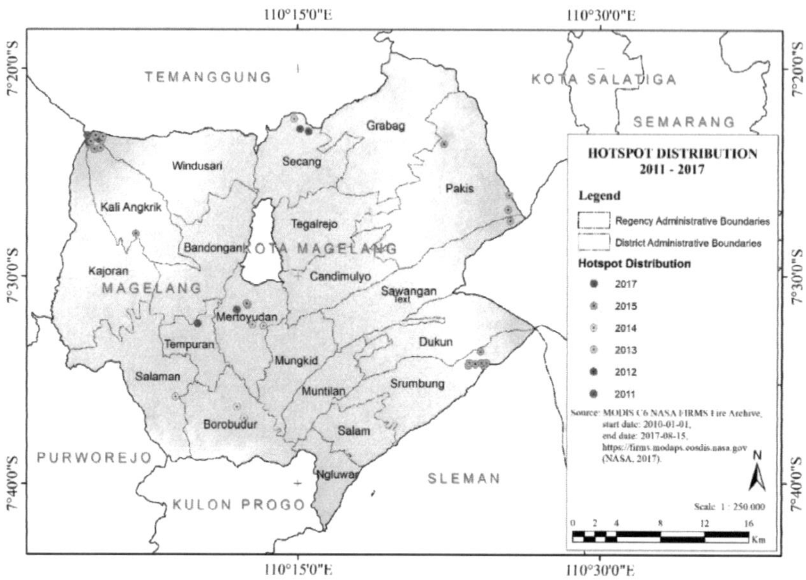

Figure 5. Hotspot Distribution Map

3.4. Priority of the forestry extension officer distribution.

Priority Of The Forestry Extension Officer Map is created from the overlay of critical land map, deforestation map and hotspot map.

Table 4. Score value for each indikator

Independent Variable			Score
Deforestation from 2000 – 2015/ (Ha)	Critical Land	Number of Hotspot from 2011 - 2017	
< 33	not critical / critical potential	< 3	1
33 s/d 66	moderate	3 s/d 5	2
> 66	critical	> 5	3

Note : To determine the classification of scores = Max value - Min value.

The next step is to reclassification based on the scores in table 4. The results of reclassification can be seen in the figure 6, 7 and 8. To determine the importance of each variable, the AHP Method is used. The weighting of each variable of the respondents can be seen from the table 5.

Table 5. Weighting of each variable of the respondents

Indicator	Respondent 1	Respondent 2	Respondent 3	Rate
Critical land	0,67	0,67	0,66	0,667
Deforestation	0,26	0,26	0,25	0,257
Hotspot	0,07	0,07	0,09	0,077

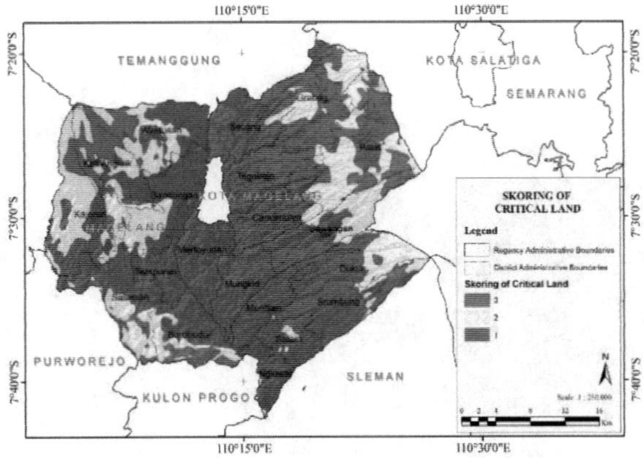

Figure 6. Skoring of critical land map

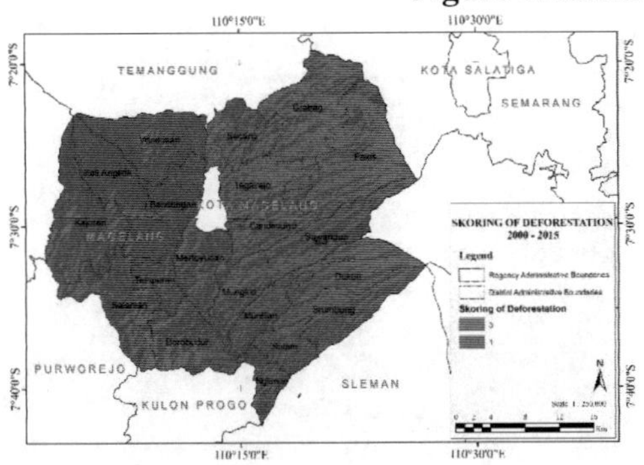

Figure 7. Skoring of deforestation map

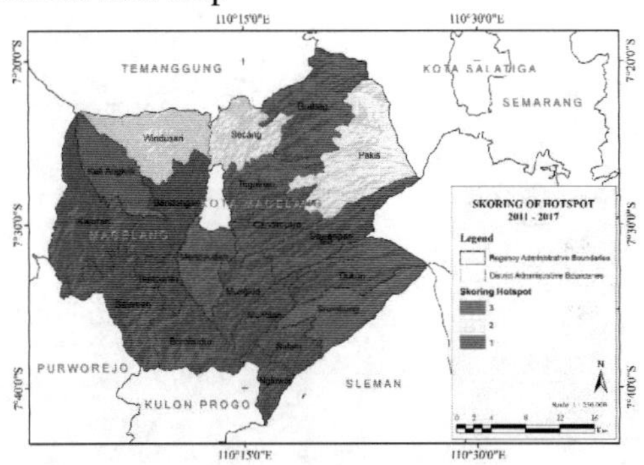

Figure 8. Skoring of hotspot map

Analysis of respondents' combined opinion shows that critical land indicator (weight value 0,667) is the most important aspect to consider in determining the priority of distribution of forestry extension officer. The following aspects to note are the indicators of deforestation (weighted value 0.257) and the hotspot indicator (weighted value 0.077).

This opinion reflects that the critical land indicator is considered very closely related to the area of forestry extension work. The preference of respondents to prioritize critical land indicators in the determination of forestry extension work area is logical because the role of extension workers as a motivator of farmers in improving the environment.

The resulting AHP rank is then converted into the SIG database for a given value. The weight of each indicator scores through the formulation of the mathematical model.

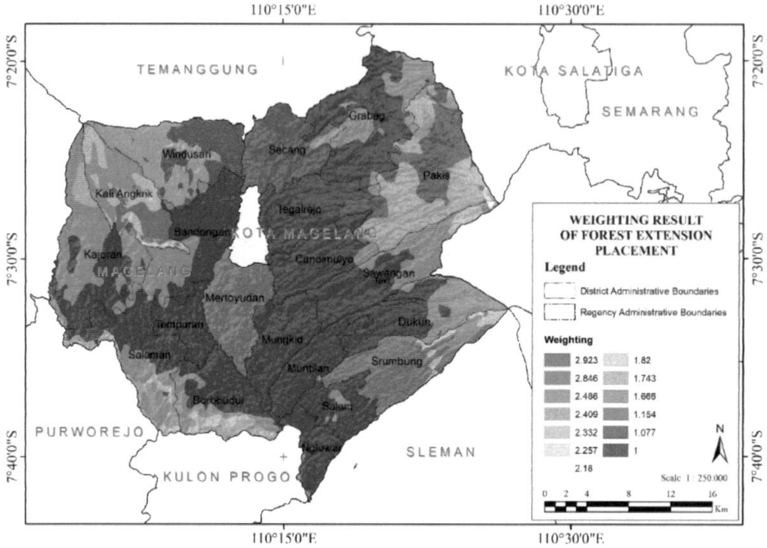

Figure 9. Weighting result of forest extension officer placement

Table 6. Area of area based on weighted results

No	Total Value of Weight	Area (Ha)
1	1.00	51.384,66
2	1.08	12.559,56
3	1.15	10.761,85
4	1.67	18.383,19
5	1.74	7.453,71
6	1.82	2.717,96
7	2.18	40,91
8	2.26	49,44
9	2.33	3.046,12
10	2.41	2.660,34
11	2.49	2.419,88
12	2.85	63,12
13	2.92	28,45
	Total	111.569,20

Note : Maks = 2.92; Min = 1.00; Mean = 1.43; Interval = 0.64

Furthermore, the scores above are grouped into three zones with the following scores: high priority has a score > 2.28, medium priority is having a score range of 1.61 - 2.28, and low priority has score value <1.641

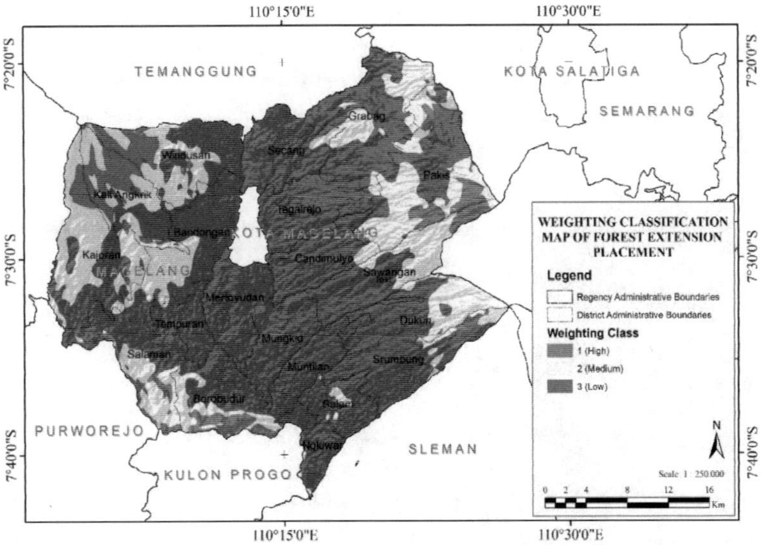

Figure 10. Weighting classification map of forest extention officer placement

The placement of forestry extension is at the district level, then the weighted value is calculated based on the smallest unit of work unit (district), thus yielding the following weighting values:

Table 7. Weighted value in each district

No	District	Weighted value	Placement Priority Forestry Extension Officer	Area (Ha)
1	Pakis	866,30	Tinggi	10,670.09
2	Grabag	483,55	Sedang	8,938.38
3	Srumbung	407,44	Sedang	5,932.57
4	Salaman	385,17	Sedang	6,853.92
5	Windusari	381,83	Sedang	7,093.02
6	Kaliangkrik	349,31	Sedang	5,365.32
7	Kajoran	257,57	Rendah	8,647.80
8	Dukun	239,27	Rendah	5,643.68
9	Sawangan	211,98	Rendah	6,548.50
10	Ngluwar	167,99	Rendah	2,287.15
11	Borobudur	130,95	Rendah	5,684.93
12	Salam	129,99	Rendah	2,843.49
13	Tegalrejo	129,66	Rendah	4,788.73
14	Bandongan	111,65	Rendah	6,567.34
15	Candimulyo	93,66	Rendah	4,439.70
16	Secang	52,87	Rendah	4,988.39
17	Muntilan	40,00	Rendah	3,017.02
18	Mertoyudan	39,24	Rendah	4,601.80
19	Mungkid	33,00	Rendah	4,101.83
20	Tempuran	25,00	Rendah	2,555.53

Note : Maks = 866; Min = 25; Median = 149; Mean = 227; Interval = 280

Priority of the forestry extension officer distribution in environment problem solutions are grouped into 3 (three) according to the weighting result, that is:
1. Low Priority (Value <305);
2. Medium Priority (305 ≤ Value ≤ 586);
3. Main Priority (Value> 586)

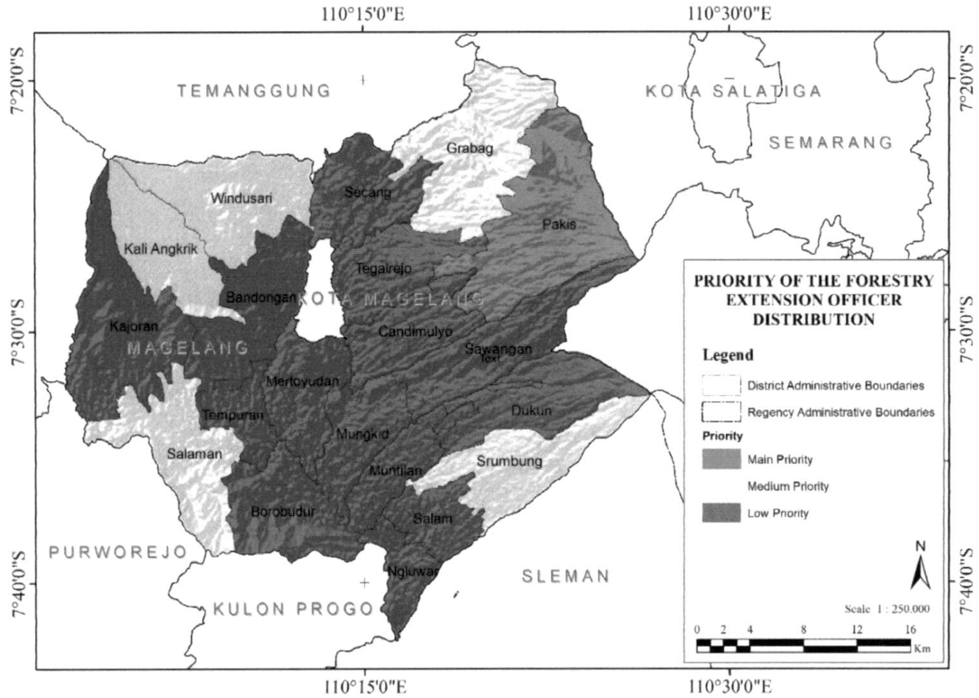

Figure 11. Priority Of The Forestry Etention Officer Distribution

Based on environmental problems that occurred in Magelang regency, there is 1 district which become the main priority of extension activity that is Pakis district, while Salaman, Srumbung, Grabag, Windusari, Kaliangkrik are districts with medium priority. Other districts are included in low priority. With a limited number of extension workers, the placement of extension workers is preferred to district with priority and medium priority. Other extension workers can be placed in districts with low priority with work area of 2-3 districts.

4. Conclusion

Priority of the forestry extention officer distribution could be created by a decisive system using AHP & GIS method. The distribution of the forestry specialist up until now is based with the principle of balance in each district. The result shows that the distribution of the forestry extention officer is not in accordance with the environmental issues occurred in Magelang regency. Work area priority is limited only to the environment condition. From the literature review and field observation, social condition could be enlisted as a parameter because the object of the extension activity is human itself.

References

[1] Kementrian Lingkungan Hidup dan Kehutanan 2015 Statistik Kementrian Lingkungan Hidup dan Kehutanan Tahun 2015 Jakarta
[2] Kementrian Lingkungan Hidup dan Kehutanan 2016 Deforestasi Indonesia Tahun 2014-2015. Jakarta : Direktorat Inventarisasi dan Pemantauan Sumber Daya Hutan

[3] Cao Q, Yu D, Georgescu M, Han Z and Wu J 2015 Impacts of land use and land cover change on regional climate : a case study in the agro-pastoral transitional zone of China. *Environ Res Lett.*;**10**

[4] Pawitan H 2014 Perubahan Penggunaan Lahan Dan Pengaruhnya Terhadap Hidrologi Daerah Aliran Sungai. Researchgate

[5] Peraturan Menteri Kehutanan Republik Indonesia No 10 Tahun 2011 Tentang 6 (enam) Kebijakan Prioritas Bidang Kehutanan dalam Program Pembangunan Nasional Kabinet Indonesia bersatu II.

[6] Peraturan Menteri Kehutanan Lingkungan Hidup dan Kehutanan Republik Indonesia No 36 Tahun 2015 Tentang Petunjuk Teknis Fungsional Penyuluh Kehutanan dan Angka Kreditnya.

[7] Alkaf M, Munibah K and Rusdiana O 2014 Model Spasial Perubahan Penggunaan Lahan Di Taman Nasional Gunung Merbabu dan Daerah Penyangganya. *Majalah Ilmiah Globe.* **16** (1):43–50.

[8] Dinas Pertanian Tanaman Pangan Peternakan dan Kehutanan 2015 Laporan Akhir : Penyusunan Rencana Kehutanan Tingkat Kabupaten (RKTK) Kabupaten Magelang Tahun 2015. Magelang

[9] Oktariadi O 2009 Penentuan Peringkat Bahaya Tsunami Dengan Metode Analytical Hierarchy Process. *J. Geologi Ind.* **4** (2) 103-16

[10] Anasiru R H 2016 Analisis Spasial Dalam Klasifikasi Lahan Kritis Di Kawasan Sub-Das Langge Gorontalo. *Informatika Pertanian* **25** (2) 261 - 72

[11] Kementrian Lingkungan Hidup dan Kehutanan 2015 Statistik Kementrian Lingkungan Hidup dan Kehutanan Tahun 2015 Jakarta

Spatial Analysis of The Correlation Between Hot Spot Distribution and Land Cover Type in Sumatra, Indonesia

Turmudi[1*], Yustisi A L G [1], Bambang Riadi[1], Yatin Suwarno[1] and Nugroho Purwono[1]

[1]Geospatial Information Agency. Jalan Raya Jakarta-Bogor Km. 46 Cibinong 16911 – Indonesia

turmudi.pokja@gmail.com

Abstract. Based on land cover data of 2015, Sumatra is dominated by shrubs, bushes, plantations with hardwoods, and plantations. It is the island with the largest peat distribution in Indonesia. Peatland forest map provided by Ministry of Environment and Forestry in 2015 represent that the total area of peatland in Sumatra is about 9.6 million hectares, mostly located in Riau, Jambi, and South Sumatra Province. Fire incidents can be indicated through hot spots distribution data. The data used is the distribution of hot spot of Sumatra in 2005-2016. The purpose of this study was to analyze the pattern of hot spot distribution and its relations with land cover changes. Overlay results indicate that 31.75% fires occur in shrubs and bushes, 20.87% in plantations with hardwoods, 16.60% in other natural cleared areas, 14.55% in forest crops, and the rest spreads in lowland crops, dry season crops, and swamp / peat forests. The highest hot spots are in 2005 and 2015 due to the El Nino effect that causes prolonged drought. Available land cover data indicates a change during 1990 through 2011. Most of peat land turn into plantation between 1990 to 2000. The result of overlay of hot spot distribution with land cover data from 2006-2015 shows that forest fires mostly occur in industrial plantation forest. This can be seen from the land cover classification where previously a plantation forest, then it is burned into a field and become an industrial plantation forest. Nevertheless, this analysis still need to be validated through ground truth to check the actual land cover and land use in field.

Keywords: Fire, Hot spot, Land cover, Sumatra

1. Introduction

Sumatra is one of five big islands in Indonesia. Based on land cover data in 2015, it is dominated by shrubs, bushes, plantations with hardwoods, and plantations. Also, it is the island with the largest peat distribution based on Ministry of Environment and Forestry in 2015. Peatland forest map provided by Ministry of Environment and Forestry in 2015 represent that the total area of peatland in Sumatra is about 9.6 million hectares, mostly located in Riau, Jambi, and South Sumatra Province. The aim of this study was to analyze the pattern of fires through hot spot distribution and its relations with land cover changes.

Fire incidents of land and forest in Indonesia in big scale started from 1982 until now. Big fire incidents happened periodically which was in 1982-1983, 1991, 1994, 1997-1998, 2006, and 2015 [1].

Forest fires can occur due to natural and / or human factors. Some experts consider that the more burnt-out area since 1991 happened as in consequence of the more natural forest changed into production forest, plantation, and so on. In 2015, fires particularly occurred in Sumatra, Kalimantan, and Papua which caused 80% of Sumatra and Kalimantan covered by thick smoke [1]. Based on research in Palangkaraya City, Central Kalimantan Province, forest and land fires densely happened in shrub and peat/swamp area at a distance about 1-2 km from residential or river and 0-1 km from the road [8]. National Agency for Disaster Management (2015) recorded that fires harmed the environment, caused economic loss, ecosystem destruction, germplasm loss, carbon emission, and the smoke spread through neighbor countries such as Singapore, Malaysia, Thailand, and South Philippine. Thereafter, the President issued an instruction about the enhancement for forest and land fire management (INPRES 11/2015). That involve more than twenty institutions coordinating to control forest and land fire through prevention, extinguish, and post-fire recovery. Furthermore, forest and land fire management had also been regulated in the Regulation of Environment and Forestry Minister (PERMENLHK) Number P.32/MenLHK/Setjen/Kum.1/3/2016 which the activity is organizing, managing the human resources and infrastructures, preventing, extinguishing, recovering, and supporting for evacuation and rescue.

The launch of NASA's Terra satellite in late 1999 marked a significant step forward in the ability to monitor fires from space. The satellite's sensor payload includes the Moderate Resolution Imaging Spectroradiometer (MODIS), an instrument having 1-km middle- and long-wave infrared bands designed specifically for the observation of actively burning fires [6]. The Terra satellite occupies a sun-synchronous polar orbit with local equatorial crossing times of 1030 (descending) and 2230 (ascending). A second MODIS instrument on NASA's Aqua satellite, launched in mid-2002, provides an additional pair of observations at 0130 (descending) and 1330 (ascending) local time [3]. The algorithm of Collection 6 MODIS active fire detection [4] is quite different with Collection 5 to solve the occurrence of false alarms caused by small forest clearings and the omission of large fires obscured by thick smoke [5].

The Aqua and Terra sensor from MODIS provide the information about hot spot and available in time series. It can be used to determine the hot spot distribution in Sumatra and to find out how the pattern changes periodically. The aim of this study was to analyze the spatial distribution of hot spot and the relation with land cover changes.

2. Method and materials

This research located at Sumatra. Data used was hot spot range from 2005 to 2016, base map, and land cover map range from 2006 to 2015. Hot spot distribution was downloaded from FIRMS using MODIS Terra/Aqua Satellite. Base map using 1:50,000 scale source from Indonesian Geospatial Information Agency (BIG). Layer that used from base map was boundary and residential building. Land cover map using 1:250,000 scale source from BIG for 2015 and Ministry of Environment and Forestry (KLHK) for 2006, 2009, and 2011.

Hot spot data was filtered based on the confidence level value to eliminate the outlier. All data over the confidence value of 90 would pass the filtering and the rest would be eliminated. The filtered data from 2005 to 2016 was statistically calculated per year to know when the hot spot mostly frequent. Then, the hot spot was plotted and the attribute table was joined with boundary layer from base map and land cover from BIG or KLHK. The overlay between hot spot, base map, and land cover map was performed for data in 2015, 2011, 2009, and 2009 separately due to land cover map availability. It would be used to perform the density value of hot spot by dividing the number of hot spot with land cover area. So, the area of land cover for each type should be calculated first and then followed by calculating the number of hot spot inside the land cover type boundary. The density value would determine which land cover type that have the most hot spot data in different year. Afterward, the statistic and density result would be used in the analysis process.

The analysis was done descriptively for all data processed by comparing the density in each year and take some sample in the overlayed map visually. The overlayed process was also done per province to find out which province experienced hot spot frequently. The result would refer whether there is a correlation or no between the region, land cover type, and the number of hot spot. The flow chart of this study is presented in the Figure 1.

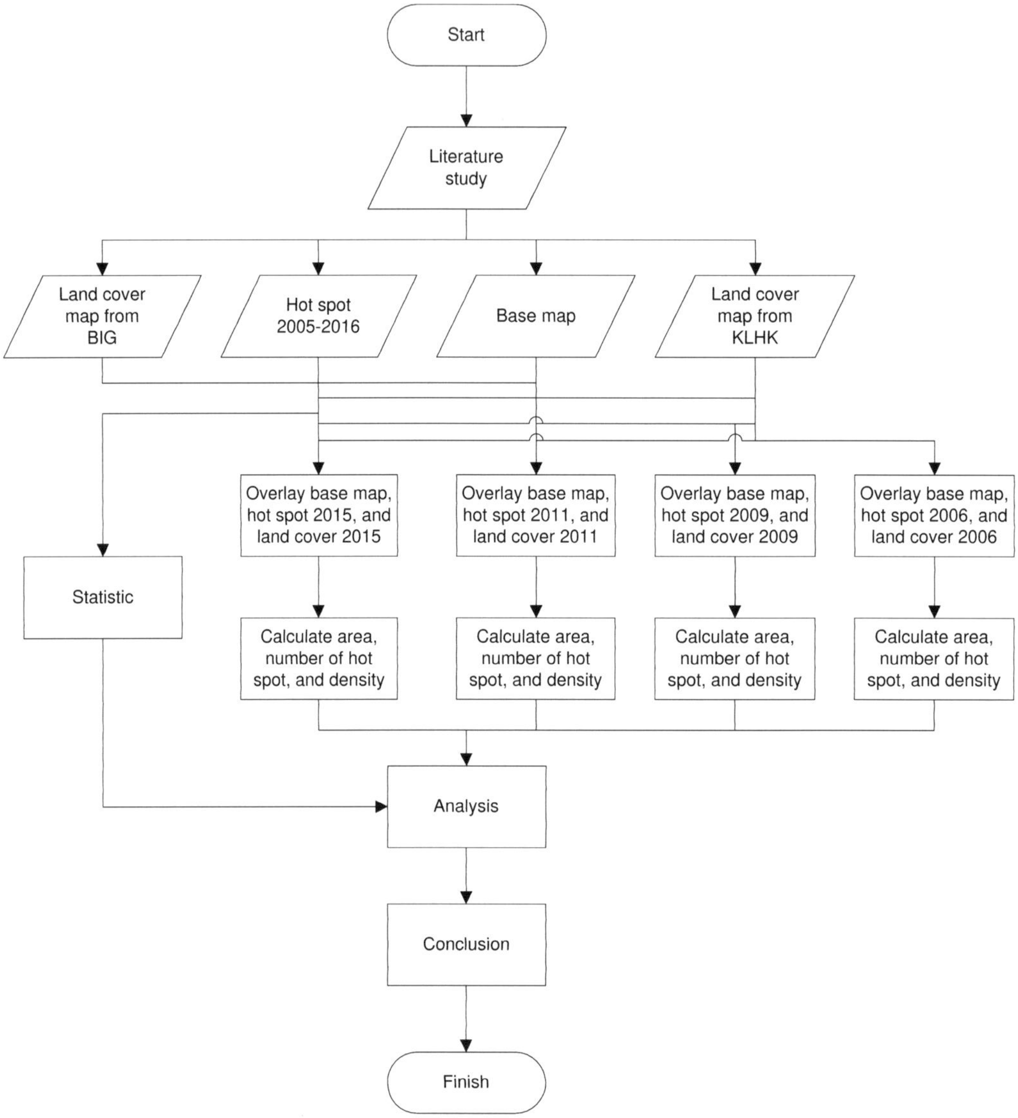

Figure 1. Flow chart of research methods

3. Result and discussion

3.1 Type of land cover in Sumatra

Indonesian Geospatial Information Agency (BIG) and Ministry of Environment and Forestry (KLHK) are government agencies which provide land cover data. BIG produces land cover into 22 classes while KLHK into 23 different classes. However, both of it have the same classification generally. The classification consists of shrub, plantation, forest, swamp/peat, open land, savanna, residential, and water.

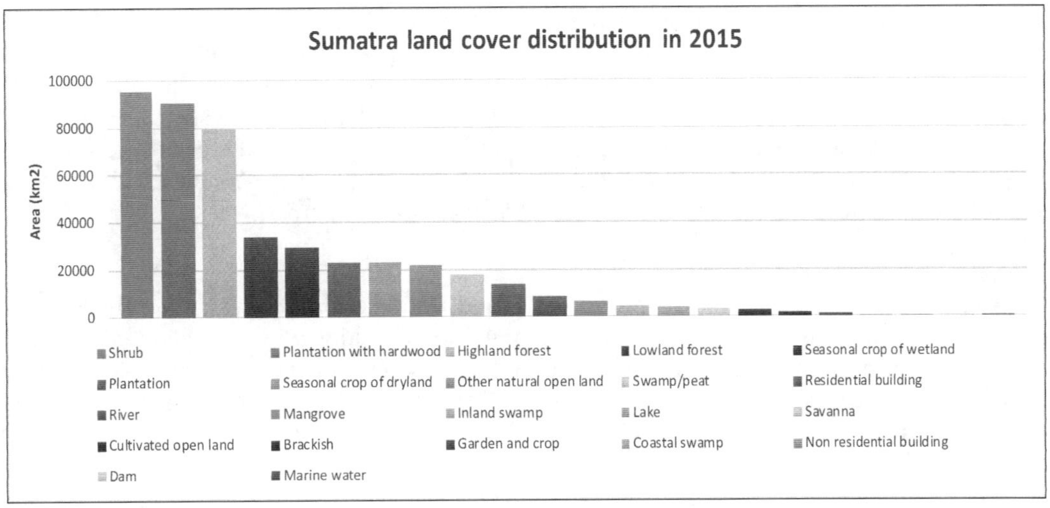

Figure 2. Type of Land cover in Sumatra

Based on land cover data of 2015, Sumatra is dominated by shrub, plantation with hardwood, highland forest, and plantation forest. Available land cover data indicates a change during 1990 through 2011. Most of peat land turn into plantation between 1990 to 2000. Then, some peat lands are classified as shrub in the land cover data.

3.2 Hot spot distribution in Sumatra from 2005 to 2016

Statistically, the number of hot spot varied each year and the most was in two periods that is 2005-2006 and 2014-2015. El Nino phenomenon was happened in those years and effected prolonged drought. It caused the land became easier to burnt so that hot spot detected by the satellite was significantly increase. The number of hot spot in 2014 increase 107.03% or more than twice from 2013.

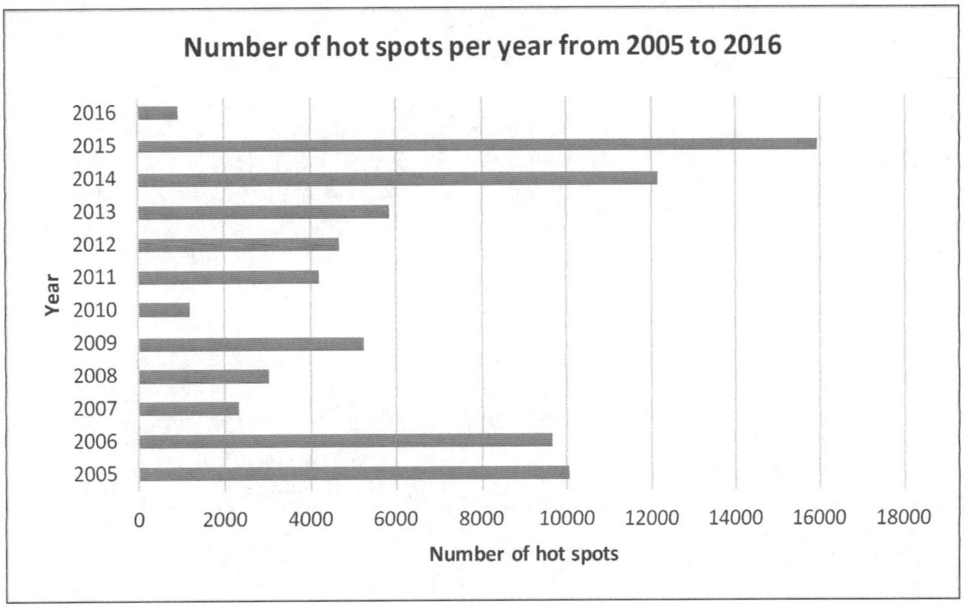

Figure 3. Distribution of hot spot per year from 2005 to 2016

IOP Conf. Series: Earth and Environmental Science **165** (2018) 012017 doi:10.1088/1755-1315/165/1/012017

Figure 3 shows that after El Nino phenomenon, the number of hot spot always decrease but increase again in some following years. If the hot spot in El Nino period was ignored, then the average number of hot spot each year detected by MODIS was 3,436 points. One of channel in satellite MODIS Terra/Aqua have resolution 250 m, it means that one hot spot point detected represent an area width 250 m x 250 m. Even though fire happen in the edge of pixel and the area is smaller than the resolution, the satellite will still detect fire in the middle of pixel as one point. If the average number of hot spot was converted into area, then the average of total area burnt was 214.75 km^2 each year.

3.3 Density of hot spot

Overlay results between land cover and hot spot in year 2015 indicate that hot spot mostly detected on plantation, shrub, other natural open land, swamp/peat, and plantation with hardwood. Meanwhile, if the density of hot spot was calculated then the most densities hot spot are on plantation, other natural open land, swamp/peat, shrub, and savanna (Table 1). The similar result was shown by overlay between land cover and hot spot in year 2006, 2009, and 2011. The highest densities hot spot was on swamp, open land, plantation forest, savanna, and shrub (Table 2).

Table 1. Density of hot spot in 2015

Land cover classification	Area (km^2)	Number of hot spots	Density (/km^2)
Plantation forest	23,234	5,286	0.228
Other natural open land	21,922	1,843	0.084
Swamp/peat	17,703	1,477	0.083
Shrub	95,088	4,937	0.052
Inland swamp	4,328	205	0.047
Savanna	3,190	112	0.035
Coastal swamp	341	11	0.032
Cultivated open land	2,690	37	0.014
Plantation with hardwood	90,372	1,212	0.013
Lowland forest	33,792	431	0.013
Seasonal crops of dryland	22,778	109	0.005
River	8,499	37	0.004
Seasonal crops of wetland	29,266	79	0.003
Lake	3,951	10	0.003
Mangrove	6,497	11	0.002
Residential building	13,642	23	0.002
Highland forest	79,607	103	0.001
Brackish	1,851	2	0.001
Plantation and crop	1,306	1	0.001
Non-residential building	66	-	-
Dam	10	-	-
Marine water	1	-	-

Table 2. Density of hot spot in 2006, 2009, and 2011

Land cover classification	Density (/km^2)		
	2006	2009	2011
Open land	0.135	0.071	0.044
Grove of swamp	0.098	0.052	0.032
Mining	0.061	0.018	0.029
Shrub	0.041	0.013	0.024
Secondary swamp forest	0.036	0.012	0.014
Savanna	0.032	0.011	0.011

Land cover classification	Density (/km^2)		
	2006	**2009**	**2011**
Mixed agricultural dryland	0.018	0.010	0.010
Swamp	0.013	0.010	0.006
Plantation forest	0.012	0.009	0.006
Plantation	0.009	0.006	0.005
Transmigration	0.009	0.004	0.003
Secondary dryland forest	0.009	0.004	0.003
Agricultural dryland	0.007	0.003	0.002
Primary swamp forest	0.005	0.002	0.001
Primary mangrove forest	0.003	0.001	0.001
Residential	0.002	0.001	0.001
Water body	0.002	0.001	0.000
Primary dryland forest	0.002	0.000	0.000
Paddy field	0.001	0.000	-
Secondary mangrove forest	0.000	0.000	-
Cloud	-	-	-
Pond	-	-	-
Airport/harbour	-	-	-

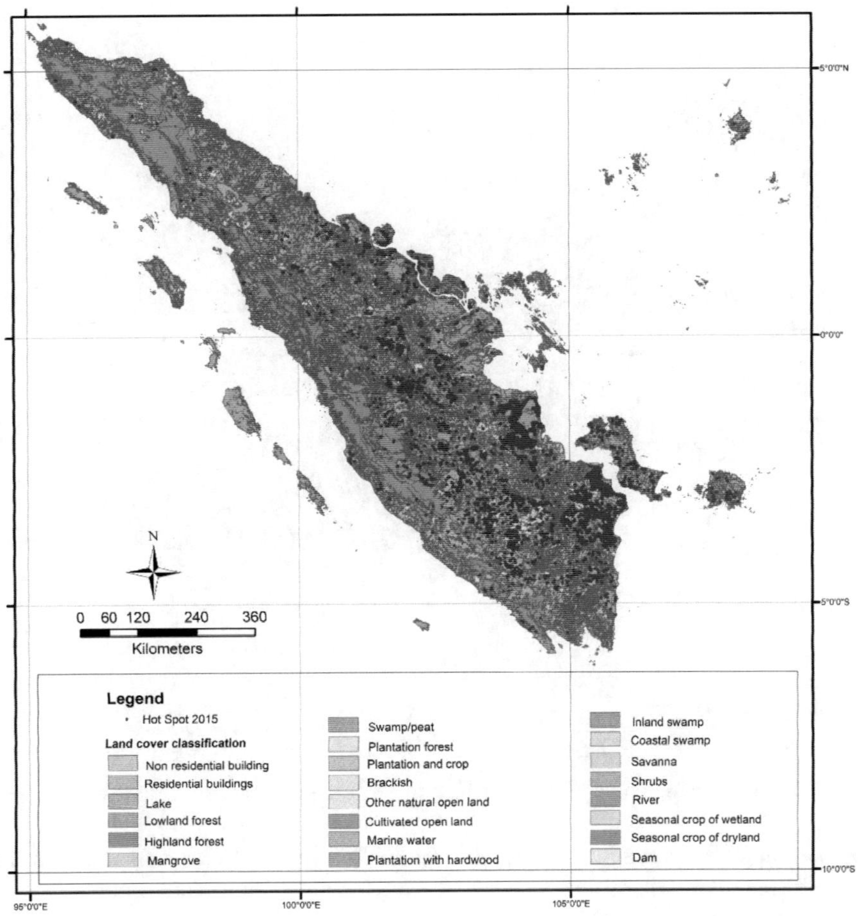

Figure 4. The overlay between land cover and hot spot in year 2015

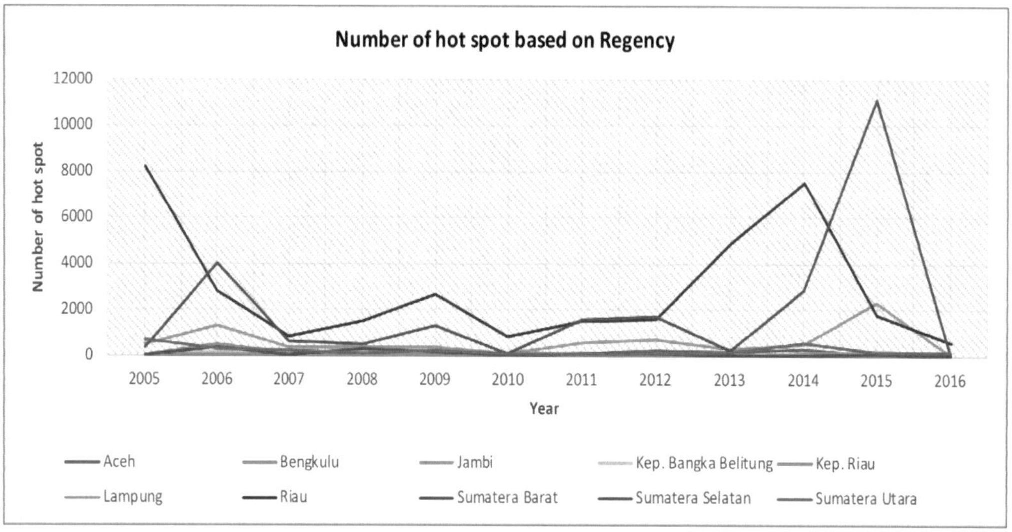

Figure 5. Distribution of hot spot based on Regency

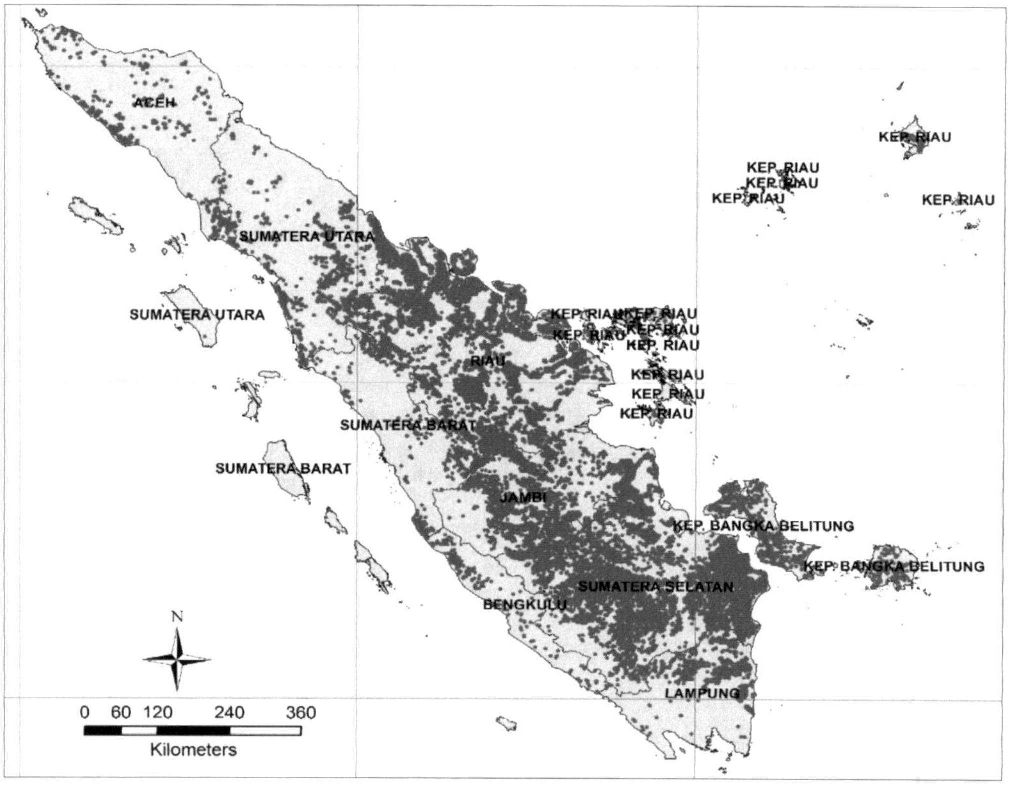

Figure 6. The spatial distribution of hot spot from 2005 to 2016

Hot spot distribution per year was clustered based on Province in Sumatra (Figure 3). The result was there are three from ten provinces in Sumatra that have the most significant number of hot spot, that is Riau, South Sumatra, and Jambi. Overlay results indicate that 31.75% fires occur in shrub, 20.87% in plantation with hardwood, 16.60% in other natural open land, 14.55% in forest crop, and the rest spreads in lowland crop, dry season crop, and swamp/peat forest. As much as 38% land cover in Riau was plantation with hardwood, 13% shrub, 12% swamp/peat, 10% plantation, 7% other natural open land. Meanwhile in South Sumatra, 30% land cover in the form of shrub, 16% plantation with hardwood, 11% plantation forest. Similar with Riau, Jambi has 22% plantation with hardwood, 17% highland forest, 13% shrub, 10% seasonal crop of dryland, 8% other natural open land.

Based on hot spot detected and land cover type, these three provinces have similarity which was hot spot mostly detected on shrub, plantation forest, other natural open land, and swamp/peat. Currently, the government has been used Satellite MODIS Terra/Aqua to monitor hot spot as initiation to detect fire. Then, the team will check on field whether the fire occurred or not. Some strategies had been applied as a quick reaction if the fire had been detected, such as waterbombing from the air.

Based on interviewed with Manggala Agni (2017), fire incident starts declining. Still, there are obstacle when detecting fire on peat area. It could not be detected through MODIS Terra/Aqua because the fire occurred in the ground and only causes smoke on the surface. Satellite MODIS Terra/Aqua do not detect smoke as hot spot and could not penetrate through thick smoke. Fire suppression on peat is also more difficult when the fire extinguisher should be injected underground. Sometimes when the fire seems to be extinguished, the fire underground has not been completely extinguished and will cause the fire to happen again.

4. Conclusion

Based on the spatial analysis, it can be concluded that fire incidents mostly occurred on plantation, open land, swamp/peat, shrub, and plantation with hardwood. Hot spot distribution in year 2006, 2009, 2011, and 2015 shows that fire happened all the time and on the similar land cover type. This indicate that fire occurred due to human factor. The result of overlay of hot spot distribution with land cover data from 2006-2015 shows that forest fires mostly occur in industrial plantation forest. This can be seen from the land cover classification where previously a plantation forest, then it is burned into a field and become an industrial plantation forest. Nevertheless, this analysis still need to be validated through ground truth to check the actual land cover and land use in field.

References

[1] Endrawati and Yusnita R 2015 *Analisis Data Titik Panas (Hotspot) Kebakaran Lahan dan Hutan Tahun 2015* Jakarta: Direktorat Inventarisasi dan Pemantauan Sumber Daya Hutan, Ditjen Planologi Kehutanan dan Tata Lingkungan Kementrian Lingkungan Hidup dan Kehutanan

[2] Fire Information for Resource Management System (FIRMS) [Online] Available: https://earthdata.nasa.gov/earth-observation-data/near-real-time/firms [Accessed 4 May 2017]

[3] Giglio L, Csiszar I, Justice C O 2006 Global Distribution and Seasonality of Active Fires as Observed with the Terra and Aqua Moderate Resolution Imaging Spectroradiometer (MODIS) Sensors *Journal of Geophysical Research*

[4] Giglio L 2015 MODIS Collection 6 Active Fire Product User's Guide, Revision A. Technical Report. University of Maryland [Online] Available: https://cdn.earthdata.nasa.gov/conduit/upload/3865/MODIS_C6_Fire_User_Guide_A.pdf [Accessed 1 September 2017]

[5] Giglio L, Schroeder W and Justice C O 2016 The Collection 6 MODIS Active Fire Detection Algorithm and Fire Products *Remote Sensing of Environtment*, 31-41

[6] Kaufman Y J, Justice C, Flynn L, Kendall J, Prins E, Giglio L, Ward D E, Menzel P and Setzer A 1998 Potential Global Fire Monitoring from EOS-MODIS *Journal of Geophysical Research*

[7] Manggala-Agni 2017 Interviewee Fire Incidents in Riau Province [Interview] 11 July 2017

[8] Mapilata E, Gandasasmita K and Djajakirana G 2013 Analisis Daerah Rawan Kebakaran Hutan dan Lahan dalam Penataan Ruang di Kota Palangkaraya, Provinsi Kalimantan Tengah *Majalah Ilmiah Globe*, 178-184

[9] Menteri Lingkungan Hidup dan Kehutanan 2016 Peraturan Menteri Lingkungan Hidup dan Kehutanan (PERMENLHK) Nomor P.32/MenLHK/Setjen/Kum.1/3/2016 tentang Pengendalian Kebakaran Hutan dan Lahan [Online] Available:

http://103.52.213.225/hukum/simppu-lhk/public/uploads/files/P.32.pdf [Accessed 30 August 2017]

[10] Peta Kesatuan Hidrologi Gambut [Online] WebGIS Kementrian Lingkungan Hidup dan Kehutanan [Online] Available: http://appgis.dephut.go.id/appgis/KHG/SUMATERA_KHGAMBUT.jpg [Accessed 28 August 2017]

[11] Presiden RI 2015 Instruksi Presiden Republik Indonesia Nomor 11 Tahun 2015 (INPRES 11/2015) tentang Peningkatan Pengendalian Kebakaran Hutan dan Lahan [Online] Available: https://www.setneg.go.id/index.php?option=com_perundangan&id=405271&task=detail&catid=5&Itemid=42&tahun=2015 [Accessed 30 August 2017]

Acknowledgment

Author would like to thank FIRMS for providing hot spot data all over Sumatra and Indonesian Geospatial Information Agency for the support and counsel so that this study can be done well.

Analysis of Orthorectification Accuracy of Pleiades and SPOT6 Images For Mapping Basic of RDTR Coastal Area (Case Study: District of Jenu, Tuban)

Bangun Muljo Sukojo[1] and Friska Melia Ode Binta[1]

[1]Department of Geomatics Engineering, Faculty of Civil, Environment, and Earth Engineering Institut Technology of Sepuluh Nopember

bangunms@gmail.com

Abstract. The coastal area of Tuban Regency in various spatial plans is planned to be one of the economic growth centers in East Java. The plan requires a detailed Spatial Plan (RDTR) map, especially the Coastal Area with high accuracy. However, the current RDTR map still lacks high accuracy. Therefore, the purpose of this research is to analyze the accuracy of high-resolution satellite image orthorectification, so that it can be used for making RDTR base map. In this research, orthorectification process is done by using satellite image data of Pleiades 1A and SPOT6 with Rational Polynomial Coefficient (RPC) method. As a supporting data are Ground Control Point (GCP) of 11 points and Independent Check Point (ICP) of 12 points and DEM Astrium Terra SAR-X data for satellite image altitude correction so that the image is upright. Based on the processing and data analysis, the accuracy horizontal image of Upright Pleiades 1A is 0.25 meters and the SPOT6 Upright image is 1.8 meters. Based on the horizontal accuracy value, the satellite images of Pleiades 1A and SPOT6 meet the horizontal accuracy standard of the base map scale of 1: 5000 and 1: 10,000 in accordance with Perka BIG No. 15 the year 2014. In addition, this research also conducted the perturbation of TSS of Pleiades 1A satellite image and SPOT6 using Laili algorithm 2015. From the result of image processing and ground truth data can be said that TSS concentration is low (<100 mg / l) according to Permen LH Number 01 of 2010. Thus can be said that marine waters in District of Jenu classified as clean and it does not affect shoreline changes. The result of correlation test of image TSS of Pleiades 1A and SPOT6 with ground truth data is very strong while image data of Pleiades 1A and SPOT6 with aerial photography data have low and very low correlation coefficient.

Keywords: Orthorectification, Pleiades 1A, SPOT6, TSS, Coastal Areas

1. Introduction

Tuban is one of the cities in the northern coastal area of Java Island which has 65 km long coastline and 22.608 km2 of sea area stretching from the east of Palang Subdistrict to the west of Bulu Bancar District[1]. Along the coast of Tuban Regency, there are many infrastructures and activity centers, including the main arterial road Pantura connecting East Java - West Java, ports, warehousing, industry, settlements, and tourism. The coastal area is planned as an integrated industrial area of East

Java with the development plan of the port, the development of fisheries and municipalities, as well as the construction of minapolitan areas[2].

As an economic and activity center, the coastal area of Tuban Regency should have good spatial use planning and in accordance with prevailing laws and regulations. Therefore, in order to avoid confusion that can cause damage to environment and ecosystem, hence needed detail map of spatial (RDTR) especially coastal area which has high accuracy. Today, however, many RDTR maps do not yet have high accuracy in accordance with the rules of the Perka BIG[3].

In making the RDTR map a high-resolution satellite image is required that has a wide coverage and can describe the physical appearance in accordance with the appearance in the field, thus providing accurate information. As is known that in the process of satellite image recording there are some error factors caused by shifts and slope and topographic variations that make coordinate information and image altitude not in accordance with the real situation. Therefore, orthorectification process is needed to reduce the various distortions caused by camera/sensor tilt and relief shift[4].

In addition to the orthorectification process also needs to be done accurate analysis orthorectification image of Pleiades and SPOT6 on field conditions. So research is needed to know the accuracy of orthorectification image of Pleiades and SPOT6 for Tuban Regency especially Jenu Subdistrict which have quite rapid development[2]. Therefore, in this final project, the authors will analyze the accuracy of orthorectification images of Pleiades and SPOT6 to give a recommendation for making the basic map of coastal area 1: 5000 and 1: 10,000 RDTR which refers to Tentang Pedoman Teknis Ketelitian Peta Dasar,2014[2].

2. Research Methodology

2.1 Location

Location of research in this final project is in District Jenu, exactly 10 km west of Tuban with an area of 81.62 km2. Geographically located at 6° 49'16 "LS - 111° 53'27" BT to 6° 53'10 "LS - 112° 3'2" BT. The administrative boundary is adjacent to the Java Sea in the north, Regency of Tuban in the east, Merakurak District in the south and Regency of Tambakboyo in the west.

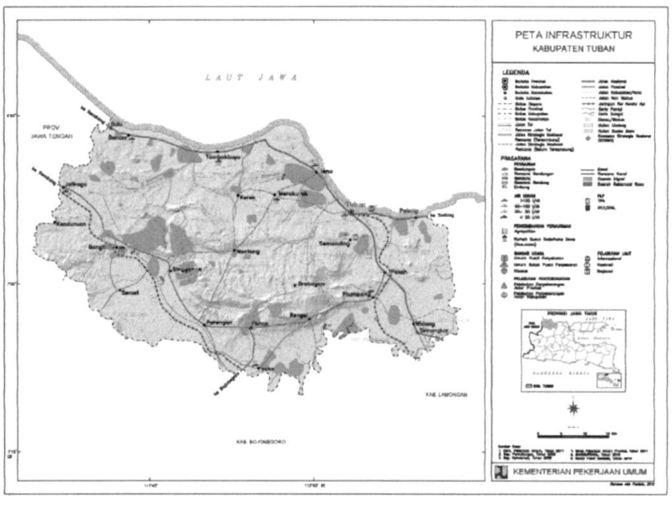

Figure 1. Location (Source: PU Tuban)

2.2 Data and Equipment

2.2.1 Data. (a) Satellite images of Pleiades 1A 2015; (b) Satellite images of SPOT6 2015; (c) DEM Astrium Terra SAR-X; (d) Map of Rupa Bumi Indonesia (RBI) 1:25000; (e) Aerial photo; and (f) Coordinate of GCP and ICP points from GPS measurement results

2.2.2 Equipment. Hardware: Geodetic GPS, Stative dan *tribrach,* Tape measure, Phantom 3 advance camera, Laptop, Stationery and survey form. Software : *Microsoft Office,* Satellite image processing software, Orthorectification processing software, Data processing software, *and* GPS data processing software

2.3 Flowchart of Data Processing

Stages of data processing in this research are as follows:

1. The data used in this final project are Pleiades 1A and SPOT6 satellite images, Astra Terra SAR-X DEM, GCP coordinate point data and ICP, RBI map 1: 25.000, aerial photo of the coastal area and sample test of TSS.
2. Pansharpening is done to obtain a high scope by combining multispectral and panchromatic channels.
3. Cropping area using RBI map 1: 25000 according to the condition of case study area of District of Jenu, Tuban.
4. The design of the net is made based on the points that have been obtained by connecting the point into a net and then calculated SoF ≤ 1.
5. Interpolate GCP to find coordinates of approaches that will be used for field measurement. GCP data from field measurements will be used in orthorectification processes previously processed using GPS software.
6. The process of orthorectification using Astrium DEM data and GCP coordinates obtained from field measurements using GPS. The orthorectification method used is the Rational Polynomial Coefficient (RPC) method with the tolerance value of RMSE ≤ 1.5 pixel [5].
7. The orthorectification result is an upright image of Pleiades 1A and SPOT6.
8. The ICP interpolation process is used to determine the coordinates of the approach to be used during field measurements. After field ICP measurements obtained field coordinates of ICP previously processed using GPS data processing software and will be used for the process of precision test.
9. The result of orthorectification is done by the accuracy test process using the coordinate of ICP point that has been measured in the field, where the accuracy test refers to Peraturan Kepala BIG Tentang Ketelitian Peta Dasar and Modul Validasi Peta Rencana Tata Ruang [3].
10. Satellite imagery TSS calculations using Laili algorithm, 2015 [6].

$$\text{TSS} = 31.42 * (\text{g}\rho\text{RED}/\log\rho\text{NIR}) - 12.719\ (\text{mg/l}) \quad (1)$$

    As the supporting data of the analysis of coastline change, then validate TSS field data from coastal air capture area and TSS sample test. The TSS sampling is done as below:
    a. Prepare bottles that have been covered with Duct tape, this is done for Reduce sun exposure directly.
    b. Bottle to be weighed for Used at a certain depth.
    c. Take a water sample (TSS) at that location has been determined.
    d. Label the bottle, not to be confused With another location.
    e. Then tested the TSS sample at Tuban DLH laboratory for getting TSS concentration value.
11. The result of TSS image calculation and TSS field validation data is done a correlation to know the relationship between the two data. The TSS consumption data is used to support an analysis of possible shoreline changes.
12. In the process of analysis, there are two data that is done analysis that is the result of horizontal accuracy test and result of correlation of TSS image and field data. Where the results of value TSS used as supporting data analysis of coastline changes.
13. The definitive map in question is a recommendation map for the RDTR basic map of the coastal area of Jenu District, Tuban 1: 5000 scale for Pleiades 1A and 1: 10,000 images for SPOT6 images.

3. Result and Analysis

3.1 Design and Strength of Figure

Determining the position of the ground control point (GCP) for geometric correction is required for the planning and calculation of the strength of the net (Strength of Figure (SoF). The smaller the net power factor number, the better the net configuration is concerned, and otherwisw[7]. The shape of the design of the net that has been formed can be seen in Figure 2.

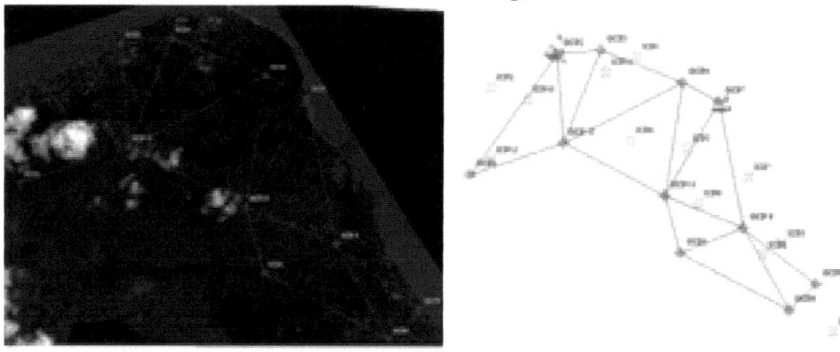

Figure 2. Design of Ground Control Point

To see the strength of the net formed, then calculated the strength of the net using the formula as follows:

Number of baselines: 31 Number of dots: 11
N_ size: Σ baseline x 3 = 93 N_ parameter: Σpoint x 3 = 33
U: (N_ size) - (N_parameter) = 60
SOF = 0.216

From the results of these calculations, it can be said that the design of the nets made has met the tolerance and considered strong[6].

3.2 Coordinates of Ground Control Point

Measurement of the ground control point in this study using GPS Geodetic with a static differential method using 3 GPS with duration of measurement ± 60 minutes. The measurement results are processed using GPS processing software including post-processing process (baseline processing) and network adjustment.

The GCP coordinates are tied to the coordinates of CORS ITSS and SRGI in the Tuban region, with SRGI coordinate difference measurements and SRGI BIG as follows dx = -4.306 meters dy = 0.391 meters. So obtained GCP coordinates as follows:

Table 1. Coordinates of Ground Control Point

Point ID	Coordinate GCP	
	X (m)	Y(m)
GCP 01	600491.450	9246507.632
GCP 02	603117.845	9251281.283
GCP 03	604623.091	9251459.875
GCP 04	607571.479	9250219.753
GCP 05	608088.481	9243562.887
GCP 06	612050.928	9241354.470
GCP 07	608909.184	9249528.496
GCP 08	612881.915	9242403.060

GCP 09	610201.187	9244564.668
GCP 10	607356.837	9245760.250
GCP 11	603632.382	9247795.181

3.3 Orthorectification

In this final project, orthorectification process is done using Rational Polynomial Coefficient (RPC) method. In the orthorectification process using data 11 coordinates GCP point and DEM Astrium Terra SAR-X as a height correction processed with orthorectification processing software so that obtained RMSE value.

From the result of orthorectification processing, the RMSE image of Pleiades 1A is 0.108 meters with details in Table 2. Where the RMSE values of both images have fulfilled the tolerance set by Perka BIG 2014[3].

Table 2. RMSE of Pleiades IA Image Orthorectification

Point ID			RMSE
	X (m)	Y(m)	(m)
GCP 01	600491.450	9246507.632	0.08
GCP 02	603117.845	9251281.283	0.04
GCP 03	604623.091	9251459.875	0.01
GCP 04	607571.479	9250219.753	0.04
GCP 05	608088.481	9243562.887	0.18
GCP 06	612050.928	9241354.470	0.21
GCP 07	608909.184	9249528.496	0.11
GCP 08	612881.915	9242403.060	0.16
GCP 09	610201.187	9244564.668	0.13
GCP 10	607356.837	9245760.250	0.11
GCP11	603632.382	9247795.181	0.1
		RMSE	0.108

While the value of the RMSE SPOT6 image of 0.642 meters with details in Table 3.

Table 3. RMSE of Pleiades IA Image Orthorectification

Point ID	**Coordinate of GCP SPOT6**		**RMSE (m)**
	X (m)	**Y(m)**	
GCP 01	600491.450	9246507.632	0,021527778
GCP 02	603117.845	9251281.283	0,024305556
GCP 03	604623.091	9251459.875	0,027083333
GCP 04	607571.479	9250219.753	0,024305556
GCP 05	608088.481	9243562.887	0,06875
GCP 06	612050.928	9241354.470	0,040277778
GCP 07	608909.184	9249528.496	0,059027778
GCP 08	612881.915	9242403.060	0,024305556
GCP 09	610201.187	9244564.668	0,022916667
GCP 10	607356.837	9245760.250	0,025694444
GCP 11	603632.382	9247795.181	0,09375
	RMSE		**0,44583333**

3.4 Test of Accuracy

In this research, there are 12 points of accuracy using ICP point according to Modul Validasi Peta Rencana Tata Ruang[4]. The distribution of ICP points can be seen in Figure 3.

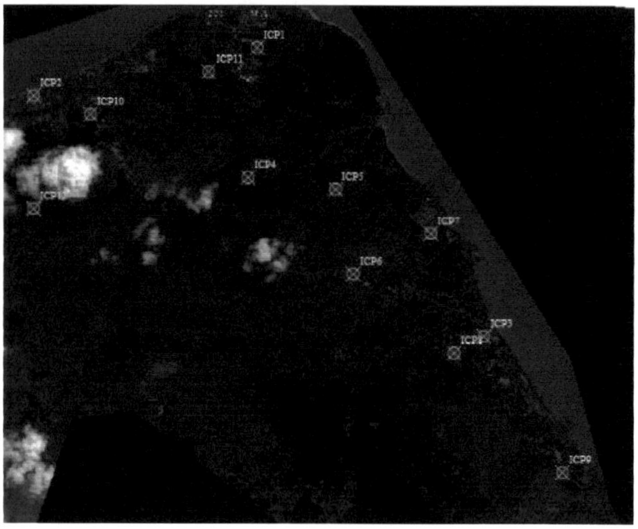

Figure 3. Distribution of ICP

The measurement of ICP point using Geodetic GPS and the method that used is rapid static with duration ± 15 minutes. The measurement result is processed using GPS processing software including post-processing process (baseline processing) and network adjustment.

These ICP coordinates are tied to the coordinates of CORS ITSS and SRGI in the Tuban region, with SRGI coordinate difference measurements and SRGI BIG as follows:

dx = -4.306 meters dy = 0.391 meters

So we get the coordinates of ICP as follows:

Table 4. Coordinate of ICP

Point	Coordinate ICP	
	X (m)	Y(m)
ICP 01	605904.661	9251229.177
ICP 02	600895.765	9249945.277
ICP 03	611581.194	9243931.582
ICP 04	605975.440	9247901.147
ICP 05	607964.143	9247642.874
ICP 06	608545.470	9245485.737
ICP 07	610194.207	9246532.696
ICP 08	610939.411	9243523.84 3
ICP 09	613761.734	9240444.200
ICP 10	602254.210	9249478.350
ICP 11	604842.409	9250578.954
ICP 12	601166.593	9247059.155

To know the accuracy of the result of orthorectification done the calculation of horizontal accuracy test using ICP point. And the value of RMSE ICP image of Pleiades 1A was 0,259 meters, with details in tables 5.

Table 5. RMSE of Pleiades

	Coordinate ICP Pleiades 1A		RMSE (m)
Point ID	X (m)	Y (m)	
ICP 01	605904.661	9251229.177	0.04
ICP 02	600895.765	9249945.277	0.31
ICP 03	611581.194	9243931.582	0.24
ICP 04	605975.440	9247901.147	0.33
ICP 05	607964.143	9247642.874	0.22
ICP 06	608545.470	9245485.737	0.2
ICP 07	610194.207	9246532.696	0.17
ICP 08	610939.411	9243523.843	0.12
ICP 09	613761.743	9240444.200	0.21
ICP 10	602254.210	9249478.350	0.11
ICP 11	6044842.409	9250578.954	0.05
ICP 12	601166.593	9247059.155	0.03
	RMSE		0.259

While the value of RMSE ICP SPOT6 image of 1,195 meters, with details in Table 6.

Table 6. RMSE of ICP SPOT6

	X (m)	Y (m)	(m)
ICP 01	605904,661	9251229,177	1,18
ICP 02	600895,765	9249945,277	2,8
ICP 03	611581,194	9243931,582	2,4
ICP 04	605975,44	9247901,147	2,51
ICP 05	607964,143	9247642,874	1,3
ICP 06	608545,47	9245485,737	0,22
ICP 07	610194,207	9246532,696	0,3
ICP 08	610939,411	9243523,84 3	1,06
ICP 09	613761,734	9240444,2	0,36
ICP 10	602254,21	9249478,35	0,31
ICP 11	604842,409	9250578,954	0,38
ICP 12	601166,593	9247059,155	1,41
	RMSE		**1,195**

In this research, the value of RMSE 0.259 meter for Upright Pleiades 1A and 1.195 meters for an Upright image of SPOT6. Based on the results of calculation of accuracy test using ICP point from field measurement result according to the rules (Perka BIG, 2014)[7], obtained RMSE 0.3 meter for Pleiades 1A vertical image and RMSE 1.8 meter for SPOT 6 upright image.

IOP Conf. Series: Earth and Environmental Science **165** (2018) 012018 doi:10.1088/1755-1315/165/1/012018

3.5 Accuracy Analysis

From the result of horizontal accuracy test of an upright image of Pleiades 1A and SPOT6 can be seen in table 7.

Table 7. Accuracy Image and Verification Value of Pleiades 1A and SPOT 6

Image Oerthorectificatified	Class 1		Class 2	
	H (m)	V (m)	H (m)	V (m)
Pleiades 1A	00.03	-	-	-
SPOT6	-	--	-	-

In accordance with Peraturan Kepala Badan Informasi Geospasial No. 15 of 2014 Tentang Pedoman Teknis Peta Dasar[7], calculation of horizontal accuracy of Upright image of Pleiades 1A and SPOT6. From the calculation of horizontal accuracy, the image value of Upright Pleiades 1A is 0.3 m, which means that the orthorectification result of the upright Pleiades 1A image is eligible and can be used for making RDTR base map of scale 1: 2,500 class 1 with 0.5m horizontal precision. While the image upright SPOT6 obtained horizontal accuracy of 1.8 m which means it has been eligible to be used for making a basic map of RDTR scale 1: 5000 class 3 with the condition of horizontal accuracy of 2.5 meters.

3.6 Calculation of Total Suspended Solid

In this research, the TSS of Pleiades 1A and SPOT6 satellite images use the algorithm from Laili 2015[5]. The TSS data is used as a supportive analysis of the possibility of shoreline changes. From the calculation, we get the TSS concentration as the detail in Table 8.

Table 8. Concentration Value of TSS on Pleiades 1A and SPOT 6

Point ID	Definition	TSS Image of Pleiades 1A (mg/L)	TSS Image of SPOT6 (mg/L)
UJI 01	0 m from coastal	17.109	12.701
UJI 02	100 m from coastal	24.066	26.121
UJI 03	200 m from coastal	20.729	22.123
UJI 04	0 m from coastal	24.737	21.03
UJI 05	100 m coastal	44.366	50.701

From the above study then it is presented in graphical form as shown in figure 4.

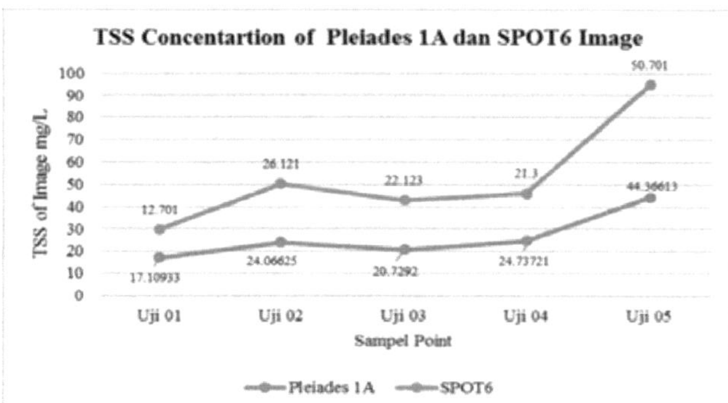

Figure 4. TSS Concentration of Pleiades 1A and SPOT 6 Image

From the above graphic can be seen that at the test point 05 located ± 100 meters from the beach has the highest TSS concentration, it is because the location of test point 05 is not far from the factory and the existence of sea transportation activity. While the lowest TSS concentration is at the test point 01 located on the edge of the beach, it is due to the location of the test point 01 remote with residential areas and estuaries.

3.7 Correlation Test of TSS Image and Groundtruth
In this study also tested the correlation between image and ground TSS value (ground truth). Imagery data used are Pleiades 1A satellite images August 21, 2015, SPOT6 June 2015 and field data April 14, 2017 with details as in Table 9.

Table 9 Concentration Value TSS Image of Pleiades 1A, SPOT6 and Ground Truth

Point ID	Value of TSS Pleiades 1A mg/L	Value of TSS SPOT6 mg/L	Value of TSS Ground Truth mg/L
Uji 01	17.109	12.701	19.8
Uji 02	24.066	26.121	23.4
Uji 03	20.729	22.123	19
Uji 04	24.737	21.3	26.3
Uji 05	44.366	50.701	45.9

The concentration of TSS from image processing of Pleiades 1A and SPOT6 with the result of ground truth needs to be done correlation test. This is done to determine the suitability of the relationship between two variables that is the result of image processing and field measurement data, so it is necessary to calculate the correlation. Graph of image correlation result of Pleiades 1A and field can be seen in Figure 5.

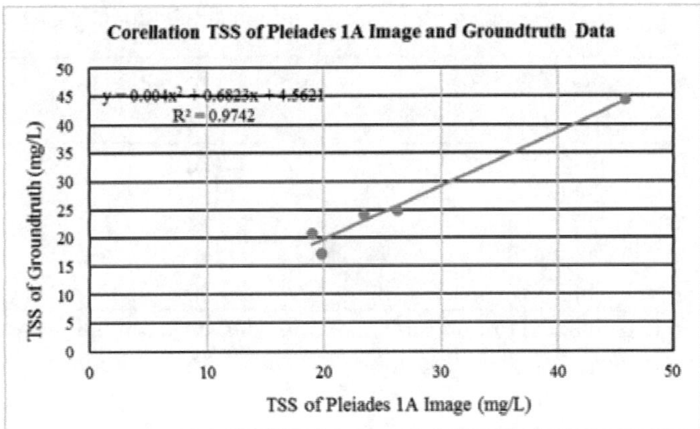

Figure 5. Correlation TSS Of Pleiades 1A and Groundtruth Data

While graph of the result of correlation calculation between SPOT 6 satellite image and field data can be seen in Figure 6.

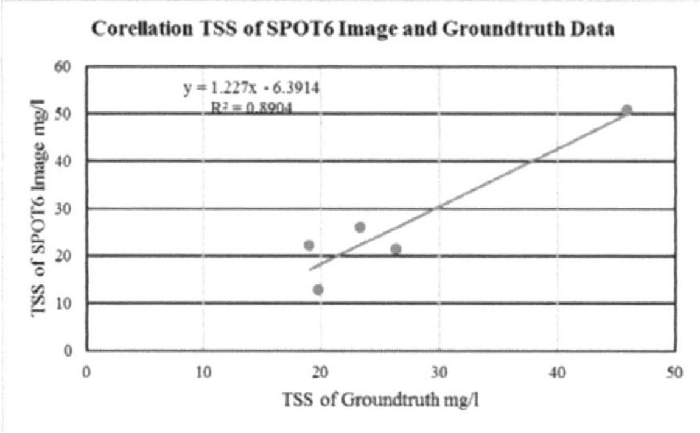

Figure 6. Correlation TSS Of SPOT6 and Groundtruth Data

The result of image correlation test of Pleiades 1A and field got correlation equal to (R) 0.98 whereas correlation test of SPOT6 and field got correlation equal to (R) 0.94. So it can be said that the correlation coefficient interpretation relation is very strong[8]. From the above calculations from both satellite imagery data and TSS ground truth data have low concentrations (<100 mg / l)[9]. So from the results of the waters of the District of Jenu classified as clean. It shows that TSS concentration does not have a big influence on the shoreline changes, especially in the marine waters of District Jenu. Because the researchers did not observe the tidal, so for the shoreline using interpretation image Pleiades 1A (09.54)[4].

3.8 Correlation Test of TSS Image dan Aerial Photo

In the research that has been done a correlation between the value of TSS Image and Air Photo, shown in Table 10.

Table 10. Concentration Value TSS Image of Pleiades 1A, SPOT6 and Aerial Photo

Point ID	Value of TSS Pleiades 1A mg/L	Value of TSS SPOT6 mg/L	Value of TSS Aerial Phpto mg/L
Uji 01	17.109	12.701	18.701
Uji 02	24.066	26.121	18.701
Uji 03	20.729	22.123	12.719
Uji 04	24.737	21.03	18.701
Uji 05	44.366	50.701	18.701

The correlation test is conducted to find out the relationship between the two data. Graphs of image calculation results of Pleiades 1A and aerial photograph can be seen in Figure 7.

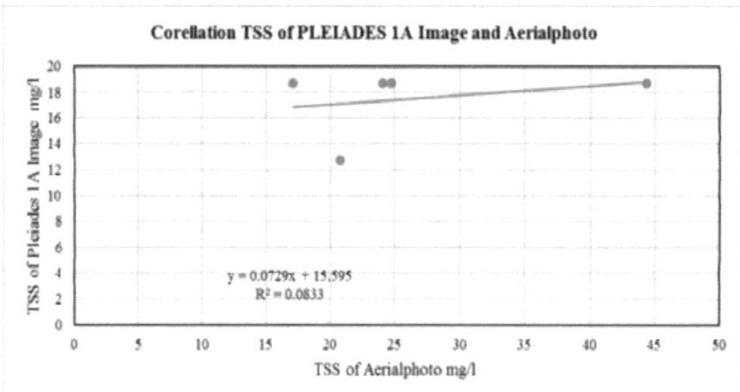

Figure 7. Graph of Correlation TSS Image Concentration Value of Pleiades 1A and Aerial Photo

IOP Conf. Series: Earth and Environmental Science **165** (2018) 012018 doi:10.1088/1755-1315/165/1/012018

From the graph above it is known that the correlation value (R) 0.28 so that it can be said that the correlation between TSS image processing Pleiades 1A and aerial photographs are low[8]. While the SPOT6 TSS image correlation test and aerial photographs can be seen in Figure 8.

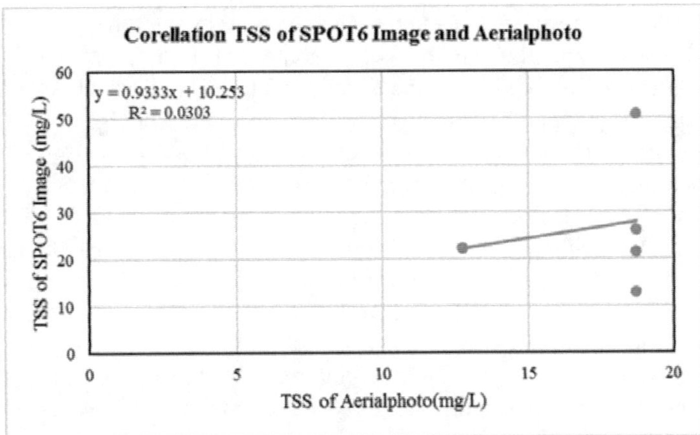

Figure 8. Graph of Correlation TSS Image Concentration Value of SPOT6 and Aerial Photo

From the graph above it is known that the correlation value (R) 0.28 so that it can be said that the correlation between TSS image processing SPOT6 and aerial photographs are very low[8].

3.9 Basic Map of RDTR District of Jenu, Tuban

The result of base map of RDTR of the coastal area of District of Jenu Tuban can be seen in Figure 9.

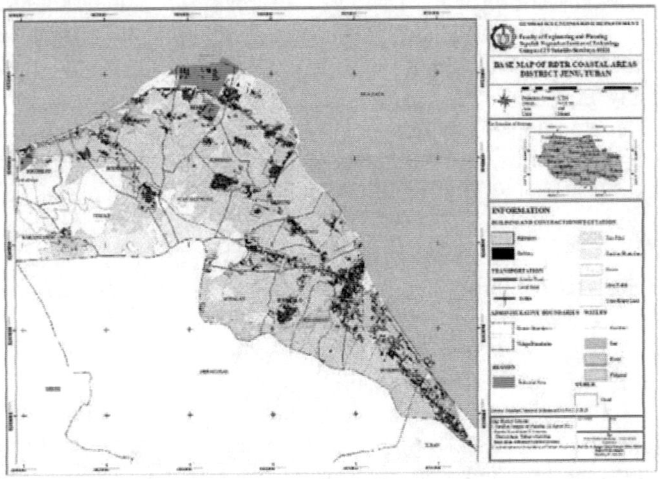

Figure 9. Base Map of RDTR Coastal Area of District Jenu, Tuban

The classification used under the rules of SNI RBI, 2010[10]. However, adjusted to the condition of the location of the case study that is District Jenu, Tuban.

4. Conclusion and Recommendation

Based on the results of the analysis that has been done in the previous discussion, it can be obtained conclusions from this study as follows:

1. In orthorectification process using 11 coordinate points of Control of Land (GCP), which got value RMSE image of Pleiades 1A 0.108 meter and SPOT6 0.642 image.
2. Based on Peraturan Kepala BIG No. 15 the year 2014[7], then the process of calculating the horizontal accuracy test using 12 coordinates of ICP point. With the horizontal accuracy of the image of Estuary Pleiades 1A 0.3 m, so the image of Estuary Pleiades 1A result of orthorectification is said to meet the requirement of RDTR base map of scale 1: 2,500 class 1. While SPOT6 has a horizontal accuracy of 1.8 m and qualified for RDTR 1 : 5000 class 3. The result will be recommended in making the base map of RDTR Coastal Area scale of 1: 5000 for the upright image of Pleiades 1A and 1: 10.000 Upright image SPOT 6.
3. The level of TSS concentration in the waters of the District of Jenu is low ie <100 mg / L[9]. So the low TSS concentration did not have a major effect on the shoreline changes in the District of Jenu. In the study of coastline, the determination is using an interpretation of the image of Pleiades 1A (09.54) which refers to Modul Validasi Peta Rencana Tata Ruang[4]. The correlation between ground truth data and image processing data of Pleiades 1A and SPOT6 is classified as very strong while correlation coefficient correlation data image of Pleiades 1A and SPOT6 and the aerial photograph is low and very low [8].

The suggestion given from the implementation of this research for subsequent research is to take the field data using aerial photographs should use aerial photographs that have a sensor Near Infra-Red (NIR) so that if processed using TSS algorithm produces a strong correlation between image TSS and TSS image processing air.

Reference

[1] Sihombing W H 2012 Kajian Kenaikan Muka Air Laut di Kawasan Pesisir Kabupaten Tuban Jawa Timur

[2] Utami V H and Pamungkas A 2013 Identifikasi Kawasan Rentan Terhadap Abrasi di Pesisir Kabupaten Tuban *Jurnal TEKNIK POMITS* **2**(2) 2337-3539

[3] Julzarika A 2009 Perbandingan Teknik Orthorektifikasi Citra Satelit SPOT 5 Wilayah Semarang dengan Metode Digital Mono Plotting (DMP) dan Metode Rational Polynomial Coefficients (RPCs) Peneliti Bidang Bangfatja Pusbangja (Jakarta: LAPAN)

[4] Badan Informasi Geospasial 2016 *Modul Validasi Peta Rencana Tata Ruang* (Bogor: Sekertariat BIG)

[5] Laili N 2015 Development of Water Quality Parameter Retrieval Algoritmhs For Estimating Total Suspended Solids and Chlorophyl-A Concentration Using Landsat-8 Imagery at Poteran Island Water

[6] Abidin H Z 2002 *Penentuan Posisi dengan GPS dan Aplikasinya* (Jakarta: PT Pradnya Paramita)

[7] Peraturan Kepala Badan Informasi Geospasial Nomor 15 2014 *Tentang Pedoman Teknis Ketelitian Peta Dasar* (Bogor: Badan Informasi Geospasial)

[8] Sugiono 2007 *Hipotesis Statistik* (Bandung: Universitas Pendidikan Indonesia)

[9] Paraturan Menteri Negara Lingkunga Hidup Nomor 01 2010 *Tentang Tata Laksana Pengendalian Pencemaran Air* (Jakarta: Menteri Negara Lingkungan Hidup)

[10] Standart Nasional Indonesia 2010 *Spesifikasi Penyejian Peta Rupa Bumi Indonesia-Bagian 2 Skala 1:25000* (Jakarta: Badan Standar Nasional)

Analysis of Accuracy Orthorectification Pleiades and QuickBird Imagery to Creation of Based Map of Detail Spatial Plan for Open Green Space (Case Study: Part of Tuban Urban Area)

Bangun Muljo Sukojo[1] **and Salwa Nabilah**[1]

[1]Department of Geomatics Engineering, Faculty of Civil, Environment, and Earth Engineering Institut Technology of Sepuluh Nopember

bangunms@gmail.com

Abstract. Appropriate Government Regulation No. 15/2010 concerning the Organization of Spatial, Spatial Plan area (RTRW) Regency/City must assign part of its territory and put together a plan of Spatial Detail (RDTR). Map of RDTR that exist today can be said to be less qualified and their use should be insufficient. One of RDTR is about Open Green Space (RTH). In Act No. 26/2007 of spatial mentions open green space should consist of 30% of the city area that is 20% and 10% of the public and the private. In making the RDTR map can be done by using very high-resolution satellite imagery previously orthorectification needed to be made in advance. Therefore, it needs to be done research on the accuracy of very high-resolution satellite imagery, satellite imagery can be known so that it can be used for making the based map of RDTR with scale 1:5000. This research was conducted in the Urban area of Tuban, East Java. To create a base map of the Spatial Detail of this Plan needs to be done orthorectification using data that are the Pleiades Imagery, QuickBird Imagery, DEM data of Astrium Terra SAR X with a resolution 9 m, the measurement point of GCP and ICP with Geodetic GPS, and limits Urban administration. GCP and ICP are used as much as 5 and 12 points. After a specified GCP point, can do calculations of SoF with results indicators of SoF $\leq$ 1, and obtained of 0.1883. Then do the process of orthorectification with the GCP has been tied to BIG's SRGI with RMSError $\leq$ 1.5 pixels. As for the value of the RMS Error of 0.548 pixels for Pleiades Image and 0.366 pixels for QuickBird Image. To test the results of orthorectification, performed the accuracy test using ICP with a horizontal accuracy of 1.5175xRMSE. Test the accuracy of obtained results of 0.463 m for the image of the Pleiades and 0.414 m for QuickBird Imagery. So it can be specified in the 1:5000 scale for the image of the Pleiades enters class 1 and for class 1 entry QuickBird Imagery.Keywords: Geographic Information System, Landsat 8 Satellite, Quickbird Satellite, Remote Sensing, SPOT Satellite, Tuban Regency.

Keywords: Accuracy Test, Open Green Space, Orthorectification, Pleiades Imagery, QuickBird Imagery.

IOP Conf. Series: Earth and Environmental Science **165** (2018) 012019 doi:10.1088/1755-1315/165/1/012019

1. Introduction

Appropriate Government Regulation No. 15/2010 concerning the Organization of Spatial, Spatial Plan area (RTRW) Regency/City must assign part of its territory and put together a plan of Spatial Detail (RDTR). RDTR should have been set at the latest 36 months as the Regional Spatial Regulation. One of RDTR is about Open Green Space (RTH).

In Law No. 26/2007 on spatial planning mentioned Open Green Space (RTH) is an elongated/line area and/or in groups, which use more open, where the plants grow, both of which grow naturally or are deliberately planted, which Consisting of 30% of urban areas should be Open Green Space consisting of 20% public and 10% private.

Tuban is one of the urban areas that do not have Detailed Spatial Plan map, especially in Open Green Space. In order for the Tuban Urban Detailed Spatial Plan map to be made soon, there is a need for a basic map as a reference for making Detailed Spatial Plan maps, especially Green Open Space which is very needed in urban areas.

Basic Detailed Spatial Plan map creation must come from very high-resolution satellite images such as Pleiades, QuickBird, WorldView, GeoEye, and IKONOS. The process of recording satellite imagery is covered from rides (satellites) moving over the Earth's surface at hundreds of kilometers, thus causing satellite imagery to have geometric distortions. To reduce the effect of geometric distortion on the image is done geometric correction that can make the object on the image more in accordance with the actual object.

Orthorectification is the process of geometric correction of satellite imagery or aerial photographs to correct geometric errors of an image originating from topographic, sensor geometry and other errors. In previous research, orthorectification of high-resolution satellite images using RPC method and using DEM ALOS PALSAR with 12.5m spatial resolution in Surabaya [3]. Orthorectification in this research using Rational Polynomial Coefficients (RPC) method by using Digital Elevation Model (DEM) Astrium Terra SAR X which has 9m spatial resolution.

Therefore, in this final project examined the accuracy of very high resolution images of Pleiades and QuickBird with Ground Control Point (GCP) and Independent Check Point (ICP) measured using geodetic GPS with static differential method through orthorectification process for recommendation of making Open Green Space of Detailed Spatial Plan based map with scale 1: 5000 in Part of Tuban Urban area.

2. Research Methodology

2.1 Research Location

The location of this research is in Part of Tuban Urban Area, Tuban Regency, East Java, which is centered on geographic coordinates 6° 53' 41" S and 112° 02' 30" E covering five districts of Tuban District, Jenu District, Palang District, Merakurak District, and Semanding District. Not all villages of the district are included in the Tuban Urban, but only a few. For Tuban District all villages, Jenu District 1 village, Palang District 6 villages, Merakurak District 1 village, and Semanding District 10 villages.

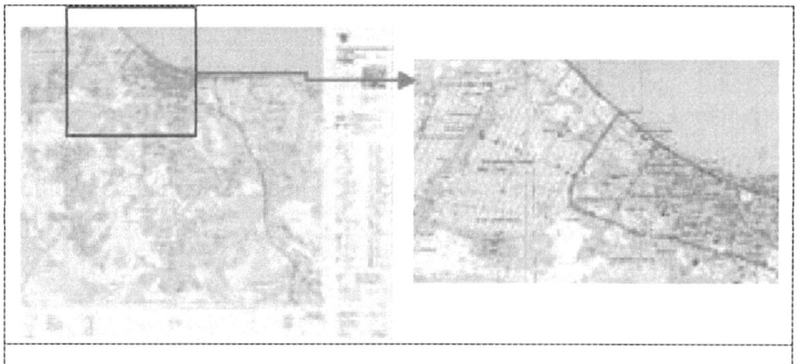

Figure 1. Research Location in Part of Tuban Urban Area

IOP Conf. Series: Earth and Environmental Science **165** (2018) 012019 doi:10.1088/1755-1315/165/1/012019

2.2. Data and Equipment

2.2.1 Data

The data used in this research are:

1. Tuban Pleiades imagery August 2015
2. Tuban QuickBird imagery October 2013
3. DEM Astrium Terra SAR X with a 9m spatial resolution
4. Data of Tuban Urban Administration boundary from RBI scale 1: 25000
5. GCP and ICP data from Geodetic GPS survey 2017

2.2.2 Equipment

Equipment used in this research are:

a. Hardware
   1. Geodetic GPS
   2. Survey Form

b. Software
   1. Geospatial processing software
   2. Satellite Imagery processing software
   3. GPS data processing software
   4. Data calculation software

2.3 Stages of Data Processing

Stages of data processing in this study are shown in Figure 2. The explanation of the flowchart as follows:

1. Data Collection
   At this stage of data collection were used that Pleiades Imagery, QuickBird Imagery, DEM Astrium Terra SAR X, Tuban Urban administrative boundary data, and the data point GCP and ICP were measured using geodetic GPS.
2. Pansharpening
   Do the process of merging panchromatic and multispectral imagery to produce very high-resolution images that have color for easier identification in the laying point GCP and ICP.
3. Making Network Design
   Make GCP and ICP of image interpretation that will be used for making networks and calculated strength are usually referred SoF value ≤ 1.
4. Field Survey
   In a field survey carried out measurement points of GCP and ICP using geodetic GPS, then obtained coordinates of GCP and ICP that were later used to orthorectification process.
5. Orthorectification
   After the obtained coordinates points of GCP and ICP and gained a good SoF value, it will be done using the orthorectification process with DEM Astrium Terra SAR-X which is useful for correcting satellite image of a geometric errors tolerance values can be seen from his error RMSE $\leq$ 1.5 pixels.
6. Accuracy Test
   The accuracy test using ICP point coordinates that have been made and which will determine the scale and grade on the resulting map. With tolerance CE90 $\leq$ 1 meter in class 1 with the scale 1:5000.
7. Cropping Area
   Satellite imagery has orthorectified then crop by the size of the corresponding region using the administrative boundaries of the RBI map with scale 1: 25000.

8. Classification of Open Green Space and Non
   After that do the classification by doing digitization. The resulting classification is the classification which is the area of green open space and non, the classification is carried out in accordance with the Regulation of the Minister of Public Works No. 05/2008
9. Based Map of RDTR for Open Green Space Creation with Scale 1: 5000
   After all, is done, the based map of RDTR for Open Green Space with scale 1: 5000, which will then be used for RTDR production, especially in the area of open green space in Part of Tuban Urban Area.

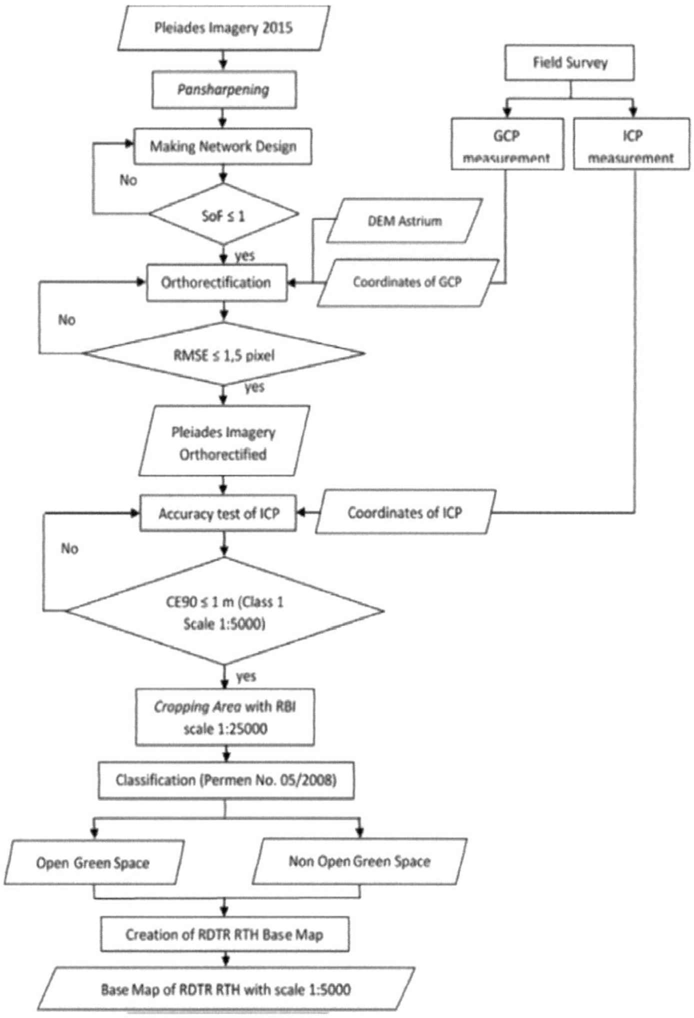

Figure 2. Flowchart of Research

3. Result and Analysis

3.1 Making Network from GCP and SoF Calculations

Good orthorectification depends on the strength of the networks that have been made. The closer to zero the better the SoF and the stronger the network. In this research, the design of networks with GCP is 5 points. From 5 GCP we get 7 baselines on the design of networks that have been made with a length of 3-4 km for a relatively flat area.

Figure 3. GCP Network Design

In this research using 5 point GCP and 12 point ICP, so Strength of Figure (SoF) got result equal to 0,1883 with calculation as follows:

Number of baseline : 7
Number of Points : 5
N size : Number of baseline x 3 = 13 x 3 = 39
Nparameters : Number of points x 3 = 5 x 3 = 15
u : Nsize - Nparameters = 39 - 15 = 24

$$SoF = \frac{\left[Trace\left([A^T] \ . \ [A]^{-1}\right)\right]}{u} \quad (1)$$

With SoF value, it can be said to enter in tolerance, where tolerance SoF ≤ 1.

3.2 Coordinates of GCP and ICP

The control point coordinate (X, Y) is obtained from satellite image interpretation and field measurements using GPS using a static method with duration of 30-45 minutes which is then processed through GPS data processing software including post processing (baseline process) and network adjustment.

The coordinates of GCP and ICP are tied to CORS ITS with ITSS code on the campus of ITS Surabaya and CTBN reference point in Tuban. Do the measurement point of CTBN then obtained coordinates then differencing with CTBN coordinates already fixed from BIG., Here is the difference obtained:

Table 1. Difference of CTBN Reference Point

CTBN	WGS84	611591.174	9243767.499
CTBN	SRGI 2013	611595.48	9243767.108
Differences		-4.307	0.391

Then from the difference is done adjustment to all coordinates GCP and ICP measurement with Geodetic GPS with using UTM coordinate system. Here is a list of GCP and ICP coordinates in SRGI:

Table 2. List of GCP Coordinates that field survey

Name of Points	East (m)	North (m)
GCP 01	611199.0355	9239278.634
GCP 02	612435.5095	9239625.138
GCP 03	612863.0395	9238136.419

GCP 04	615038.3985	9239216.128
GCP 05	615117.9465	9237055.242

Table 3. List of ICP Coordinates

Name of Points	East (m)	North (m)
ICP 01	611790.178	9239035.116
ICP 02	612775.385	9239699.464
ICP 03	612341.793	9238991.136
ICP 04	612482.56	9239267.626
ICP 05	613082.787	9239145.672
ICP 06	612764.403	9238295.941
ICP 07	613260.634	9238479.201
ICP 08	614717.44	9238872.364
ICP 09	614333.647	9239344.962
ICP 10	614437.088	9237593.941
ICP 11	614963.046	9238119.501
ICP 12	614932.938	9237005.816

3.3 Orthorectification

Once the GCP coordinates are known, then the image orthorectification process can be done. The orthorectification process uses RPC methods that require DEM data. The DEM used is DEM Astrium Terra SAR X with 9m spatial resolution.

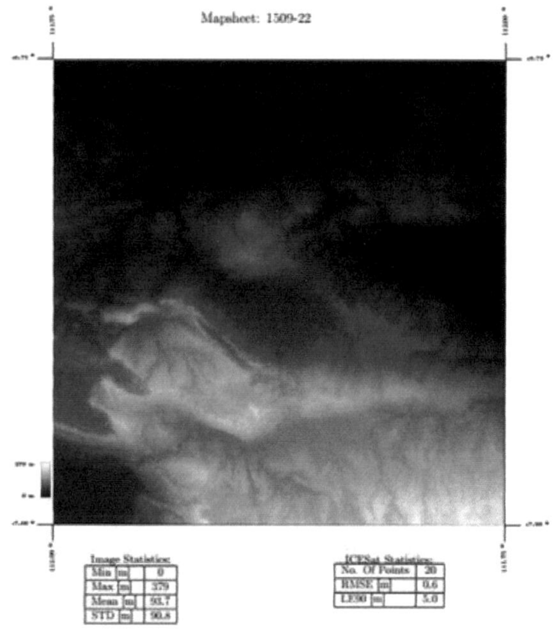

Figure 4. DEM Astrium Terra SAR X

From orthorectification process then got RMSE value which have fault tolerance RMSE ≤ 1.5 pixels.

$$RMSE_{horizontal} = \sqrt{\frac{D^2}{n}} = \sqrt{\frac{\sum[(x_{data}-x_{cek})^2 \ + \ (y_{data}-y_{cek})^2]}{n}} \tag{2}$$

Where :
n = Total number of checks on the map
D = The difference between the coordinates measured in the field with the coordinates on the map
x = Coordinate value on the – X axis
y = Coordinate values on the – Y axis
By using the above calculation, obtained the following results:

Table 4. RMSE of Pleiades Image Orthorectification

Name of Points	East (m)	North (m)	Residual (m)
GCP 01	611199.0355	9239278.634	0.57
GCP 02	612435.5095	9239625.138	0.24
GCP 03	612863.0395	9238136.419	1
GCP 04	615038.3985	9239216.128	0.61
GCP 05	615117.9465	9237055.242	0.32

Table 5. RMSE of QuickBird Image Orthorectification

Name of Points	East (m)	North (m)	Residual (m)
GCP 01	611199.0355	9239278.634	0,49
GCP 02	612345.5095	9239625.138	0,015
GCP 03	612863.0395	9238136.419	0,43
GCP 04	615038.3985	9239216.128	0,1
GCP 05	615117.9465	9237055.242	0,66
		RMSE	0,366

RMSE value obtained on Pleiades image of 0.548 and QuickBird image of 0.366 so it can be said to enter tolerance.

3.4 Accuracy Test

The accuracy test in this research is conducted by using 12 points of Independent Check Point (ICP) measurement coordinates according to the requirements given by BIG on spatial map validation module in 2016.

Table 6. RMSE of Pleiades Image Accuracy Test

Name of Points	East (m)	North (m)	Residual (m)
GCP01	611199.04	9239278.63	0.285
GCP02	612435.51	9239625.14	0.12
GCP03	612863.04	9238136.42	0.5
GCP04	615038.39	9239216.13	0.305
GCP05	615117.95	9237055.24	0.16
ICP01	611790.178	9239035.116	0.45
ICP02	612775.385	9239699.464	0.345
ICP03	612341.793	9238991.136	0.405
ICP04	612482.56	9239267.626	0.43
ICP05	613082.787	9239145.672	0.275
ICP06	612764.403	9238295.941	0.355
ICP07	613260.634	9238479.201	0.205
ICP08	614717.44	9238872.364	0.36
ICP09	614333.647	9239344.962	0.38
ICP10	614437.088	9237593.941	0.42
ICP11	614963.046	9238119.501	0.26
ICP12	614932.938	9237005.816	0.36
RMSE (meter)			0.305

Table 7. RMSE of QuickBird Image Accuracy Test

Name of Points	East (m)	North (m)	Residual (m)
GCP01	611199.04	9239278.63	0,221527778
GCP02	612435.51	9239625.14	0.098
GCP03	612863.04	9238136.42	0,19375
GCP04	615038.39	9239216.13	0.065
GCP05	615117.95	9237055.24	0,297916667
ICP01	611790178	92390351.16	0,234722222
ICP02	612775385	9239699464	0,078472222
ICP04	612482.56	9239267626	0,197222222
ICP05	613082787	9239145672	0,184027778
ICP06	612764403	9238295941	0,224305556
ICP 07	613260634	9238479201	0,17708333
ICP09	614333647	9239344962	0,119444444
ICP10	614437088	9237593941	0,15625
ICP11	614963046	9238119501	0,170138889
ICP12	614932938	9237005816	0,102083333
	RMSE (meter)		0,273

The horizontal accuracy test can be calculated using the calculations determined by BIG for the creation of a base map:
Horizontal accuracy test = 1.5175 x RMSE Using the above calculations, obtained 0.463 m for Pleiades and 0.414 m for QuickBird.

Table 8. Geometry RBI Map accuracy

Scale	**RBI Map Accuracy on Horizontal Position**		
	Class 1	Class 2	Class 3
1: 5000	1	1.5	2.5
1: 2500	0.5	0.75	1.25
1: 1000	0.2	0.3	0.5

From those value it can be said that Pleiades can be used as base map of scale 1: 5000 class 2, while QuickBird scale 1: 5000 class 1.

3.5 Base Map of RDTR for Open Green Space in Part of Tuban Urban Area
Below is a based map of RDTR for open green spaces in Part of Tuban Urban Area according to Minister of Public Works Regulation No. 05/2008.

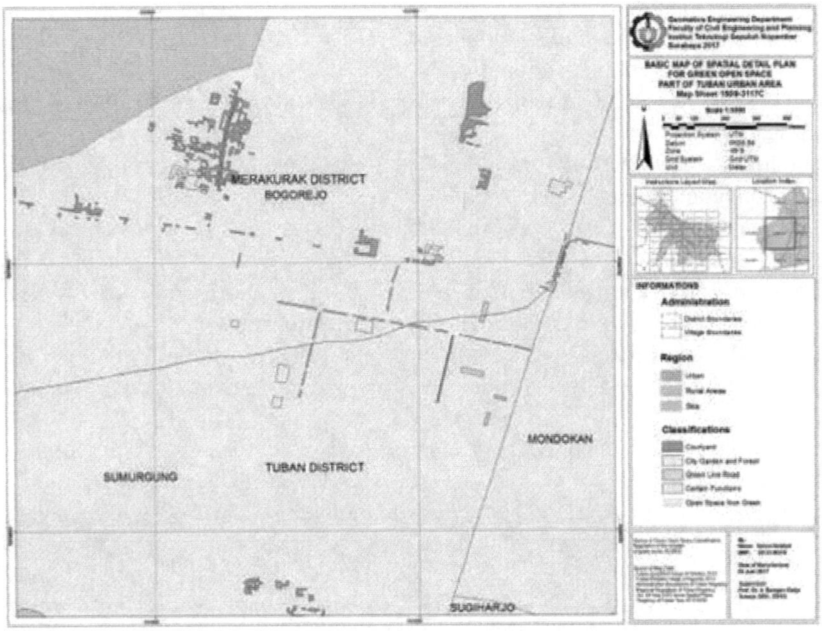

Figure 5. Based Map of RDTR for Open Green Space in Part of Tuban Urban Area

4. Conclusion and Suggestion

4.1 Conclusion

From data processing to analysis, researchers can conclude as follows:

1. With this final project, research has got the base map of RDTR for open green space in the Urban area of Tuban.
2. From orthorectification using Ratio Polynomial Coefficient (RPC) method with 5 point GCP, we get a RMSE value of 0.548 pixels for Pleiades Imagery and 0.366 pixels for QuickBird Imagery. With a tolerance of RMSE $\leq$ 1.5, it means that RMSE gained tolerance. The accuracy test using 12 ICP points, using the provisions of BIG obtained a result of 0.463 m for 1: 5000 scale Pleiades imagery entering class 1 and 0.414 m for 1: 5000 scale QuickBird imagery entering class 1.
3. In making the based map of RDTR for the open green space in the Part of Tuban urban area is required Very High-Resolution Satellite Imagery such as Pleiades and QuickBird Imagery which have 0.5 m and 0.65 m spatial resolution and DEM with high resolution also like DEM Astrium Terra SAR X with 9 m spatial resolution.

4.2 Suggestion

Suggestions from researchers for further research should:

1. When do the measurement point of SRGI (CTBN) on the ground that it was not possible to quantify then it can be done by doing the measurement around the point with a point of help, then the distance between the point SRGI (CTBN) and point help can be differenced.
2. If the satellite image data used in the study there were 2 that one of its data is lacking, then the determination of GCP on both the satellite images are distinguished in accordance with widely available satellite image data so that the image can be orthorectified with entirely.

IOP Conf. Series: Earth and Environmental Science **165** (2018) 012019 doi:10.1088/1755-1315/165/1/012019

References

[1] Peraturan Pemerintah Republik Indonesia No 15 2010 *Tentang Penyelenggaraan Penataan Ruang* (Jakarta: Republik Indonesia)

[2] Undang-Undang Republik Indonesia No 26 2007 *Tentang Penataan Ruang* (Jakarta: Republik Indonesia)

[3] Sumarsono M 2016 Analisa Ketelitian Orthorektifikasi Citra Pleiades untuk Pembuatan Peta Rencana Detail Tata Ruang

[4] Terbuka Hijau (Studi Kasus: Kota Surabaya) Tugas Akhir 1-3 (Surabaya: Jurusan Teknik Geomatika ITS)

[5] Peraturan Daerah Kabupaten Tuban No 09 2012 *Tentang Rencana Tata Ruang Wilayah Kabupaten Tuban Tahun 2012-2023* (Tuban: Bupati Tuban)

[6] Peraturan Menteri Pekejaan Umum No 05 2008 *Tentang Pedoman Penyediaan dan Pemanfaatan Ruang Terbuka Hijau di Kawasan Perkotaan* (Jakarta: Kementrian Pekerjaan Umum)

[7] Badan Informasi Geospasial 2016 *Modul Validasi Peta Rencana Tata Ruang* (Bogor: Badan Informasi Geospasial)

[8] Peraturan Kepala BIG No 15 2014 *Tentang Pedoman Teknik Ketelitian Peta Dasar* (Bogor: Badan Informasi Geospasial)

IOP Conf. Series: Earth and Environmental Science **165** (2018) 012020 doi:10.1088/1755-1315/165/1/012020

Geographic Information System For Mapping Potential of Traffic Congestion (Case Study : Setiabudi Street - Perintis Kemerdekaan Street, Semarang City)

Nurhadi Bashit[1*], Arwan Putra Wijaya[1] and Yudo Prasetyo[1]

[1]Remote Sensing and Photogrammetry Lab, Geodesy Engineering Dept, Faculty of Engineering, Jl. Prof. Soedarto SH, Tembalang, Semarang, 50277, Central Java, Indonesia

nurhadi.bashit@live.undip.ac.id

Abstract. Rapid population growth in major cities Indonesia causes transportation problems such as traffic congestion. This happens because the rapid population growth rate is not balanced by the development of road infrastructure. Unbalanced transportation growth with road construction causes traffic congestion such as in Setiabudi Street, Semarang City. One solution related to traffic congestion and traffic accident is the need for policy related to traffic arrangement. Therefore, we need accurate data and an integrated information system between the factors causing the traffic congestion with spatial data. So this research purpose is creating the map of potential traffic congestion. Field surveys were conducted by collecting information on road conditions, volume of transportation and the causes of traffic congestion in each segment of the road. The results of the survey were analyzed by Geographic Information System (GIS) using clustering method to make the potential map of traffic congestion. The analysis of this study resulted in the conclusion that the intersection of Sukun to the inner direction of the city experiencing traffic congestion along 600 meters while the intersection that does not experience traffic congestion is at the intersection of Ngesrep.

Keywords : Traffic congestion, GIS, Clustering

1. Introduction

Rapid population growth in major cities Indonesia always accompanied by transportation problems such as traffic congestion. This happens because the rapid population growth rate is not balanced by the development of road infrastructure. Other factors causing traffic congestion such as existing road conditions are getting damaged due to weather factors and vehicles crossing roads exceeding capacity. Increased traffic congestion in big cities causes the potential for greater traffic accidents. Traffic accidents are frequent in Semarang City, according to Satlantas Polres Semarang, during the year 2015 has occurred as many as 491 traffic accidents in Semarang District, especially the Semarang - Solo Street [4]. While in the city of Semarang, in the year 2013 recorded as many as 164 people killed by traffic accidents [7].

One type of transportation that is suspected to often cause traffic congestion and traffic accidents is a freight transport. Freight transport with a load that exceeds capacity causes the vehicle to run very slowly to give impact to other road users during crowded conditions. It certainly can cause to traffic congestion and prone to traffic accidents. Traffic accidents that occurred in the city of Semarang recently, especially in Sub-district Ngaliyan, Prof. Hamka Street, there has been a successive accident that starts from a container truck that can not control its speed on the road derivatives, so that crashed into other vehicles around it [5]. One solution related to traffic congestion problems and traffic accidents is the need for policies related to traffic arrangements so that complete information is needed related to factors that cause traffic congestion and traffic accidents. Information related to factors causing traffic congestion and traffic accidents will be easier to use if based on geospatial information because of not only textual or historical data but using spatial data.

Geographic Information System (GIS) is a system used to capture, store, manipulate, analyze, organize and display all types of geographic data [1]. In this case, this method can be used to store, analyze, organize and display geographic data related to the length of roads experiencing traffic congestion so that it can be used to take policies related to traffic management.

2. Materialis and Methods

2.1 Materials

In this paper using primary and secondary data as follows:

Table 1. Data.

NO	Data	Data Source	Type Of Data
1	QuickBird Image	Bappeda	Digital
2	The Semarang Urban Land Use Plan	Bappeda	Digital
3	Field Survey	Field Survey	Manual and Digital

2.2 Methods

2.2.1. Remote Sensing. Remote sensing is a science, technique, art to get information about an object, area or phenomenon through analysis of data obtained without direct contact with the object [3]. Simple sense of remote sensing is a person doing the process of reading the text on the book to get the information contained in the text [3]. The process of acquiring remote sensing data using sensors placed on satellite. Remote sensing data has resolution on remote sensing data depending on satellite. Resolution on satellite imagery is very important to be understood by the user of the image to know the data to be used as needed. The condition and quality of the image can be viewed based on image resolution information such as spatial resolution, spectral resolution, temporal resolution and radiometric resolution. The resolution contained in satellite imagery has characteristics generated by the sensor system on the satellites [8]. Remote sensing satellites experience the development of sensor capability in obtaining high resolution image data. High resolution satellite imagery is an image to obtain information of object detailed on the earth's surface. High resolution image can be used to classify land use from the process of visual interpretation and classification image. Visual interpretation and classification image of high resolution image can produce good classification accuracy [9].

2.2.2. Geographic Information System (GIS). Geographic information systems are computer-based systems used to store and manipulate geographic information. Geographic Information System (GIS) is designed to collect, store, and analyze objects and phenomena in which geographical location is an important or critical characteristic to be analyzed. Thus, GIS is a computer system that has the following four capabilities in handling geo-referenced data [6]:

IOP Conf. Series: Earth and Environmental Science **165** (2018) 012020 doi:10.1088/1755-1315/165/1/012020

1. Input Data.
This sub-system is responsible for collecting, preparing, and storing spatial data and its attributes from various sources. This subsystem is also responsible for converting and transforming the original data formats into a format (native) that can be used by the GIS device.

2. Output Data.
This subsystem is responsible for displaying or generating output (including exporting it to the desired format) all or part of the database (spatial) in both softcopy and hardcopy forms as well as tables, graphs, reports,and maps.

3. Data Management.
This subsystem organizes both spatial data and attribute data in the form of tables into a database system so that it is easy to be recalled, updated, and modified.

4. Data Manipulation and Analysis.
This subsystem determines the information that can be generated by the GIS. in addition, this subsystem also manipulates (evaluation and use of mathematical functions and operators and logic) and data modeling to produce expected information. Illustration of the GIS subsystem can be seen in the figure 1.

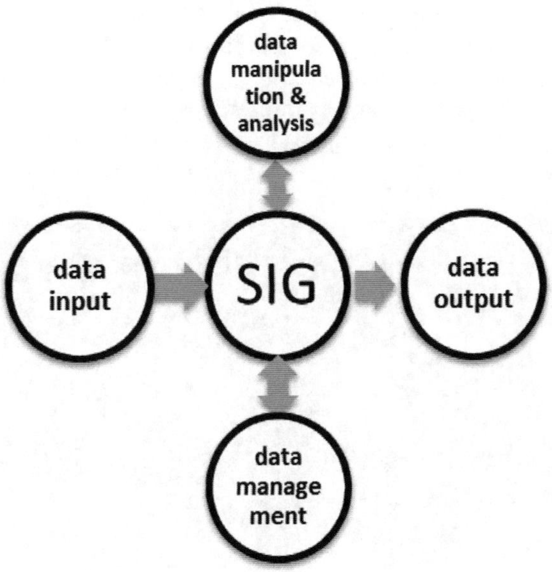

Figure 1. Ilustration GIS subsystem [6]

2.2.3. Metode Clustering. Cluster analysis is an algorithm used for grouping data by viewing objects that have intrinsic characteristics or similarities. Samples of data on a valid cluster have the same characteristics of each other than the different cluster samples. The clustering methodology is particularly suitable for the exploration of the interrelationships between samples to make an initial assessment of the sample structure. Human capability is competitive with automatic grouping procedures in one, two, or three dimensions, but most real problems involve grouping in higher dimensions. It is very difficult for humans to intuitively interpret data embedded in high dimensional [2].

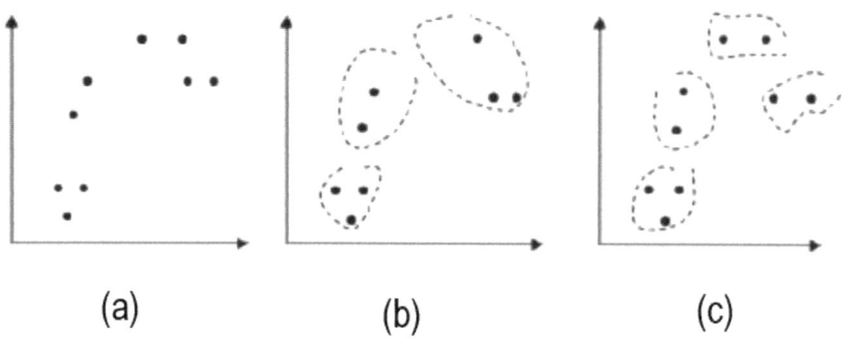

Figure 2. Cluster analysis at point 2D; (a) initial data, (b) three data clusters, (c) four data clusters

3. Results and Discussion

3.1. Visual interpretation and digitization

Satellite images used in this study are Quickbird satellite images, so it is expected that the resulting map in this study can be generated on a large scale. Quickbird imagery can display objects on the surface of the earth with detail so easy to do manual interpretations. The process of satellite image processing is done by manually digitization road segment and land use that exist around the road. Digitization is done manually aimed to produce good results. In this case, the digitization is done on the Setiabudi Street – Perintis Kemerdekaan Street because it is the study area in this study. Digitization is performed only on roads and land use adjacent to the road. The land use around the road is one of the factors causing traffic congestion on a road segment. The result of digitization on screen can be seen in figure 4.

Figure 3. Digitization on screen

3.2. Mapping of traffic congestion by clustering method

Results from field surveys and digitization were spatially treated by clustering method, to obtain a potential congestion map. The classification itself will be divided into three clusters, which are very

IOP Conf. Series: Earth and Environmental Science **165** (2018) 012020 doi:10.1088/1755-1315/165/1/012020

traffic congestion, medium traffic congestion, not traffic congestion. The classification is based on the characteristic approach of each sample obtained in the field survey.

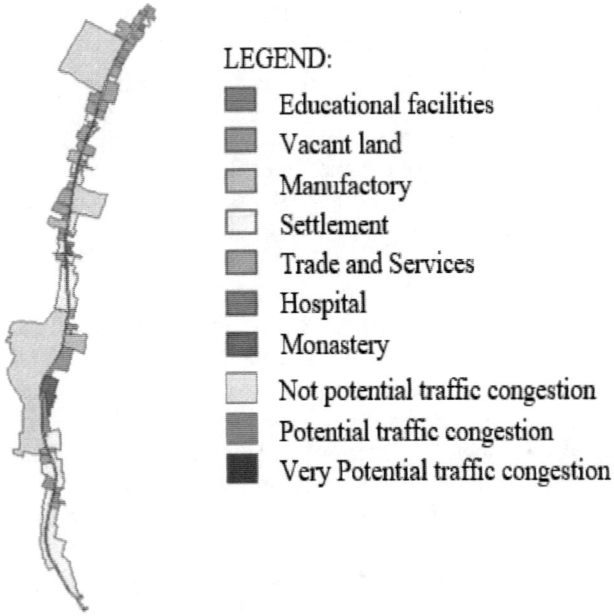

Figure 4. Map of traffic congestion

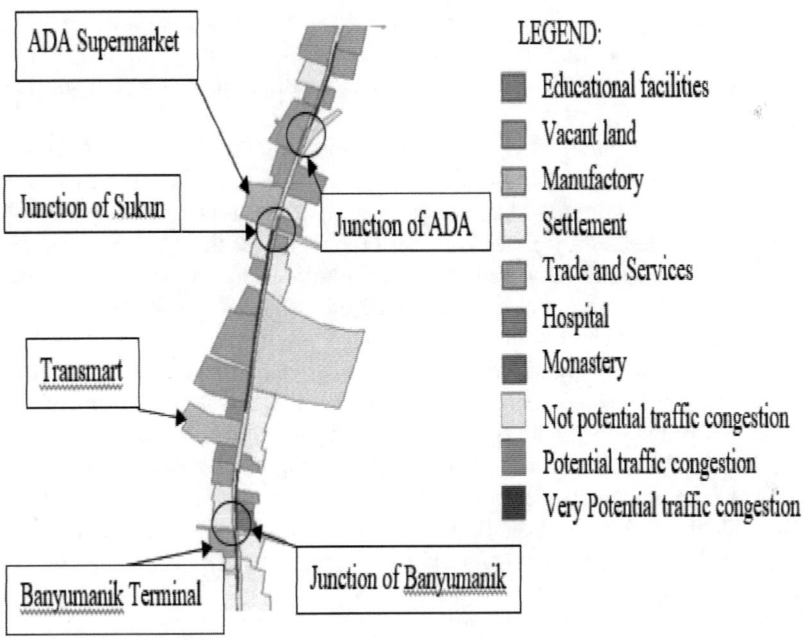

Figure 5 Road segment has traffic congestion

Figure 4 shows a potential map of traffic congestion created by a geographic information system in order to directly see the visual form of a congested road segment. Road segments experiencing traffic congestion with very congestion class are found at the intersection of Sukun in the direction of entry into the city with a congestion length of about 600 meters. The length of traffic congestion is due to an increase in cars volume during the morning and afternoon. The intersection of sukun has increased the volume of cars as many as ± 300 cars on 2 lane road. Traffic congestion that occurs at the intersection of sukun is caused by the improvement of vehicle volume due to traffic lights and cars parked on the street and vehicles that want to get into the Transmart as well as other trades and services located in the area. Ngesrep junction is a road segment that does not experience congestion in the direction of entry into the city. In the morning and afternoon on Ngesrep junction is not experiencing congestion. The potential map of traffic congestion can be seen in figure 5.

3.3. Traffic Congestion Analysis on Road Conditions
Geometrically, the path of Perintis Kemerdekaan Street until Setiabudi Street has an average width of 15 meters road without roadside with 4 lanes 2 direction. In terms of pavement structure, road segment using pavement in the form of hotmix construction with good road pavement condition. Based on the road conditions have good enough condition and decent by passed so it can be called as the primary arterial road because according to road function. The condition of the road is not the main factor causing traffic congestion. According to the primary arterial road function, transportation past the road should be at high average speed as they pass but in reality, there is a traffic congestion on the Perintis Kemerdekaan Street to Setiabudi Street. Road conditions are not the main factors causing traffic congestion because road conditions are good enough and worth to pass.

3.4. Traffic Congestion Analysis on landuse
Transportation passing the Perintis Kemerdekaan Street to Setiabudi Street comes from within the city and the Banyumanik Toll Highway so there is a buildup of vehicles at the junction ADA Supermarket. In the road segment is often experiencing congestion due to the cumulation of public transportation in the form of inter-city buses that reduce passengers and raise passengers on the road as well as public transport vehicles parked. Based on the results of the field survey on Perintis Kemerdekaan Street to Setiabudi Street shows the use of land around the area of the road is a trade, services, military, and settlement.

Land use in the vicinity of the road is one of the factors causing the potential for traffic congestion. The use of land which is the main factor of traffic congestion is trade and services because vehicles that want to enter or exit the building takes time according to the type of vehicle. In this case, Junction of ADA Supermarket is often a traffic congestion on the road into the city or out of town. The buildup of vehicles is caused by parking of vehicles on the roadside causing traffic congestion. The Sukun Junction is congested due to the buildup of cars and motorcycles that want to enter into Transmart. Traffic congestion caused by vehicles that want to enter Transmart cut off the flow of roads so that the buildup of vehicles in front of Transmart. This causes traffic congestion at the junction of Sukun. Traffic on Perintis Kemerdekaan Street to Setiabudi Street needs to be arranged to avoid potential traffic congestion is more severe. Potential of traffic congestion on the road is always troubling road users who want to pass the road because of the growing number of vehicles that continue to increase each year. Traffic arrangements on the road need to be done so that road users feel comfortable when crossing the road. There needs to be a regulation on the curbing of vehicles parked along the road so as not to disturb other road users while crossing the road.

3.5. Traffic Congestion Analysis on the Semarang Urban Land Use Plan
Based on the the Semarang Urban Land Use Plan, Banyumanik is the main region of development to the south (Semarang District), this region serves as settlement areas, tourism, and military. The direction of development in Banyumanik area is caused the density of settlement in Semarang City so that the direction of settlement development is moving towards South Semarang (Semarang District).

Therefore, development settlement in Banyumanik Subdistrict continues to increase as well as other causes south of Banyumanik Subdistrict is the industrial area (Bawen Subdistrict). However, Banyumanik Subdistrict is expected to support the provision of settlement areas.

The development settlement of Banyumanik Subdistrict caused the primary arterial road like Perintis Kemerdekaan Street to Setiabudi Street experiencing the traffic congestion. In the morning and afternoon, there is always traffic congestion on the primary arterial road is caused by the volume of vehicles passing through the road is very crowded. Based on the Semarang Urban Land Use Plan, Perintis Kemerdekaan Street to Setiabudi Street is a network of roads with primary arterial road function because this road is the entry of Semarang City that connects with other city limits such as Ungaran District. Therefore, Perintis Kemerdekaan Street to Setiabudi Street has a transportation system linking urban service activities centers such as markets, offices, trade and services, education and settlement.

Perintis Kemerdekaan Street to Setiabudi Street is a road connecting Banyumanik District development center (settlement, industrial, trade, and services) with the center of Semarang City which serves as a center of the office, trade, public service, education and densely populated residential center. Based on observations in the field of traffic flow on Perintis Kemerdekaan Street to Setiabudi Street are a mixed flow of private transport, public transport and heavy transport and motorcycles that quite a lot. Therefore, the accumulation of transport that pass by in the morning and afternoon causing traffic congestion on the road.

4. Conclusion

In this paper can be taken some conclusions on the process that has been done:
Based on the map of the traffic congestion on the Perintis Kemerdekaan Street to Setiabudi Street, traffic congestion divided into three clusters, which are very traffic congestion, medium traffic congestion, not traffic congestion. The class of roads experiencing traffic congestion with very traffic congestion class is located at the intersection of Sukun in the direction of entry into the city with a long congestion of about 600 meters and the class of roads does not experience congestion is located at the intersection of Ngesrep in the direction of entry into the city.

The potential for traffic congestion on Perintis Kemerdekaan Street to Setiabudi Street is more due to land use that is located around of the road. The factors such as road conditions and vehicle volume are not the main factors of potential traffic congestion. Land use in the vicinity of Perintis Kemerdekaan Street to Setiabudi Street by the Semarang Urban Land Use Plan is a settlement but in fact, the use of land has been largely a trade and services. Supervision is necessary to supervise the development trade and services around of the Perintis Kemerdekaan Street to Setiabudi Street so that the potential for traffic congestion can be reduced.

Reference

[1] Irwansyah, Edy 2013 *Sistem Informasi Geografis : Prinsip dasar dan Pengembangan Aplikasi* (Yogyakarta: Digibooks)
[2] Kantardzic, Mehmed 2011 *Data Mining : concepts, models, methods, and algorithms* (New Jersey : John Wiley and Sons)
[3] Lillesand T. M. Kiefer, R.W. and Chipman J.W. 2004 *Remote Sensing and Image Interpretation* (USA: John Wiley and Sons)
[4] Munir, Syahrul 2016 *Tahun Lalu, 130 Orang tewas akibat kecelakaan di Jalur Semarang – Solo.http://regional.kompas.com/read/2016/01/15/20424921/Tahun.Lalu.130.Orang.Tewas.akibat.Kecelakaan.di.Jalur.Semarang-Solo* Akses date on 04 – 04 – 2017
[5] Pamungkas R T 2017 *Tabrakan beruntun di Ngaliyan, Tiga korban dirawat, Sopir kontainer belum ditemukan. http://jateng.tribunnews.com/2017/02/27/tabrakan-beruntun-di-ngaliyan-tiga-korban-dirawat-sopir-kontainer-belum-ditemukan* Akses date on 05 – 04 – 2017
[6] Prahasta Eddy 2009 *Sistem Informasi Geografis: Konsep-Konsep Dasar (Perspektif Geodesi dan Geomatika)* (Bandung: Informatika)

IOP Conf. Series: Earth and Environmental Science **165** (2018) 012020 doi:10.1088/1755-1315/165/1/012020

[7] Setiawan, Eka 2013 *10 Bulan 839 Kasus kecelakaan di Semarang https://daerah.sindonews.com/read/809146/22/10-bulan-839-kasus-kecelakaan-di-semarang-1385148691* Akses Date on 04 – 04 – 2017
[8] Smith R B 2012 *Introduction to Remote Sensing* (Nebraska: Microimages.inc.)
[9] Sudaryanto 2013 Studi Penggunaan Lahan Di Kecamatan Umbulharjo Kota Yogyakarta Berdasarkan Interpretasi Citra Quickbird *Magistra* 86 Th. XXV 112–118

IOP Conf. Series: Earth and Environmental Science **165** (2018) 012021 doi:10.1088/1755-1315/165/1/012021

ESDM One Map Indonesia Indonesia: Opportunities and Challenges to Support One Map Policy based on Applied Web-GIS

H A Setyowati[1*], M P Dwinugroho[2], Sigit Heru Murti B S[3], A Yulianto[4], N E Ajiwihanto[5], J Hadinata[6] and Sanjiwana A K[3]

[1]GIS Officer, Center for Data & Information Technology, Ministry of Energy & Mineral Resources

[2]Head of Center for Data & Information Technology, Ministry of Energy and Mineral Resources, Republic of Indonesia

[3]Lecturer, Departement of Cartography and Remote Sensing, Faculty of Geography, Universitas Gadjah Mada

[4]Web Programmer, Center for Data & Information Technology, Ministry of Energy & Mineral Resources, Republic of Indonesia

[5]Head of Data Management, Center for Data & Information Technology, Ministry of Energy and Mineral Resources, Republic of Indonesia

[6]Data and Information Manager, Center for Data & Information Technology, Ministry of Energy and Mineral Resources, Republic of Indonesia

heratania@gmail.com

Abstract. The law number 4 : 2011 and Presidential Decree number 9 : 2016 policy in Indonesia regulates the management of standardization and unification of geospatial data called One Map Policy to achieved one geospatial information which is needed across the institution and ministries. Ministry of Energy and Minerals Resources (MEMR) of Republic Indonesia have managed diverse data which are dynamic and periodically changing, completed with their spatial aspect. To manage those datasets to follow the one map policy, Center of Data and Information Technology unit with task is managing data and information technology developed a Web-GIS platform called ESDM One Map Indonesia that can be used for displaying, analyzing, and monitoring energy and mineral resources. ESDM One Map Indonesia is expected to support various stakeholders in their decision and policy making process, especially in the case of territorial issues. The data in this system is stored in PostgreSQL that exist in an enterprise geodatabase which then published as map services in WMS, WFS, or KML format using Enterprise GIS. Overall, this Web-GIS application could help in monitoring the data development and performing spatial analysis which are can be used for supporting the decision maker. ESDM One Map Indonesia is strong efforts -from MEMR of Republic Indonesia in the form of Web-GIS application for supporting the Indonesian One Map Policy which requires- support and data collaboration from another Ministry/Organization. The willingness from MEMR still needs a support especially keep up with the advancement of Web-GIS based application so that the service provided by ESDM One Map Indonesia can be continuously maintained.

Keywords : Web-GIS Application, MEMR of Republic Indonesia, ESDM One Map Indonesia, Decision Support System, One Map Policy

1. Introduction

Nowadays, the need of space-based analysis continually increases, especially in the fields of regional development and planning. The problems of space and land utilization, which trigger conflicts on land-use overlap, accentuate the requirement of spatial data. Act No. 4/2011 on Geospatial Information becomes one of many emphases on the necessity of geospatial information management. Also, the issuance of the Presidential Regulation No. 9/2016 on the Acceleration of One Map Policy affirms the urgency of spatial data provision. A real-time, spatially performed resource inventory can assist leaders in recognizing changes, trends, and developments in an area as well as provide a complete, accurate, valid, reliable, and up-to-date data for decision making. Regulations on the management of geospatial data standardization and integration, as elaborated in the Acceleration of One Map Policy, are expected to accommodate every existing geospatial information and similar data required by institutions and ministries.

From the perspectives of standard and quality, the existing data is still in need of updates and widely accepted standards to avoid spatial inaccuracy when converted to the standardized data. Moreover, many sectors in ministries or institutions are divided further into several subsectors, resulting in different data formats. The study on One Map Policy, conducted by Hasyim et al. from the Geospatial Information Agency [11], revealed a considerable number of incomplete spatial planning legislations in Indonesia, as listed in **Table 1**. This finding implies that the arrangement of spatial data has not been fully resolved. **Table 2** shows that the geospatial information and topographic maps used as reference in the maps of Basic Geospatial Information are also incomplete. The topographic maps scale 1:250,000 have been finished and available since 2013. In 2014, the topographic maps scale 1:50,000 were available for 72% of the existing national territory, and the topographic maps scale 1:25,000 were limited to Java, Bali, Nusa Tenggara, and Sulawesi. Meanwhile, the provision of topographic maps in the other areas is afoot.

Table 1. Status of the spatial planning legislation at provincial district and city
(Source: www.penataan ruang.net (update May 26,2015) in Hasyim,2016 [11])

Status	Province	District	City
Not Finished	9	70	9
Already Finished	25	329	84
Formasi Total	34	399	93
Precentage of completion	73.5%	82.5%	90.3%

Table 2. Availability of topographic maps
(Source: Center for topographic mapping and toponymy, Geospatial Information Agency (BIG) in Hasyim,2016 [11])

Scale	Number of Sheet			Percentage
	National Coverage	Already available	Not yet available	
1:5.000	379.012	539	378.473	0.14%
1:10.000	91.547	1074	90.473	1.17%
1:25.000	13.020	3894	9126	29.91%
1:50.000	3899	2837	1.062	72.76%
1:250.000	309	309	0	100%

Puyam S. Singh [18] found a growing need of web-based GIS for easy and fast information sharing and dissemination, which can significantly assist in the decision-making process related to natural resource-based analysis. The rapid development of GIS in the previous decades is an interesting phenomenon, but it also is a new challenge for large-data management and processing [20].

The Center for Data and Information Technology, under the Indonesian Ministry of Energy and Mineral Resource, is responsible- for 18 out of the whole 85 thematic maps in the Acceleration of One Map Policy. The diversity of the data scattered in each unit is a challenge for this center in providing complete, accurate, and latest data. As a network node of spatial data, this institution is developing an Enterprise Geographic Information System, termed as ESDM One Map Indonesia (Energy and Mineral Resource-One Map Indonesia). This system is designed to integrate all data in the sector of energy and mineral resource and to collaborate data management with different parties. ESDM One Map Indonesia displays both reference and thematic maps that come directly from relevant ministries or institutions. Also, it can demonstrate, analyze, and monitor energy and mineral resources that are temporally dynamic and spatially diverse.

A.A. Alesheikh [2] explains that the distribution of geospatial information on the internet allows different layers of society to access it and provides them with media for processing it without location restrictions. Web-based GIS evolves from a variety of web maps and client-server architecture into a wide distribution. Therefore, the internet reshapes every role of an information system, including the functions to collect, store, retrieve, analyze, and visualize data. The high costs of GIS, a specific database system, and system-improving software development fade after the introduction of web-based GIS. The dissemination of spatial information through the internet improves the decision-making process. Both web development and internet expansion have greatly assisted geoscientists in their works, as well as society in general [2], because, first, the web allows for visual interaction with data. Due to the development of web server, clients can now produce maps conveniently. Moreover, since maps and graphics are available on the internet, they can observe any updates directly and hence, fasten the evaluation process. Second, the internet is ubiquitous so that geospatial data can be widely accessed. Consequently, clients can access and acquire necessary data quickly; which enables them to work from almost any locations.

The application of ESDM One Map Indonesia, developed by the Center for Data and Information Technology, is in line with the One Map Policy, i.e., one geospatial reference, one standard, one database, and one geoportal. Furthermore, it conforms with the Economic Policy Package Vol. 8 that elaborates the President's *Nawacita* Program (literally the nine hopes, agendas, or expectations) on the One Map Policy. This policy is actualized in the Presidential Regulation No. 9/2016 as an effort to facilitate the Acceleration of One Map Policy (scale 1:50,000) [6]. Moreover, it gains the support of the Corruption Eradication Commission during the data collection from all ministries or institutions through the coordination and supervision activities.

This WebGIS application develops along with its advantages and disadvantages [2]. The ESDM One Map Indonesia was developed in 2015 with a standard ArcGIS license [9] that allows for simple analyses but not for the complex ones. It also has a speed problem. Because GIS associates with a variety of graphical data, a slow internet connection impedes the appearance of the map service. Alesheik et al. [2] found this condition non-equivalent to the complexity of a specific GIS program like ArcView and ArcInfo or MapInfo in the future. Furthermore, various bugs often appear and potentially disrupt the development of this program. Another problem may emerge from the outdated technology of the application. It is developed using Flex, whereas the technology of WebGIS application is based on Javascript [19]. Emphasizing on the necessity of this application system, the Center for Data and Information Technology develops a better ESDM One Map Indonesia with an updated system. On May 31, 2017, the ESDM One Map Indonesia was finally launched with an updated WebGIS system. This day has, therefore, become a new momentum for GIS development in Indonesia. Qualified infrastructure and architecture are expected to support ESDM One Map Indonesia. The nowadays ESDM One Map Indonesia has already been integrated not only with the

IOP Conf. Series: Earth and Environmental Science **165** (2018) 012021 doi:10.1088/1755-1315/165/1/012021

internal sector of the Indonesian Ministry of Energy and Mineral Resources (MEMR) but also with different ministries and institutions. Maps with various themes can help users to identify the potentials and conditions of energy resources.

This research explores ESDM One Map Indonesia from the making step until the development, as well as its opportunities and challenges. In the future, if MEMR wants to optimize the data and the application in ESDM One Map Indonesia, then it will require a full support for up-to-date technology and permanent development. Consequently, ESDM One Map Indonesia is not only the name of an application but also a mature concept of data and information management, especially for spatial data, which is actualized in the form of WebGIS application that can function as a Decision Support System (DSS). The commitment to continuously develop the capacity, capability, and content of the ESDM One Map Indonesia to assist stakeholders in the decision-making process is inevitable. This application is expected to be a reliable source of information and the basis of decision and policy-making processes to support the One Map Policy in national level.

2. Research Area

The coverage of the mapped area on each data contained in One Area Indonesia's EMR ie the entire territory of Indonesia, as presented in **Figure 1** below. Some maps that have coverage of almost all areas of Indonesia, namely Map of Oil and Gas Working Areas, Mining Business License, Geological Map, Water Basin Map, Solar Energy Potential Map, and several other maps, while maps that have not covered all regions of Indonesia such as maps, geological disaster maps that are only located in certain locations.

Figure 1. The Area Coverage of ESDM One Map Indonesia
Source: Portal ESDM One Map Indonesia, geoportal.esdm.go.id, 31 May 2017

3. Materials and Methods

3.1. Data

Fuad Hasyim [11], in his paper 'One Map Policy (OMP) Implementation Strategy to Accelerate Mapping of Regional Spatial Planning (RTRW) in Indonesia', explains that the demand on geospatial information in the preparation stage of spatial planning can be divided into two major groups. The first group is Basic Geospatial Information (*Informasi Geospasial Dasar*-IGD) in the form of Indonesian Topographic Map (*Rupa Bumi Indonesia*-RBI). The second group is Thematic Geospatial Information (*Informasi Geospasial Tematik*-TGT) that covers thematic network and basic infrastructure. However, the local governments have to face particular obstacles in mapping the spatial plans, namely the availability of geospatial information, guidelines, and cross-border human resource and technology.

The data contained in ESDM One Map Indonesia is collected from various sectors in MEMR and other ministries or institutions. All of these data are arranged and stored in the form of geodatabase. This database requires the same coordinate system. Therefore, different sets of data need coordinate system change or transformation [15]. These data are stored in one database according to their themes. Each theme is saved in one map package (.mxd) before being published. The status of the data in ESDM One Map Indonesia is currently divided into two, namely the data coming from the internal sectors of MEMR (presented in **Tables 3-7**) and those from the collaboration of different ministries and institutions (listed in **Table 8**).

The Ministry of Energy and Mineral Resources is a part of JIGN (The National Geospatial Information Network) and is also a member of the One Map Acceleration Policy activity (PKSP). Both are also in line with ESDM One Map Indonesia so that the Map on the ESDM One Map is similar to the one in the acceleration activity of one map. It's just that not all the maps in the ESDM One Map is in the list of map fulfillment on the activity of one policy Acceleration Map in other words the map in ESDM One Map is more varied. In this PKSP Activity, the ESDM Ministry is obliged to fulfill 17 Thematic Maps of the existing themes and the Map in EMR One Map Indonesia is used for the internal needs of the ESDM ministry while for PKSP activities for public interest with certain standards.

Table 3. List of the Maps Acquired from the Oil and Gas Sector in ESDM One Map Indonesia (Source: Portal ESDM One Map Indonesia, geoportal@esdm.go.id, 31 May 2017)

No	Maps	No	Maps
1	Pipelines	20	Oil Import Lane
2	Custody Transfer Points	21	Oil Distribution Lane
3	Platforms	22	Secondary Oil Distribution Lane
4	Lifting	23	Gas Transmission Network
5	HNK Gas and Oil Working Area	24	Gas Transmission Network Towers
6	Conventional Oil and Gas Working Area	25	Gas Distribution Network
7	Current Oil and Gas Working Area	26	Gas Distribution Network Towers
8	Wells	27	Existing LNG Processing Plants
9	Premium Oil and Solar Agents (APMS)	28	Existing LPG Processing Plants
10	Aircraft Filling Depots (DPPU)	29	Existing Upstream Pipes in 2015
11	Gas Filling Stations (SPBG)	30	Existing Open Access Pipes in 2015
12	Subsidized Vehicle Fuel Stations (SPBKB)	31	Existing Distribution Pipes in 2015
13	Diesel Filling Stations (SPDN) – Fueling Stations (SPBN) for Fishers	32	On-going Infrastructures of Open Access Pipes in 2015
14	LPG Bulk Filling and Transport Stations (SPPBE)	33	Areas Reached by Gas Networks
15	Public Filling Stations (SPBU)	34	Oil and Gas Production in 2015
16	Oil Terminals	35	Oil and Gas Production in 2016
17	Gas Processing Plants (Refinery)	36	Wells
18	Oil Processing Plants (Refinery)	37	SEISMIC_2D
19	Gas Supply Lines	38	SEISMIC_3D

Data format is also incredibly significant while using web-based GIS system. The formats are vector and raster; both of which have advantages and disadvantages [2]. A raster data, such as satellite image, depicts an object according to reality. Even though it is highly visible and identifiable, it is, however, lack of cartographic aspects. Moreover, identifying a single object is improbable because raster data is composed of pixels. A contact to the server is required per request by the clients when loading a large raster data, for instance, during data zooming. This type of data also has a limited appearance in different zoom levels. The dimension of raster data recording changes when users try to obtain more details from it. The data appears in blurs and becomes less informative, prompting the need to provide multiscale raster data with different accuracy or to deliver good quality image in various scale range.

Table 4. List of the Maps Acquired from the Mineral and Coal Sector in ESDM One Map Indonesia (Source: Portal ESDM One Map Indonesia, geoportal@esdm.go.id, 31 May 2017)

No	Maps
1	Mining Business License Area (WIUP) for Metal Mineral
2	Mining Business License Area (WIUP) for Non-metal Mineral and Stone
3	Mining Business License Area (WIUP) for Coal
4	Smelters
5	Coal Terminals
6	Mining Areas
7	Coal Formations
8	Mineral Formations

Table 5. List of the Maps Acquired from the Electricity Sector (Source: Portal ESDM One Map Indonesia, geoportal@esdm.go.id, 31 May 2017)

No	Maps	No	Maps
1	Geothermal Power Plants (PLTP)	11	Electricity Infrastructure in Bangka
2	Micro Hydropower Plants (PLTM/PLTMH) <10MW	12	Electricity Infrastructure in Belitung
3	Solar Power Plants (PLTS) <10MW	13	Electricity Infrastructure in Java Island
4	Off-grid Generators (national budget)	14	Electricity Infrastructure in Kalimantan Island
5	Bioenergy-based Generators	15	Electricity Infrastructure in Sumatra Island
6	Substations	16	The Project Locations in 2,500 Villages
7	Distribution Substations	17	Electrification Ration in 2014-2016 per Regency
8	Transmission Network	18	Electrification Ration in 2014-2016 per Province
9	Distribution Network	19	Recapitulation of Installed Capacity
10	Distribution Line Network	20	35.000MW Electricity Generator Program

Table 6. List of the Maps Acquired from the New Renewable Energy and Energy Conservation Sector (Source: Portal ESDM One Map Indonesia, geoportal@esdm.go.id, 31 May 2017)

No	Maps	No	Maps
1	Geothermal Potentials	12	The Blending Facilities of PT. AKR
2	Oil Palm Potentials	13	The Blending Facilities of PT. Pertamina
3	Potentials of Waste Power Plants (PLTSA)	14	Biodiesel
4	Micro Hydro Potentials	15	Bioethanol
5	Sea Current Potentials	16	Biogas Plant Constructions
6	Sea Current Conversion	17	Distribution of Industries with Energy Consumption > 6000 TOE
7	Distribution of Methane Content	18	Compact Fluorescent Lamp Company with Energy Efficient Label
8	Distribution of Gas Types	19	Existing Geothermal Working Area
9	Distribution of Gas Indications	20	Specified Geothermal Working Area

No	Maps	No	Maps
10	Wind Energy Potentials	21	Assignment Area for Preliminary Survey
11	Solar Potentials		

Vector data has a lighter load and, therefore, it transmits faster than raster. Users can do more spatial analysis on vector data rather than on raster, which requires initial extraction. In different zoom levels, the dimension of the object in vector data remains. Furthermore, a single, specific object is identifiable with an excellent detail. Vector data does not need a contact to the server in every execution in the browser because it has a light data load. However, the disadvantage of vector data includes a rigid appearance, which is unlike the real world. This data largely depends on the processing and the making of the data itself, the accuracy of data processing, and the adequacy of the existing information. Moreover, the distribution of vector data may also compromise copyright rules. The choice of data format (either vector or raster) varies according to the existing application and infrastructure [2].

Table 7. List of the Maps Acquired from the Geology Sector in ESDM One Map Indonesia (Source: Portal ESDM One Map Indonesia, geoportal@esdm.go.id, 31 May 2017)

No	Maps	No	Maps
1	Geology	11	Non-metal Mineral and Stone Potentials
2	Hydrogeology	12	Potentials of Seabed Surface Sediment Distribution
3	Sediment Basin	13	Potentials of Marine Minerals
4	Groundwater Basin	14	Groundwater Drilling (2003-2014)
5	Karst Landscape Area	15	Distribution of Gravity Core Data in the Sea
6	Coal Potentials (2016)	16	Volcanic Disaster-prone Areas
7	Solid Bitumen Potentials	17	Earthquake-prone Areas
8	Coal Bed Methane (CBM) Potentials	18	Tsunami-prone Areas
9	Peat Potentials (2011)	19	Mass Wasting Susceptibility Zones
10	Metal Mineral Potentials (2015)		

Table 8. List of the Maps Acquired from Other Ministries and Institutions in ESDM One Map Indonesia (Source: Portal ESDM One Map Indonesia, geoportal@esdm.go.id, 31 May 2017)

No	Maps	No	Maps
1	Provincial Administration Boundaries	11	Navigation Points
2	Administrative Boundaries of Regencies	12	Crossing Points
3	Administrative Boundaries of Districts	13	Seaports
4	Protected Forest Area	14	Railway Stations
5	Conservation Forest Area	15	Sea Lanes
6	Production Forest Area	16	Railway
7	Limited Production Forest Area	17	Master Plan for the Development of National Industries
8	Convertible Production Forest Area		
9	Special Purpose Terminals		
10	Private Interest Terminals		

3.2. Methods

3.2.1 The Development of ESDM One Map Indonesia .The data in ESDM One Map Indonesia is stored in an enterprise geodatabase using the DBMS (Database Management System) PostgreSQL and, then, published as a map service in the format WMS, WFS, or KML. This WebGIS application is developed using Javascript and HTML5. It is also called as a dynamic web browser [1]. A dynamic map allows users to select any features displayed on it, such as scale and location [1]. In this type of application, the map is depicted according to predefined parameters and, then, the web browser displays it as a picture. Users can change these parameters and thereby, observe and produce a new map. However, since this application contains geographic data, GIS software, and an interface program, it causes a heavy load on the server. It also requires a program to provide a map service. GIS-based planning support system enables planners and citizens to create and test scenarios of development alternatives quickly and efficiently as well as determine the potential impact of a particular land use pattern along with the population and employment trends in the future. Therefore, it assists public officials to make an informed decision for future planning. Furthermore, most urban, community, and regional planning strategies are complicated, which then require the involvements of local, provincial, and national governments [10].

A.A. Alesheikh explains that the main purpose of the development stage of WebGIS is to determine 'how' WebGIS performs the applications required in database planning, depicts an object into graphics (e.g., color, size, and symbol), creates the structures of graphic files and non-graphic attribute files, activates the layers, chooses the scale on which the map layer is opened, and displays the GIS products. Also, the management and security restrictions during data access by users are determined at this stage. This development can be performed by selecting the sources (e.g., documents, maps, and digital files), preparing the actual database designs (logical/physical design), determining the procedures for data conversion from source media to the compatible database, and defining the procedures for database management and maintenance [2]. The structure of the spatial database is designed to accommodate (i.e., to save, manage, and access) various types of data, including basic geographic data, field observation data, and remote sensing information. ESDM One Map Indonesia is an integrated spatial data management system built using ArcGIS technology. ArcGIS provides software for spatial data processing (ArcGIS for Desktop), data storage system (Geodatabase), software for spatial data sharing (ArcGIS for Server), and a multi-purpose spatial data platform (Portal for ArcGIS) for data utilization. Geoserver is used to generate and publish web-based map service and web feature service, keyhole markup language, and all of the things necessary to the development of WebGIS application [7] [19].

3.2.2 The Management of ESDM One Map Indonesia

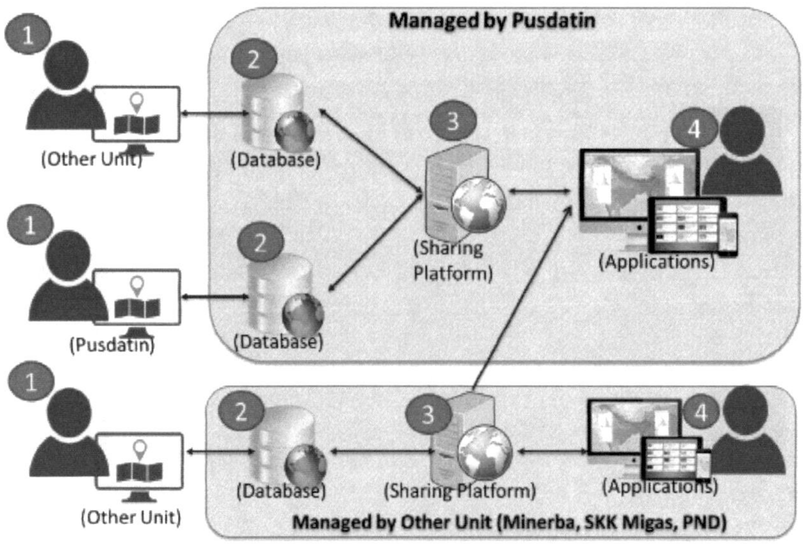

Figure 2. The Management of ESDM One Map Indonesia
Source: Center for Data and Information Technology MEMR, 2017

The management process of ESDM One Map Indonesia is performed in two ways. First, the units (i.e., Directorate-Generals and Agencies) that have the same GIS application and infrastructure can integrate their maps through a map service in ESDM One Map Indonesia directly. Relying on this system, the Center for Data and Information Technology can obtain the latest updates from the units without the complexity of data processing. Second, the units that have no GIS application and infrastructure receive supporting facilities to collaborate with other units in data management in the ESDM One Map Indonesia. In this way, the data can be managed by the respective Directorate-Generals and Agencies. **Figure 2** illustrates the data integration flows in ESDM One Map Indonesia, which include:

1. The Persons in Charge (PICs) of the data in each unit manages their data by themselves, while the Center for Data and Information Technology manages the cross-sector data and monitors the data from the other units.
2. The data is saved in a Geodatabase Server that is managed by the Center for Data and Information Technology or every unit.
3. The data is published using ArcGIS Server that is organized by the Center for Data and Information Technology.
4. The data is displayed in various applications and shared with relevant parties.

3.2.3 The Future Management of ESDM One Map Indonesia. Surya Herjuna et al., in a scientific article titled Pemanfaatan Aplikasi SIG Berbasis Web Untuk Penataan Sektor Pertambangan Indonesia (The Utilization of Web-based GIS Application for Structuring the Mining Sector in Indonesia) [23], stressed on the constant development of web-based GIS application, which aims to optimize monitoring process and synergize cross-institution data (between different ministries and institutions). The process of data synergy involves ministries and other institutions to be able to obtain a view of the systems and the processes used in web-based GIS system. Surya Herjuna et al. also divided the development of web-based GIS application into short, mid, and long-term. Each term consists of several strategic plans, for instance, to develop the following hardware, software, database, and brainware [23]:

a. Hardware development includes the provision of server, computer, firewall system, internet network, and others.
b. Software development on map application, attribute data input technique, and tools for data analysis.
c. Database development includes the cooperation of ministries or other institutions in data exchange to strengthen national database, the reconciliation of license data in the regions, and the input of attribute data in every company.
d. Brainware development consists of training for central administrators and local governments.

In the future, ESDM One Map Indonesia will be developed further by optimizing the role of every management position presented in **Figure 3**.

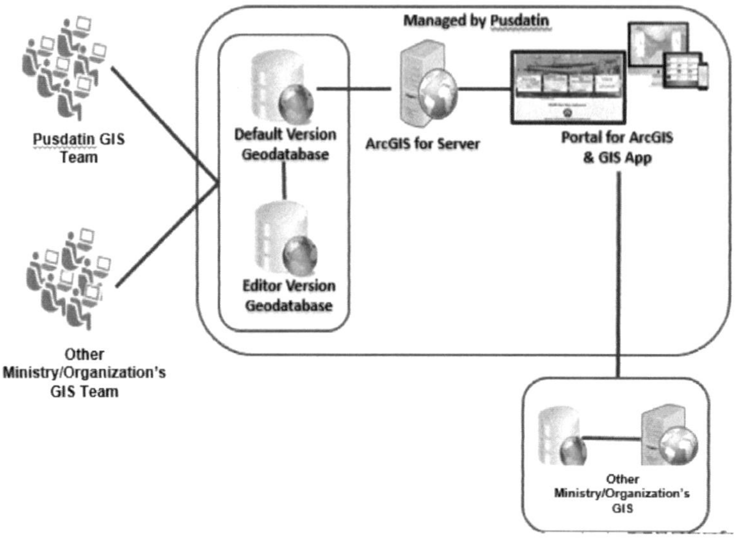

Figure 3. The Future Management of ESDM One Map Indonesia
Source: Center for Data and Information Technology MEMR, 2017

4. Result, Analysis, and Discussion

4.1 One Map Policy

According to the Act No.4/2011 on Geospatial Information Article 2 [5], geospatial information has to be organized by legal certainty, integrity, openness, sophistication, accuracy, usefulness, and democracy. Article 3 explains the objectives of this act, namely (b) to actualize an effective and fruitful implementation of geospatial information through a cooperation in coordination, integration, and synchronization; and (c) to encourage the use of geospatial information in governance and various aspects of the society's life. The Coordinating Minister for Economic Affairs who read out the Economic Policy Package Vol. 8 states that the development of an area or an infrastructure often collides with some problems related to space and land utilization. This conflict is very difficult to resolve because the Thematic Geospatial Information overlaps to each other. Therefore, the One Map Policy, referring to one geospatial reference, one standard, one database, and one geoportal, is expected to accelerate the implementation of national development, which is one of the government's priorities. The same basis for map reference will also increase the reliability of information on the locations of various economic activities for business certainty. The diverse information compiled in this one map can also be used to design various simulations, e.g., disaster mitigation. Following the Presidential Regulation on the Acceleration of the Implementation of One Map Policy [6], ministries and institutions will prepare thematic maps scale 1:50,000 before the deadline, i.e., in the year 2019, according to their respective action plans.

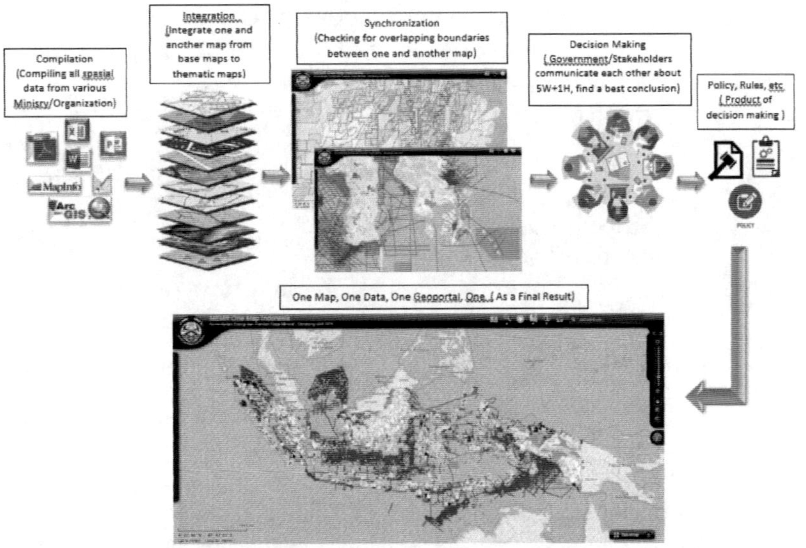

Figure 4. The Illustration of One Map Policy in ESDM One Map Indonesia
Source: Center for Data and Information Technology MEMR, 2017

The Acceleration of One Map Policy is expected to become a reference for spatial data improvement, spatial planning accuracy, and the accuracy of policy formulation and decision making through the following steps: the compilation, integration, and synchronization of 85 thematic maps from 19 ministries/institutions in 34 provinces in Indonesia. **Figure 4** illustrates the One Map Policy in ESDM One Map Indonesia as a collaboration between GIS and web, which also employs compilation, integration, and synchronization steps in producing a widely accessible portal. A.A. Mustaffa [4] reveals that the compilation of GIS techniques is widely used in many sectors, especially those that require a lot of data management. The concept of One Map Policy using web-based GIS is in line with Yu P. Yuronen [25], which explains that the data in GIS supports the current information technology and informatics and provides an opportunity to evaluate the status of the natural environment in many regions in real time. There are various definitions and objectives of web-based GIS [4]. According to the field study of this research, it is defined as a database system on a computer designed to support data or information acquisition, storage, analysis, and presentation in a selected location. Through GIS the required information can be accessed quickly and accurately. Another advantage is that GIS offers the ability to connect spatial data with attribute data so that the GIS data becomes more useful for analytical purposes. Consequently, web-based GIS can assist leaders to make an informed decision

As mentioned previously, the data contained in ESDM One Map Indonesia is collected from different sectors of MEMR and other ministries or institutions, and then arranged and stored in a geodatabase. The data used in this research are mostly vector, in the forms of point, line, and polygon. The GIS data collected from the above sectors and ministries/institutions have different formats. Therefore, these data have to be transformed into one reference system [1] according to the standard of the One Map Policy (**Tables 9-11**) before they are combined into a single entity. The datum also conforms to the same standard, i.e., datum WGS 84 with GCS projection. The standard for the data format adopted in the Acceleration of One Map Policy is as follows.

Table 9. The Parameter of Geospatial Information Standard for One Map Policy
Source: *Modul Integrasi Informasi Geospasial Tematik* (Module of Integrated Thematic Geospatial Information), The Acceleration of One Map Policy Secretariate, 2016

Information Type	Units	Description
Geospatial Thematic Information's Name	-	Name of Geospatial Thematic Information
Data Format	Shape file (.shp) and Geo database (.gdb)	Type of data which is used in data making
Coordinate System	Geographic	Coordinate system that used is Geographic Coordinate System
Datum	WGS (World Geographic System) 1984	The datum information based on WGS 1984
Map Scale	1:10.000 up to 1:1.000.000	The scale for making Geospatial Thematic Information

Table 10. Geospatial Information Quality Parameter (position accuracy) for One Map Policy
Source: *Modul Integrasi Informasi Geospasial Tematik* (Module of Integrated Thematic Geospatial Information), The Acceleration of One Map Policy Secretariate, 2016

Types of Information	Units	Descriptions
Position Accuracy		
The conformity of IGT* elements with the shoreline in IGD*	Yes/No/Incomparable	The conformity of object depiction in IGT with shoreline in IGD
The conformity of IGT elements with the regional boundaries in IGD	Yes/No/Incomparable	The conformity of object depiction in IGT with regional boundaries in IGD
The conformity of the geometric elements of IGT with the other IGD elements and/or satellite images	Yes/No/Incomparable	The conformity of object depiction in IGT with shoreline in IGD

Notes: *IGT= Thematic Geospatial Information; IGD= Basic Geospatial Information

Table 11. Geospatial Information Quality Parameter (consistency) for One Map Policy
Source: *Modul Integrasi Informasi Geospasial Tematik* (Module of Integrated Thematic Geospatial Information), The Acceleration of One Map Policy Secretariate, 2016

Types of Information	Descriptions
Consistency	Improvement activity process includes primary field, attribute consistency, and consistency
Database	Improvements based on the standards of IGT mapping
a. Primary field	The improvements of several main pieces of information (fields) contained in the database
b. Attribute Consistency	The improvements of geospatial attribute
c. Standard Consistency	
c.1 Classification-based standard consistency c.2. Standard consistency according to geodatabase	Improvements based on the standards of IGT mapping

Types of Information	Descriptions
Topology a. Point	The improvements refer to 1) a point inside a polygon and 2) no duplication at the same location
b. Line	The improvements refer to 1) no overshoot and undershoot lines, 2) no intersection between different lines, and 3) no stacking lines
c. Area	The improvements refer to 1) no overlapping areas and 2) no abutting empty areas

4.2. GIS dan Web-GIS

Geographic Information System (GIS) is a tool used to acquire, manage, analyze, and present every information related to space or spatial properties. GIS is capable of displaying various complicated data on a map and presenting an easily readable, accessible, and understandable information [1]. GIS grows along with the integration of web system and the internet and hence, it can synchronize spatial information in real time. With the internet, the distribution process of a considerable number of data becomes more efficient compared to manual dissemination.

In the development, GIS and WebGIS-based analyses are increasingly adopted for a variety of purposes, such as environment and resource analysis [11], [15], [18], [21], [26]; rainfall distribution analysis [3], mining analysis [16], [19]; monitoring activities [7], [22], [25]; disaster analysis [1], [4], [17], [28]; and urban planning [12], [13], [14], [27]. Stancovic et al. [19] demonstrated the benefits of WebGIS system for the management of abandoned mining sites in Serbia. In line with this study is Herjuna et al. [23] that employed WebGIS to structurize the mining sector in Indonesia.

Aye et al. [28] studied the utilizations of a collaborative WebGIS to manage the risks of hydrometeorological hazards in different countries in Europe as well as a WebGIS-based spatial application like jecami.eu (developed in Switzerland) to monitor the ecological conditions of the Alpine [11]. These two studies are examples of the developments of GIS and WebGIS utilizations in urban and resource planning. All of them used data that were obtained from relevant authorities. In other words, the data used in their studies meet the specific standards and quality or, in other words, they are accountable. Consequently, web-based GIS application becomes significant in supporting the actualization of One Map Policy.

Nowadays, many research and development projects apply web services to the GIS systems in various companies and institutions [27]. The ability to store and process the necessary data that refer to an area or space induces this system's wide utilization as a tool for solving the problems in management and planning. The mapping method employed by this system gives users a considerable amount of information visually, which can function as a reference in the predictions of many objects and events [21]. GIS is a suitable tool for planners to analyze their target areas.

WebGIS application improves the efficiency of information dissemination to stakeholders, particularly decision makers. Siti Maneesa Amiruddin [22] studied the web-based GIS application like ESDM One Map Indonesia and its functions in assisting staffs to organize and analyze relevant data and to publish information in the forms of maps, diagrams, and reports. The benefits of applying a real-time WebGIS to resource monitoring [22] include te reduction of monitoring cost, the acceleration of decision-making process (due to the ability to make decisions based on real-time information), and the extensive accessibility (everyone can access it via the ubiquitous internet).

GIS allows its users to conduct a more detailed spatial analysis in a region [16]. There have been various benefits of a collaborative GIS in ESDM One Map Indonesia.

a. Every working unit responsible for data management can erase the conflicts that emerge due to different versions of the data used as a map reference.
b. It is effective and time- and resource-efficient.

IOP Conf. Series: Earth and Environmental Science **165** (2018) 012021 doi:10.1088/1755-1315/165/1/012021

c. It guaranteed that the accessed data is the latest (up-to-date) version
d. The availability of a complete, accurate, and up-to-date data provides an opportunity for stakeholders to increase their investment and production of energy and mineral commodities.

GIS is a computer system used to capture, store, check, and display data related to positions on the Earth's surface [3]. It can show various types of data on one map, which then allows people to observe, analyze, and understand patterns and relationships more easily. With GIS technology, people can compare the locations of many objects and events to find how they relate to each other [12]. Overall, GIS is a suitable tool for organizing information and involving the following operations, i.e., data planning, observation, collection, storing, management, analysis, to produce usable information in the decision-making process [3]. Web-based GIS enables people, even those with no technical GIS background, to use it as a decision-making tool and to observe as well as modify real-time information [8] for analysis of energy and mineral resource potentials [17].

4.3. ESDM One Map Indonesia

ESDM One Map Indonesia is a web-based information system capable of displaying various thematic geospatial information (maps) in the field of energy and mineral resource online (webGIS). As illustrated in **Figure 5**, the Minister of EMR, Ignasius Jonan, officially launched ESDM One Map Indonesia on May 31, 2017. This application presents a variety of thematic geospatial information obtained from different sectors of MEMR, namely Oil and Gas, Mineral and Coal, New Renewable Energy and Energy Conservation, Electricity, Geology, and Research Development Sectors. The information includes, for instance, the maps of Oil and Gas Working Area, Mining Business License Area, Geothermal Working Area, Electricity Infrastructure, Oil and Gas Infrastructure, and Potentials (Resources and Reserves) of New Renewable Energy and Geology. This system also provides thematic geospatial information from other sectors in different ministries or institutions.

Figure 5. The Launching of ESDM One Map Indonesia on May 31, 2017. The Minister of EMR gave a direction during the launching event (a), the Head of the Center for Data and Information Technology assisted the Minister and Secretary General of EMR to try the application of ESDM One Map Indonesia (b), the Minister of EMR signed the map in ESDM One Map Indonesia (c), and committees assisted the guests to try the analytical features in ESDM One Map Indonesia. Source: Center for Data and Information Technology MEMR, 2017

The initial purpose of ESDM One Map Indonesia was to create a regionally-arranged data inventory in MEMR. However, this application develops into a platform for the decision makers in MEMR to plan EMR infrastructures and analyze mining licenses and existing EMR potentials.

Figure 6. (a) The Home Page and (b) After Log-In Window in the Portal
Source: Portal ESDM One Map Indonesia Indonesia, geoportal@esdm.go.id, 2017

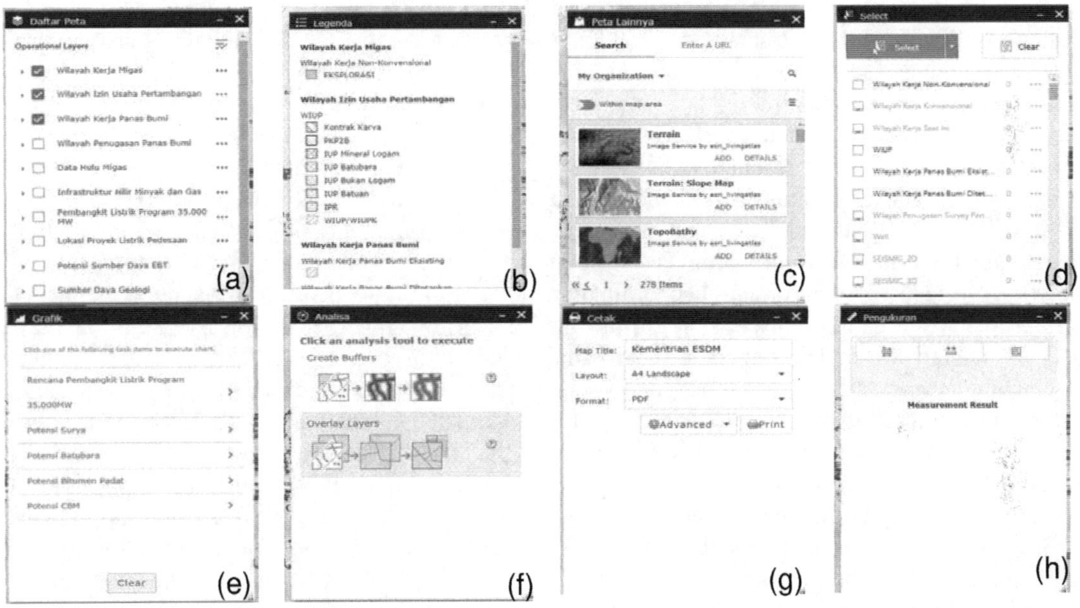

Figure 7. Various Menus in ESDM One Map Indonesia, i.e., (a) Map List, (b) Legends, (c) Search for Other Maps, (d) Selecting and Statistic Tools, (e) Graphics, (f) Analysis Tools, (g) Print, and (h) Measurement Source: Portal ESDM One Map Indonesia Indonesia, geoportal@esdm.go.id, 2017

Figure 6 shows the home page and after log-in window in the Portal of ESDM One Map Indonesia, while **Figure 7** shows the menu panel in the portal. The top right side in the after log-in page offers several options for ArcGis basemap or Openstreet Map, viz. topographic map, road network map, and satellite image, that facilitate various perspectives and interests in some analyses. The various menus in ESDM One Map Indonesia accommodate the diverse needs of users. The 'Map List' shows the default maps in the portal, while the other maps can be searched using 'Search for Other Maps'. Every map in this list automatically displays the description of the objects on the menu 'Legend'. For a simple analysis like distance and area, users can use 'Measurement'. Users can also find a simple statistic of the selected data from 'Selecting and Statistic Tools'. Moreover, they can display the graphics of various selected maps using the menu 'Graphics', further analyze some particular data (e.g., land use overlap) using 'Analysis Tools', and print the resultant map from the menu 'Print'.

Currently, ESDM One Map Indonesia has been integrated with other geographic information systems from some units of MEMR, namely the Special Work Unit of Oil and Gas and the Directorate-General of Mineral and Coal. The Portal of ESDM One Map Indonesia has organized hundreds of data from all units of MEMR, including the Oil and Gas Working Area, Data of Upstream Oil and Gas Activities, Mining Business License Area, Electricity Infrastructure, Geothermal Working Area, Potential of New Renewable Energy and Energy Conservation, and Natural Resource and Geological Disaster. It also displays the data obtained from other ministries. These data are Administrative Boundaries, Transportation Infrastructures, and Forest Areas. This system can be accessed publicly or by using an account (username and password) for data editing purpose and internal use of MEMR units. All data presented in ESDM One Map Indonesia has gone through layers of verification and validation process by either the data management team at the Center of Data and Information Technology or the sources (units or institutions) of related data.

4.4. ESDM One Map Indonesia in One Map Policy

The development of ESDM One Map Indonesia is the manifestation of one objective of the Energy Coordination and Supervision Activities initiated by the Corruption Eradication Commission and MEMR. The objective is to develop an integrated data system as an implementation of One Map Policy in MEMR. The EMR data has a strategic role in decision making. The available spatial data of energy and mineral resource is distributed in different working units in MEMR, business entities, and related stakeholders. Furthermore, the spatial data is dynamic, i.e., changing over time. Also, there is a high demand from the leaders in MEMR and the public on open access to the data and information about energy and mineral resource. The need of easily acquired, complete, and accurate data is the reason why ESDM One Map Indonesia becomes necessary and in line with One Map Policy.

In the National Geospatial Information Network (*Jaringan Informasi Geospasial Nasional* -JIGN), the Center for Data and Information Technology acts as a network node in MEMR. It performs the functions of geospatial information coordination, integration, and dissemination between the Data Representatives (i.e., those responsible for the data) in MEMR and other ministries/institutions. It also acts as a Network Node Connector (Geospatial Information Agency). The role of this center in organizing thematic spatial data about energy and mineral resource is affirmed in the Act of Geospatial Information (Act No. 4/2011) [4] and the National Geospatial Information Network (as a network node). Regarding the integration and dissemination of EMR thematic geospatial information, the existence of ESDM geoportal (ESDM One Map Indonesia) is strongly significant in providing the tools that can meet the demand for thematic geospatial information online. ESDM One Map Indonesia, built by the Center for Data and Information Technology, has also met the standard for geoportal application specified by the Geospatial Information Agency as the Network Node Connector.

ESDM One Map Indonesia provides a variety of information on EMR potential and infrastructure. Some of which can be used to identify the distribution of filling stations, which support the implementation of one fuel price program, to display the statistical potential of EMR and attract investors to develop EMR infrastructures, to analyze regional zonation, and to analyze land-use overlap in spatial planning, as presented in **Figure 8**.

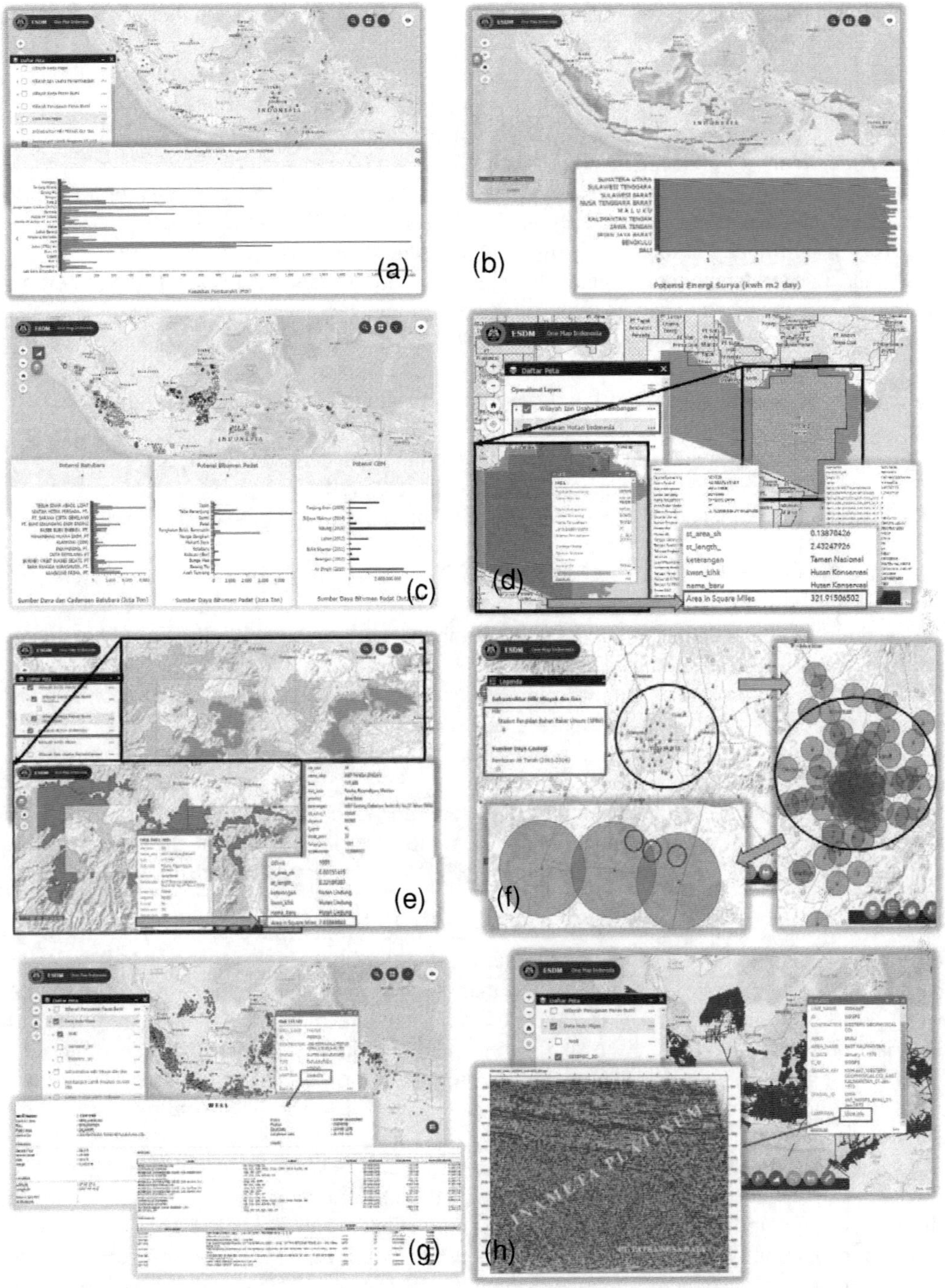

Figure 8. (a) Analyses of the Potent Resource of Power Generator, (b) Solar Energy Potential, (c) Geological Resource Potential, (d) Land Use Overlap between Mining Areas and Conservation Forest Areas, (e) Land Use Overlap between Geothermal Working Area and Protected Forest, (f) Buffer Distance between Filling Stations and Groundwater Drilling Sites, (g) Integration of Non-spatial Data (Detailed Data of Wells) and Well Distribution, (h) Integration of Non-spatial Data (Detailed Seismic Data) and Well Distribution
Source: Center for Data and Information Technology MEMR, 2017

ESDM One Map Indonesia is expected to provide solutions for a collaborative Geographic Information System, namely:

1. Eliminating any problems emerging due to different versions of the data used as a reference and giving a guarantee that the accessed data are the latest (up-to-date) version.
2. Increasing the effectiveness and efficiency of time and resources in data management and dissemination. Working units are responsible for their own data.
3. Supporting the leaders in the decision-making process and providing a vast opportunity for stakeholders to increase their investments and productions of EMR commodities.

The benefits of ESDM One Map Indonesia in One Map Policy are as follows:

a. To facilitate the preparation of large-scale space utilization planning using an integrated spatial plan document.
b. To facilitate and accelerate the settlement of land use conflicts, including the ones in customary lands.
c. To accelerate the implementation of the development programs, including the development of areas and infrastructures.
d. To facilitate and accelerate the completion of the regional boundaries in Indonesia.
e. To facilitate and accelerate the issuance of land use-related permits.
f. To facilitate the implementation of any simulations that require maps, such as disaster mitigation, environmental preservation, and defense.
g. To improve the reliability of information related to various economic activities, which will provide a much-needed business assurance at this time.

Apart from the many benefits mentioned above, some critical notes on ESDM One Map Indonesia must be considered in the future, such as:

a. The data owners, who have submitted their data to the Center for Data and Information Technology, are less responsive and participative in monitoring the results on ESDM One Map Indonesia, especially regarding the precision of object location (e.g., the inland generators that are still located in the sea or the locations of any companies that shift after data conversion).
b. The data format used by the units responsible for the same resources is different. The unification of the data attribute tends to take much time (e.g., the electricity data has different formats for resource inventory). The variety of data formats affects the data attribute in ESDM One Map Indonesia)
c. The differences in the application or platform used in the units (internal sector of MEMR) and other ministries/institutions. The diversity complicates the integration process. Until recently the data that can be integrated directly with the map service are those coming from Elnusa (PND), Special Work Unit of Oil and Gas (SKK Migas), Mineral and Coal Sector, and some other ministries or institutions like Geospatial Information Agency (which provides the data of administrative boundaries). However, since the application of ESDM One Map Indonesia relies strongly on internet connection, a network interference on data sender's location causes the map service to not appear.

ESDM One Map Indonesia enables accessibility and a collaborative interaction between different stakeholders [28]. Regarding legal responsibility, a real collaborative decision making is not always possible and is beyond the capacity of the decision support system. This platform supports a collaborative interaction between different stakeholders to actualize a better and transparent decision-making environment. The frame of this platform adopts a common design so that it is applicable and highly flexible in different regions. The types of users and the levels of involvement and interaction in the platform depend on the institutional setting, role, and responsibility of each user in a particular area.

ESDM One Map Indonesia is a significant effort of MEMR, along with the supports and data collaborations from other ministries or institutions, to actualize the One Map Policy with WebGIS application. The willingness of MEMR still requires many supports especially to advance the Web-

GIS application constantly and to be able to provide services through ESDM One Map Indonesia continuously.

5. Conclusion

ESDM One Map Indonesia can be used for various regional analyses, for instance, the analyses of resource potential, land use overlap, and the integration of non-spatial and spatial data. It can be accessed using computers and mobile phones from anywhere by anyone and at any time as long as internet connection is available. It can also function as a source of information and the basis of policy and decision-making process, all of which promote the actualization of One Map policy on a national level.

References

[1] A Garagon Dogru, T Selcuk, H Ozener, O Gurkan and G Toz Developing A Web-Based Gis Application For Earthquake Information
[2] A A Alesheikh, H Helali and HA Behroz 2002 Web GIS: Technologies and Its Applications. In : *Symposium on Geospatial Theory, Processing and Applications,* Ottawa.
[3] A A Kadir, M Kaamin, N S Azizan, S Sahat, S M Bukari, M Mokhtar, N Ngadiman and N B Hamid 2015 Application of Geographic Information Systems (GIS) in Analysing Rainfall Distribution Patterns in Batu Pahat District. In : Soft Soil Engineering International Conference 2015. *IOP Conf. Series: Materials Science and Engineering* 136, doi:10.1088/1757-899X/136/1/012064
[4] A A Mustaffa, M F Rosli, M S Abustan, R Adib, M I Rosli, K Masiri and B Saifullizan 2015 A Study of Flood Evacuation Center Using GIS and Remote Sensing Technique. In : Soft Soil Engineering International Conference 2015. *IOP Conf. Series: Materials Science and Engineering* 136. doi:10.1088/1757-899X/136/1/012078
[5] Anoni mous 2011 Undang- Undang Nomor 4 Tahun 2011 tentang Informasi Geopsasial, Jakarta
[6] Anonimous 2016 Peraturan Presiden No. 9 Tahun 2016 tentang Percepatan Kebijakan Satu Peta
[7] Arun Kumar Pratihast, Ben DeVries, Valerio Avitabile, Sytze de Bruin, Martin Herold and Aldo Bergsma 2016 Design and Implementation of an Interactive Web-Based Near Real-Time Forest Monitoring System. *PLOS ONE,* DOI:10.1371/journal.pone.0150935
[8] Dou Changyong, Guo Huadong, Han Chunming, and Liu Ming 2014 An Open Source Software and Web-GIS Based Platform for Airborne SAR Remote Sensing Data Management, Distribution and Sharing. In : 35th International Symposium on Remote Sensing of Environment. *IOP Conf. Series: Earth and Environmental Science* 17. doi:10.1088/1755-1315/17/1/012255
[9] ESRI 2012 ArGIS 10 Functionality Matrix. USA
[10] ESRI 2006 GIS Solutions for Urban and Regional Planning (Designing and Mapping the Future of Your Community with GIS). USA
[11] Fuad Hasyim, Habib Subagio, and Mulyanto Darmawan 2016 One map policy (OMP) implementation strategy to accelerate mapping of regional spatial planing (RTRW) in Indonesia. In : 8th IGRSM International Conference and Exhibition on Remote Sensing & GIS, Kuala Lumpur, Malaysia. *IOP Conf. Series: Earth and Environmental Science* 37, doi:10.1088/1755-1315/37/1/012054
[12] Kamil Faisal, Ahmed Shaker and Suhaib Habbani 2016 Modeling the Relationship between the Gross Domestic Product and Built-Up Area Using Remote Sensing and GIS Data: A Case Study of Seven Major Cities in Canada. *International Journal of Geo-Information ISPRS Int. J. Geo-Inf.* 2016, **5**, 23; doi:10.3390/ijgi5030023. www.mdpi.com/journal/ijgi
[13] Kathryn Clark and Suresh Muthukrishnan. Development and Application of a WebGIS based Facilities Management and Decision Support System. Department of Earth and Environmental Sciences, Furman University, Greenville, SC 29613

[14] Martin Loidl, Gudrun Wallentin, Rita Cyganski, Anita Graser, Johannes Scholz and Eva Haslauer 2016 GIS and Transport Modeling—Strengthening the Spatial Perspective. *International Journal of Geo-Information ISPRS Int. J. Geo-Inf.* 2016, **5**, 84; doi:10.3390/ijgi5060084 . www.mdpi.com/journal/ijgi

[15] Noor Yasmin Zainun, Ismail Abdul Rahman and Rosfazreen Azwana Rothman 2016 Mapping Of Construction Waste Illegal Dumping Using Geographical Information System (GIS). In : International Engineering Research and Innovation Symposium (IRIS). *IOP Conf. Series: Materials Science and Engineering* 160. doi:10.1088/1757-899X/160/1/012049

[16] O Alipbeki, G Kabzhanova, G Kurmanova and Ch Alipbekova 2016 Application of GIS technologies to monitor secondary radioactive contamination in the Delegen mountain massif. In: 8th IGRSM International Conference and Exhibition on Remote Sensing & GIS. *IOP Conf. Series: Earth and Environmental Science* 37 doi:10.1088/1755-1315/37/1/012080, Kuala Lumpur, Malaysia

[17] Qianna Wang, Martin Mwirigi M'Ikiugu, Isami Kinoshita and Yanyun Luo. 2016. GIS-Based Approach for Municipal Renewable Energy Planning to Support Post-Earthquake Revitalization: A Japanese Case Study. *Sustainability* 2016, **8**, 703; doi:10.3390/su8070703. www.mdpi.com/journal/sustainability

[18] Puyam S. Singh, Dibyajyoti Chutia, and Singuluri Sudhakar 2012 Development of a Web Based GIS Application for Spatial Natural Resources Information System Using Effective Open Source Software and Standards . *Journal of Geographic Information System*, 2012, **4**, 261-266. http://dx.doi.org/10.4236/ jgis.2012.43031 . Published Online June 2012 (http://www.SciRP.org/journal/jgis)

[19] Ranka Stankovi´c , Nikola Vulovi´c, Nikola Lili´c, Ivan Obradovi´c, Radule Tošovi´c and Milica Peši´c-Georgiadis 2016 A WebGIS Decision Support System for Management of Abandoned Mines. www.mdpi.com/journal/energies. *Energies* 2016, **9**, 567; doi:10.3390/en9070567

[20] Ravi Dhungel 2012 Web Mapping and Application Towards a Cloud: Enabling a WebGIS Prototype in an Open Source Environment. Research Paper. Department of Geography and Environmental Resources In the Graduate School Southern Illinois University Carbondale.

[21] S M Bukari, M A Ahmad, T L Wat, M Kaamin and N. Alimin 2015 Spatial Analysis in Determination Of Flood Prone Areas Using Geographic Information System and Analytical Hierarchy Process at Sungai Sembrong's Catchment. Soft Soil Engineering International Conference 2015, *IOP Conf. Series: Materials Science and Engineering* 136. doi:10.1088/1757-899X/136/1/012039

[22] Siti Maneesa Amiruddin 2016 Real-time Web GIS to monitor marine water quality using wave glider. In: 8th IGRSM International Conference and Exhibition on Remote Sensing & GIS. *IOP Conf. Series: Earth and Environmental Science* 37. doi:10.1088/1755-1315/37/1/012074

[23] Surya Herjuna, Dimar Wahyu Anggara and Bayu Raharja Pemanfaatan Aplikasi Sig Berbasis Web Untuk Penataan Sektor Pertambangan Indonesia. *Simposium Nasional Sains Geoinformasi IV* 2015

[24] The Swiss National Park 2014 JECAMI (Joint Ecological Continuum Analysis and Mapping Initiative) A Booklet on What Jecami is and how to use it. Swiss EU.

[25] Yu P Yuronen, E A Yuronen, V V Ivanov, I V Kovalev and P V Zelenkov 2015 The concept of creation of information system for environmental monitoring based on modern GIS-technologies and earth remote sensing data. TIAA2015. *IOP Conf. Series: Materials Science and Engineering* 94, doi:10.1088/1757-899X/94/1/012023

[26] Yudo Prasetyo, Setyo Ardi Gunawan and Zia UI Maksum. 2016. Determination of The Water Catchment Area in Semarang City Using a Combination of Object Based Image Analysis (OBIA) Classification, InSAR and Geographic Information System (GIS) Methods Based On a High-Resolution SPOT 6 Image and Radar Imagery. In : 2nd International Conference

of Indonesian Society for Remote Sensing (ICOIRS) 2016, Yogyakarta, Indonesia. *IOP Conf. Series: Earth and Environmental Science* 47, doi:10.1088/1755-1315/47/1/012027

[27] Xiaolin Lu 2006 Develop Web GIS Based Intelligent Transportation Application Systems with Web Service Technology. 6th International Conference on *ITS Telecommunications Proceedings.* 0-7803-9586-7 IEEE.

[28] Zar Chi Aye, Teresa Sprague, Vivian Juliette Cortes, Kathrin Prenger-Berninghoff, Michel Jaboyedoff and Marc-Henri Derron 2015 A collaborative (web-GIS) framework based on empirical data collected from three case studies in Europe for risk management of hydro-meteorological hazards. *International Journal of Disaster Risk Reduction* **15** (2016) 10-23. http://dx.doi.org/10.1016/j.ijdrr.2015.12.001 2212-4209/& 2015 The Authors. Elsevier Ltd.

Acknowledgment
Authors would like to thank the Corruption Eradication Commission for their support in coordinating and supervising data collection from various sectors, ministries, or institutions. Authors would also like to express their gratitude to all ministries/institutions that collaboratively contribute to ESDM One Map Indonesia, Mrs. Mira Muliati for her assistance in information technology and system, all staffs of the Center for Data and Information Technology (the Indonesian Ministry of Energy and Mineral Resource) especially those of Data Management Section, and all users of ESDM One Map Indonesia.

IOP Conf. Series: Earth and Environmental Science **165** (2018) 012022 doi:10.1088/1755-1315/165/1/012022

The Impacts Analysis of Pre And Post Merapi Mount Eruption on Residential Areas Using Sentinel 1, ALOS Palsar and Landsat Satellite Images Combination in 2009-2015

Yudo Prasetyo[1*], Riska Pratiwi[1] and Nurhadi Bashit[1]

[1]Remote Sensing and Photogrammetry Lab, Geodesy Engineering Dept, Faculty of Engineering, Diponegoro University, Jl. Prof. Soedharto SH, Tembalang-Semarang, Central Java, Indonesia, 50277

yudo.prasetyo@ft.undip.ac.id

Abstract. Indonesia is a country located in the presence of three large tectonic plates like the Indo-Australian, Eurasian and Pacific plates. The path of plate movement approaching or steering away each other leads to high volcanic and earthquakes activity. As a proof of the active path of this plate movement can be seen from the series of active volcanoes lies from the island of Sumatra, Java, Nusa Tenggara into Papua. The impact of active volcano activity will be felt in the densely populated area where the island of Java became one of the islands with the largest residential area in Indonesia. This impact needs to be anticipated with an integrated high-tech technology and volcanic disaster mitigation system. One of the most active volcanoes in the island of Java is Merapi mount which is experienced the last major eruption peak on October 26th, 2010. This volcanic eruption is effusive eruption type where magmatic gas pressure in the crater is not too strong and magma eruption is just flowing out past the slopes of the Merapi mount area. However, magmatic gas pressure and magma volume still result in deformation changes that have a direct impact on residential areas throughout the Merapi mount area. Therefore, this research performs deformation calculation using DInSAR (Differential Interferometry Synthetic Aperture Radar) method on Sentinel 1 SAR and ALOS PALSAR 1 satellite images between 2005-2015 acquisition year. Then the deformation patterns obtained from the DInSAR method will be spatially correlated with residential areas throughout the Merapi mount area. Residential areas were obtained through supervised classification process from Landsat 5 and 8-satellite imagery in the 2005-2015 acquisition year. The reason of observation year selection is based on pre and post eruption concept to get pattern of Merapi's mountain body change through deformation analysis. The result of the combination of deformation pattern and residential area changes was further analysed to obtain spatial correlation. The value of this spatial correlation will be an indicator of how's far the deformation changes due to volcanic activity will have an effect on the residential areas in the Merapi mount area. The output of this research is a map of Merapi mount deformation (due to volcanic activity) impacts on the residential area. The results are expected to contribute to the volcanic disaster mitigation system and the development of residential area plans in Merapi mount area.

Keyword: ALOS, DInSAR, Disaster, Sentinel, Volcanic

1. Introduction

Indonesia is a country located Indonesia is a country located at the meeting of three large tectonic plates of the Indo-Australian, Eurasian and Pacific Plate [6]. The movement of the approaching plates will plunge into the bottom of the younger plate.

Behind the plate subduction path will form a series of magmatic and volcanic activities. This causes the Indonesian territory has many volcanoes scattered from the island of Sumatra, Java, Nusa Tenggara to Papua. Java Island has several volcanoes that are still active, one of which is Mount Merapi. Mount Merapi is located between several districts in the Central Java Province. The south side slopes are in the administration of Sleman Regency and the Special Province of Yogyakarta, Magelang Regency on the west side, Boyolali Regency on the north and east side and Klaten Regency on the southeast side. For geographical area of research lies in latitude position 110°14'60" E- 110°32'30" E and in longitude position 7°29'47" S-7°47'53" S and can be seen clearly in **Figure 1.**

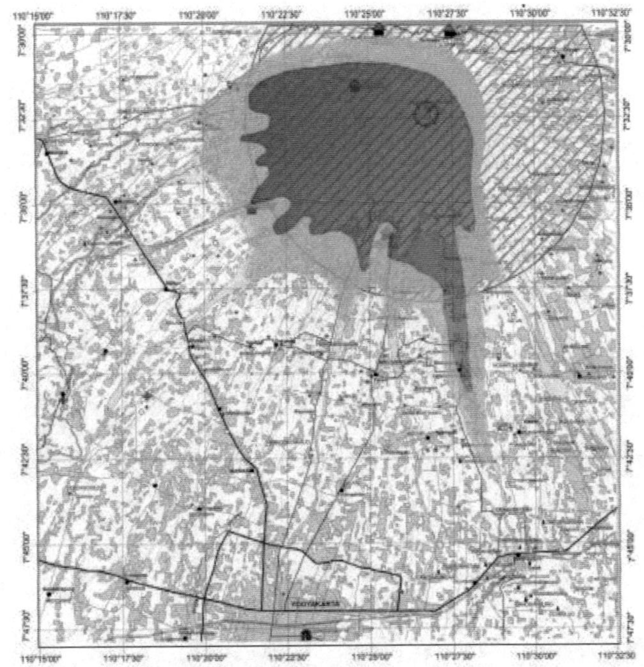

Figure 1. Research Area on Merapi Volcanic Mount (BNPB, 2016)

Mount Merapi has a topography height around 2,930 meters above mean sea level. Mount Merapi eruption occurred on October 26th, 2010. Merapi volcano eruption is a type effusive eruption in the form of lava melts where the pressure of magmatic gas is not so strong. For the flow of magma coming out of the cavity just flows through the slopes on the top of the mountain. Although only a type effusive eruption, this eruption causes many physical, environmental and economic losses. The flow of lava flow will affect the land cover around the area, so there will be changes in land cover before and after the eruption of Mount Merapi. This causes significant deforestation (deformation) before the eruption. Volcanic deformation occurs due to volcanic activity in the form of magma movements below the surface that affect the changes in pressure on the pocket of magma. The presence of submarine motion is an early indication of eruption and an increase in pressure that will result in ground deformation. This research uses remote sensing technology using active sensor data and passive sensor. Active sensor data used is ALOS PALSAR and Sentinel-1, which is used to analyse pre and post eruption using DInSAR (Differential Interferometry Synthetic Aperture Radar) method. This research will analyse the pattern of deformation that occurred at Mount Merapi in the time before and after the eruption. Includes an analysis of the impacts on changes in land cover in the research area.

2. Materials and Methods

2.1. Research Methodology

For research data used consist of spatial and attribute data where the data research details can be seen in **Table 1.**

Table 1. Research Data

No.	Description	Acquisition Time
1.	ALOS PALSAR Images	2009-2010
2.	Sentinel 1 1W Images	2015
3.	Landsat 5	2009
4.	Landsat 8	2015
5.	SRTM DEM image	2010
6.	TerraSAR DEM image	2011
7.	Administration Map of Yogyakarta Province and Central Java	2016
8	Merapi Volcanic Hazard Map	2009-2016

In this research, there are several research stages that have been done; the outline of research methodology is described in **Figure 2** and DInSAR processing stage is described in **Figure 3**

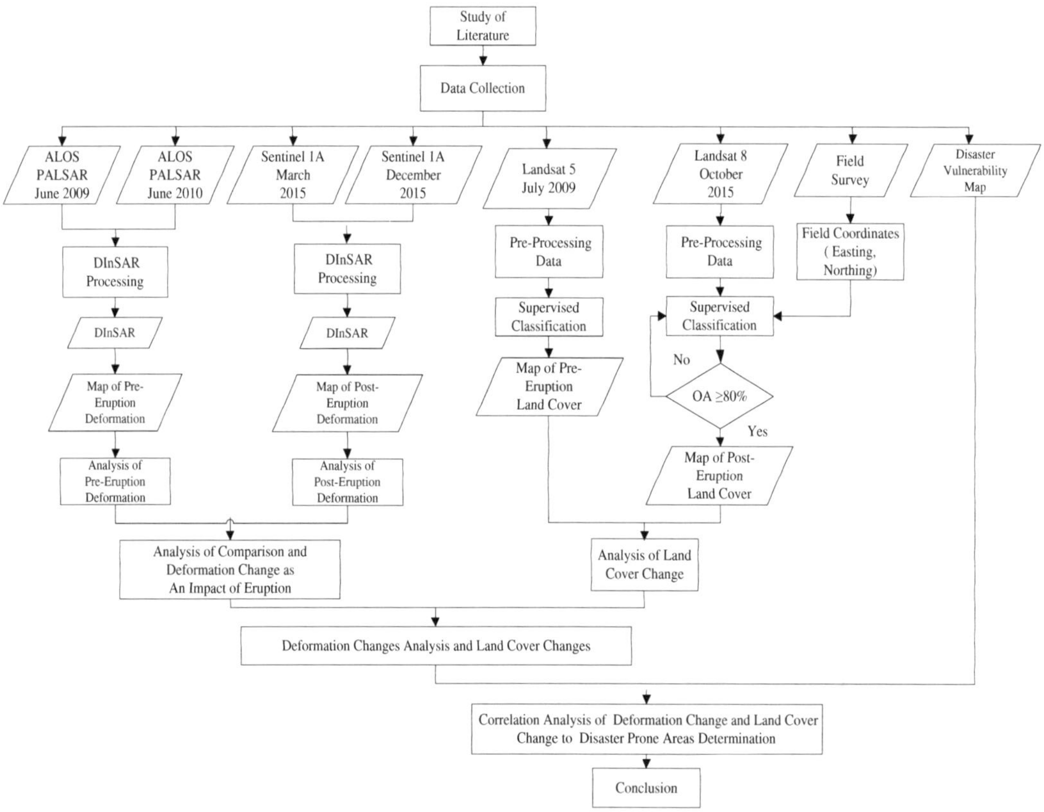

Figure 2. Research Methodology

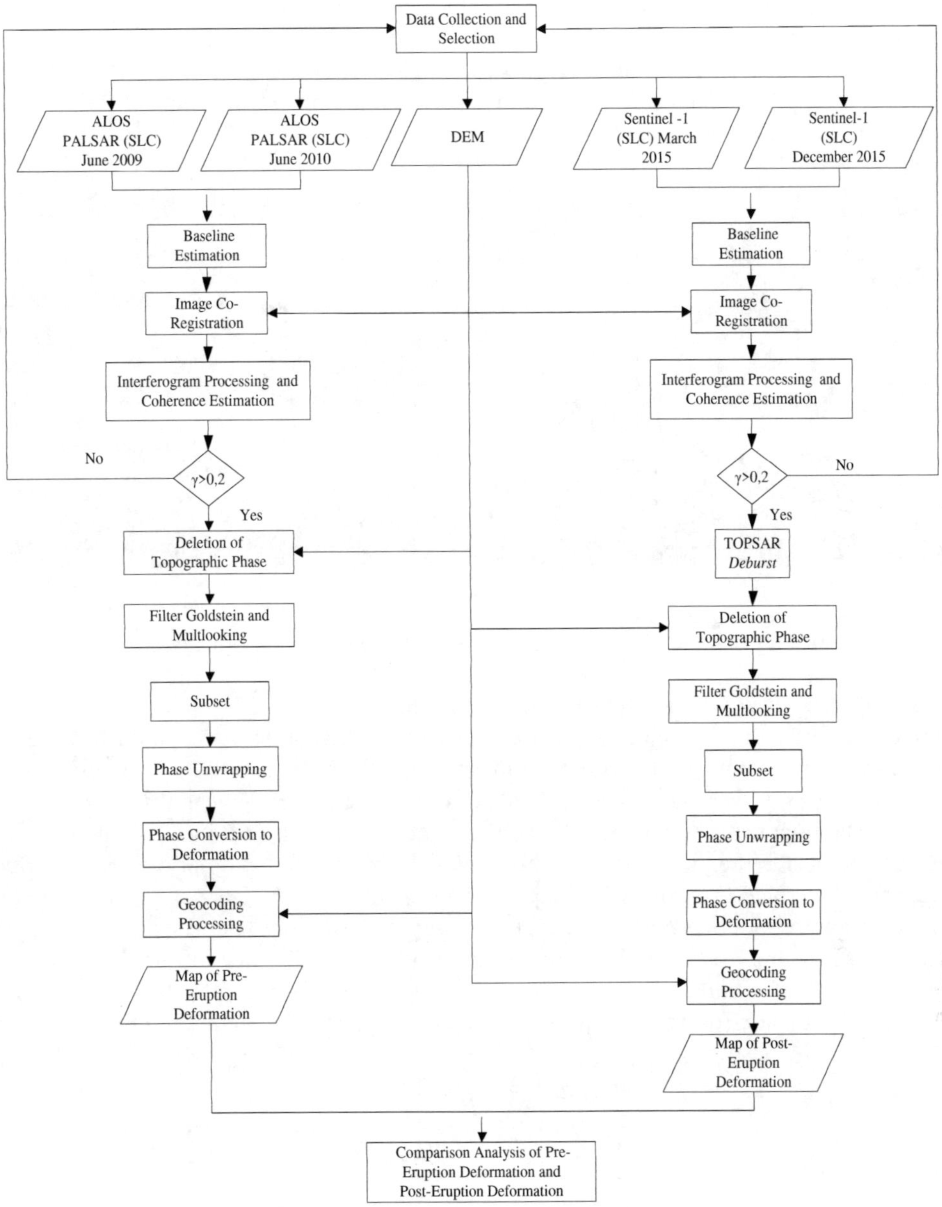

Figure 3. DInSAR Processing Stage

2.2. Deformation on Volcanic Mount

Land surface deformation occurs because of changes in magma pressure or magma intrusion [3]. Usually deformation is expressed by displacement of horizontal, vertical and tilting direction [8]. In volcanic areas there is often a change in topography caused by lava flows or eruptions. These changes may be additional materials. Another case, sometimes the magma activity in the magma bag does not produce additional material on the surface. The result of all of the magma activity in the volcanic mountain is deformation on land surface. In principle the deformation of the volcano body can be an increase in soil surface (uplift) or a decrease in soil surface (subsidence). Deformation in the form of inflation generally occurs because the process of magma movements to the surface that presses the surface of the soil above it. Maximum deformation is usually observed shortly before volcanic eruptions take place. Deformation of deflation generally occurs during or after eruption. At that time the magma pressure inside the volcano's body has weakened, causing the surface to tend to return to its original position. Symptoms of volcanic deformation will cause a shift in the position of a point in the body of a volcano. This position shift can occur in both horizontal and vertical directions. According [7], the value of this shift can reach tens of meters on a silicone volcano that forms a lava dome.

(a) (b)

Figure 4. (a) Explosive Type (b) Effusive Type (Explore Volcanoes, 2008)

2.3. DInSAR (Differential Interferometry Sythetic Aperture Radar)

Differential Interferometry Synthetic Aperture Radar (DInSAR) is an acquisition technique of two paired SAR images in combination of complex image data at the same spatial position (differential SAR) or slightly different position (terrain height InSAR) in the same area by multiplication of multiple conjugations. The final result is a digital elevation model (DEM) or a shift in a surface of the earth [2]. DInSAR is often used to monitor changes (deformation) of an area to the order of order cm. Such order of thoroughness, obtained from InSAR data processing methods in different way. In this method used several pairs of interferogram at once to detect the topographic surface changes with very high accuracy [5]. In the interferogram, the phase difference is described as the distance difference measured by the line of sight (LOS) radar and includes topography, orbit shifts, surface deformation and atmospheric effects [1]. The following **Equation 1** can show the phase difference equation in DInSAR method.

$$\Delta\varphi = \Delta\varphi_{topo} + \Delta\varphi_{defo} + \Delta\varphi_{atm} + \Delta\varphi_{orb} \quad (1)$$

In:

$\Delta\varphi$ = Phase Differential
$\Delta\varphi_{topo}$ = Topography Phase
$\Delta\varphi_{defo}$ = Deformation Phase
$\Delta\varphi_{atm}$ = Atmospheric Phase
$\Delta\varphi_{orb}$ = Orbital Phase

There are two ways to implement DInSAR method in this research. One is with an external elevation model (DEM) and by using extra radar imagery to form topography model. The method using an external DEM requires only two radar images, assuming the deformations that occurred during the acquisition and known by the two-pass method. The second method does not require data other than radar data. The topographic reference here is made with additional radar data obtained, three or four images (used) and sequentially called by the three-pass and four-pass method.

3. Results and Discussion

3.1. Deformation analysis for volcanic pre-eruption in Merapi Mountain

The internal RMSe value of pre-eruption deformation processing results is 0.0044 ± 0.1933 m/yr. The result of pre-eruption deformation can be seen in **Figure 5.**

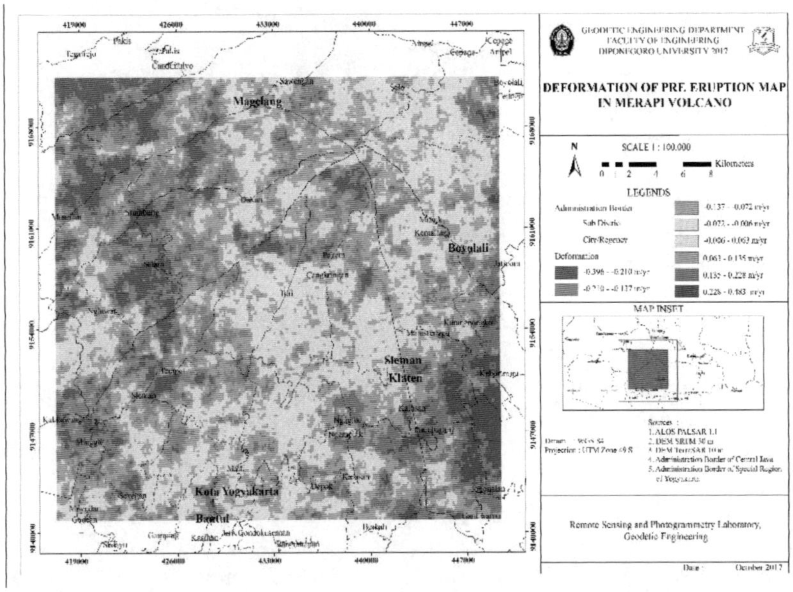

Figure 5. Deformation of Pre-Eruption in Merapi Volcanic Mount

In **Figure 6** it can be seen that the highest land subsidence in land surface occurred in Candimulyo District, Magelang Regency with decrease reach -0.2109±0.067 m/yr. The land subsidence rate calculation is earned by searching for the highest value of land subsidence for each districts administration. For the land subsidence rate is obtained by finding the average value of land subsidence from the DInSAR processing data. This land subsidence phenomenon is common happened in Magelang and Sleman districts. This is because the two districts are directly adjacent to the peak of Mount Merapi, so it can cause a land subsidence in the ground surface when there is an unstable magma movement before the eruption. This land subsidence may also occur due to dense development activities in both districts.

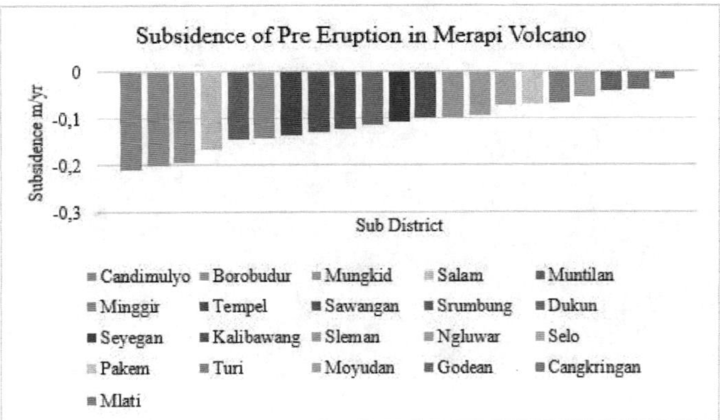

Figure 6. Land Subsidence in Pre-Eruption Phase in Merapi Volcanic Mount

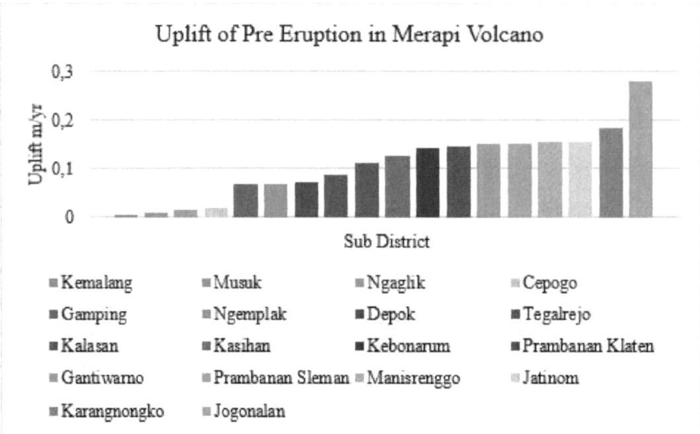

Figure 7. Uplift in Pre-Eruption Phase in Merapi Volcanic Mount

The highest rise of the land subsidence and uplift of land surface in **Figure 6** and **Figure 7** is located in Jogonalan Sub-district, Klaten Regency with uplift on land surface around 0.2788±0.0803 m/yr. The cause of this land surface rise is the lack of perfect phase unwrapping process, so there are still many artefacts that have not disappeared and cause the final result to be less suitable, especially in the southern part of the study area located in Klaten regency.

3.2. Deformation analysis for volcanic post-eruption in Merapi Mountain

The internal RMSe value of post-eruption deformation processing is -0.0346 ± 0.0441 m/yr. The result of post-eruption deformation can be seen in **Figure 8**. In **Figure 9** the highest land subsidence in land surface reaches -0.0871 ± 0.0024 m/year in Tegalrejo Sub-district, Jogjakarta City. This land subsidence is common in Sleman and Magelang regencies. This is because the two districts are directly adjacent to the peak of Mount Merapi, so that it can be affected directly during the eruption. The immediate impact is the flow of lava that flows and can cause a decrease in the face of the soil. This land subsidence may also occur due to dense development activities in both districts.

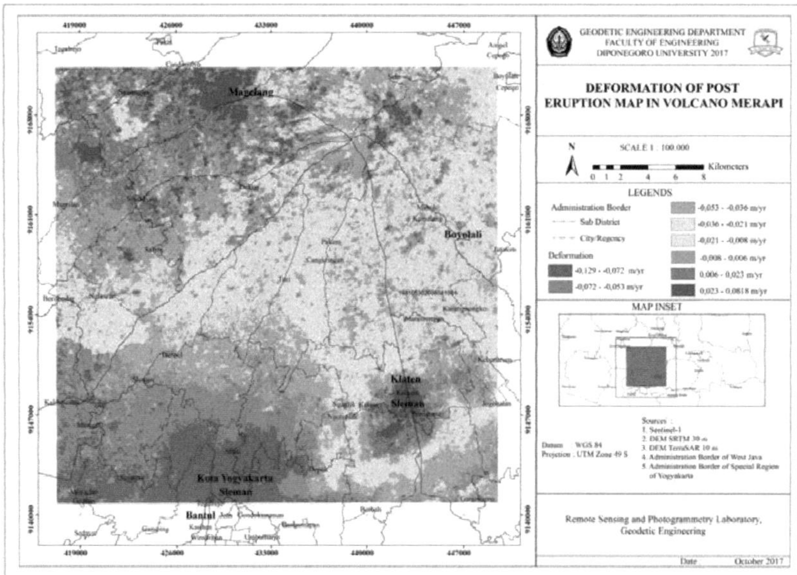

Figure 8. Deformation of Post Eruption in Merapi Volcanic Mount

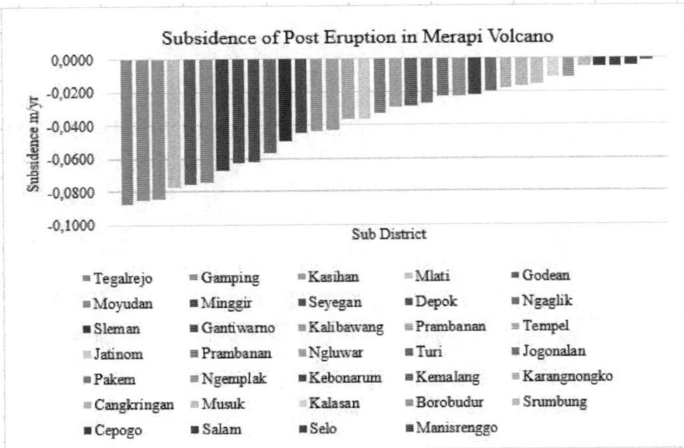

Figure 9. Land Subsidence in Post Eruption Phase in Merapi Volcanic Mount

The highest uplift is in Sawangan District, Magelang Regency with an increase of 0.0148 ± 0.0184 m/year can be seen in **Figure 10**. One of the causes of post-eruption soil eruption is the process of phase unwrapping is not perfect, so there are still many artefacts that have not been lost. The deformation change is obtained from the pre and post eruption volumetric values. The changes are classified into 3 classes, namely low, medium and high classes. This classification is based on the highest and lowest values then divided equally. The lowest deformation change value was 0.0187 m/year and the highest was 0.3056 m / yr. The lowest grade has a deformation change value of 0.0187 m/year up to 0.0749 m / yr. The moderate class has a deformation change value of 0.0749 m/year up to 0.1667 m / t and the highest grade has a deformation change value of 0.1667 m/year up to 0.3056 m/yr.

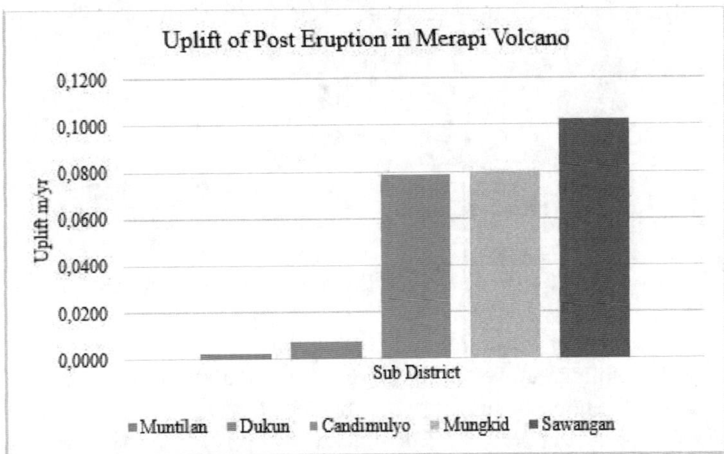

Figure 10. Uplift in Post Eruption Phase in Merapi Volcanic Mount

The results of the classification resulted in 16 sub-districts entering the lower classes, 12 sub-districts entering the middle class and 11 sub-districts entering the highest class. The highest rate of deformation change occurred in Jogonalan Sub-district, Klaten regency with a change value of 0.3056 m/year. The lowest rate of deformation change occurred in Moyudan sub-district, Sleman District with a change value of 0.0187 m/yr. One of the causes of high deformation changes in the southern areas of the study area is the process of unwrapping processing is less than the maximum, both on the result of pre-eruption deformation. The comparison of pre and post eruption deformation values and the deformation change map is overlaid as shown in **Figure 11** and **Figure 12**

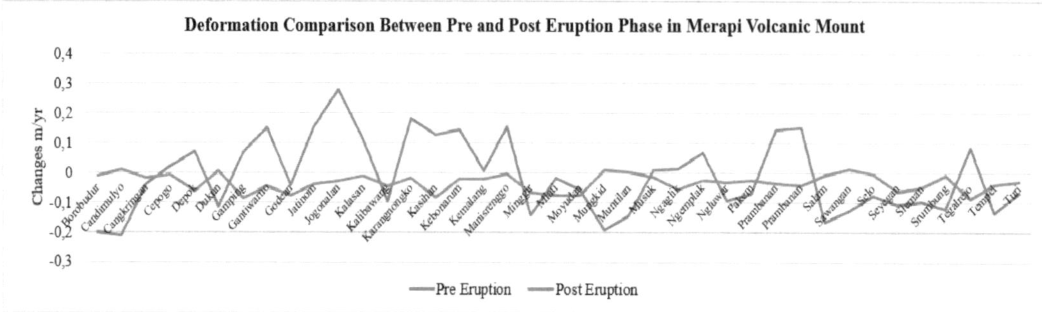

Figure 11. Deformation Comparison Between Pre and Post Phase in Merapi Volcanic Mount

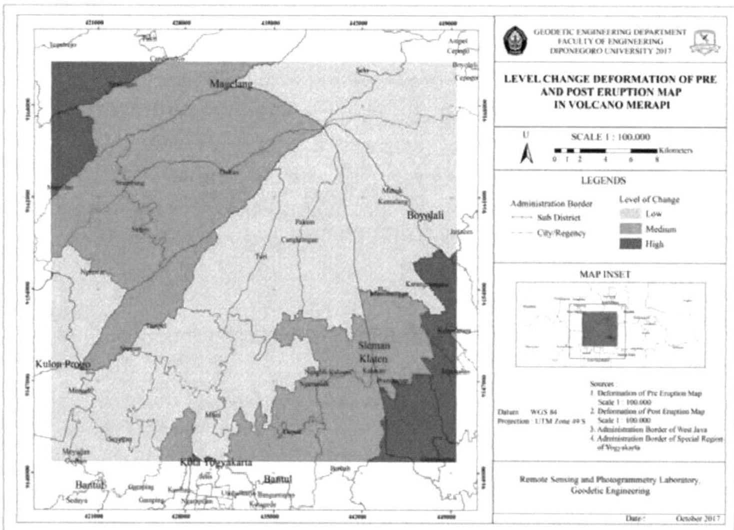

Figure 12. Deformation Changes Between Pre and Post Phase in Merapi Volcanic Mount

3.3. Volumetric analysis based on pre and post eruption in Merapi Mountain

Volumetric pre and post eruption can be obtained from the calculation of area and distance shift from the surface of the earth to the point of shifting. The distance of this shift is obtained from the speed of decline and the increase calculated per year. The objective of research to find the volumetric deformation value from radar image observation is to observe the pattern of land surface change in the period pre and post-eruption. The change pattern will be seen and observe to know the influence into land cover change pattern in the research area.

In **Table 2** and **Table 3** it can be seen that the pre-eruption volume is 117,803,878,288 m³ or 164,925,429.6 tons and the volume of post-eruption is 29,960,942,643 m³ or 41,945,319.7 tons. From these volumetric results can be seen that there is a volumetric decrease from pre-eruption to post eruption of 87,842,935.645 m³ or 122,980,109.9 tons.

Table 2. Volumetric Calculation for Pre-Eruption Phase

No	Width (m²)	Velocity (m/year)	Time (Year)	Distance (m)	Volume (m³)
1	74430266,6117	-0,3954 -0,2100	1	0,3024	22509886,545
2	16095160,2800	-0,210 - -0,1371	1	0,1733	27884875,365
3	20834611,8200	-0,1371- -0,0716	1	0,1044	21741857,243
4	17797532,4830	-0,0716 - -0,0062	1	0,0389	6923906,8846
5	137465200,1480	-0,0062 - 0,0627	1	0,0283	3886060,3099
6	123928131,9090	0,0627 - 0,1351	1	0,0989	12254923,997
7	74237052,7870	0,1351 - 0,2280	1	0,1816	13478627,815
8	25662850,5688	0,2280 - 0,4830	1	0,3555	9123740,1283
		Total			117803878,288

Table 3. Volumetric Calculation for Pre-Eruption Phase

No	Width (m²)	Velocity (m/year)	Time (Year)	Distance (m)	Volume (m³)
1	70954856,5955	-0,1296 - -0,0716	1	0,1006	7139713,4536
2	98892397,8844	-0,0716 - -0,0525	1	0,0621	6137969,5900
3	119145335,8040	-0,0525 - -0,0359	1	0,0442	5271002,3449
4	172359218,0000	-0,0359 - -0,0210	1	0,0285	4909819,9441
5	178655849,6640	-0,0210 - -0,0078	1	0,0144	2570887,7802
6	180346915,2730	-0,0078 - 0,0063	1	0,0007	127856,4045
7	123503417,0110	0,0063 - 0,0229	1	0,0146	1806929,6487
8	38137024,9132	0,0229 - 0,0818	1	0,0524	1996763,4770
		Total			29960942,643

This change can occur because of volcanic activity in the form of magma movements below the surface that affect the pressure changes in the pocket of magma. The land subsidence movements are an early indication of eruptions and increases in pressure that will result in ground deformation.

Deformation in the form of inflation generally occurs because the process of movement of magma to the surface that presses the surface of the soil above it. This causes the maximum deformation usually observed shortly before the volcano eruption that is at the time of pre-eruption, while deflation deformation generally occurs during or after the eruption. At that time the magma pressure inside the volcano body has weakened, so that the ground surface tends to return to its original position. This causes post-eruption deformation more stable than pre-eruption, so the volume due to deformation is reduced.

3.4. Impact analysis on residential area based on deformation results

The results of land cover classification pre and post eruption can be seen in **Figure 13** and **Figure 14.**

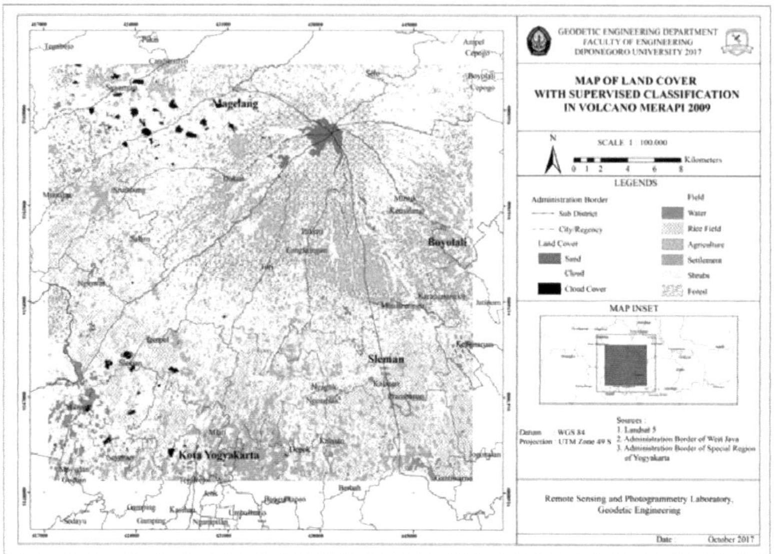

Figure 13. Land Cover Classification in Pre-Eruption Phase

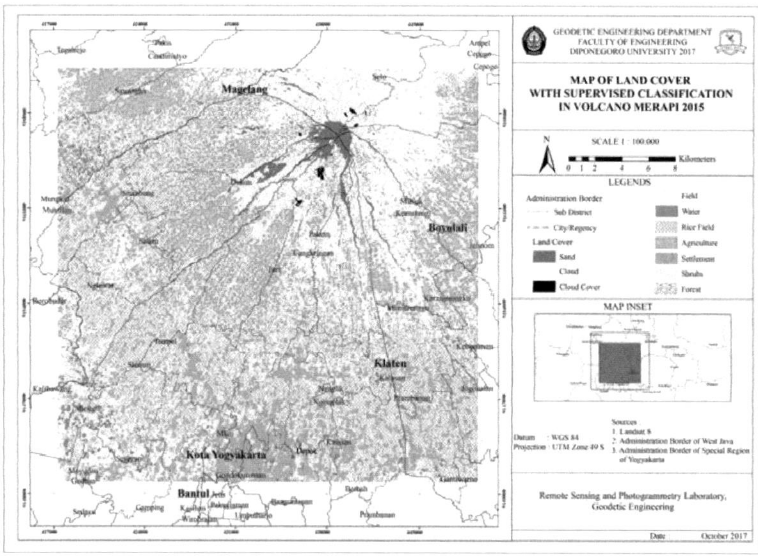

Figure 14. Land Cover Classification in Post Eruption Phase

Table 4. Land Cover Classification in Pre and Post Eruption

No	Description	Pre Eruption (Ha)	Post Eruption (Ha)
1.	Sand	902.982	1,846.726
2.	Bush	6,258.923	5,070.847
3.	Forest	14,880.213	5,967.533
5.	Field	43,111.097	35,284.768
7.	Water Body/River	1,987.325	1,253.26

No	Description	Pre Eruption (Ha)	Post Eruption (Ha)
8.	Rice Plant	12,215.235	16,186.164
9.	Farm	8,546.549	15,729.437
10.	Residential	9,014.236	16,160.846

In **Table 4** it can be seen that any land cover that undergoes changes, either increase or decrease values. This uplift land surface during post eruption can impact to the increasing number of people every year, thus affecting the increase of settlements, rice fields and plantations. Based on calculation result of land cover at condition of pre and post eruption got difference value equal to 7,146.610 Ha. The subsidence in land cover during pre-eruption can be impact to the forests, fields, bush and water bodies area. All this area is widely used for the development and establishment of more useful and durable land such as plantations. Forests and shrubs located around the peaks of Mount Merapi can be severely damaged as they become the path of the flow of magma that occurs during the eruption.

4. Conclusion

Based on land surface changes in pre and post-eruption phase, volcanic deformation is classified into 3 levels (low, medium and high). The highest rate of deformation change occurred in Jogonalan Sub-district, Klaten Regency with a change value of 30.56 cm/year. And the lowest deformation rate occurred in Moyudan Sub-district, Sleman Regency with a change value of 1.87 cm/yr. In addition to deformation changes, this research has found a land volumetric pre-eruption amounted to 164,925,429.6 tons and a land volumetric post-eruption amounted to 41,945,319.7 tons. The result of volumetric calculation can be seen that there is a volumetric decrease from pre-eruption to post eruption of 122,980,109.9 tons. A significant pre and post eruption deformation also seen in pre-eruption deformation ranged from -21.09±6.7 cm/year to 27.88±8.03 cm/year. Likewise, a post eruption ranged from -8,71±0.24 cm/year to 1.48±1.84 cm/year. This is due to the movement of the magma is less stable and can expand the volcanic dome during pre-eruption of Mount Merapi.

For the analysis of the level of land cover change during the pre and post eruption period, some areas were affected by the volcanic activity of the Merapi Mount. The highest rate of land cover change occurred in Turi Sub-district, Sleman Regency with 70.187% change value. And the lowest land cover change rate occurred in Prambanan Sub-district, Sleman Regency with change value 21,83%. Related to the impact of the pre-eruption activity in Merapi Mount to residential area gained a result around 9,014.236 Ha and in the post-eruption phase around 16,160.846 Ha. Thus, obtained significant value of the impact of deformation due to volcanic activity on Mount Merapi to residential areas amounted to 7,146.610 Ha.

References

[1] Castaneda *et al* 2011 Dedicated SAR interferometric analysis to detect subtle deformation in evaporite areas around Zaragoya, NE Spain. *International Journal of remote sensing* **32**(7):1861-1884.

[2] Cumming I G *et al* 2005 *Digital Processing of Synthetic Aperture Radar Data.* Artech House Remote Sensing Library.

[3] Danoedoro P 1996 *Pengolahan Citra Digital.* Yogyakarta. Gadjah Mada University Publisher

[4] Sharav A 2003 *Differential SAR Interferometry for Crustal Deformation study. The* Netherlands: International Institute For Geo-Information Science And Earth Observation Enschede : The Netherlands.

[5] Prasetyo Y, Fahrudin and Islam L F J 2011 Integrated analysis of differential interferometric synthetic aperture radar (DInSAR) and geological data for measuring deformation movement of Kaligarang fault, *AIP Conference Proceedings* 1857: 090003. doi: 10.1063/1.4987102

[6] Purwadhi F S H 2001 *Interpretasi Citra Digital* Jakarta. Grasindo Publisher

[7] Van der Laat R 1996 *Ground-Deformation Methods and Results- In Monitoring and Mitigation of Volcano Hazards.* Springer Verlag, Berlin, pp. 147 - 168.
[8] Vidyan Y *et al* 2013 Pemanfaatan Metode TLS untuk Pemantauan Deformasi Gunung Api, Studi Kasus : Kerucut Sinder Gunung Galunggung, Jawa Barat. *Environment and Geology Disaster Journal* **4** (1): 49-69.

Acknowledgment
The authors would like to thank the reviewers for their valuable comments. The authors would also like to thank at Badan Nasional Penanggulangan Bencana (BNPB) for provide ALOS PALSAR raw images and all research assistant in Remote Sensing and Photogrammetry Laboratory in Diponegoro University for the assist of radar satellite image data processing.

Landslide Hazard Analysis In Tuban Regency Using Multilevel Satellite Imagery Processing For Landslide Potential Mapping

Bangun Muljo Sukojo[1]* **and Citra Prastika**[1]

[1]Department of Geomatics Engineering, Faculty of Civil, Environment, and Earth Engineering,Institut Technology of Sepuluh Nopember

bangunms@gmail.com

Abstract. Landslide is a kind of land or rock movement, come from a place that has a higher slope as a consequence of disturbed land stability. The landslide can cause huge damage around us, but there are so many ways that can decrease landslide risk assessment, for example, to keep away from landslide area, spread a warning around the people, and apply safety standard when entering a landslide area. This research has purpose to know landslide potential that influences in hazard mitigation strategy in Tuban Regency. This research result is landslide potential map based on the land characteristic, rainfall, slope, and land use. That four parameters pass by an analysis processing an will be sure by remote sensing data that is medium resolution satellite imagery of Landsat 8, high-resolution satellite imagery of SPOT, and very high-resolution satellite imagery of Quickbird. This research result is a map of landslide susceptibility that places Tuban Regency in 5 classifications of it, where Tuban Regency domination in the low stage of susceptibility with 106.618 ha, and for very high of susceptibility with 175.5 ha of areas.

Keywords: Geographic Information System, Landsat 8 Satellite, Quickbird Satellite, Remote Sensing, SPOT Satellite, Tuban Regency.

1. Introduction

Hydro-meteorology hazard which happens in Indonesia frequently is caused by high rainfall and tropical climate. One of hydro-meteorology hazard that can cause other risk and danger to happen is flood and dryness, where National Board for Disaster Management have noted that flood that followed by the landslide is the highest amount of hydro-meteorology hazard in Indonesia.

Tuban Regency is one of the places that have the potency to lose a lot when there is an impact from landslide hazard. Strategy and effort are needed to reduce the amount of landslide hazard that could happen, for example, by issuing the landslide susceptibility area from landslide potential mapping.

The danger of landslides can be identified through geographic information systems and remote sensing by processing the parameters of landslides such as rainfall, soil type, slope, and land use, and then strengthened by using remote sensing data, multilevel satellite imagery. The landslide parameter along with the supporting data is then processed and analyzed using Remote Sensing based technology and Geographic Information System.

IOP Conf. Series: Earth and Environmental Science **165** (2018) 012023 doi:10.1088/1755-1315/165/1/012023

This research was conducted with the purpose to know the potential landslide area, as well as the parameters that influence the occurrence of the landslide in Tuban Regency with the output in the form of landslide potential map wherein its preparation refers to some rules of landslide potential map. In its completion, this research is expected to provide information to the government and the public about the area that has the potential of the landslide in Tuban Regency, especially to the people around the area with the potential of landslide.

2. Research Methodology

2.1. Research Sites

The location of this research is in Tuban Regency with an area of about 183,994,562 Ha. Location of astronomy Tuban at coordinates 111°30' - 112°35' longitude and 6°40' - 7°18' Latitude.

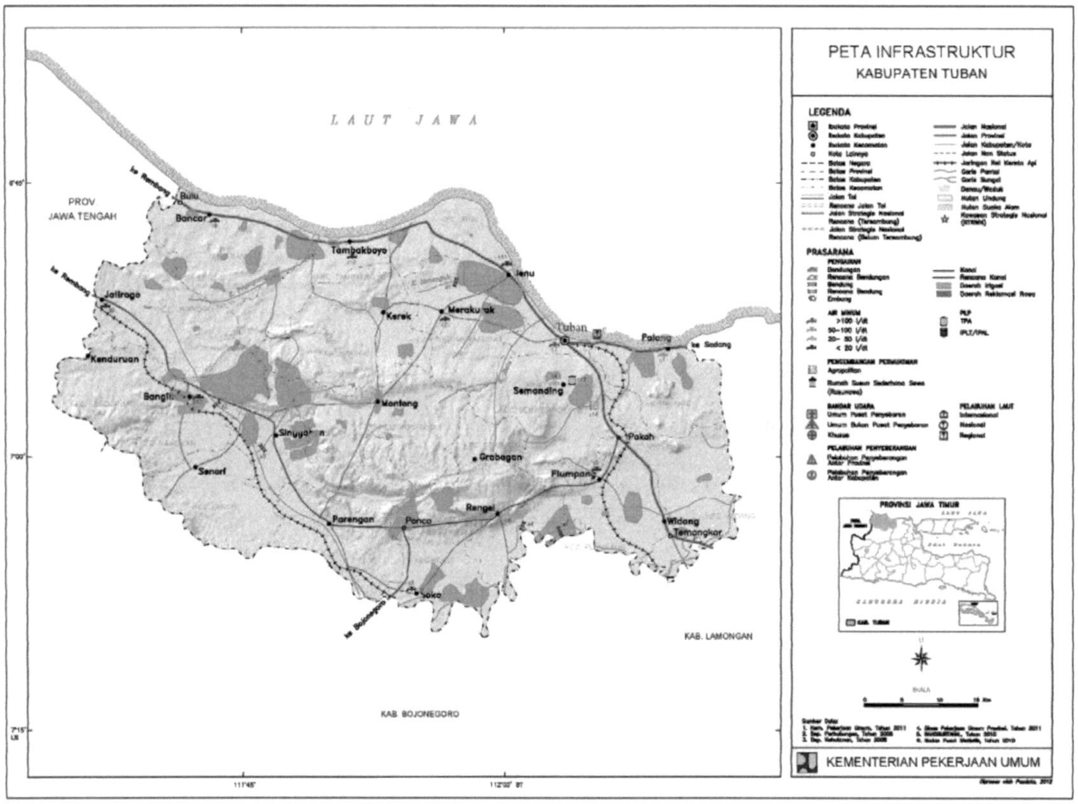

Figure 1. Research Sites Data and Equipment
(Source: PU Tuban)

2.2. Data and Equipment

2.2.1. Data. The data used in this research are:

1. Rainfall data of 2015 (source: Public Works Service Tuban)
2. DEM Satellite Astra Terra SAR X Tuban Regency in 2012 (source: Bappeda Tuban)
3. Soil type data of 2008 (source: Bappeda Tuban)
4. Slope data of 2008 (source: Bappeda Tuban)
5. Medium Resolution Satellite Imagery of Landsat 8 Path 119 Row 65 with acquisition date on 13 October 2015 (source: http://earthexplorer.usgs.gov)
6. High-resolution satellite imagery of SPOT with acquisition on 2016 (source: LAPAN)

7. Very high-resolution satellite imagery of Quickbird with acquisition on 2013 (source: Bappeda Tuban)
8. Administrative boundary (source: RBI Map 1:25.000)
9. Aerial Imagery from UAV (Unmanned Aerial Vehicle)
10. GCP and ICP data from GPS Survey

2.2.2. Stages of Data Processing Equipment. The equipment used in this study include:
1. Hardware
   a. GPS Geodetic dual frequency
   b. Phantom 3 Advance
   c. Laptop
2. Software
   a. Software geospatial processing
   b. Software satellite imagery processing
   c. Software GPS processing
   d. Software orthorectification processing
   e. Software statistic processing

2.3. Stages of data processing in this study are as follows:
1. Making the Rainfall Map
   Interpolation in rainfall data then classification and scoring on the interpolation result.
2. Making the Type of Land Map
   Classifying and scoring on available data
3. Slope Creation
   Classifying and scoring on available data
4. Making the Landuse Map
   Perform radiometric correction and digital classification on Landsat 8 using Supervised Classification: Maximum Likelihood of confusion matrix and calculate the classification results and do the scoring.
5. Overlay, scoring, and reclassification
   Overlay and score on all parameters. The scoring used in this study refers to some references from similar studies as shown in the following table.

Table 1. Parameter Scoring

Score/ Parameter	1	2	3	4	5
Rainfall (mm/year)	<1500	1500-2000	2000-2500	2500-3000	>3000
Soil Type	Alluvial, Glei	Latosol	Brown Foret, Mediteran	Andosol, Grumosol, Podsol	Regosol, Litosol, Organosol
Slope (%)	0 - 8	8 - 15	15 - 25	25 - 45	> 45
Landuse	Water	Forest	Garden	Tackle, rice field, settlement	-

Reclassifications are made up into 5 classes of vulnerability to landslides and then finding the determination of the interval.

6. High and very high-resolution image processing
   Do mosaics and pansharpening so that the image can be seen clearly. Interpolation of GCP and ICP coordinates is required for field measurements to be calculated by the strength of the net by calculating the SoF. Coordinate results that meet GCP and ICP standards are used in orthorectification and accuracy test processes.
7. Vegetation Map Creation

Doing the calculation algorithm Normalized Difference Vegetation Index (NDVI).

8. Wetness Map
   Doing the calculation algorithm Normalized Difference Water Index (NDWI).
9. Groundtruth Parameter Processing
   Comparing the data obtained from the image and obtained from the field.

3. Result and analysis

3.1. Corrected Satellite Imagery

Geometric correction is performed on each image to be able to produce the area corresponding to the situation in the field.

1. Landsat 8
   Geometric correction is performed to produce a rmse of 0.005m which is then continued by radiometric correction using FLAASH method using image data processing software.
2. SPOT
   SPOT Satellite Imagery is rendered using image processing software to obtain a rmse of 0.88 m. It shows that the process of rectification is done can be said to meet the standards, namely <1m as stated in PerkaBIG [1].
3. Quickbird
   Quickbird high-resolution satellite images are used in the manufacture of vegetation maps and wetting maps. Begins with a mosaic and pansharpening followed by interpolating the coordinates of the image of a point GCP and ICP for use in the measurement field.

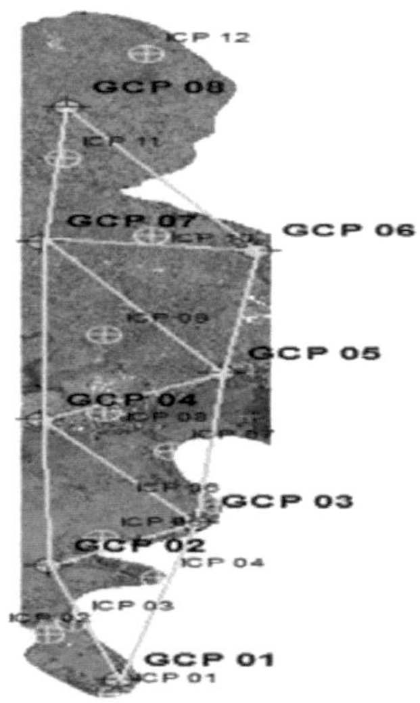

Figure2. Tentative Map

The choice of location and distribution point GCP and ICP Module Validation refers to the Spatial Plan Map BIG [2] where the dot placement must be clearly identified in the image. There are 8 points GCP and 12 ICP points used in this study. SoF value obtained for 0.2115 in accordance with the following calculations.

Σbaseline	: 22	
Σpoint	: 8	
N_u	: Σbaseline x 3	= 66
N_p	: Σtitikpoint x 3	= 24
U	: N_u – N_p	= 42

The smaller the SOF value the better the net configuration, and vice versa. SOF calculations performed have resulted in values <1 so that the design nets that have been made can be considered to be strong [3]. Data collection in the form of coordinate of control point (GCP) and accuracy test point (ICP) is done by using dual frequency geodetic GPS with static method in duration of measurement 30-45 minutes at each point. Measurement data processing is done using GPS data processing software that includes the post-processing and network adjustment.

Coordinates obtained GCP and ICP has been tied with CORS ITS S and has been corrected by SRGI BIG to obtain dx = -4.306m and dy = 0.391m for subsequent use in the process of orthorectification using image processing software to obtain the required RMSE <1.5 Pixel that is equal to 0.7875 pixels. Here is the RMSE of the processed coordinates.

Table 2. RMSE Quickbird Imagery (SRGI)

Point Name	Residual (*pixels*)	X (m)	Y (m)
GCP01	0.493	611415.062	9212055.260
GCP02	0.391	610493.765	9215153.548
GCP03	0.207	612407.164	9216339.387
GCP04	0.644	610388.390	9219113.217
GCP05	0.22	612762.528	9220378.888
GCP06	0.784	613208.440	9223709.584
GCP07	0.597	610415.784	9223937.052
GCP08	0.448	610728.314	9227575.156
ICP01	0.628	611176.463	9211689.356
ICP02	0.222	610453.878	9213282.668
ICP03	0.303	610779.393	9213638.819
ICP04	0.293	611867.692	9214775.977
ICP05	0.216	611103.617	9215903.052
ICP06	0.194	612664.987	9216831.495
ICP07	0.37	612078.452	9218234.885
ICP08	0.439	611217.088	9219342.436
ICP09	0.397	611224.687	9221431.059
ICP10	0.626	611717.011	9224112.681
ICP11	0.381	610678.329	9226189.472
ICP12	0.257	611768.987	9229012.678

Image result from the process of orthorectification is then performed accuracy test to determine the value of accuracy of the image. The accuracy test is performed by calculating the RMSE of all ICP points first then calculated by the formula in accordance with the following Basic Maps Accuracy Technical Guidelines.

The results of test calculations obtained precision horizontal accuracy of 0.9105 so it can be said to meet the standards by including the Class 1 on a scale of 1: 5000.

3.2. Image Data Correlation

The use of multilevel satellite imagery in this research needs to be tested strength / suitability relationship between image data. The relationship is obtained by applying the algorithm NDVI Landsat and QuickBird, as well as the use of the band Blue and Red on SPOT and Aerial Imagery (Foto) from UAV (Unmanned Aerial Vehicle).

Table 3. NDVI on image

Point Name	NDVI			
	Foto	SPOT	Landsat	Quickbird
GT-01	0.04	0.026	0.28	0.4
GT-02	0.063	0.3	0.4	0.51
GT-03	0.056	0.06	0.29	0.62
T-04	0.028	0.06	0.25	0.26
GT-05	0.03	0.26	0.46	0.28

Here is an illustration that shows the relationship between data on the image obtained from NDVI processing of each image with wavelength approach in image and SPOT image.

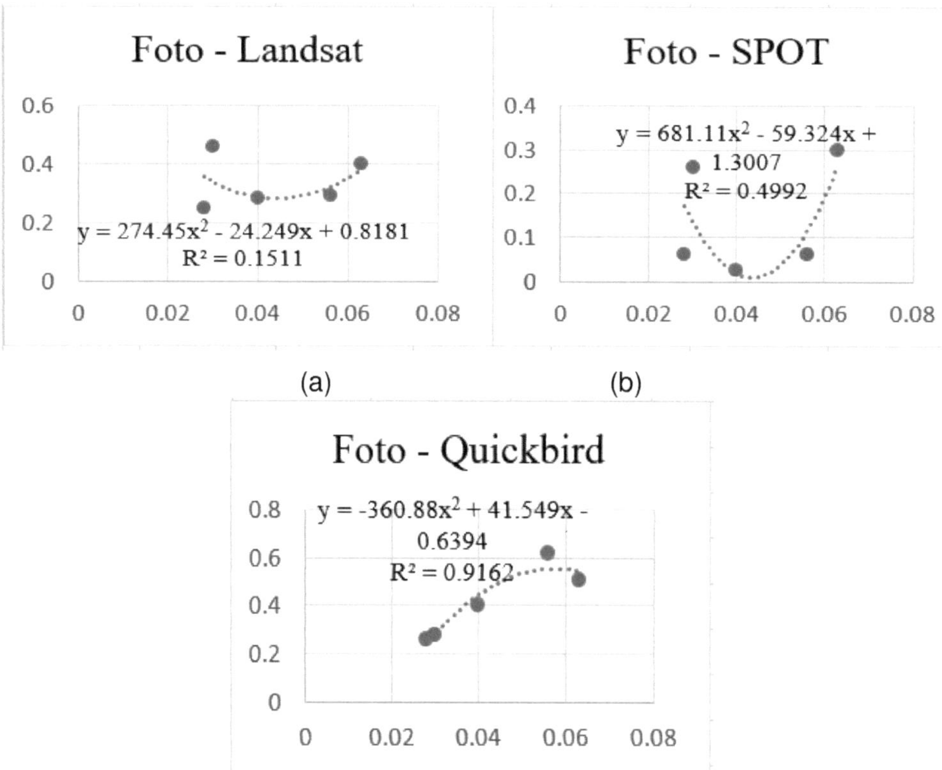

(c)

Figure 3. (a) Image Correlation with Landsat (b) Image Correlation with SPOT (c) Image Correlation with Quickbird

From Figure 3 it can be seen that the lowest correlation to the strongest is a photo with Landsat of 0.0945, a photo with a SPOT of 0.2677, and a photo with Quickbird of 0.912. This is possible because of the spatial resolution of each image. Quickbird image has a very high resolution of 0.6m so it has a stronger relationship with the photo when compared with the SPOT of 1.5m and Landsat by 30m.

Therefore, in this study the parameters of groundtruth are processed by focusing on the use of Quickbird image.

3.3. Rainfall Map
Rainfall data obtained in tabular form includes 26 rainfall stations spread in Tuban Regency. Classification is done with reference to the Puslitbangtanak in Isti [4], followed by a comprehensive layouting to get in each class as follows rainfall.

Table 4. Rainfall

Rainfall	Area (ha)	Percentage (%)
<1500 (mm/year)	173785	87.84
2000-2500 (mm/year)	24056.1	12.16

3.4. Soil Type Map
Data types of soil found in shapefile format (vectors) that include all districts in Tuban. Classification is done with reference to the Sugianti [5] to obtain the area of each class as follows.

Table 5. Soil Type

Soil Type	Area (ha)	Percentage (%)
Mediteran/Rezina	70695.4	35.76
Litosol	22976.4	11.62
Regosol	5273.62	2.67
Grumosol	65923.5	33.35
Alluvial	31773.3	16.07
Rawa	1033.63	0.52

3.5. Slope Map
Slope data in shapefile format (vectors) with classification based on Sugianti [5]. The area of each classification results as follows.

Table 6. Slope Wide

Slope	Area (ha)	Percentage (%)
0 - 8 %	169429	85.71
8 - 15 %	18428.2	9.32
15 - 25 %	6561.86	3.32
25 - 45 %	2824.31	1.43
> 45 %	432.704	0.22

3.6. Landuse Map
Landsat 8 imagery can product a landuse map which processed by radiometric correction as the first, and then followed by a digial interpretation methods called Supervised Classification: Maximum Likelihood with 75,38% of confusion matrix. Classification processed based on Kesaulya and Permen PU No.22 tahun 2007 [6]. The result of landuse area as follows.

Table 7. Landuse Area

Landuse	Area (ha)	Percentage (%)
Water	2054.1	1.04
Forest	138040.0	69.79
Garden	10984.8	5.55
Tackle, rice field, settlement	46700.8	23.61

3.7. Landslide Hazard Map

In this study used Storie method that requires rainfall data, soil type, slope inclination, and land cover in the preparation, where the determination of these parameters refers to the Guidelines for Preparation of Spatial Planning of Landslide Disaster Prone Areas of Ministry of Public Works No.22 / PRT / M / 2007 [7].

Data is given a score as shown in Table 1 and then be overlaid and applied tools for dilajutkan intersect the scoring process and the reclassification in accordance with the intervals determined by reference to Susanto [8] as the following formula.

$$\text{Interval} = \frac{\text{maximum score - minimum score}}{\text{class}} \quad (1)$$

From the calculation of the interval got the range of values of each class along the 2.2 to obtain classification results in 5 classes as follows.

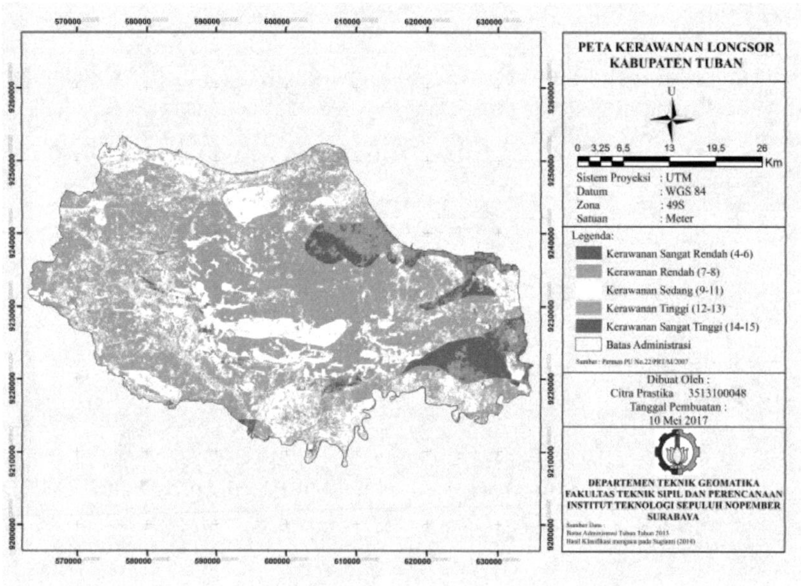

Figure 4. Map of Landslide Hazard

From the result of classification that has been done got wide of each class of landslide vulnerability like following table.

Table 8. Area of Landslide Hazard

Category	Area (ha)	Percentage (%)
Very Low	11,854.200	6.0
Low	104,618.000	53.1
Medium	75,639.300	38.4
High	4,691.870	2.4
Very High	175.502	0.1

3.8 Correlation in Internal Parameters of Landslide Hazard

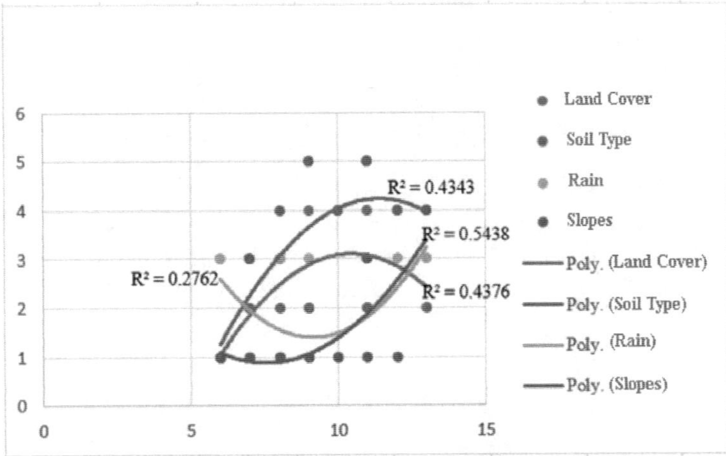

Figure 5. Correlation Parameter of Landslide Hazard

From Figure 5 it can be seen that the magnitude of landslide parameter correlation with landslide vulnerability level from low to high is rainfall of 0.06, land cover is 0.49, land type is 0.59, and slopes is 0.63.

3.9. Groundtruth

Five points of groundtruth are placed on each myander of Bengawan Solo River. The sample soil (for soil moisture content and texture) were done on disturb land (disturbed soil / ground) by using the pipe which is plugged deep around 30cm on land taken about 1kg in order to do the testing on soil samples. Photo image acquisition is done by using Phantom 3 Advance flown around 10-15m above the groundtruth point. It is then being adjusted to the results of the classification and to form such a landslide susceptibility map as shown in figure 4. From table 9 can be seen that Groundtruth point is at a level of vulnerability.

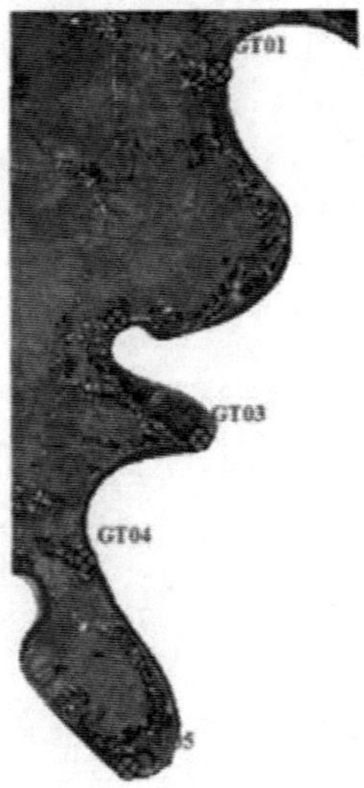

Figure 6. Groundtruth Point

IOP Conf. Series: Earth and Environmental Science **165** (2018) 012023 doi:10.1088/1755-1315/165/1/012023

Table 9. Classifixation Groundtruth Points

Point Name	Classification
GT-01	Very High
GT-02	Low
GT-03	Very Low
GT-04	Medium
GT-05	Very Low

3.10. NDVI and NDWI (Quickbird)

Corrected satellite image that has been generated is then processed with the NDVI algorithm for known greenishness. Generated class 5 in accordance with the Regulation of the Minister of Forestry of the Republic of Indonesia No.P.12 / Menhut-II / 2012 [8] to generate a map as follows

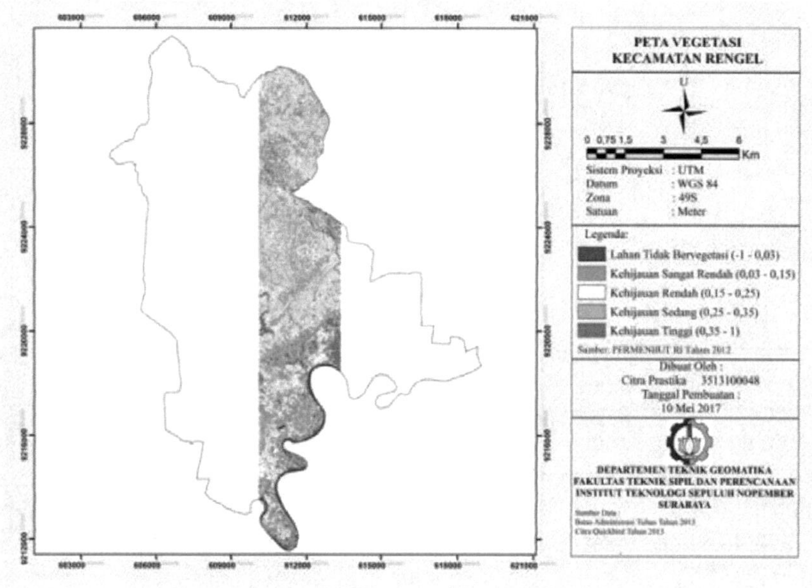

Figure 7. Vegetation Map

Corrected satellite image is further processed by NDWI algorithm. The classification process carried out by referring to Xu (2006, referring to MNDWI) [9] to be produced as follows.

IOP Conf. Series: Earth and Environmental Science **165** (2018) 012023 doi:10.1088/1755-1315/165/1/012023

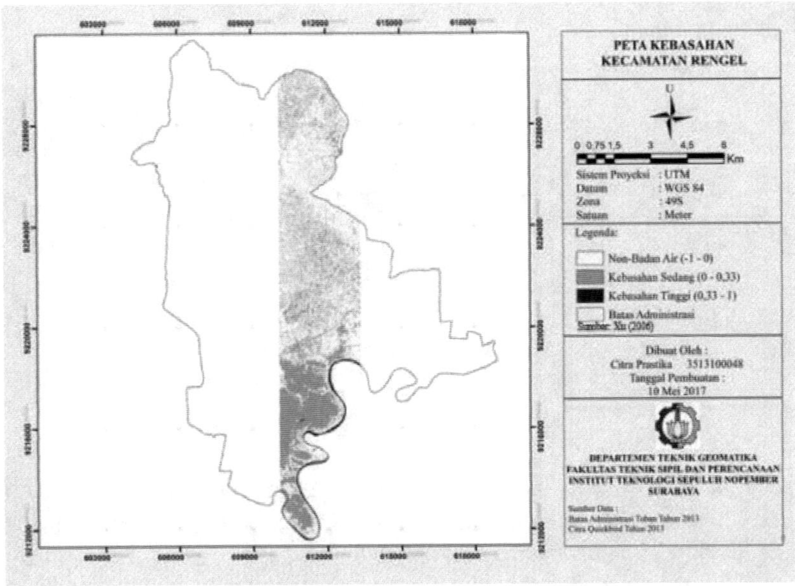

Figure 8. Wetness Map

3.11. *Laboratory Tested*

Testing the moisture content in the soil was done at the Environmental Service Laboratory of Tuban Regency. Tests carried out on the soil texture Soil and Rock Mechanics Laboratory of Civil Engineering - ITS using a sieve and hydrometer which refers to the Guidelines for Basic Research and Testing Soil for Road Works Department of Public Works No.003-03 / BM / 2006 [10]. The soil testing was conducted based on grain size analysis to find out the fraction of the soil.

The test results of water content and texture of the soil that has been combined with NDVI values and NDWI through Quickbird image processing at each point Groundtruth can be seen in the following table.

Table 10. Results of parameter Groundtruth

Point	NDWI	NDVI	Moisture	Textrure		
				Sand	*Silt*	*Clay*
GT-01	-0.13	0.27	75.22	80.2	13.9	5.9
GT-02	-0.05	0.20	57.17	71.1	11.2	17.6
GT-03	-0.20	0.45	42.55	79.4	14.2	6.4
GT-04	-0.12	0.27	37.17	81.9	12.2	5.9
GT-05	-0.30	0.56	37.67	70.7	15.2	14.1

Soil texture shows the composition of the soil constituents expressed as a ratio of the proportion (%) relative to the sand fraction *(sand)* in diameter from 2.00 to 0.20 mm, dust *(silt)* in diameter from 0.20 to 0.002 mm, and clay *(clay)* (<2 micrometer). Soil sampling is at the *disturb land* (about 20 cm from the soil surface to the inside) so there are some very clear material with sand of different sizes (by naked eye) like a stone / gravel, snails, plant roots, and so on. In soil testing the material is counted in the category> 2.00 mm.

3.12. *Correlation of Groundtruth Parameters*

The dominance of the sand fraction will cause the formation of few macro pores, so that the grasp of the water is very weak. This condition causes the water easily in and out of the ground and only a few are retained[11]. it is evidenced by the acquisition of a correlation of 0.1 to moisture and sand.

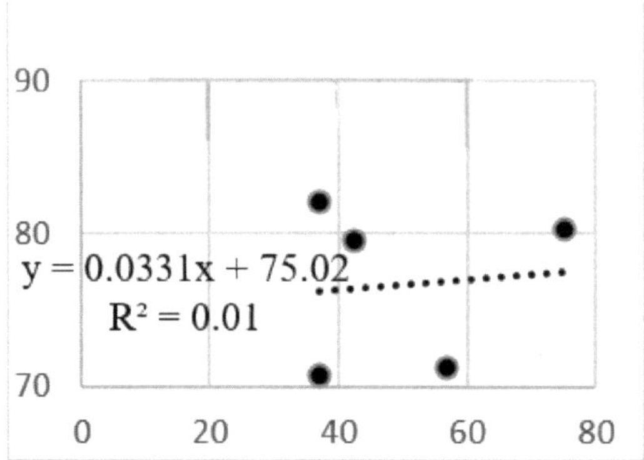

Figure 9. Correlation of Water and Sand Content

The dominance of the clay fraction will lead to the formation of many micro pores, so that the grasp of the water is very strong. This condition causes water to enter as soon trapped, as evidenced by the acquisition of R^2 of 0.95 against water (NDWI) and clay.

The dominance of the dust fraction will lead to the formation of meso pores in moderation, resulting in a strong hold of the water. This causes the water and air quite easily in and out of the ground, some water will be restrained. Dust-textured soils are generally more fertile than sand-textured soils [11].

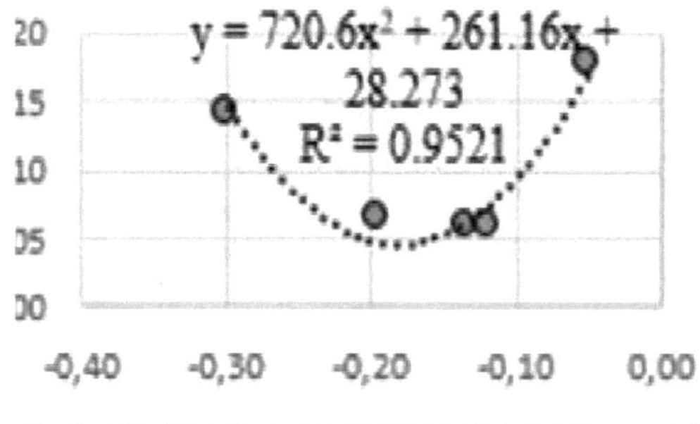

Figure 10. NDWI and Clay Correlation

In addition to soil texture and moisture content in the soil, vegetation also plays an important role in controlling soil stability. The existence of trees along the cliffs are very affected the stability of the cliff through rooting function that protects the soil thus affecting shear resistance *(shear strength)* of land. The presence of organic matter on the soil will affect the increase of water holding capacity, the dissolution of nutrients from minerals, as well as the increasing activity of decomposition of organic matter. This is supported by the acquisition of R^2 of 0.95 to vegetation (NDVI) and water (NDWI)

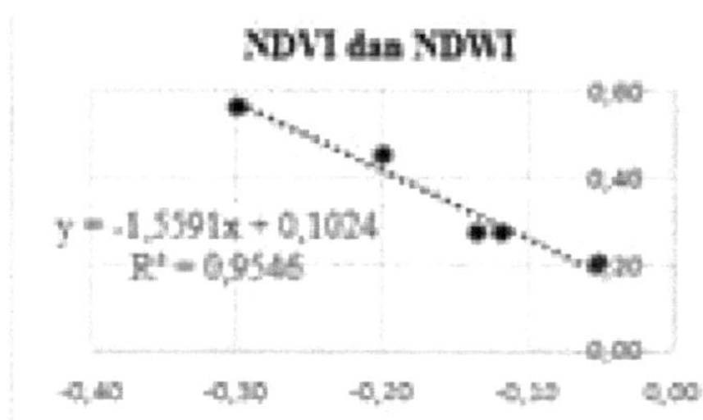

Figure 11. NDVI and NDWI correlations

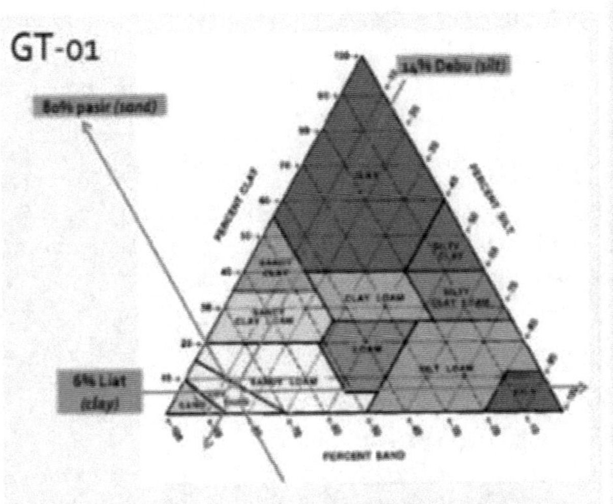

Figure 12. Land Texture According to The USDA Triangle

When viewed from the physical properties of the soil (soil texture and water content) as well as the vegetation on *Groundtruth point,* the GT-01 has the potential of being so large (compared to other *Groundtruth* point) to occur an avalanche. With sand fraction content of 80.2%, 13.9% dust, and clay 5.9%, referring to the USDA (1999) [13], land on the GT-01 has a category of argillaceous sand. With NDWI value of -0.13 which means that there is water content in GT-01 so that there is a possible vegetation growing at that point which is also reinforced with NDVI value of 0.27. Water content in GT-01 has a fairly high category of 75.22% which is a trigger for avalanches. If there is an increase in the water content in the soil (such as the absorption of rainwater, overflowing river water, leaking water / ponds), there will be accumulation of water which further stretches the bonds between grains of soil which will eventually push the soil grains to landslides. Here is the data related parameters that have been obtained.

Table 11. Groundtruth Parameter

Points	NDWI	NDVI	Moisture	Soil Texture			Category (USDA)	Classification
				Sand	*Silt*	*Clay*		
GT-01	-0.13	0.27	75.22	80.2	13.9	5.9	Loamy sand	High
GT-02	-0.05	0.20	57.17	71.1	11.2	17.6	Sandy loamy	Low
GT-03	-0.20	0.45	42.55	79.4	14.2	6.4	Loamy sand	Very Low
GT-04	-0.12	0.27	37.17	81.9	12.2	5.9	Loamy sand	Medium
GT-05	-0.30	0.56	37.67	70.7	15.2	14.1	Sandy loamy	Very Low

4. Conclusion

Based on the processing and data analysis has been done, here are the conclusions that the authors get in completion of this final task.

1. There are 5 classes of landslide vulnerability that is very low, low, medium, high, and very high.
2. Tuban Regency is dominated by low level of vulnerability with an area of 104,618 ha, while for very high level of vulnerability is only 175.5 ha.

3. Of the 5 point *Groundtruth* studied, the point of the GT-01 has a higher avalanche vulnerability compared with other *Groundtruth* point. GT-01 is at a moderate level of vulnerability where it is one to two levels higher than GT-02, GT-03, GT-04, and GT-05.

References

[1] Peraturan Kepala Badan Informasi Geospasial Nomor 15 2014 Tentang Pedoman Teknis Ketelitian Peta Dasar (Bogor: Badan Informasi Geospasial)
[2] Badan Informasi Geospasial 2016 *Modul Validasi Peta Rencana Tata Ruang* (Jakarta: Badan Informasi Geospasial)
[3] Abidin H Z 2002 *Penentuan Posisi Dengan GPS dan Aplikasinya* (Jakarta: PT. Pradnya Paramita)
[4] Isti Adhina Yura 2013 Pemetaan Zonasi Kerentanan Banjir di Kabupaten Sragen Menggunakan Teknologi Penginderaan Jauh dan Sistem Informasi Geografis (Universitas Gajah Mada: Penginderaan Jauh dan SIG Sekolah Vokasi)
[5] Sugianti Khori 2014 *Pengklasan Tingkat Kerentanan Gerakan Tanah Daerah Sumedang Selatan Menggunakan Metode Storie* (Bandung: Pusat Penelitian Geoteknologi LIPI)
[6] Kesaulya Hertine 2016 Perencanaan Mitigasi Bencanan Longsor di Kota Ambon *Jurnal Perencanaan Wilayah dan Kota Universitas Sam Ratulanggi Manado* 228-235
[7] Dinas Pekerjaan Umum 2007 Peraturan Menteri Pekerjaan Umum Tentang Pedoman Penataan Ruang Kawasan Rawan Bencana Longsor (Indonesia: Jakarta)
[8] Susanto Aditya Dhani 2014 Analisis Tingkat Rawan Kekeringan Lahan Sawah Dengan Pemanfaatan Penginderaan Jauh dan Sistem Informasi Geografis di Kabupaten Sragen (Surakarta: Universitas Muhammadiyah)
[9] Menteri Kehutanan 2012 Peraturan Menteri Kehutanan Tentang Perubahan Kedua Atas Peraturan Menteri Kehutanan Nomor P.32/Menhut-Ii/2009 Tentang Tata Cara Penyusunan Rencana Teknik Rehabilitasi Hutan dan Lahan Daerah Aliran Sungai (Rtk RHL-DAS) Jakarta
[10] Departemen Pekerjaan Umum 2006 *Pedoman Penyelidikan dan Pengujian Tanah Dasar untuk Pekerjaan Jalan Departemen Pekerjaan Umum* (Jakarta: Departemen Pekerjaan Umum)
[11] Ali Hanafiah Kemas 2005 *Dasar-dasar Ilmu Tanah* (Jakarta: Rajagrafindo Persada)
[12] Xu H 2006 Modification of normalized difference water index (NDWI) to enhance open water features in remotely sensed imagery *Int. J. Remote Sens.* 2006 27 3025–3033
[13] USDA 1999 *Soil Taxonomy: A Basic System of Soil Classification for Making and Interpreting Soil Surveys* (Washington DC: U.S. Government Printing Office Washington, DC)

Acknowledgments:
This research was supported by Department of Geomatics Engineering, Faculty of Civil, Environment, and Earth Engineering, Institut Technology of Sepuluh Nopember (ITS), Surabaya Indonesia and Institut of Research and Public Service, Institut Technology of Sepuluh Nopember (ITS), Surabaya Indonesia.

IOP Conf. Series: Earth and Environmental Science **165** (2018) 012024 doi:10.1088/1755-1315/165/1/012024

Accuracy Assessment of Land Cover Fraction Map Derived from Landsat 8 Imagery Using Linear Spectral Mixture Analysis Method

B Susilo[1], R Sulastri[1]* and F Khikmah[1]

[1]Department of Geographic Information Science, Universitas Gadjah Mada, Bulaksumur, Sinduadi, Mlati, Sleman, DI Yogyakarta 55281

riesa.sulastri@mail.ugm.ac.id

Abstract. Accuracy assessment is conducted to assess how well a classification method is. One of the land cover classification method for mixed pixel problem is Linear Spectral Mixture Analysis (LSMA). Error matrix is commonly used accuracy assessment of categorical data. But accuracy assessment for fraction map as result of LSMA processing is not really examined yet. LSMA method produces map fraction which contains land cover proportion. This experiment aims to examine the effect of sampling unit on land cover fraction map accuracy assessment result. 70 samples were taken using stratified random sampling. Geoeye 1 image is used as reference data corresponding to ground truth data. Accuracy assessment was performed in two scenario sampling unit which is single pixel area and a cluster of pixels area. The accuracy of LSMA model was influenced by position error between classified map and the reference data. Selection the right sampling unit is a one way to minimize the deviation caused by positional error corresponding reference data.

Keywords: Accuracy Assessment, Land Cover, Fraction Map, Linear Spectral Mixture Analysis

1. Introduction

Remote sensing technology has become a data source of land cover/ land use mapping because of easy data acquisition compared to field surveys. The land cover information and thematic map can be extracted through image classification methods [1]. The classification method commonly used in land cover assessments is pixel-based classification, including per-pixel classification and sub-pixel classification. The per-pixel classification assumes that one pixel has the one class. However, the reality showed that one pixel contains information more than one type of land cover. Medium resolution imagery provides low accuracy in this methods, especially used for heterogeneous areas [2]. This is because the existing of mixed pixel forced classified into one class. The development of remote sensing data processing techniques introduces a classification of sub-pixels that classifies one pixel into several different classes [3]. Sub-pixel classification is able to overcome the lack of per-pixel classification by providing predictions of the proportion of object classes in a single pixel [4].

IOP Conf. Series: Earth and Environmental Science **165** (2018) 012024 doi:10.1088/1755-1315/165/1/012024

One of the commonly used methods in sub-pixel classification is Linear Spectral Mixture Analysis (LSMA). This method is widely used to overcome mixed spectral problems which produce land cover fraction map [5]. The accuracy of land cover fraction map needs to assessed to knowing the reliance of the LSMA classification result. The reliance of LSMA method to be applied is showed by its accuracy [6]. The accuracy assessment method of sub-pixel classification is different from the per-pixel classification. The confusion matrix is widely used to assess the accuracy of categorical data/ per-pixel classification [7]. Meanwhile, the previous study has been used error measurements to assess subpixel classification, like expected sub-pixel class overlap constrained to the unmatched sub-pixel fraction [8]. Wu (2004) used two type of error measurement, root-mean-square error (RMSE) and systematic error (SE) to evaluate urban composition [9]. Weng et al. (2009) calculate the accuracy of land cover fraction map using RMSE and mean average error (MAE) [10]. Yang et al (2010) employed RMSE to assess the impervious surface area for four different methods – SMA, NSMA, MESMA, and PNMESMA[11]. The RMSE were used in this study to calculate the accuracy of land cover fraction map.

Low accuracy might happen because of an external factor such as an error in reference data. Generally, the accuracy results can be affected by error sources that caused by the positional error between classified map and reference data. The positional errors between the references and map affected on the low accuracy of the thematic map. This issue needs to be a consideration in determining the sample for accuracy assessment. Therefore, the influence factors on accuracy assessment are very important to identify to improve the objectivity of the assessment.

There are some very important considerations in determining the sample of accuracy assessment, including the appropriate sampling unit, number of samples, and sample scheme [7]. The sampling unit is a small part of the map used for the accuracy assessment, it can be a single pixel and a cluster of pixels. The sampling unit is an important aspect to consider the possible errors [12]. The errors indicate the classification errors on the result of the map. This study aims to map the land cover fraction and calculate the accuracy assessment using two scenarios of sampling unit.

2. Materials and Methods

2.1. Materials

This study used Landsat 8 OLI Level 1TP as the main data sources and GeoEye Image level 2A as a reference for maps. Processing of the 1TP level has gone through image calibration which is radiometric and geometric corrections. Landsat 8 OLI Level 1TP show pixel value at level surface reflectance value. Therefore, the radiometric correction was not performed in this study. The study used Landsat Image 8 multispectral channel consisting of seven bands. The wavelengths used in the analysis, in more detail are described in Table 1.

Table 1. Specification of multispectral bands of Landsat 8 OLI imagery

	Bands	Ranges of Wavelength (μm)
Band 1	Coastal/aerosol	0,435-0,451
Band 2	Blue	0,452-0,512
Band 3	Green	0,533-0,59
Band 4	Red	0,636-0,673
Band 5	Near infrared	0,851-0,879
Band 6	SWIR 1	1,566-1,651
Band 7	SWIR 2	1,566-1,651

The image acquisition on June 23, 2015, cut into 450x390 pixels covering the urban area of Surakarta. Selection of the study area considering the heterogeneity of land cover. Recording result in medium resolution imagery such as Landsat with 30 meter resolution produced dominated by mixed pixels. The study area located in the 49 M zone UTM 472884 - 486320 mT and 9158020 - 9169533 mU, presented in Figure 1 below.

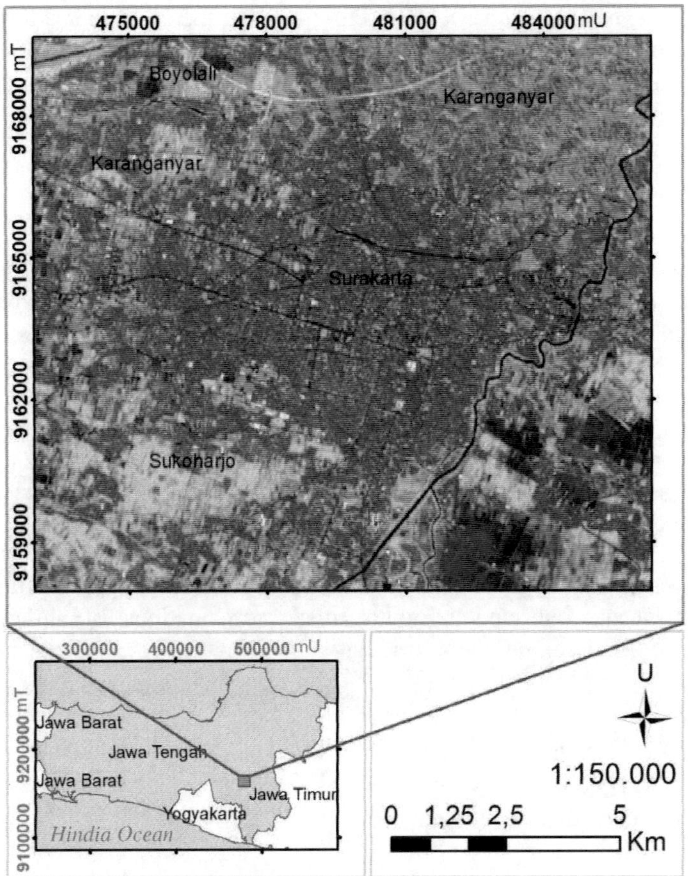

Figure 1. Study Area (Composite Band 452)

The GeoEye imagery with processing level standard 2A has been orthorectification a using a rough DEM. The imagery is used as a reference in geometric correction and reference data for accuracy assessment. The pan-sharpening process applied in GeoEye imagery to increase the resolution to 0.5 meters.

Image processing in this study uses several tools including a laptop with i3 processor and Envi 5.3 software and Idrisi Selva 17.0. Envi 5.3 software is used for calibration, rectification, and transformation. The LSMA processing is done on the Idrisi Selva software because it is able to produce the total value of land cover fraction which constraint is equal to one. Fieldwork utilizes the Global Navigation System (GPS) as a navigation tool.

2.2. Methods

This study consists of two stages, such as mapping of land cover fraction using Linear Specific Mixture Analysis method and classification accuracy assessment. Land cover is divided into four classes, ie, built up area, vegetation, bare land, and water body. Each land cover class extracts the pure spectral reflectance through image, called endmember. The accuracy assessment of LSMA done by calculating the RMS Error.

2.2.1. Land Cover Fraction Maps. There are three stages to produce the land cover fractions map, include image pre-processing, endmember determination, and Linear Spectral Mixture Analysis (LSMA) process. The extraction method of land cover using LSMA is described in detail as follows.

2.2.1.1 Geometric Correction. Geometric correction is performed to equalize the position of Landsat 8 OLI imagery with the correct position. The reference used in this correction is GeoEye's imagery that has assumed to have coordinates corresponding to the field. This geometric correction is done to reduce position error between GeoEye Image as the reference of accuracy assessment and Landsat Image 8. Geometric correction is done until the RMS Error reaches 0,3265, where there is error position as far as 10 meters.

2.2.1.2 Endmember Selection. Endmember selection is the essential process in the LSMA method. Endmember is the pure object whose the spectral reflection value will be used as a reference in LSMA process. Endmember obtained from the image itself. It used to extract the land cover map includes built-up area, bare land, water bodies, and vegetation. The process to determine the endmember is through a Minimum Noise Fraction transformation and Pixel Purity Index (PPI). The next process is manual selection on a pure pixel of PPI results by checking land cover on GeoEye imagery. The endmember for water bodies is taken from Cengklik Reservoir. The pure pixel of vegetation is found in the north of Surakarta City and rubber plantations. Endmember of bare land is taken from the ground before construction and rice fields that have not been planted. The high density of built-up area was found in the center of Surakarta City.

2.2.1.3 Linear Spectral Mixture Analysis (LSMA). LSMA is a technique to estimate the reflectance recorded on pixels. It can be modeled with linear combinations of summations of multiple endmembers [13]. The linear combination, formulated mathematically as follows [14].

$$DN_{ij} = Ef_{ij} + e_{ij} \quad (1)$$

Where f_{ij} is *endmember* fraction pixel in position ij, E is the pattern matrix of endmember spectral reflectance, DN_{ij} is the reflectance value of a pixel ij, e_{ij} is the error between the reflectance value of model and brightness value of measured band.

Formula (1) assumed that the number of fraction of all endmember is sum to 1.

$$\sum_{l=1}^{L} f_l = 1 \quad (2)$$

Beside that, each fractions must be positive, where

$$f_l \geq 0 \quad (3)$$

The LSMA method was used inputs of Landsat 8 image that has been corrected radiometric to surface reflectance value and endmember. In the LSMA processing, each mixed pixel will be separated into four land cover fractions. This analysis yielded land cover fraction maps of the built up area, vegetation, bare land, and RMSE. These maps show the percentage composition of each land cover on pixels whose value ranges from 0 to 1 or 0% to 100%.

2.2.2. Accuracy Assessment of Land Cover Fraction Map. Accuracy assessment was conducted by comparing the proportion of land cover with accurate results with the proportion of land cover in the field. The formula used in the test is the root-mean-square error (RSME) as follows.

$$RMSE = \sum_{i=1}^{N} (v'_i - v_i)^2 / N)^{\frac{1}{2}} \quad (4)$$

Where v'_i is sample *i*' percentage of land cover map fraction, v_i *is* sample *i*' percentage of actual land cover from reference images, *N* sum of sample members

The proportion of land cover in the field was obtained from the extraction of pan-sharpening GeoEye imagery. Reference image was classified into four classes of land cover – built up area, bare land, vegetation, and water body using visual interpretation. For each sample site, the sampling units are single pixel and cluster of pixels. Then, the percentage of land cover were extracted from each area of the sample unit. Ground truth percentage of land cover within sample unit can be calculated through a dividing area of each land cover by the total area of sample unit (900 m2 or 8100 m2). This study used 70 samples which distribute in the entire study area. It was referred to Jensen (1996) which the minimum of sufficient samples for the variance-covariance matrix is 10 times the number of bands used. The sample determined using stratified random sampling method based on fraction class of each land cover map yields. Sample site was chosen in the homogenous area. This study examined the effect of sampling unit on the results of the accuracy assessment. The sampling unit used was single pixel and cluster of pixels with 3x3 pixel square. The figure 2 and figure 3 show how the sample unit is utilized.

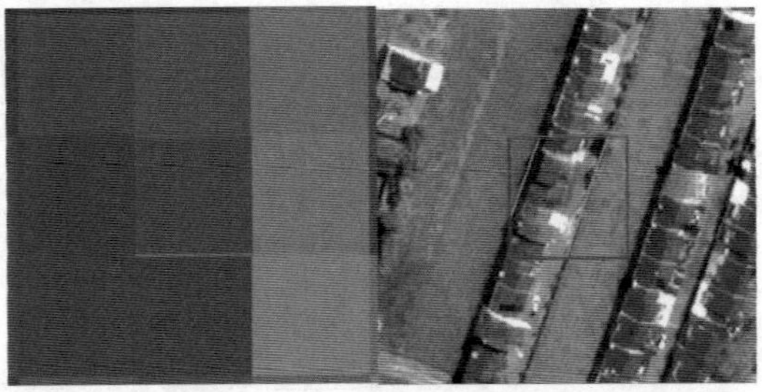

(a) (b)

Figure 2. Single pixel sampling unit (on Landsat 8 OLI image (b) polygon with red line indicates digitized land cover.

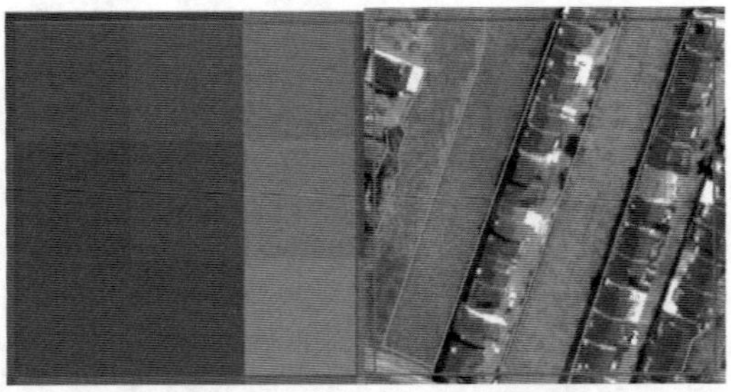

(a) (b)

Figure 3. Cluster of pixels sampling unit (on Landsat 8 OLI image (b) polygon with red line indicates digitized land cover.

The methods used in this study is summarized by the flow chart in Figure 4.

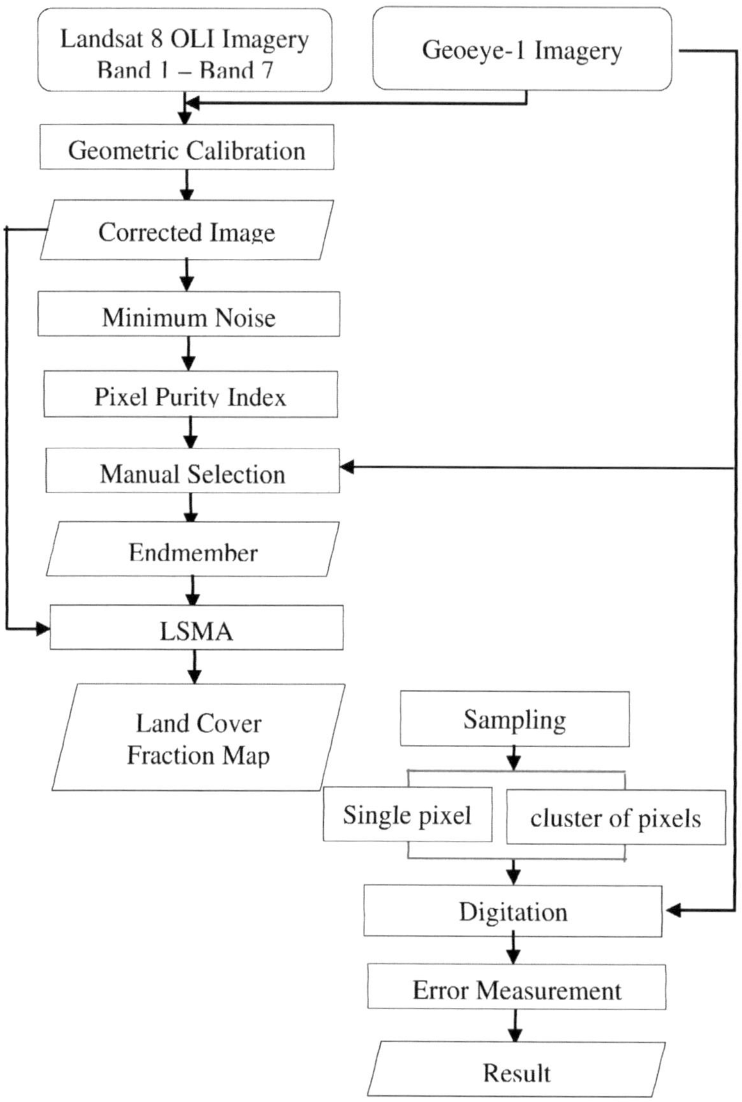

Figure 4. The flow chart of study.

3. Result and Discussion

3.1. Land Cover Fraction Maps

LSMA is a frequently used method to solve mixed pixel problems. This study applied the LSMA method to estimate the proportion of land cover class; vegetation, bare land, water body, and built up area in each pixel. The distribution of land cover fraction was presented in Fig. 5 to Fig. 8. The land cover fraction maps were presented using color gradations from red to dark blue. The color that closer to the red color indicated the low proportion of land cover. Meanwhile, the color which closer to the dark blue shown the high proportion of land cover.

Land cover was dominated by the class range in the proportion of 0.7 - 0.8. Most of the built-up land fraction was located in the urban areas dominated by buildings. The dominant of vegetation fraction was located in the urban fringes used as the paddy field. The highest of the bare land fraction

IOP Conf. Series: Earth and Environmental Science **165** (2018) 012024 doi:10.1088/1755-1315/165/1/012024

was in the range of 0.4 to 0.5. The bare land was relatively low due to few bare areas in the urban areas of Surakarta. Most of the water bodies were located in the river and irrigated fields.

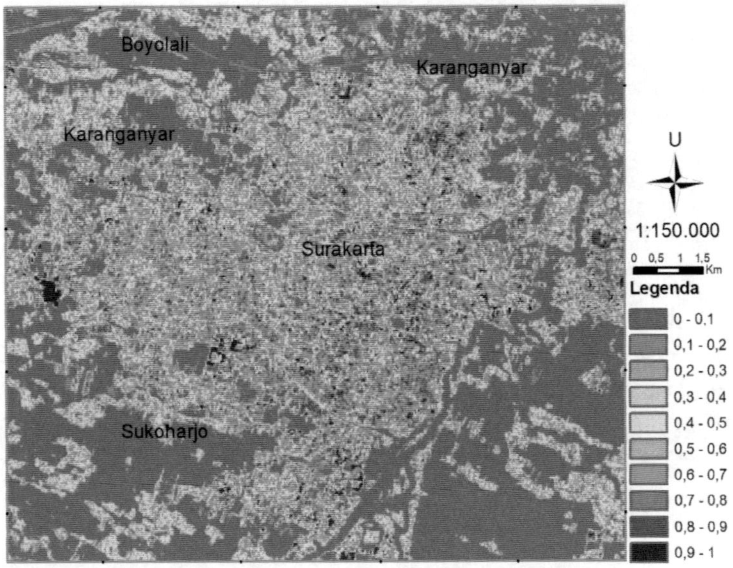

Figure 5. Built up Land Fraction Map

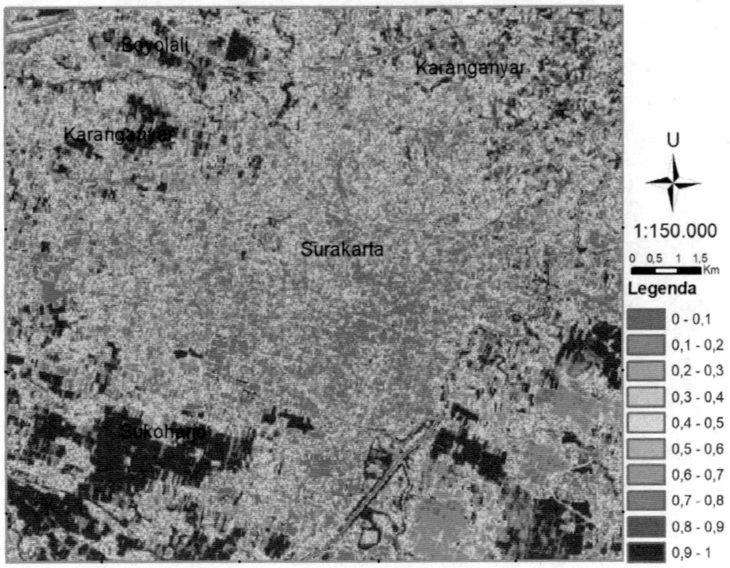

Figure 6. Vegetation Fraction Map.

IOP Conf. Series: Earth and Environmental Science **165** (2018) 012024 doi:10.1088/1755-1315/165/1/012024

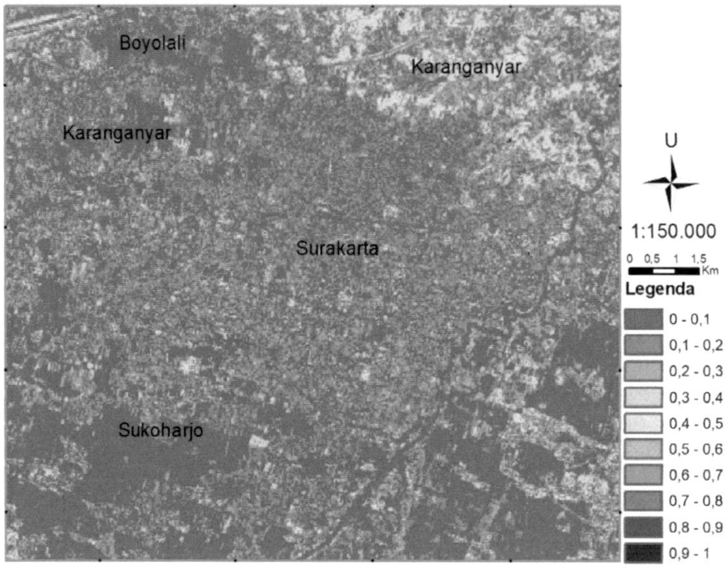

Figure 7. Soil Fraction Map

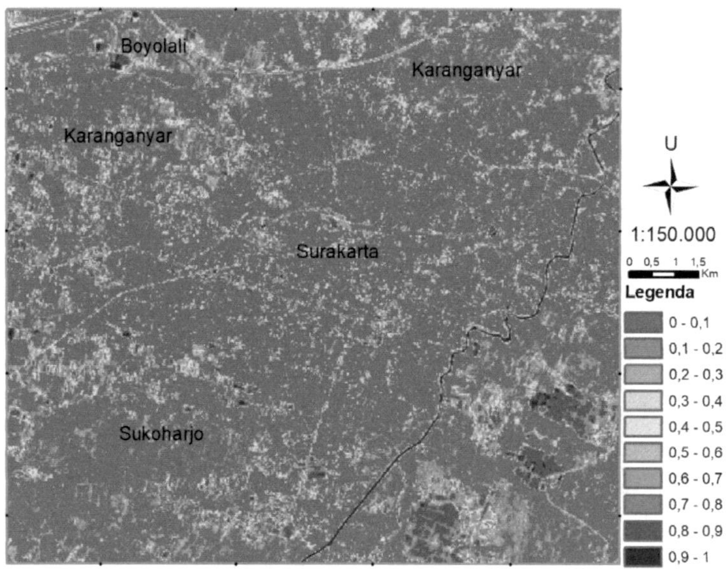

Figure 8. Water Bodies Fraction Map

The LSMA processing generated residual imagery as shown in Figure 9. The residual value represents the difference between the modeled brightness value and the measured brightness value of the band. The building made from concrete had the high residual value. This is caused by the reflectance value of the building higher than average reflectance of built-up-area endmember.

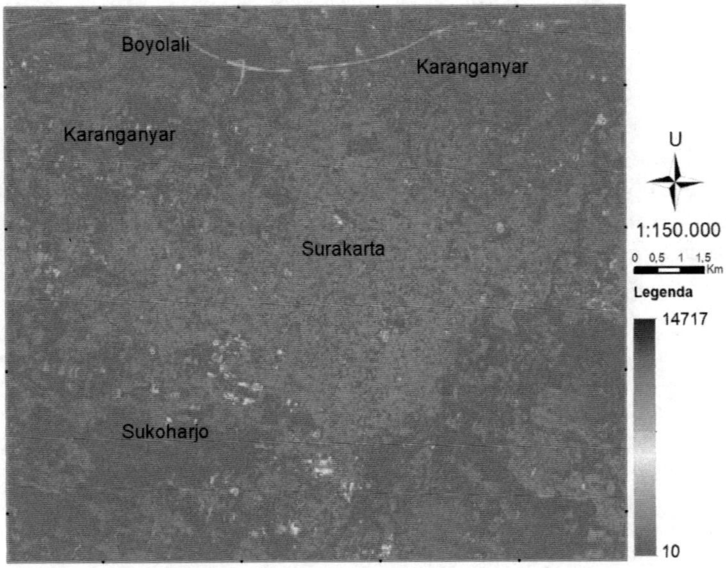

Figure 9. Residual Map

3.2. Accuracy Assessment

To evaluate the accuracy of land cover fraction map, 70 samples were used. Geoeye pansharpening image with 0.5 meter resolution was used as a reference data corresponding ground truth to assess the accuracy of the land cover map fraction as LSMA result. Two scenarios of sampling unit were used to assess the model accuracy. Land cover fraction map use pixel as the analysis unit, so sampling unit could be pixel or cluster of pixels.

Table 2. The Accuracy of Land Cover Fraction in 3x3 pixels Sampling unit

Land Cover	Land Cover				Overall Accuracy
	Built up Area	Vegetation	Bare Land	Water Body	
RMSE	26.93	15.21	24.39	28.09	23.66
Accuracy (%)	73.07	84.79	75.61	71.91	76.34

Table 2 shows the result of accuracy assessment of cluster of the pixel as the sampling unit. Overall accuracy derived from a cluster of pixels sampling unit is 76.34. Waterbody fraction map has the highest error among others map with RMSE 28.09. Low albedo impervious surface are similar to water body reflectance. Waterbody was estimated higher on low albedo with asphalt material.

Table 3. The Accuracy of Land Cover Fraction in single pixel Sampling unit

Land Cover	Land Cover				Overall Accuracy
	Built up Area	Vegetation	Bare Land	Water Body	
RMSE	30.59	21.66	34.95	30.53	29.43
Accuracy (%)	69.41	78.34	65.05	69.47	70.57

The result of accuracy assessment for single pixel sampling unit was presented in Table 3. The accuracy of land cover fraction map using single sample unit is 70.57. The RMSE were 30.59, 21.66, 34.95, and 30.53 of built-up area fraction map, vegetation fraction map, bare land fraction map, and water body fraction map respectively. Bare land fraction map has the highest error among the others. The bare land was predominantly confused with the impervious surface. Dry soil has similar reflectance to high albedo.

The sample unit with 3x3 pixels square has higher overall accuracy than the single pixel. A pixel with 90 m2 area may have less relation with the actual area in the ground. Besides that, low accuracy might happen because of positional error corresponding reference image and land cover map fraction. Meanwhile, using a cluster of pixels, the positional error can be minimized where the center pixel is measured precisely.

The previous study conducted by Powell (2007) used several windows size for sampling unit. Windows size is block of Landsat pixels square [15]. The size of sampling unit was varied from a single Landsat pixel to 17x17 pixel square. Sampling unit variation was used to investigate impact size of sampling unit on accuracy assessment based correlation between model and reference from high-resolution imagery. Based on the correlation between modeled fraction and reference, a single pixel has low correlation. As the size of sampling unit increased, the correlation increased.

4. Conclusion

The accuracy of LSMA model was influenced by the positional error between Landsat 8 imagery and the reference imagery. Selection the right sampling unit is a one way to minimize the deviation caused by position error corresponding reference data. In this study, other factors might have the influence on accuracy result.

References

[1] Mare Marina R R and Mihai B A 2016 Mapping Land Cover Using Remote Sensing Data and GIS Techniques: A Case Study of Prahova Subcarpathians Procedia Environmental Sciences, 32 244

[2] Tran T V, Julian J P and de Beurs K M 2014 Land Cover Heterogeneity Effects on Sub-Pixel and Per-Pixel Classifications ISPRS *International Journal of Geo-Information* **3** 540

[3] Helmi and Badwai 2012 Sub pixel Accuracy Analysis of Phase Corelation Shift Measurement Methods Applied to Sattelite Imagery *International Journal of Advanced Computer Science and Appliation* **3** 202

[4] Aplin P and Atkinson P M 2010 Sub pixel Land Cover Mapping for Per Field Classification *International Journal of Remote Sensing* **22** 2853

[5] Lu D Batistella and M Moran E 2002 Linear Spectral Mixture Analysis of TM Data for Land use and Land Cover Classification in Rondonia, Brazilian Amazon *Proc Symposium on Geospatial Theory, Processing and Application* Ottawa

[6] Danoedoro P 2015 Pengaruh Jumlah dan Metode Pengambilan Titik Sampel Penguji Terhadap Tingkat Akurasi Klasifikasi Citra Penginderaan Jauh Conf. accessed from www.researchgate.net

[7] Congalton R G and Green K 2009 *Assesing the Acuracy of Remotely Sensed Data: Principle and Practices* Second Edition (Boca Raton: CRC Press, Taylor & Francis Group) p 55

[8] Wang L and Silvan-Cardenas J L 2008 Sub-pixel confusion–uncertainty matrix for assessing soft classifications *Remote Sensing of Environment* **112** 1081

[9] Wu C 2004 Normalized Spectral Mixture Analysis for Monitoring Urban Composition using ETM+ Imagery *Remote Sensing of Environment* **93** 480

[10] Weng Q, Hu X and Liu H 2009 Estimating Impervious Surface Using Linear Spectral Mixture Analysis with Multitemporal ASTER Images *International Journal of Remote Sensing* **30** 4807

[11] Yang F, Matsushita B and Fukushima T A 2010 A Pre-screened and Normalized Multiple

IOP Conf. Series: Earth and Environmental Science **165** (2018) 012024 doi:10.1088/1755-1315/165/1/012024

Endmember Spectral Mixture Analysis for Mapping Impervious Surface Area in Lake Kasumigaura Basin, Japan ISPRS *Journal of Photogrammetry and Remote Sensing* **65** 479

[12] Foody G M 2008 Sample Size Determination for Image Classification Accuracy Assesmet and Comparison Proc. of the 8th International Symposium on Spatial Accuracy Assessment in *Natural Resourches and Environmental Science,* vol **25** p 154

[13] Weng F and Pu R 2013 Mapping and Assessing of Urban Impervious Areas using Multiple Endmember Spectral Mixture Analysis: a Case Study in the City of Tampa Florida *Geocarto International* **28** 594

[14] Schowengerdt R A 2007 *Remote Sensing Models and Methodes for Image Processing* (United Kingdom : Elsevier, Inc) p 437

[15] Powell R L, Roberts D A, Dennison P E and Hess L L 2007 Subpixel mapping of Urban Land Cover using Multiple endmember Spectrak Mixture Analysis Manaus Brazil *Remote Sensing of environment* **106** 253

Acknowledgments

This research was supported by Faculty of Geography, Universitas Gadjah Mada. We also thanks to Digital Globe who provide Geoeye-1 imagery.

IOP Conf. Series: Earth and Environmental Science **165** (2018) 012025 doi:10.1088/1755-1315/165/1/012025

Applied Geo-Eye Imagery and GIS for Estimating Crop Production Based On Land Suitability in Karangasem Regency, Bali

H A Setyowati[1*], S H Murti B S[2], R Nurani[1], E Susanti[1], S Puspitasari[1], R Ariyani[1], R Esti P[3], and A A Firdausi[3]

[1]Graduated student from Departement of Cartography and Remote Sensing, Faculty of Geography, Universitas Gadjah Mada, Sekip Utara, Sleman, Yogyakarta 55281
[2]Lecturer from Departement of Cartography and Remote Sensing, Faculty of Geography, Universitas Gadjah Mada, Sekip Utara, Sleman, Yogyakarta 55281
[3] Student from Departement of Cartography and Remote Sensing, Faculty of Geography, Universitas Gadjah Mada, Sekip Utara, Sleman, Yogyakarta 55281

heratania.a.s@mail.ugm.ac.id

Abstract. Farming practices are vital to agrarian countries like Indonesia. Over the years, regional development process, coupled with lack of information on agricultural production, has induced human interventions on land that markedly convert agricultural into non-agricultural land use. Therefore, estimating crop production becomes necessary for a deeper understanding on the farming potential of a region. The study was conducted in various physical environments in Karangasem Regency, Bali. It aimed to 1) estimate both crop production and productivity using land suitability approach; 2) calculate the crop production and productivity; and 3) map the type of cropping pattern and schedule in the study area. Crop production was estimated with remote sensing and Geographic Information System (GIS) techniques and based on land suitability for rice, maize, and cassava production. The results show that remote sensing and GIS are essential in the estimation of crop production and productivity that employs land suitability approach and matching method. The estimated productions of rice, maize, and cassava are 540,173.840 kg/year, 1,843,857.202 kg/year, and 903,416.517 kg/year, respectively. In addition, the cropping pattern in Karangasem Regency varies between land suitability units, as determined by *subak* irrigation system, weather variation, and the economic condition of farmers.

Keywords: Agriculture, Crop Production, Land Suitability, Remote Sensing, GIS

1. Introduction

As an agrarian country, Indonesia depends largely on agricultural sector. Developing countries are typically defined with a vast agricultural sector in which, statistically, 75% of the populations engage in agriculture, more than 50% of the national incomes also originate from the same sector, and almost all of the exports include farming materials [10]. However, the existing agricultural potential lacks the inducement for the Indonesians to develop this sector. The paucity of information on crop production and productivity in every region creates limited knowledge on the beneficial outputs of farming activities. Furthermore, as the region develops, agricultural to non-agricultural land use conversion becomes preferable. Therefore, the estimation of crop production is important for describing the significance of farming potential in a region.

Bali, part of Sunda Kecil Islands, is 153 km in length, 112 km in width, and approximately 3.2 km from Java Island. It lies between 8°25′23″ S and 115°14′55″ E and, thereby, has tropical climate. Even though Bali is famous of its tourism sector, the agricultural practices also contribute significantly to its development. Food availability in Bali is determined, for instance, by the agricultural production level. Meanwhile, the food demand is affected by the increasing trend of population growth over the years. Accordingly, the need for access to information on the accurate crop production estimates grows rapidly. Applying remote sensing and Geographic Information System makes the estimations faster and more accurate than conducting terrestrial surveys. The estimation of crop productivity in Karangasem Regency, Bali aimed to estimate the crop production and productivity using land suitability approach in the study area, to calculate the crop production and productivity in the study area, and to understand the type of cropping pattern and schedule in the study area. Crop production was estimated with remote sensing and Geographic Information System (GIS) techniques and based on land suitability for rice, maize, and cassava production. The results show that remote sensing and GIS are essential in the estimation of crop production and productivity that employs land suitability approach and matching method.

2. Materials and Methods

2.1. Data

The study employed land suitability approach with matching method to estimate the crop production and productivity in Karangasem Regency. The field surveys used GPS, soil test kit, and abney level. Meanwhile, the analysis was conducted on a set of data, viz. GeoEye satellite image in 2014, the land use maps of Karangasem Regency (scale 1:25,000), the Reconnaissance Soil Map of Bali Province (scale 1:250,000), rainfall data, and the table of land suitability for crop productions. The estimation of crop production in this study was based on land units, with various parameters of land suitability, i.e. soil, climate, slope, rainfall, and agricultural land (paddy field and dryland), which were 'matched' with the table of land suitability for every commodity.The study obtained information on agricultural and non-agricultural land use from GeoEye satellite image because of its high spatial resolution. The image also accentuated certain objects that were useful for extracting the physical characteristics of land, such as slope, soil feature, and water potential, and thereby simplified the identification process.

2.2. Methods

2.2.1 Study Area Selection. Karangasem Regency was selected because it represents regions that depend on vast agricultural areas with diverse cropping patterns and commodities. It also has extensive natural resource potentials for agriculture, fishery and marine sector, and mining. Its economic development is mainly supported by farming activities, as shown by the agricultural share of the Gross Regional Domestic Product (GRDP) in 2009, i.e. 28.96%. The major dryland farming in the study area is due to the diverse topography including an old volcano (Mt. Agung) on one of its side. The estimation of crop production, as sampled according to land units, considered the physical environments of the study area. The development of cropping calendar significantly promotes the process of maximizing crop productivity especially in the presence of diverse cropping patterns for various types of farming commodities in Karangasem Regency.

2.2.2 Pre-field Preparation

2.2.2.1 GeoEye Image Correction .The 2014 GeoEye image of the study area shows that the landform change in the corresponding year was not significant. The image was corrected geometrically, i.e. restoring the position of pixels in a way that the transformed digital image can visualize the sensor-recorded objects on the Earth's surface [3]. Geometric correction can be conducted in two ways: 1) by referring to data from spacecraft and knowledge on internal distortion; and 2) by conducting Ground Control Point (GCP)-based transformation [3].The geometric correction in this study used image-to-image registration method, i.e. correcting one image with other geo-referenced satellite images in the same area. The coordinates from the same area were then adopted based on the GCPs as the coordinate reference of the study area. GeoEye satellite image have high spatial resolution so its can obtained information on agricultural and non-agricultural land use. The satellite image was identified using visual interpretation, i.e. on-screen digitation on the corrected satellite image to classify the cropland such as rice, maize, and cassava. The result of identified image using visual interpretation were verified by comparing with field survey data.

2.2.2.2 Agricultural and Non-agricultural Land Classification. Agricultural land was one of the main factors in the crop estimation process that served as the object of this study. It represented the extent of the cultivated area in the regency. It was identified using visual interpretation, i.e. on-screen digitation on the corrected satellite image. The study used on-screen digitation because visual delineation, using elements of interpretation, offers a good accuracy. The object identification provided the distribution of agricultural and non-agricultural land in the study area

*2.2.2.3 Agro-ecosystem.*The regional potential for the development of agricultural and forest commodities is essentially determined by the nature of the physical environments, which includes climate, soil, topography, hydrology, and several conditions for crop growth [7]. Meanwhile, land suitability is shaped by several parameters such as slope, rainfall intensity, soil feature, and water potential. Slope defines the distribution of water in an area and, at the same time, water availability becomes the main factor in determining farming commodities and cropping patterns. Slope was extracted from contour data presented in the topographical map (RBI) and, then, used for providing information on terrain units. Meanwhile, the characteristics of rainfall determine the corresponding crop types. Rainfall data in the last 10 years were interpolated in order to obtain the rainfall intensity in the study area. The interpolated rainfall also provided the relative classification of rainfall intensity and its interval, as compared to other regions with similar rainfall intensity. In addition, slope profile was extracted from the soil map of Bali (scale 1:25,000). This map was processed to increase its details, using the slope map, i.e. by classifying the slope and associating it with soil type. As a determinant of land suitability, water potential was analyzed based on the presence of river.

The physical parameters like landform, slope, soil feature, and water potential were combined to shape the characteristics of terrain units. Afterwards, these units were also merged with agricultural land use data in order to divide the study area into land units that were used as sampling reference during field survey. In addition to land units, the study also created agro-ecosystem units as the representations of existing agricultural ecosystem in which productivity is defined as a balanced presence of soil, nutrient, sunlight, humidity, and organism that jointly creates a healthy cropping environment and sustainable yield [1]. Agro-ecosystem units combined a set of data on land units, which had been matched with the table of land suitability for crops, and agricultural land use. This matching method was used for land evaluation, i.e. comparing the criteria of land suitability and the exact data on land quality [4]. Meanwhile, land suitability assessment is defining a specific land based on its existing qualities and characteristics that shape certain conditions for crop growth; in other words, the evaluated land has specific suitability for certain crops [9]. The study also conducted a field check for agro-ecosystem units.

2.2.2.4 Field Survey. The field survey was conducted in order to acquire primary data and check the results of pre-field data processing. The results of the field survey were used for checking the accuracy of land unit approach and sampling on land units indicated that every land unit was represented by several field samples that met the objective of the study and the extent of land units. The wider the land units, the more the samples. Field samples were selected effectively in order to represent the real condition. Field survey was conducted with direct observation and interview. Direct observation aimed to understand the types of commodity, while interview aimed to find more information on cropping pattern, crop yield, and agricultural productivity.

2.2.2.5 Post-field Analysis. Post-field activities focused on field data analysis in order to map the crop production estimates. The calculating crop production. Field survey aimed to acquire a set of data on commodity, productivity, and cropping pattern as well as to evaluate the pre-field image interpretation of agricultural area. Purposive random calculation of the estimation mainly attempted to acquire information on agricultural production in Karangasem Regency. The estimation process included the multiplication of every type of commodity in every identified land unit by agricultural productivity per area. It also involved the cropping pattern in the study area in determining the annual crop production. The estimation produced the annual production of every commodity in Karangasem Regency.

3. Result, Analysis, and Discussion

3.1 Pre-field.

The first step of the study was pre-field activities, i.e. creating rainfall map, drainage map, soil map, and land suitability maps for maize, rice, and cassava production. Afterwards, these base maps were used for creating field survey maps, namely slope map, land use map, tentative map of agro-ecosystem, and map of sampling locations. Slope map is defined as the change in elevation over a given horizontal distance between two points. It transformed the contour data presented in the topographical map (RBI scale 1:25,000) into DEM and defined the slope based on the classification of [12] as shown in Figure 1. slope determines land suitability in a way that a steep area has different water availability, as well as different land suitability, from a gently sloping area. However, in Karangasem Regency, slope is not an issue in land suitability due to the presence of terraces as land modification.

The land use in Karangasem Regency is relatively homogeneous, as shown in Figure 2., i.e. dryland and dry agricultural area. Land suitability mapping for estimating crop production refers to two types of land use, namely agricultural and non-agricultural land use shown in Figure 3. In this case, remote sensing provided satellite images, while the elements of interpretation provided a means of acquiring data on the aforementioned land use types from satellite images. Agricultural land appears as an area with low vegetation and, thereby, with smooth texture, as compared with other vegetation appearances. Analysis of the application of remote sensing for estimating crop production refers to agricultural land use only because the cultivation of rice, maize, and cassava, as the focus of the study, presents in agricultural land use. The study assumed that the wider the agricultural land, the higher the crop production.

Soil characteristics presented in land suitability map includes soil surface texture, soil permeability, pH, and soil fertility. Four of these characteristics determine the suitability of agricultural land. As shown in Figure 4. based on the derivative information of soil maps, the soil types in the study area are grayish-brown alluvial, brown latosol, Litosols, regosol with humus presence, brown regosol, yellowish-brown regosol, grayish-brown regosol, and gray regosol. Regosols, mainly found in Karangasem Regency, are young soils that show no or little soil horizon development. They are characterized by sandy texture, normal pH, and medium fertility, and originated from pyroclastic materials; hence, their spatial associations with slopes of young volcanoes. Soil texture and permeability are the two intercorrelated soil properties that determine water availability for crop growth. Texture defines the amount of stored and transmitted water. Therefore, the more coarse or silt the texture, the higher the infiltration capacity. On the contrary, the finer or more clay-sized particles the soil has, the lower the infiltration capacity. Water infiltration capacity determines the rate of water movement in a soil (permeability); therefore, the slower the water transfer rate, the more the land suitability for crop production. Soil pH represents the acidity of a soil. A 'good' soil has normal pH (5-7); in other words, the lower the pH, the lower the land suitability for crop production. Soil fertility is determined by the nutrient contents of a soil. The more the nutrient contents, the more fertile the soil. Furthermore, the more fertile the soil, the better the crop growth.

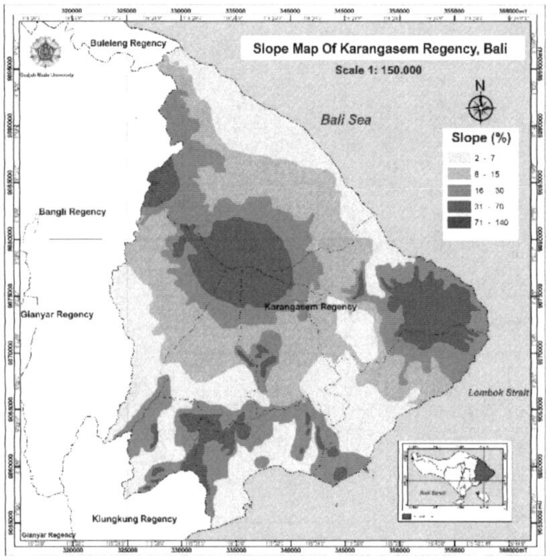

Figure 1. Slope Map of Karangasem Regency

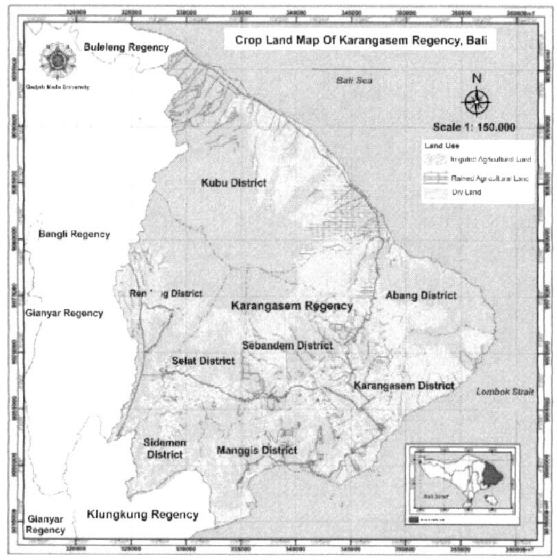

Figure 2. Crops Land Map of Karangasem Regency

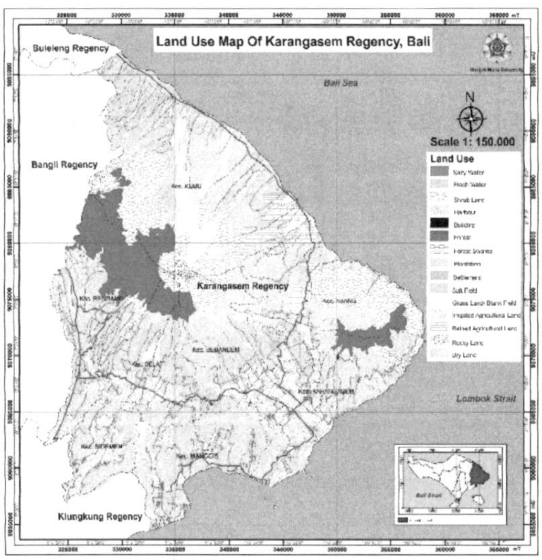

Figure 3. Land use Map of Karangasem Regency

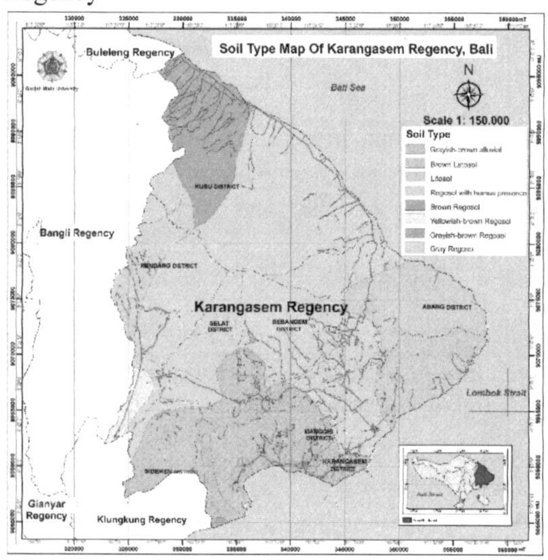

Figure 4. Soil Types Map of Karangasem Regency

Rainfall intensity in the study area is categorized into 7 classes, ranging from the lowest intensity, i.e. 1,300 mm/year, to the highest intensity, i.e. 3,300 mm/year. It indirectly represents the water availability of an area, as stored either in surface water or in groundwater. Even though maize and cassava require less water than rice, the productions of these three commodities imply that a better land suitability for crops is associated with higher rainfall intensity, and vice versa. The implication refers to the relationship between higher rainfall intensity and better drainage. However, such conclusion needs to consider the role of slope and soil profile as well. As shown in **Figure 5.** the drainage condition in the study area is divided into 3 classes, namely good, fair, and poor drainage. Poor drainage is commonly found in steep slope with sandy soil, while good drainage is typically found in a gently sloping to flat area.

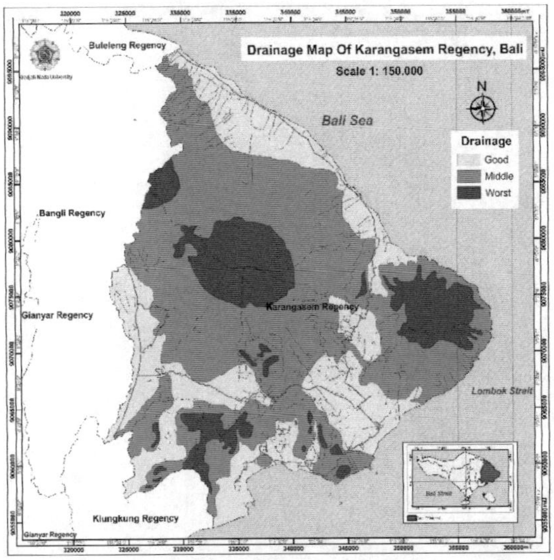

Figure 5. Drainage Map

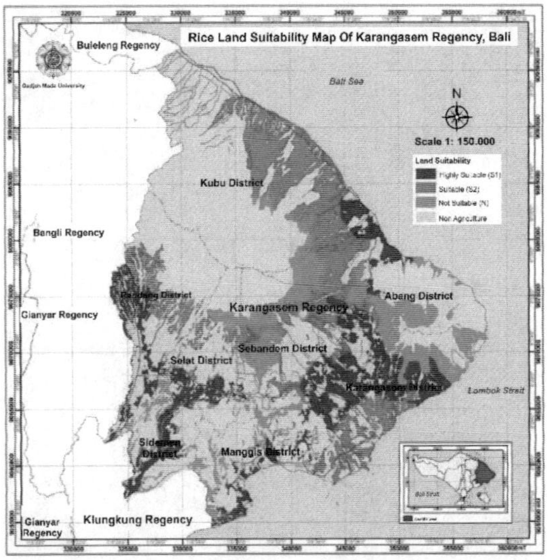

Figure 6. Rice Land Suitability Map

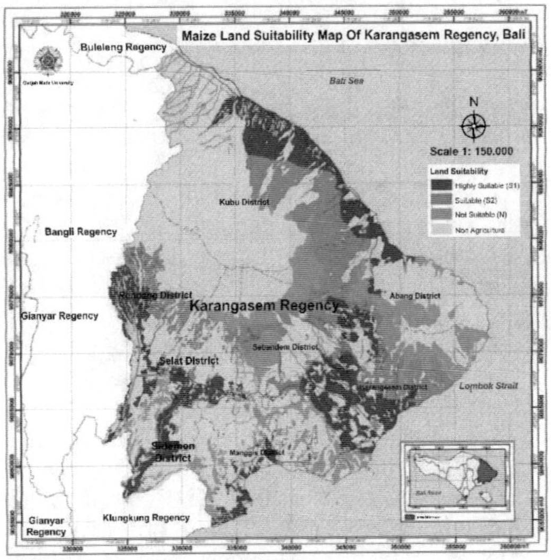

Figure 7. Maize Land Suitability Map

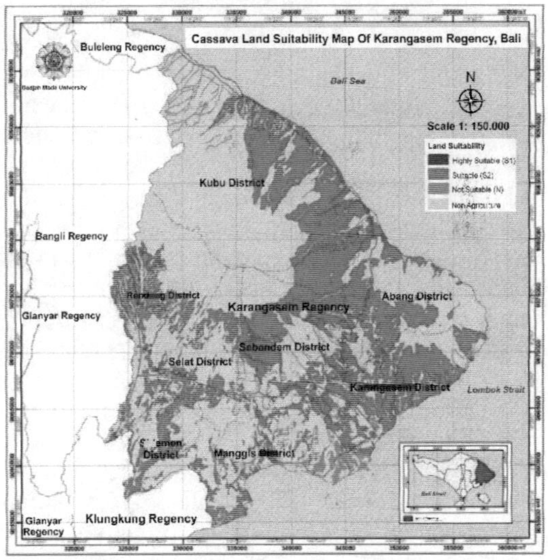

Figure 8. Cassava Land Suitability Map

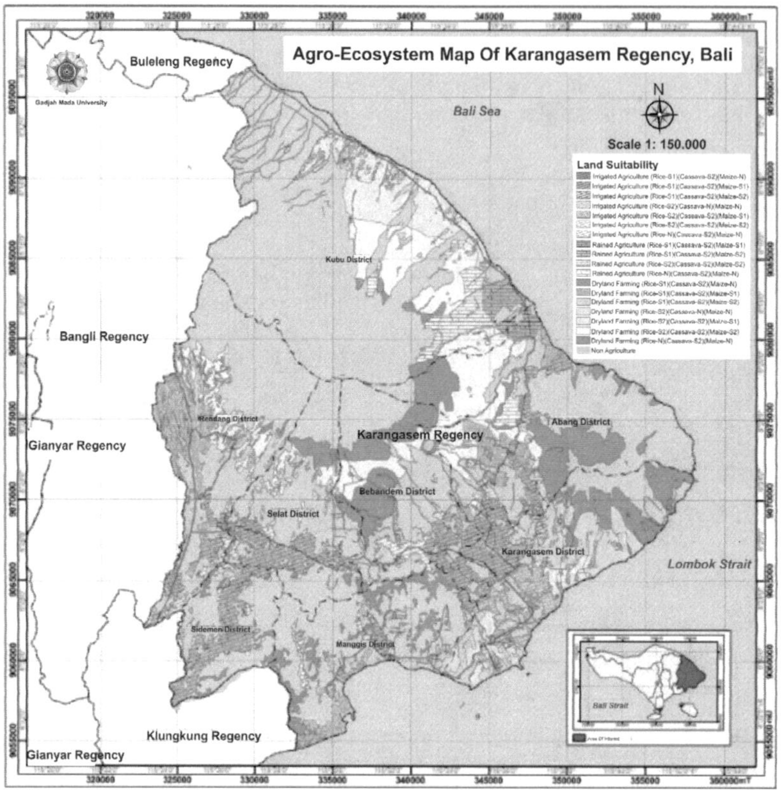

Figure 9. Agro-ecosystem Map

Land suitability is determined by several parameters, namely rainfall, drainage, temperature, effective soil depth, soil pH, and slope. The combination of these parameters lead to different land suitability, viz. highly suitable (S1), suitable (S2), and not suitable (N) (Hardjowigeno and Widiatmaka, 2011). In addition, land suitability for every crop is different. As shown in **Figure 6.,7.,** and **8.** the conditions for rice growth compose the criteria in every class of land suitability: a highly suitable land for rice production (S1) has rainfall intensity >1,500 mm/year, poorly drained, effective soil depth >50, pH >5.5-7, and slope <3%; suitable land (S2) has rainfall intensity 1,200-1,500 mm/year, poorly drained, effective soil depth >40-50, pH >7-8 or 5-5.5, and slope 3-8%; and not suitable land (N) has low rainfall intensity (<800 mm/year), excessively drained, effective soil depth 20-25, and slope >15-25%.

Meanwhile, the land suitability for maize growth is as follows: highly suitable (S1) comprises rainfall intensity >1,200 mm/year, poorly to fairly drained, effective soil depth >60, pH 6-7, and slope <3%; suitable (S2) comprises rainfall intensity between 900-1,200 mm/year, imperfectly drained, effective soil depth >40-60, pH >7-7.5 or 5.5-<6, and slope 3-8%; not suitable (N) comprises low rainfall intensity, no drainage, effective soil depth 20-24, pH >8-8.5 or 4-4.5, and slope >15-24%. Land suitability data for the three commodities are presented in the agro-ecosystem map, which consists of 18 types of land suitability for all commodities, as shown in **Figure 9** Based on this agro-ecosystem map, 22 samples were selected using random sampling technique.

3.2 Field Survey

Field survey included physical measurements and interviews. The physical measurements used abney level for slope measurement and soil test kit for soil pH measurement. Field survey was conducted because not every parameter of land suitability can be extracted from satellite images or secondary data during pre-field activities. Besides, field survey provides a means to check the accuracy of pre-field interpretation results like slope. Interviews were conducted to farmers who managed the sampled agricultural land. These interviews used questionnaires for acquiring data on productivity, cropping pattern, as well as land preparation, seeding, and fertilizer application techniques. **Figure 10.** shown field appearance of rice, maize, and cassava.

Figure 10. Field Appearance of Rice, Maize, and Cassava

3.3 Post-field Analysis

Post-field analysis is the last step in estimating the crop production in the study area. The analysis included crop production and productivity data management, map of crop production estimates, and cropping calendar. A set of data obtained from direct measurements and interviews shows that the highest crop productivity is found in irrigated agricultural land (Rice-N) (Cassava-S2) (Maize-N) and rainfed agricultural land (Rice-N) (Cassava-S2) (Maize-S2) for rice and cassava. Meanwhile, maize is mostly found in irrigated agricultural land (Rice-S1) (Cassava-S2) (Maize-N), irrigated land (Rice-S1) (Cassava-S2) (Maize-S1), and dry agricultural land (Rice-S2) (Cassava-N) (Maize-N).

3.4 Data on Crop Production

The assessment of crop production has been conducted with various methods including land suitability approach, commonly referred to as crop production estimation. This study estimated the productions of rice, maize, and cassava and resulted in 18 land suitability classes with different production and productivity. Irrigated agricultural land, rainfed agricultural land, and dry agricultural land have respectively seven, four, and seven land suitability classes. Here are three examples of land suitability and corresponding crop production found in the study area: 1) the land suitability of irrigated agricultural land for (Rice-N) (Cassava-S2) (Maize-N) produces 24,381.267 kg/are of rice and 48,762.533 kg/are of cassava; 2) despite the land suitability of rained agricultural land for (Rice-N) (Cassava-S2) (Maize-N), field survey showed that this specific land was used for rice production, i.e. 266,867,520 kg/are; and 3) dryland with land suitability for (Rice-N) (Cassava-S2) (Maize-N) produces 573,876 kg/are of maize and 143,469 kg/are of cassava. The agricultural productivity in the study area are presented in detail in **Table 1**.

Table 1. Table of Production per Land Suitability (Source: Field Survey and Data Processing, 2014)

Land Suitability	Productivity			Total Production (kg/are)		
	R	M	C	R	M	C
Irrigated Agricultural Land - (Rice-N)(Cassava-S2)(Maize-N)	2,000	0	2,000	24,381	0.000	48,762
Irrigated Agricultural Land - (Rice-S1)(Cassava-S2)(Maize-N)	950	500	0	4,803	3,792.044	0.000
Irrigated Agricultural Land - (Rice-S1)(Cassava-S2)(Maize-S1)	950	500	0	277,080	218,748.088	0.000
Irrigated Agricultural Land - (Rice-S1)(Cassava-S2)(Maize-S2)	1,472	0	0	48,210,592	0.000	0.000
Irrigated Agricultural Land - (Rice-S2)(Cassava-N)(Maize-N)	0	0	40	527,170	0.000	6,589
Irrigated Agricultural Land - (Rice-S2)(Cassava-S2)(Maize-S1)	200	200	0	63,947	31,973.833	0.000
Irrigated Agricultural Land - (Rice-S2)(Cassava-S2)(Maize-S2)	1,266	0	0	289,847,814	0.000	0.000
Rained Agricultural Land -(Rice-N)(Cassava-S2)(Maize-N)	1,600	0	0	266,867,520	0.000	0.000
Rained Agricultural Land -(Rice-S1)(Cassava-S2)(Maize-S1)	0	0	20	0.000	0.000	76,042
Rained Agricultural Land -(Rice-S1)(Cassava-S2)(Maize-S2)	0	40	40	0.000	2,941.040	1,470
Rained Agricultural Land -(Rice-S2)(Cassava-S2)(Maize-S2)	0	0	20	0.000	0.000	97,939
Dry Land -(Rice-N)(Cassava-S2)(Maize-N)	0	40	20	0.000	573,876	143,469
Dry Land -(Rice-S1)(Cassava-S2)(Maize-N)	0	87	0	0.000	18,786	0.000
Dry Land -(Rice-S1)(Cassava-S2)(Maize-S1)	0	87	0	0.000	267,903	0.000
Dry Land -(Rice-S1)(Cassava-S2)(Maize-S2)	0	40	40	0.000	122,379	61,189
Dry Land -(Rice-S2)(Cassava-N)(Maize-N)	180	500	180	9,758.082	40.658	4,879
Dry Land -(Rice-S2)(Cassava-S2)(Maize-S1)	0	40	40	0.000	206.128	103,064
Dry Land -(Rice-S2)(Cassava-S2)(Maize-S2)	0	18	18	0.000	375.610	372,095
Total				605,833,068	1,862,796	915,502

Rice is a commodity with the highest production because Karangasem Regency relies on *subak* irrigation system for its production. This system divides water that flows from the upstream into rice fields periodically in order to provide a good rice growth and sustainable rice yield rate. Maize production results in a higher yield than cassava production because the former is perceived as a commodity with higher selling price; therefore, maize is widely grown and highly produced. Every land suitability results in different production due to differences in land characteristics. Consequently, the land capability for crop growth and the crop production also vary. Information on crop production is essential not only for understanding land suitability but also for estimating the crop production in every district. The results of crop production in every district of Karangasem Regency were obtained from the calculation of productivity, harvest age, and agricultural land size. The annual production was assessed using the multiplication of average productivity by harvest age. From the annual production, the total production was also known and, then, used for estimating the production in the whole regency, which was divided again into crop production in every district. The highest rice production is found in Karangasem District, i.e. 128,728.428 kg, while the lowest production is found in Kubu District, i.e. 25,482.400 kg. Maize is highly produced in Abang District, i.e. 420,600.491 kg, and least produced in Sidemen District, i.e. 65,912.957 kg. Meanwhile, cassava is mostly produced in Kubu District, i.e. 355,946.087 kg, and least produced in Manggis District, i.e. 13,807.417 kg. The crop production in every district is presented in detail in **Table 2**.

Tabel 2. Production per Commodity per District in Karangasem Regency (Source: Data Processing, 2014)

District	Production per Commodity (kg)		
	Rice	Maize	Cassava
Abang	69,665.613	420,600.491	204,638.574
Bebandem	80,353.590	221,084.407	79,217.263
Kubu	25,482.400	324,557.023	355,946.087
Manggis	67,293.266	127,711.610	13,807.417
Karangasem	128,728.428	364,512.098	113,078.060
Rendang	31,324.283	214,561.226	75,929.621
Selat	64,129.358	104,917.390	39,234.722
Sidemen	73,196.902	65,912.957	21,564.773
Total	**540,173.840**	**1,843,857.202**	**903,416.517**

3.5 Map of Agricultural Production Estimates

Karangasem is one the regencies in Bali that has a great agricultural potential. According to [2] this regency produces 77,870.18 tons of rice, 22,838.94 tons of maize, and 107,903.97 tons of cassava from the widely found agricultural land in 8 districts in Karangasem Regency. This statistics is slightly different from the crop production estimates in 2014, which were calculated based on land suitability evaluations and interviews. However, the classification of land suitability required further field check because the results of image interpretation were different from the actual condition. Such difference occurred due to the influence of several factors like the *subak* irrigation system. Every commodity has different criteria for land characteristic and suitability.

Every district has various classes of land suitability for rice production. **Figure11** shows that the highest rice production is in Karangasem District with a total yield of 128,728.428 kg, the second highest productions is in Bebandem and Sidemen with total yields of 80,353.590 kg and 73,196.902 kg, respectively. The land in these three districts is mostly used for irrigated agricultural land, rainfed agricultural land, and dry agricultural land with 'high suitability' (S1) for rice. Despite the vast dry and rainfed agricultural lands in Bebandem District, the district is still able to produce a significant amount of rice due to the presence of *subak* irrigation system. This system ensures reliable water supply in dry seasons. Kubu and Rendang Districts are mostly non-agricultural land and drylands with very low productivity, i.e. 25,482.400 kg and 31,324.283 kg, respectively. Even though Rendang District has some irrigated agricultural lands, its productive lands have limited suitability for rice production. Mountainous area and dry climate are the factors that significantly influence the rice production in this district. The government is developing various programs that aim to improve national food security, particularly for national-level subsistence farming. One of the programs is enhancing the quality of intensification, optimization, and extensification of agricultural land in order to ensure the annual availability of rice [8].

Maize is the second greatest commodity after rice, which is usually grown in the transition period between two rice cultivation seasons. This cropping rotation aims to maintain the soil quality and break the chain of pests like rats and leafhoppers. **Figure12** shows the maize production estimates in Karangasem Regency. High maize production often associates with the capability of maize to grow in either dry or wet land. The highest maize production is found in Abang District with a total yield of 420,600.491 kg. This district mostly produces rice; but maize is also suitable for the mountainous topography, dry climate, and limited access to water supply in the district. The second highest maize yield is found in Kubu and Karangasem Districts that produce 324,557.023 kg and 364,512.098 kg, respectively. Kubu District has the most suitable land (S1) for maize. Bebandem and Rendang Districts produce the third highest maize yield, i.e. 221,084.407 kg and 214,561.610 kg, respectively. Meanwhile, Selat and Manggis Districts have lower maize yields, i.e. 104,917.390 kg and 127,711.610 kg, respectively. In Selat District, maize is produced in lands that are suitable for irrigated agriculture (Rice-S1) (Cassava-S2) (Maize-S1) and three types of dryland farming, namely (Rice-N) (Cassava-S2) (Maize-S2), (Rice-S1) (Cassava-S2) (Maize-S1), and (Rice-S2) (Cassava-S2) (Maize-S2). Meanwhile, in Manggis District, maize is grown mostly in lands that are suitable for irrigated agriculture (Rice-S1) (Cassava-S2) (Maize-S1). Sidemen District has the lowest maize production in the regency, i.e. 65,912.957 kg. In this district, maize is grown in lands that are suitable for irrigated agriculture (Rice-S1) (Cassava-S2) (Maize-S1) and dryland farming (Rice-N) (Cassava-S2) (Maize-N). Maize is also cultivated in hilly to mountainous areas where the few productive lands are beyond the reach of *subak* irrigation system.

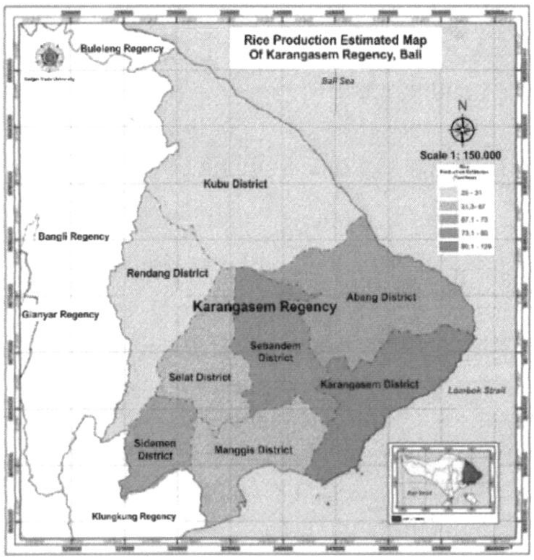

Figure 11. Rice Production Estimated Map (Source: Data Processing, 2014)

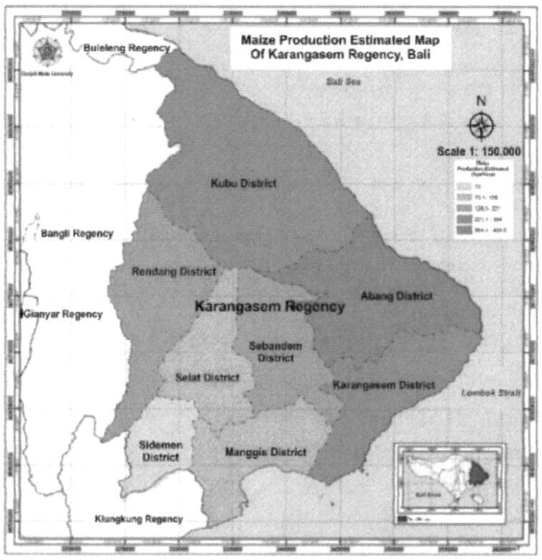

Figure 12. Maize Production Estimated Map (Source: Data Processing, 2014)

Figure 13. Cassava Production Estimated Map (Source: Data Processing, 2014)

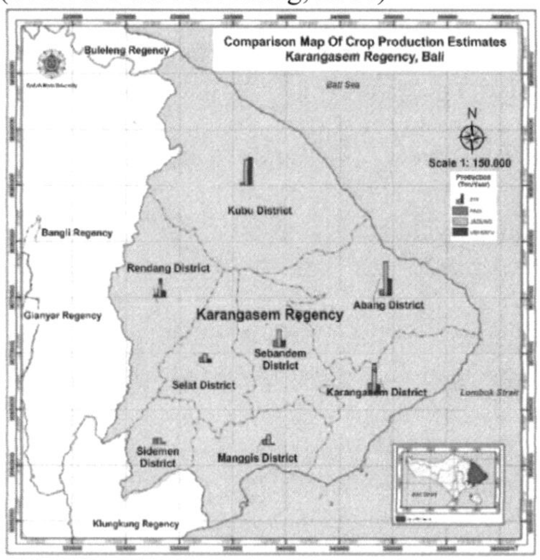

Figure 14. Comparison Map of Crop Production Estimates (Source: Data Processing, 2014)

Cassava is an alternative crop during the transition period between the cultivations of other crops. It is usually grown in dryland, barren land, and water-deficit area. **Figure 13.** shows that Kubu District produces the highest cassava yield, i.e. 355,946.087 kg. Abang District also has a high cassava production, i.e. 204,638.574 kg. Meanwhile, Rendang, Bebandem, and Karangasem Districts have lower cassava yields, ranging from 39,100 kg to 113,000 kg. The cassava productions in Karangasem, Bebandem, and Rendang District yield 113,078.060 kg, 79,217.263 kg, and 75,929.621 kg, respectively. The lowest cassava productions are found in Selat District, which vary between 21,000 ton/year and 39,000 ton/year, and Sidemen and Manggis Districts, which vary between 14,000 ton/year and 21,000 ton/year. Cassava is categorized as the highly suitable crop for every rainfed agricultural land, irrigated agricultural land, and dryland in every district because it can grow almost everywhere in the regency, i.e. in areas with no or little water and flat to mountainous topography.

The comparison between these three crop production estimates in Karangasem Regency is presented in a map, as shown in **Figure14**. The map shows that Kubu District has the highest cassava yield, as compared to its maize and rice productions. The land characteristics of this district are suitable for cassava productions. Meanwhile, the proportion of maize production in every district is mostly higher than the proportions of the other two commodities. Therefore, maize production is highly suitable and productive in Karangasem Regency, Bali. The highest maize yield is found in Abang District and, followed by, Karangasem District. Maize stays productive throughout the year in spite of its various cropping patterns. The opposite is true for rice production, which has the lowest yield among the commodities. The *subak* irrigation system has an open-and-close mechanism that becomes both opportunity and challenge for rice growth due to the strong dependence of rice on water availability. This irrigation system significantly promotes the rice production in agricultural lands located close to the water resource or, in other words, with high probability to obtain water. On the contrary, this system becomes the primary obstacle of rice productions in agricultural lands that rarely receive irrigation water. Consequently, instead of planting rice, farmers tend to cultivate crops with water-deficit tolerance like maize and peanuts in order to sustain land productivity.

3.6 Map of Cropping Calendar

Cropping calendar is a schedule for the potential cropping and harvesting time of food crops. This study provided cropping calendars for rice, maize, and cassava in Karangasem Regency, Bali, as shown in **Figure 15**. Creating the map of cropping calendar aimed to understand the cropping schedule in one year based on the land suitability of the study area. Cropping pattern and rotation is spatially determined by land suitability and water availability or *subak* irrigation system. Bali is historically known to have unique cultural tradition and religious commitment, one of which is in the form of community organization named *Subak*. *Subak* is a group of indigenous farming communities that is based on Balinese tradition and culture, i.e. the Hindu philosophy "Tri Hita Karana". It functions as a social organization in the field of local water resource management for agricultural lands. *Subak* is an irrigation sub-system whose main function is to regulate irrigation water utilization so that farmers can irrigate their fields sufficiently, fairly, and equally. So far, *Subak* has provided a very effective and strategic role in water resource management, particularly for irrigation, by ensuring water availability and utilization in Bali.

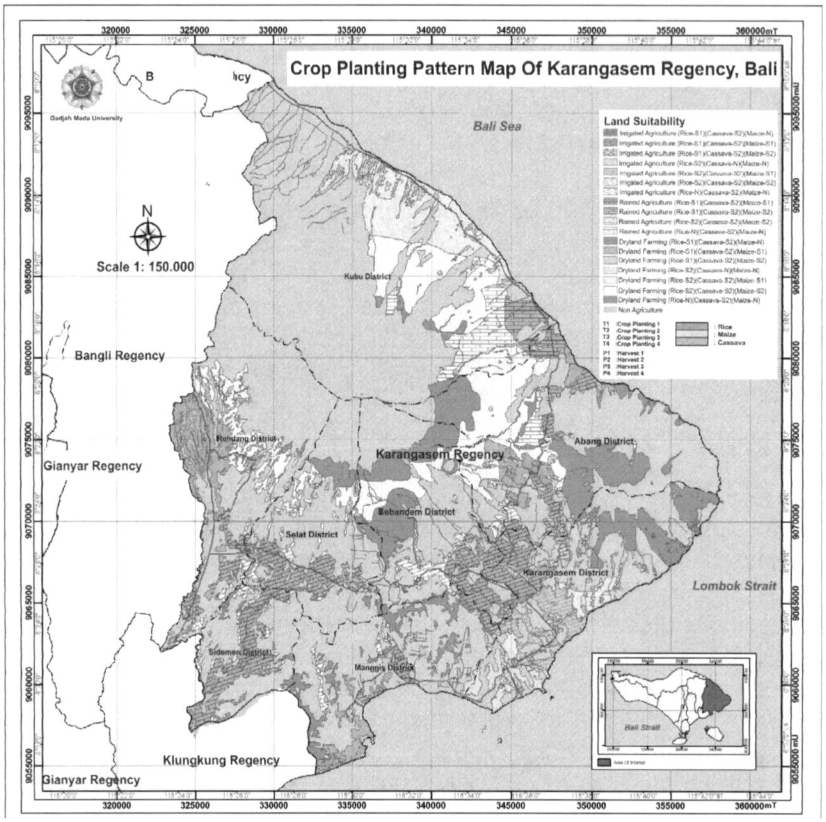

Figure 15. Crop Planting Pattern in Karangasem Regency (Source: Data Processing, 2014)

The cropping patterns may be diverse even within the same land suitability because of the influence of *subak* irrigation system. The crops grown in an area are possibly different from which crops the area is suitable for. As an example, the suitable land for the irrigated farming of (Rice-N) (Cassava-S2) (Maize-N) in the study area is actually cultivated for rice and cassava. In this agricultural land, rice is grown once a year in December and harvested in March, while cassava is grown twice a year in February-July and August-January. In addition, the suitable land for the rainfed (Rice-N) (Cassava-S2) (Maize-N) farming is actually cultivated for rice and cassava, even though (Rice-N) means that this specific land is not suitable for rice production. The difference between the results of land suitability and actual cropping pattern in the study area is likely caused by *subak* irrigation system. Both cropping pattern and calendar are influenced not only by the characteristics of land suitability but also by other factors like *subak* irrigation system, weather, and economic condition. As a farming community, not only does *subak* maintain the regulation of irrigation water, but it also provides information on crop seed, cropping schedule, harvest time, and even agricultural land management. The irrigation water regulation in *subak* is different according to dry and wet months. In wet months, *subak* does not give any significant influence on farming practices; therefore, farmers can conveniently grow any crops. However, in dry months, farmers have to comply with the open-and-close system of *subak*. In other words, *subak* irrigation system affects the cropping patterns in dry months.

4. Conclusion

According to the study results and the previously formulated objectives, remote sensing and Geography Information System are essential to estimate the crop production and productivity of, for instance, rice, maize, and cassava in Karangasem Regency, Bali using land suitability approach and matching technique. The total crop production estimates in Karangasem Regency are 540,173.84 kg/year of rice, 1,843,857.20 kg/year of maize, and 903,416.52 kg/year of cassava. The cropping pattern and schedule in every land suitability class are different due to the influence of *subak* irrigation system, weather, and the economic condition of farmers in Karangasem Regency.

References

[1] Alteri M A 1994 *Biodiversity and Pest Management in Agroecosistems* (New York: Haworth Press)

[2] BPS Karangasem 2009 *Karangasem Dalam Angka Tahun 2009* (Bali:BPS Kabupaten Karangasem)

[3] Danoedoro P 2012 *Pengantar Penginderaan Jauh Digital* (Yogyakarta:ANDI Yogyakarta) p 167

[4] Hadmoko D S 2012. *Evaluasi Sumber Daya Lahan, Prosedur dan Teknik Evaluasi Lahan : Aplikasi Teknik Skoring dan Matching* (Yogyakarta: Gadjah Mada University Press)

[5] Hardjowigeno S and Widiatmaka 2011 *Evaluasi Kesesuaian Lahan dan Perencanaan Tataguna Lahan* (Yogyakarta: Gadjah Mada University Press)

[6] Mather P M 2004 *Computer Processing of Remotely-Sensed Images, Third Ed.*, (England: John Widley & Sons)

[7] Prasetyo T H, T M Basuki and Y Mile 1997 Kesesuaian Lahan Calon Areal HTI Trans Berdasarkan Sifat-Sifat Tanah; Studi Kasus di PT. Indo Kayu. Surakarta: *Buletin Pengelolaan DAS*. Journal number IV.1.1997 e.g 53-71

[8] Sahara B and Idris 2006 Kontribusi Penggunaan Faktor Produksi dan Analisis Produksi terhadap Pendapatan Petani Rice Sawah. *Jurnal Pengkajian dan Pengembangan Teknologi Pertanian Balai Pengkajian Teknologi Pertanian*

[9] Sitorus S 1998 *Evaluasi Sumber Daya Lahan*. (Bandung: Tasito)

[10] Todaro M 2000 *Economic Developmnet,* Seventh edition (New York University: Addison Mesley)

[11] Wahyunto W and Heryanto B 2006 *Pendugaan Produksi Tanaman Rice Sawah melalui Analisis Citra Satelit*. (Warta Informatika Pertanian) Vol. **15** pp 853-869

[12] Zuidam Van R A and F I Van Zuidam Cancelado 1979 *Terrain Analysis and Classification using Aerial Photographs*. (ITC: Textbook of Photointerpretation) Vol. **VII-6** p 348

Estimating Oil Palm Production Using Sentinel 2: By-age Separation Method

Muhammad Adnan Shafry Untoro[1*], Lina Indriani[1], Bagus Wiratmoko[1], Ahnaf Hampar Sasongko[1], Balada Wisma Sandika[1], Donna Rizky Amora[1], Nur Lely Wahyunissa[1] and Sigit Heru Murti B S[1]

[1]Department of Geographic Information Science, Faculty of Geography, Gadjah Mada University, Yogyakarta 55281, Indonesia

muhammad.adnan.s@mail.ugm.ac.id

Abstract. Multispectral imagery can be used for estimating production of oil palm fruit in a plantation. One of the multispectral imagery that capable for production estimation is Sentinel 2. Sentinel 2 is a free imagery that have great 10 m medium spatial resolution bands, which are red, green, blue, and near-infrared. These bands can be used to generate several generic vegetation index such as Ratio Vegetation Index (RVI) and normalized difference vegetation index (NDVI) also non-generic vegetation index such as modified soil adjusted vegetation index (MSAVI) and atmospherically resistant vegetation index (ARVI). Those vegetation index applied for estimating the production of oil palm fruit in PT. Perkebunan Nusantara VII's plantation at Rejosari, Natar. By-age separation method demonstrate that it has a better accuracy which is 87.5 % rather than normal method that fully transformed without being separated by-age. The fully transformed method accuracy are 61.1 % for RVI, 59.1 % for NDVI, 55.3 % for MSAVI, and 62.1 % for ARVI.

Keywords: Estimation of crop production, Oil palm, Vegetation Index, Sentinel 2

1. Introduction

Oil palm or *Elaeis guineensis jacq* is one of the most widely produced and consumed commodities around the world. Oil palm is also the most profitable oil crops commodities because it's gives highest ratio of production per hectare [4]. Indonesia is the biggest palm oil producer and exporter in the world that in 2015 the total area of oil palm plantation in Indonesia reaches 11.3 million hectares and producing 31.28 million tons of palm oil [1]. East coast of Sumatera and Aceh is the area of first oil palm plantation taking place [6] and then it spread out to South Sumatera, West Sumatera, Riau, Bengkulu, Jambi, Lampung, West Java, West Kalimantan, Central Kalimantan, East Kalimantan, Maluku, Sulawesi, and Papua [8]. PT Perkebunan Nusantara is one of company that carry out oil palm plantation of Indonesian government in 3 provinces.

Currently, the calculation of the estimated amount of palm oil production is mostly done terrestrially. The terrestrial method certainly takes a lot of time and energy because the oil palm plantations usually are so large that this method is considered inefficient. Remote sensing can be used to assist the calculation of production estimation. The estimation of palm oil production in remote sensing requires data of tree's age, vegetation index, and heavy sample data of palm oil in the field [7]. The data is processed together with the vegetation index transformation of the remote sensing imagery used, which is Sentinel 2. Therefore, calculation of production estimation is no longer done by census in the field (terrestrial) but uses sampling technique and then integrated with remote sensing imagery. However, estimating production of oil palm using remote sensing can be challenging because oil palm tree physical appearance is almost identical at age of 5 – 25, but they have different level of production. Therefore, a method which can differentiate oil palm tree physical appearance and relate it with their level of production is needed.

2. Study Site

The research location of palm oil production estimate is done at oil palm plantation owned by PT. Perkebunan Nasional VII Lampung Rejosari Unit. The plantation is divided into four smaller part called *afdeling,* a Dutch word, which can be translated into section. Each section divided into smaller part called block. The plantation is located on Rejosari, Lampung Selatan. PT. Perkebunan Nasional VII oil palm plantation has variety in age. The oldest trees planted on 1992 and the youngest was planted on 2010.

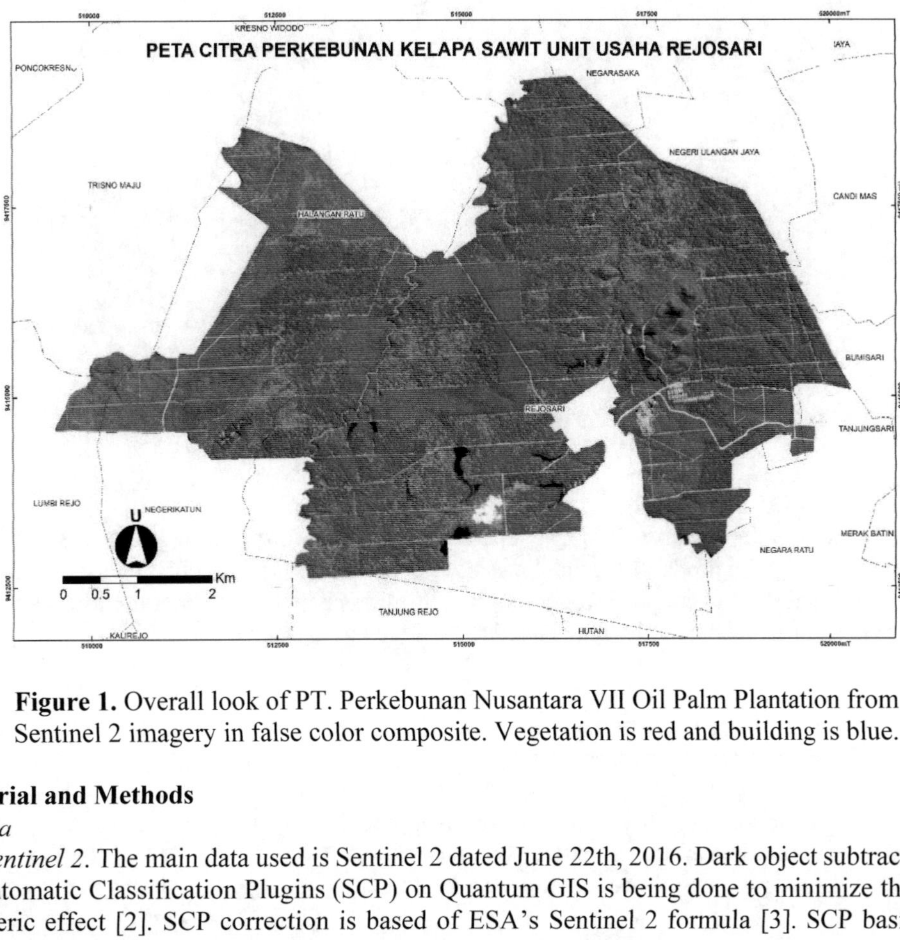

Figure 1. Overall look of PT. Perkebunan Nusantara VII Oil Palm Plantation from Sentinel 2 imagery in false color composite. Vegetation is red and building is blue.

3. Material and Methods

3.1. Data

3.1.1. Sentinel 2. The main data used is Sentinel 2 dated June 22th, 2016. Dark object subtraction using Semi-Automatic Classification Plugins (SCP) on Quantum GIS is being done to minimize the noise of atmospheric effect [2]. SCP correction is based of ESA's Sentinel 2 formula [3]. SCP basically is a shortcut to help people doing atmospheric correction for the imagery.

The imagery then masked into only study area site. Corrected and masked image then transformed with vegetation index. The goal of this transformation is to highlight the information of vegetation objects on the surface of the earth. The vegetation index used in this paper divided into two group, which are generic and non-generic index. The generic index used are generic vegetation index such as *Ratio*

Vegetation Index (RVI) and *normalized difference vegetation index* (NDVI). The non-generic vegetation index used are *modified soil adjusted vegetation index* (MSAVI) to reduce soil background and *atmospherically resistant vegetation index* (ARVI) to reduce atmospheric effect [5].

3.1.2. Plantation Time Map. Used to mask Sentinel 2 Imagery and separate the oil palm blocks based on different planting ages. The reason for the separation is because the age difference of oil palm is have different production capability [4]. The image transformation of vegetation index is explained into 3 classes of planting year i.e. 1992-1998, 1999-2005, and 2006-2010.

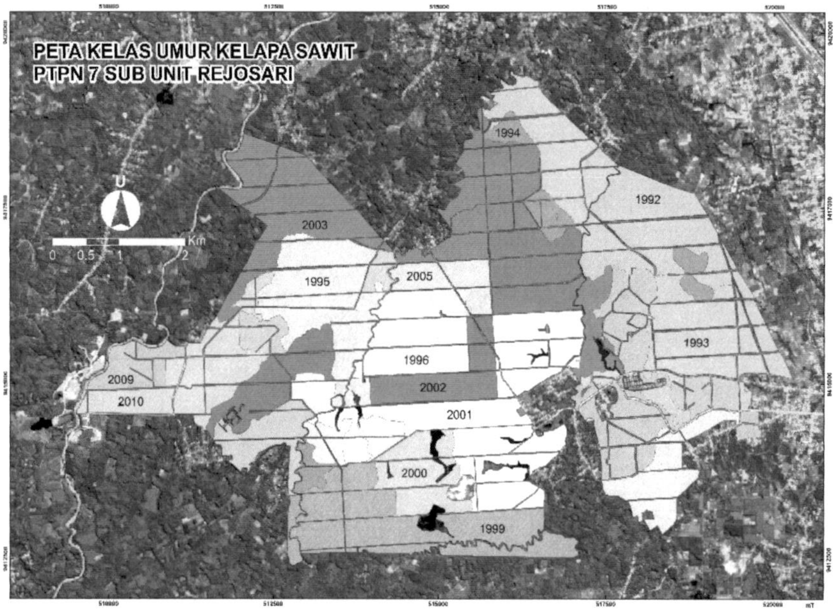

Figure 2. PT. Perkebunan Nusantara VII Plantation Time Map, oldest trees planted on 1992 and the youngest was planted on 2010.

3.2. Methods

3.2.1. By-age separation method. This age separation method is inspired from the work of Setyowati (2015) that build oil palm production model based on a multivariate regression analysis from age of the tree, vegetation index value, and the production itself. In this paper, the age of the tree separated from the regression analysis so the modelling is need to be done age-by-age.

3.2.2. Regression analysis. Regression analysis used is linear regression analysis which is done on each age classification, so that yielded some regression equation and some image of model produced. The simple linear regression algorithm used is as follows [10]:

$$Y = a + bX \tag{1}$$

Y = Production value according to empirical model (dependence variable)
a and b = Coefficients of values generated at model
X = Image pixel value used to build the models (independence variable)

3.2.3. Sample determination and field sampling. Sample determination was done by using stratified random sampling method. The data taken in the field is the data weight of fruit in one rotation in one unit of analysis in the form of 1.5 times bigger than the pixel of Sentinel 2 to avoid GPS location error. The number of samples that were planned before the field were 30 samples but due to time constraints then the sample can be measured when in the field as many as 18 samples.

3.2.4. Accuracy test. The accuracy test of production estimation result is based on the difference between the total productions of the modelling results from Sentinel 2 image compared to the production data of PT. Perkebunan Nusantara VII Rejosari Business Unit in six months where the production is start to declining because when the samples taken is when the production declined. Six months is time required for oil palm to be formed [4]. February-July 2016 is the time when PT. Perkebunan Nusantara VII oil palm production is declining. The accuracy test method refers to Susetyo (2010) production estimation and production of reality which can then be calculated the level of accuracy. Estimation test formula estimation model used is as follows:

$$Accuracy = 100\% - ((\frac{estimation - reality}{reality})100\%) \tag{2}$$

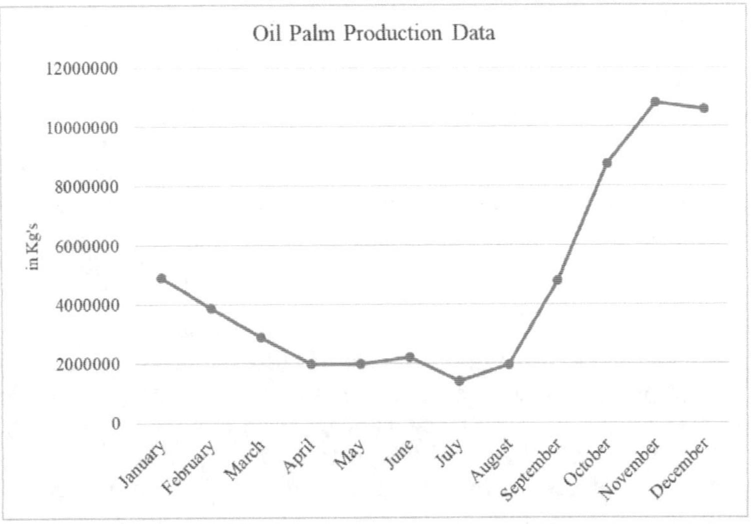

Figure 3. Production Data of Oil Palm in PT. Perkebunan Nusantara VII.

4. Results

4.1. Normality test for data samples

The samples is needed to build the production model. Model development is done to correlate between indices generated from image with existing production in the field. Komogolrov-Smirnof method is used to testthe normality of samples data by determining the value of S(n), Z-factor, F(x) dsn difference on a sample training data set. Normality test is done by comparing the maximum difference value (Dn = max) with the maximum difference resul from KomogolrovSmirnof test (Dn, a). From 18 samples acquired a Difference Maximum value is 0.61396, while the difference value is of 0.307 so it can be concluded that the overall data is normally distributed.

Table 1. Production Data of Oil Palm in PT. Perkebunan Nusantara VII

Sample number	X (m)	Y (m)	Afdeling	Block	Plantation time	Age classification
1	519379	9416486	4	3	1992	1 (1992 -1998)
2	519335	9416475	4	3	1992	
3	512482	9414570	2	4	1993	
4	514470	9415941	3	2	1996	
5	514209	9416012	3	3	1996	
6	515376	9414751	1	2	2001	2 (1999-2005)
7	515379	9414770	1	1	2002	
8	515525	9415781	3	1	2002	

Sample number	X (m)	Y (m)	Afdeling	Block	Plantation time	Age classification
9	515531	9415764	3	1	2002	
10	512483	9414640	2	3	2002	
11	515097	9416836	3	4	2005	
12	517665	9414350	4	1	2009	3 (2006-2010)
13	517722	9414181	4	1	2009	
14	519247	9414666	4	2	2009	
15	519279	9414723	4	2	2009	
16	511851	9414945	2	1	2010	
17	511805	9414936	2	2	2010	
18	512260	9415363	2	2	2010	

4.2. Oil Palm Production Model

Production model is built based on the regression formula. Each model is compared with PT. Perkebunan Nusantara's production data to justify the best index and method. There are two methods of regression. The first method is a method that does not pay attention to the planting time so that regression is done using all sample data. Estimated oil palm production on all vegetation indices regardless of planting age group overestimate. Secondary data shows that the production of palm oil plantation in February – July 2016 was 14,300.9 tons; while the estimation result using RVI vegetation index was 19,860.5 tons; NDVI 20,139.7 ton; MSAVI 20,691 tons; and ARVI 19,715,3 tons. The second method, by-age separation, implies that oil palm have different reaction to each vegetation index based on their age, so the best way to predict oil pam production is to use different vegetation index for each age class. This method done by separating the plantation site into four age classes; 1992-1996, 2001-2005 and 2009-2010 then each age classes will be modelled using each indices regression independently. The most accurate model from each age classes then merged into one imagery model so called by-age separation model. This method result is 16,228.9 tons of oil palm fruits from February to July 2016.

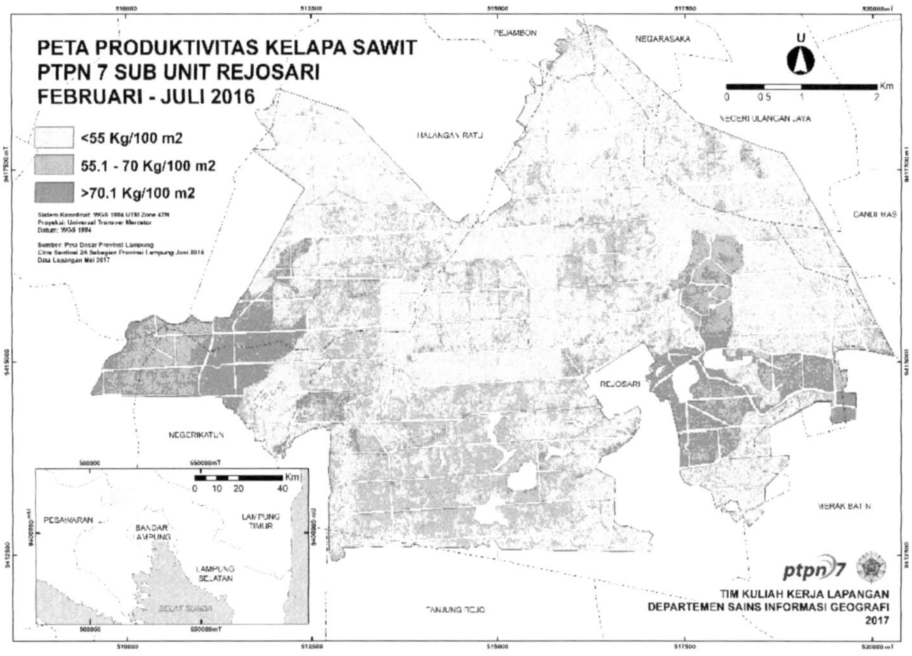

Figure 4. By-age separation method production model. This model is built from best estimation model in each age classes and predicted 16,228.9 tons of oil palm fruits from February to July 2016.

5. Discussion

The number of samples taken on the field to build the production models is actually 22 samples, while used in the model builder to determine the production is only 18 samples. There are 4 samples removed because it was too different from the existing training data.

Based on the treatment with the by-age separation method, the RVI index has the highest R square value in the first group (1992-1996) and the second (1999-2005) with the values of 0.4592 and 0.4787 while in the third group (2009-2010) The best index is ARVI with a value of 0.0344. The second method, the method that does not pay attention to age group, the highest R square value is at ARVI index with value 0.2495. Regression conducted using field data and image data, where imagery data will affect the level of productivity of oil palm in the field. The graph below shows the best regression results between the samples and the vegetation index image in each age group.

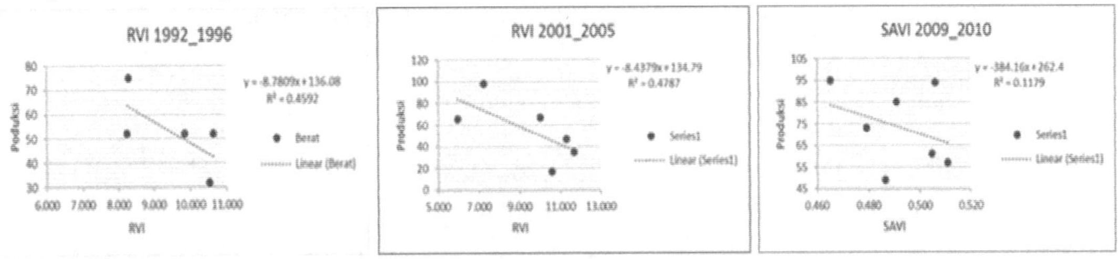

Figure 5. Best vegetation index in each age group to build by-age separation model

The best transformation index for the 1992-1996 planting age group was ARVI with 65.9% accuracy and estimated production of 5,961.8 tons. The best transformation index for the 1999-2005 planting year was RVI with an accuracy of 97.9% and an estimated production of 8,059.0 tons. The best transformation index for the 2009-2010 planting group is NDVI with an accuracy of 63.81% and an estimated production of 2,208.1 tons.

The accuracy test of production estimation result is done based on the difference between the total production of the modelling result from Sentinel 2 imagery compared with the production data of PTPN 7 Rejosari Business Unit in February-July 2016. The result of accuracy test shows that the best accuracy is on the modelling with separation of planting age group with level Accuracy of 87.5%.

6. Conclusion

By-age separation method is better than the fully transformed without being separated by-age. It is because which can differentiate oil palm tree physical appearance and relate it with their level of production. By-age separation method implies that oil palm have different reaction to each vegetation index based on their age. The best transformation index in the estimation of oil palm production planting year 1992-1996 is ARVI, in 1999-2005 planting year that is RVI, and in planting year 2006-2010 that is NDVI. The combined production estimates for each age group resulted in a total production estimate of 16,228.9 ton/semester with an 87.5% modeling accuracy.

References

[1] Badan Pusat Statistik 2015 *Statistik Kelapa Sawit Indonesia* (Jakarta: Badan Pusat Statistik)
[2] Chavez P S 1996 *Image-based Atmospheric correction-revisited and improved (Photogrammetric Engineering & Remote Sensing vol 62)* p 1025-36.
[3] Congedo Luca 2017 *Semi-automatic classification plugin documentation*
[4] Corley R H V and P B Tinker 2003 *The oil palm 4th edition* (Victoria: Blakwell)
[5] Danoedoro P 2012 *Pengantar penginderaan jauh digital* (Yogyakarta: ANDI Yogyakarta)
[6] Fauzi Yan Y.E. Widyastuti I. Setyawibawa and R. Hartono 2002 *Kelapa Sawit: Budidaya, Pemanfaatan Hasil Limbah, Analisis Usaha, dan Pemasaran (Revisi)* (Jakarta: Penebar Swadaya)
[7] Setyowati H A 2015 *Aplikasi citra SPOT-6 berbasis transformasi indeks vegetasi untuk estimasi produksi kelapa sawit (Elaeis guneensis jacq.) (kasus perkebunan kelapa sawit PT. Tunggal*

IOP Conf. Series: Earth and Environmental Science **165** (2018) 012026 doi:10.1088/1755-1315/165/1/012026

Perkasa Plantations, Air Molek, Kabupaten Indragiri, Hulu, Provinsi Riau, Sumatera.) Essay (Yogyakarta: Fakultas Geografi UGM)
[8] Sunarko 2007 *Petunjuk praktis budidaya dan pengolahan kelapa sawit* (Jakarta: Agromedia Pustaka)
[9] Susetyo Imam 2012 *Pemodelan estimasi produksi tanaman karet (Havea brasiliensis mell.-Arg) berbasis citra satelit, potensi genetik, dan data satuan medan (studi kebun Getas, PTPN IX Salatiga, Kabupaten Semarang, Jawa Tengah)* Thesis (Yogyakarta: Fakultas Geografi UGM)
[10] Pranowo 1982 *Statistik praktis* (Yogyakarta: Ananda Yogyakarta)

Acknowledgements
This research was part of Department of Geographic Information Science's 3rd field work and supported by PT. Perkebunan Nusantara VII who allow us to done our field work in their plantation and provided secondary data. We also like to thank Faculty of Geography, Universitas Gadjah mada who help us to fund this publication.

Assessing WorldView-2 Satellite Imagery Accuracy for Bathymetry Mapping in Pahawang Island, Lampung, Indonesia

L Hakim[1*], W Lazuardi[1], I S Astuty [1], A Al Hadi [1], R Hermayani[1], D Noviandias[1] and A C Dewi[1]

[1]Departement of Geographic Information Science, Faculty of Geography, Universitas Gadjah Mada, Yogyakarta, Indonesia 55281

lukman.pro@mail.ugm.ac.id

Abstract. Bathymetry is one of a vital marine geospatial data which is required in many hydrographic analysis. The extent of remote sensing gives a possibility of advancement even onto bathymetry modeling. The visible electromagnetic spectrum gives different reflectance intensity in various sea depth due to attenuation processes. This peculiar spectral responses is then used as the basis of bathymetry modeling using satellite imagery. This paper's goal is to test the WorldView-2 ability for bathymetry mapping in Pahawang Island, Lampung, Indonesia. The WorldView-2 Image of Pahawang Island is corrected into top of atmospheric (ToA) reflectance. It is then being analyzed using bivariate statistic (linear, exponential, and polynomial regression) along with the tide-corrected sea depth data which is acquired from field survey using GPSMAP 840xs single beam in a sampling framework. We then perform a standard error method to determine the most effective result of bathymetry model by its accuracy value.

Keywords : Bathymetry, WorldView-2, ToA, Regression, Standard Error

1. Introduction

Bathymetry is the technique used to measure and explore the depth of the water bodies covering the majority of the earth surface. The earliest systematic bathymetry data acquisition was conducted in 1845 by U.S. Coast Survey in Gulf Stream while the first bathymetry map was published by Maury in the 1853 Wind and Current Charts of the North Atlantic Ocean [1]. The seafloor mapping plays important role in oceanographic study. Those early bathymetry mapping has lead to many important discovery such as the exposal of Mid-Atlantic Ridge and the indication of Mariana Trench. Later on, bathymetry data acquisition also becomes a trend when dealing with the shallow water environtment analysis.

The basic method of bathymetry data acquisition was conducted by using a manual-match transect sampling framework in several decided line. The depth data was retreieved by weighted flax rope which was dropped over the side of a vessel and the length of the line recorded once the weight reached the bottom. Despite the difficult fieldwork activity, this traditional bathymetry data acquisition method has a big issue regarding the accuracy. The internal current force causing the flax to drift and resulting in the depth overestimate. As the technology develops, this problem was also tackled by the development of digital bathymetry data acquisition method. It is now possible to conduct an efficient bathymetry survey using acoustic, radar, and electromagnetic instrument [2].

Pahawang Island in Lampung, Indonesia is located between 5°41'53"-5°39'02"S and 105°11'44"-105°14'59" E. It has approximately 3,3 km^2 area surrounded by quite massive shallow water ecosystem. It is mainly consisted of coral reefs, sea grass, algae, and countless benthic organism [3]. In the framework of sustainable marine conservation, bathymetry becomes one essential data to monitor and evaluate ecosystem condition. But, some believe that conventional bathymetry survey (through traditional or acoustic method) is considered ineffective because it takes a lot of time and work to do the field survey.

Meanwhile, WorldView-2 satellite imagery as a product of electromagnetic remote sensing indicates a great potential to do this task. The spectral resolution containing coastal (400-450 nm), blue (450-510 nm), green (510-580 nm), yellow (585-625 nm), red (630-690 nm), red-edge (705-745 nm) and two near infrared (770-895 and 860-1040 nm) has a great capability to perform various technique involving several band combination. Besides, satellite imagery pixel contain unique and important values which could be derived into so many type of data with a relatively wide area. Not to mention WorldView-2 has a high resolution pixel that has a good chance in modelling the bathymetry with reasonable accuracy. It is also efficient to assess a lot type of data with a minimum requirement of fieldwork. This paper's focus is to produce an equation to transform digital number in WorldView-2 raster into depth data through several image processing and statistical procedure. We also perform a certain method to assess the final bathymetry result to justify the relevance and show the error margin.

2. Data and Method

2.1 Data

We use standard equipment for bathymetry survey with echosounder GPS map 2108 as the main tool. Meanwhile, we use WorldView-2 imagery provided by The DigitalGlobe Foundation as the modeling base to produce bathymethry maping. WorldView-2 imagery that used in this research is recorded on August, 3rd 2015. This imagery has six visible bands, which is coastal (400-450 nm), blue (450-510 nm), green (510-580 nm), yellow (585-625 nm), red (630-690 nm), and red-edge (705-745 nm) and two near infrared (770-895 and 860-1040 nm). WorldView-2 imagery has 16-bit radiometric resolution and 1.8 m spatial resolution. To handle WorldView-2 imagery, we use ENVI software by Harris Geospatial. Acording to [4], log-transformed band have high accuracy value at Bathymetry modeling. This fact was also proved by [5] which stating that the best accuracy value of their bathymetry modelling was produced by log transformed blue/green band.

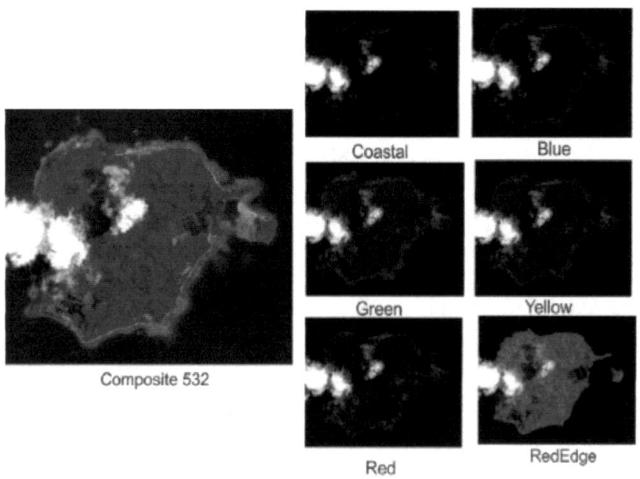

Figure 1. Visible Band of WorldView-2 Imagery

2.2 Method

2.2.1 Image Correction. Image correction used in this study is atmospheric correction. This correction useful for improving the pixel value from atmospheric disturbance in order to match the pixel value with the object reflectance value. The radiometric corrections should be done by considering atmospheric disturbance factors as the main source of error [6]. The atmospheric correction is done by converting the WV2 pixel value to the at sensor value reflectance [7] and then removes the radiance path by the Dark Subtract method [8].

2.2.2 Single Band Method. Single band method that we use to build bathymetry empirical model depends on water depth and has been corrected until atmospheric level. The attenuation of waves will stronger along the increase of water depth. Water depth variation in study area and wavelength must be noticed by observer. Band that most effective to attenuate energy uses to build bathymetry model. After that, exponent empirical model of single band will be compared with log-transformed single band to see linearization process. Spectral reflection and water depth linearization will effect and increase bathymetry model accuracy.

2.2.3 Ratio Band Method. Electromagnetic wave energy at water colomn influenced by depth variable because the energy will fall of with increasing depth. Variation of existing cover at the bottom of water does not have an effect on depth variable [9], but influences to difference in spectral reflection. In a sense, independent depth information with other variables, so on bathymetry modeling band ratio method becomes a solution to accommodate this fact ([10]; [11]; [4]). In this study, band ratio method was performed on three bands original that is coastal band, Blue band, green band and log-transformed band of WorldView-2 Imagery, the total there are 11 bands.

2.2.4 Empirical Modeling. Empirical bathymetry modeling is done by using R_w WorldView 2 imagery and Z value with looking for relationship between both of them. Z value are obtained from field survei measurement using echosounder. The total sample from field survei were 247, 122 were used for test accuracy and 125 were used for modeling sample. Modeling begins with correlation analysis to obtain correlation coefficient (r) value. All of bands as a materials for input in empirical modeling using regression analysis. To do so, we use regression as a kind of bivariate statistical analysis technique there are three type of regression for bathymetry modeling : *linear regression, power regression,* and *exponential regression*. The dependent variable for those regression is pixel value in 11 band from 3 single band, 4 ratio band and 4 log transformed band, while the independent variable is the training sample consisted of bathymetry survey.

2.2.5 Accuracy Test. Coeficient value (r) and regression value (R^2) canot be evaluate acuracy value, so that why standart Error of Estimate (SE) used for evaluate acuracy from bathymetry modeling. The sample used for accuracy test is an unused sample in modeling. Sample used for test accuracy were 122 from 247 total sample field survey.

IOP Conf. Series: Earth and Environmental Science **165** (2018) 012027 doi:10.1088/1755-1315/165/1/012027

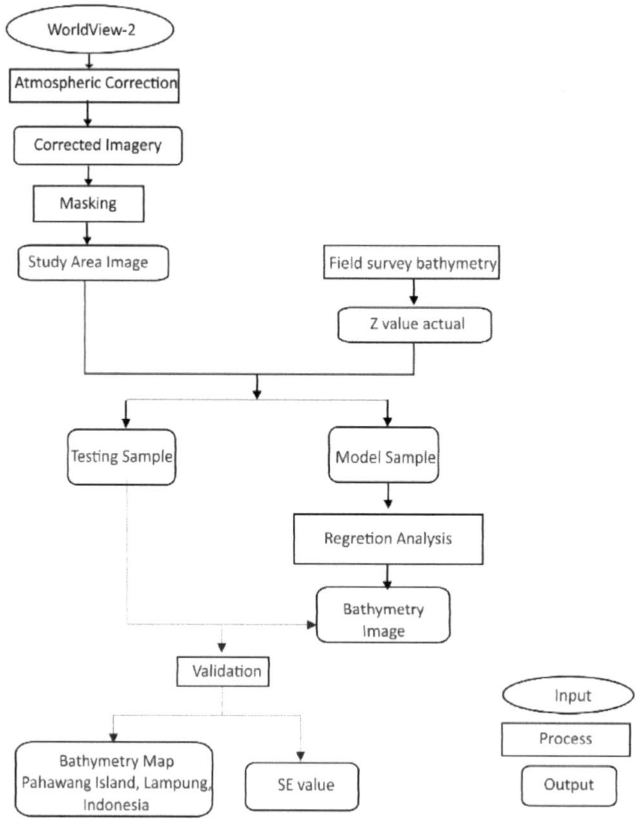

Figure 2. *Flow Chart*

3. Result and Discussion

3.1 Regression Model

The change of spectral value to depth data uses formula referred to [4]. The equation is equal to y = ax + b, where *mi* is the value of *a*, *m0* is the value of *b*, and the value of *x* is the spectral band to be used.

$$Z = m1\ Rw \times m0 \tag{1}$$

Table 1. Bathymetry Regression Model

Band		Linier	R^2	Exponen	R^2	Power	R^2
Single Band	*Coastal*	y = -232.23x + 43.721	0.0652	$y = 47510e^{-57.13x}$	0.0306	$y = 6E\text{-}08x^{-9.976}$	0.031
	Blue	y = -66.679x + 13.956	0.0272	$y = 61.6e^{-20.72x}$	0.0204	$y = 0.0048x^{-3.355}$	0.0213
	Green	y = -187.2x + 25.7	0.4418	$y = 8758e^{-69.22x}$	0.4692	$y = 4E\text{-}08x^{-8.387}$	0.4766
Ratio Band	*Coastal/ Blue*	y = 7.6492x - 4.9376	0.0083	$y = 0.0601e^{3.333x}$	0.0122	$y = 1.6681x^{3.62=24}$	0.0119
	Blue/Co astal	y = -9.3552x + 11.988	0.0084	$y = 84.255e^{-3.932x}$	0.0116	$y = 1.6676x^{-3.625}$	0.0119
	Blue/Gr een	y = 34.186x - 41.709	0.7888	$y = 1E\text{-}07e^{12.641x}$	0.8375	$y = 0.0234x^{16.628}$	0.8393
	Green/B lue	y = -57.739x + 47.287	0.7591	$y = 4E{+}07e^{-21.79x}$	0.8395	$y = 0.0234x^{-16.63}$	0.8394
Log Transfor	*(ln) Coastal/ Blue*	y = -15.505x + 18.196	0.0071	$y = 1421.3e^{-6.746x}$	0.0105	$y = 1.6792x^{-6.451}$	0.0106

Band		Linier	R^2	Exponen	R^2	Power	R^2
m Ratio Band	*(ln) Blue/Coastal*	y = 13.892x - 11.161	0.007	$y = 0.0035e^{6.1804x}$	0.0108	$y = 1.6776x^{6.4676}$	0.0107
	(ln) Blue/Green	y = -113.02x + 101.88	0.7819	$y = 4E+16e^{-42.86x}$	0.873	$y = 0.0131x^{-37.44}$	0.8731
	(ln) Green/Blue	y = 86.995x - 96.498	0.7958	$y = 1E-16e^{32.696x}$	0.873	$y = 0.0131x^{37.453}$	0.8733

3.2 Single Band

Bathymetric modeling using single band analysis shows that the WorldView-2 visible band has a negative relation with the depth value, the spectral reflectance value would be decreased as the depth increases. The results of bathymetric modeling using single band has the highest determinant coefficient (R^2) value of 0.4766 on the green band.

This corresponds to [4] research, that the green band has a good capability in penetration in shallow water. Therefore the use of green bands can improve accuracy in shallow water applications (bathymetry) compared to other single bands. Bathymetry modeling using a single band has problems, due to the influence of various bottom object reflectance, so the correlation between single band pixel value with the depth value becomes low [12].

Table 2. Accuracy and SE value of Single Band

Band		Liniear		Exponent		Power	
		Accuracy %	SE	Accuracy %	SE	Accuracy %	SE
Single Band	*Coastal*	31.798	3.130	24.271	3.476	24.031	3.487
	Blue	31.788	3.131	23.837	3.496	24.082	3.484
	Green	50.505	2.272	48.763	2.352	50.108	2.290

Regression function in single band shows best linear equation on green band with value of SE 2.272 and accuracy 50.505% with equation y = -187.2x + 25.7. The value of SE 2,272 shows the difference of 2,727 meters between bathymetric modeling with actual depth condition.

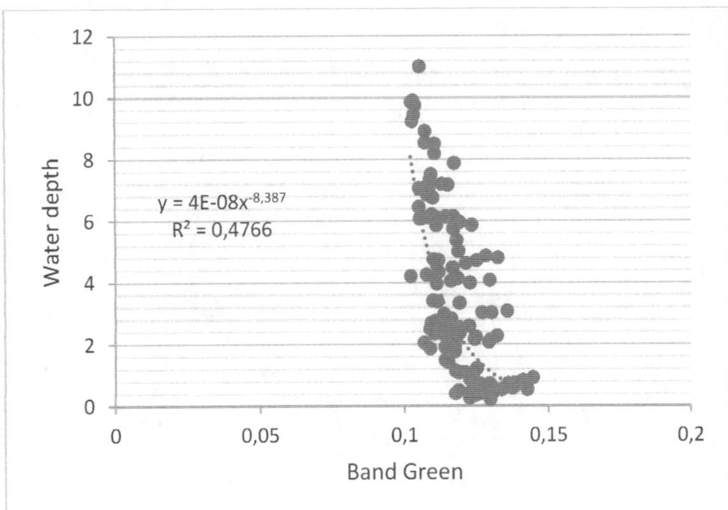

Figure 3. Regression Model of Band Green

3.3 Ratio Band

The application of band ratios is used to normalize the reflections of the water column due to the effects of the various bottom object reflectance. The Ln (Log-transform) function in each image is used to linearize relation between the depth value with the image pixel value. The combination of band ratio used for modeling has a relationship with depth value. The result of bathymetric modeling with the highest determinant coefficient (R^2) value in Band ratio of original band is 0.8395 using Band Green / Band Blue, and the highest determinant coefficient (R^2) value in Band ratio of log transform band is 0.873 in Ln Blue / Ln Green band. In band ratio of original band, has highest accuracy or lowest SE value which obtained through the ratio of log transformed ratio band of Ln band blue / Ln band green that is 72.313% and 1,271 meters. The highest accuracy bathymetric modeling result, obtained through the log transformed ratio band Ln band blue / Ln band green of 72.675% and with SE of 1,254 meters.

Table 3. Accuracy and SE value of Ratio Band

Band		Liniear		Exponent		Power	
		Accuracy %	SE	Accuracy %	SE	Accuracy %	SE
Band Ratio	*Coastal/Blue*	31.378	3.150	23.446	3.514	23.448	3.513
	Blue/Coastal	31.371	3.150	23.451	3.513	23.448	3.513
	Blue/Green	70.056	1.374	62.654	1.714	71.868	1.291
	Green/Blue	67.673	1.484	72.313	1.271	71.847	1.292
Log Transform Band Ratio	*(ln) Coastal/Blue*	31.323	3.152	23.413	3.515	23.409	3.515
	(ln) Blue/Coastal	31.324	3.152	23.480	3.512	23.409	3.515
	(ln) Blue/Green	68.542	1.444	72.675	1.254	72.455	1.264
	(ln) Green/Blue	69.652	1.393	70.465	1.356	72.429	1.265

Regression function in band ratio shows best equation on log transformed ratio band Ln band blue / Ln band green with value of SE 1.264 and accuracy 72.455% with power regression equation $y = 0.0131x^{-37.44}$. The value of SE 1.264 shows the difference of 1.264 meters between bathymetric modeling with actual depth condition.

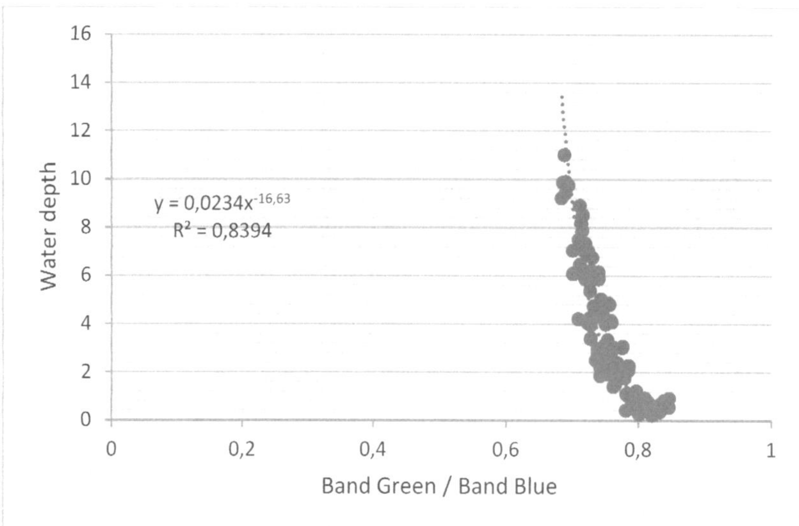

Figure 4. Ratio Band of original band (Band Green/Band Blue)

The curve above show that ratio band green/ blue have negative relationship between the pixel value and depth value, same as bathymetry model using single Green band.

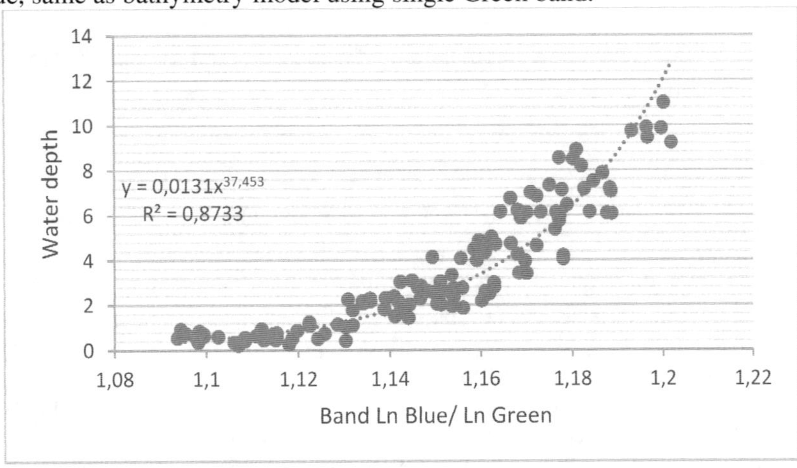

Figure 5. Log transformed ratio band Ln band blue / Ln band green

The log-transformed ratio band regression curve above proves that log transformed band can linearize relation between depth value and exponential pixel value.

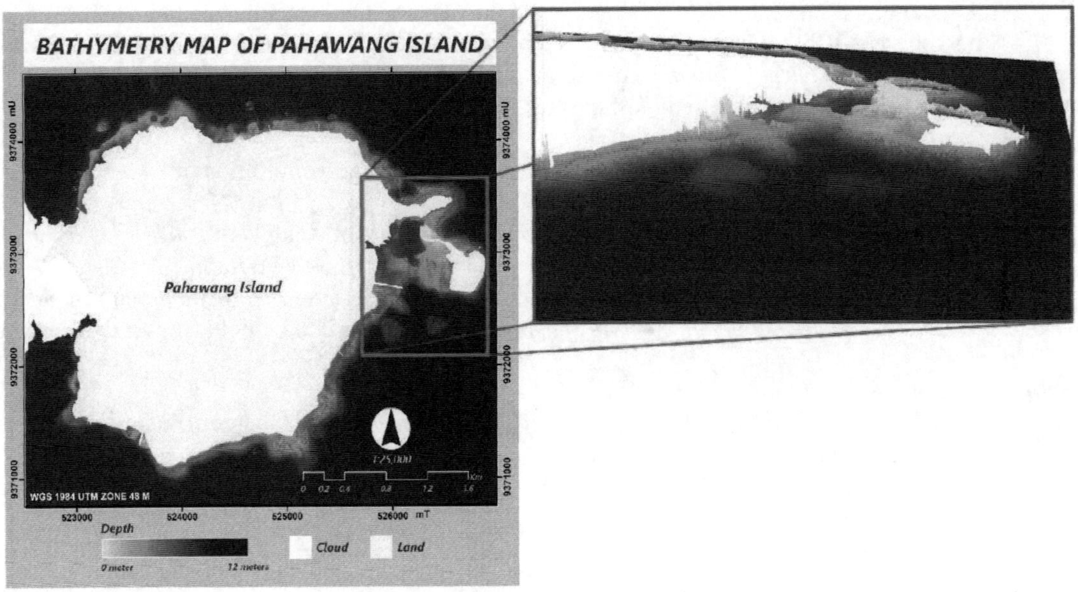

Figure 6. Bathymetry Map

4. Conclussion

The basic method of bathymetry data acquisition was conducted by using weighted flax rope which was dropped over the side of a vessel and the length of the line recorded once the weight reached the bottom. Despite the difficult fieldwork activity, this traditional bathymetry data acquisition method has a big issue regarding the accuracy. The internal current force causing the flax to drift and resulting in the depth overestimate. As the technology develops, this problem was tackled by the development of digital bathymetry data acquisition method including the satellite imagery. We try to assess WorldView-2 image capability for bathymetry modelling in Pahawang Island, Lampung, Indonesia. We perform standard atmospheric correction to gain pixel ToA value. Method used for bathymetry modeling is single band and ratio band, the highest accuracy 50.505% using single band method for green band, highest accuracy for band ratio is Band Green / Band Blue and log transformed ratio band Ln band blue / Ln band green of 72.675% . Bathymetry modeling in Pahawang Island Lampung can show the depth variation up to + 14 meters.

References

[1] Heidi M Dierssen and Albert E. Theberge 2014 *Bathymetry: History of Seafloor Mapping* (Encyclopedia of Natural Resources vol II – Water and Air) ed Yeqiao Wang (Taylor & Francis Group) pp 1-6

[2] Harper Jeremy 2015 *Bathymetry Concepts and Aplication* (US: Callisto Reference)

[3] Lazuardi W, Hakim L, Astuty I S, Al Hadi A, Hermayani R, Pratama D N and Dewi A C 2017 Coral Health Assessment Using WorldView-2 Satellite Imagery In Pahawang Island Lampung Indonesia (Seminar Nasional Geomatika 2017)

[4] Wicaksono P 2010 Integrated Model of Water Column Correction Technique for Improving Satellite-based Benthic Habitat Mapping, A Case Study on Part of Karimunjawa Islands, Indonesia. Universitas Gadjah Mada (Yogyakarta: Faculty of Geography)

[5] Nurkhayati R and Khakhim N 2013 Pemetaan Batimetri Perairan Dangkal Menggunakan Citra Quickbird di Perairan Taman Nasional Karimun Jawa Kabupaten Jepara Jawa Tengah. *Jurnal Bumi Indonesia*

[6] Danoedoro P 2012 *Introduction to Digital Remote Sensing* (Yogyakarta: Penerbit Andi)

[7] Updike T and Comp C 2010 *Radiometric Use of WorldView-2 Imagery* (Longmont, Colorado: DigitalGlobe®)

[8] Armstrong R A 1993 Remote sensing of submerged vegetation canopies for biomass estimation *International Journal of Remote Sensing*. vol **14** p 10-16

[9] Lyzenga D R 1978 Passive Remote-Sensing Techniques for Mapping Water Depth and Bottom Features. *Applied Optics* vol **17** p 379-383.

[10] Hogrefe K R 2005 Deriving Shallow Water Bathymetric Data from Ratios of Blue and Green λ Radiance Values. Oregon: Department of Geosciences (Oregon State University)

[11] Stumpf R P, Holderied K and Sinclair M 2003 Determination of water depth with high-resolution satellite imagery over variable bottom types. *Limnology and* Oceanography Vol **26** p 547-556.

[12] Jupp D L 1988 Background and extensions to depth of penetration (DOP) mapping in shallow coastal waters *Proc., Symp., Remote Sensing of the Coastal Zone.* (Gold Coast : Queensland)

Acknowledgements

We thank The DigitalGlobe Foundation for freely providing WorldView-2 imagery covering the whole Pahawang Island and fully support our research in Lampung, Indonesia.

IOP Conf. Series: Earth and Environmental Science **165** (2018) 012028 doi:10.1088/1755-1315/165/1/012028

The Ability of Lyzenga's Algorithm for Seagrass Mapping using Sentinel-2A Imagery on Small Island, Spermonde Archipelago, Indonesia

M S Thalib[1,2*], N Nurdin[1,2] and A Aris[2]

[1] Department of Marine Science, Hasanuddin University, Jl. Perintis Kemerdekaan Km. 10, Makassar, 95245, Indonesia

[2] Research Center for Regional Development and Spatial Information, Hasanuddin University, Jl.Perintis Kemerdekaan km.10, Makassar, 95245. Indonesia.

msafahthalib@gmail.com

Abstract. The existence of the value of depth variation in the study area is also a requirement in the application of this algorithm. The purpose of this study is to determine the ability of the use of algorithms in the mapping of seagrass by comparing the results of classification between imagery using the Lyzenga's algorithm with imagery without using the Lyzenga's algorithm. The result of imagery classification applying the Lyzenga's algorithm shows the visibility of baseline water objects that are more easily recognizable using Depth Invariant Index (DII) value format as evidenced by the increase of 2 classes of seagrass cover. In this study there are five seagrass cover classes of shallow water in Pajenekang Island, Spermonde-Indonesia Islands using Lyzenga's algorithm, namely class are 1) Very good: > 75%; 2) Good: 50,5-75,4%; 3) Rather Good: 25,5-50,4%; 4) Bad: 5,5-25,4%; and 5) Very Bad: <5.5%.

1. Introduction

The availability of spatial information of coastal aquatic benthic becomes very important as awareness of environment-based management. Various methods have been undertaken and developed to obtain coastal aquatic benthic information, one of which is the application of the Lyzenga's algorithm.

Based onits Morphology, the Seagrass consists of leaves, rhizomes that have roots, flowering and seeds, and to obtain energy through photosynthesis process. Seagrass are also defined as flowering aquatic plants (*angiospermae*). However, this plant species is different from strong plants in general, owned stems are used to overcome the force of gravity on the substrate habitat to remain upright. It also has an air cavity used to create buoyancy under flexible conditions when exposed to current waves [1].

As one of the coastal and marine ecosystems, seagrass beds have an irreplaceable ecological function. The ecological functions of the seagrass ecosystem include, as a residence for juvenile and adult biota, as a shelter, feeding ground, nursery ground, and spawning ground. In addition, the seagrass ecosystem also becomes important because as a balance factor of marine ecosystem. The total area of seagrass beds in Indonesia was originally estimated at 30.000 km^2, but is now estimated to have decreased by 30-40% due to human activities. So that the seagrass getting less attention to its existence [2].

IOP Conf. Series: Earth and Environmental Science **165** (2018) 012028 doi:10.1088/1755-1315/165/1/012028

Indonesia as a tropical region has seagrass species that generally grow together, forming mixed meadows consist of 2-8 species. While in temperate, seagrass beds are often dominated by only one single species [3].

That the Spermonde-Pangkep archipelago, mainly in Sabangko Island, the condition of the seagrass ecosystem is spread on several sides of the island under varying conditions and has a seagrass ecosystem with medium-term categorical condiments Braun-Blanquet [4]. One of the islands is also found in the islands of Spermonde-Pangkep is the island of Pajenekang. But the potential of this island has not been revealed, especially the seagrass ecosystem. This is the reason to choose the research topic and research location.

Research on mapping and monitoring of shallow marine ecosystems (Coral, Mangrove and Seagrass) has been done using satellite imagery. Some researches have succeeded to map coastal aquatic using Landsat satellite imagery on small Island-Spermonde archipelago [5-8]. As for the type of imagery ALOS AVNIR [9,10].

The use of *Advanced Land Observing Satellite* (ALOS) satellite imagery with 10 m spatial resolution was used to perform multispectral classification in coral and seagrass mapping on small island-Spermonde archipelago of Indonesia. In this study also used the method of Depth Invariant Indext stating that DII value is not significantly different with radian value [10].

The spatial resolution of the 10-meter ALOS imagery is similar to that of Sentinel-2 which has a multispectral instrument with 13 spectral channels of visible light channels,near infrared, as well as shortwaveinfrared. The planned satellite lasts for 7 years, has 4 spectral bands at 10 meter spatial resolution, 6 bands at 20 meter spatial resolution (four for vegetation spectral) and 3 spectral bands at 60 meter spatial resolution [11]. So that Sentinel-2 imagery allows to be used in the classification of seagrass cover, since 4 bands can be used for spectral vegetation.

Various studies have proven that the Lyzenga's method is often used to map the basic shallow water base closure material. One is to lower the Lyzenga's algorithm for mapping the shallow seagrass closure material (<20 m) with the research method wich is used for merging operation of two visible channels TM1 and TM2 that can penetrate into the water body up to a certain depth, so it can be used to identify the object in shallow marine waters[12]. Meanwhile implemented the Lyzenga's algorithm for the imagery that gives clearer appearance of the picture of the basic types of waters [9]. Another study was also conducted by Jaelani who used high resolution imagery to be used in the method of extracting seagrass information using the Lyzenga's algorithm and without algorithm [13].

Where this study aims to determine the effect of the use of algorithms to mapping substrate baseline waters, especially seagrass by looking at the results of comparison of spatial accuracy between the image classification results Lyzenga algorithm and without Lyzenga image algorithm. So this method is expected to be done on the type of Sentinel-2A imagery as study material and research author.

2. Material and methods

This section discusses are the location of the research, the type of data used and how the data is processed.

2.1. Study area

The study was conducted in March-August 2017 in shallow waters of Pajenekang Island-Spermonde Islands of Pangkep, Indonesia with an area of 104.57 Ha geographically located at 119° 19'39.92 "E longitude and 4° 58'09, 42 "S (Figure. 1).

2.2. Data collection

The main data used in this study is Satellite Sentinel-2A Imagery acquisition 01-March-2017 which is downloaded on the link: https: //scihub.copernicus.eu/dhus/#/home. As the specification of this imagery data can be viewed in Table 1. While for data processing mostly use ArcGIS 10.5 software.

Other data used in the form of secondary data. The secondary data types are ocean watershed condition data from previous studies and prediction of tidal data obtained from tidal book prediction during 2017 published by DISHIDROS (Hydrographic-Oceanographic Service) of the Indonesian Navy.

Table 1. Data and imagery specification

Description	Specification
Multyspectral	Sentinel-2A
Spatial Resolution	10 meters
Temporal Resolution	5 days
Date of Acquisition	01 March 2017

2.3. Processing Imagery

Sentinel-2 Multi-Spectral Instrument (MSI) has 13 spectral bands from which are visible and Visible and Near Infrared (VNIR) to Short-Wave Infrared (SWIR), where this showing four spectral bands at 10 m are classic blue (490 nm), green (560 nm), red (665 nm) and near infrared (842 nm); six bands in 20 m are four bands in spectral vegetation (705 nm, 740 nm, 783 nm and 865 nm) and two large SWIR bands (1610 nm and 2190 nm); and three bands at 60 m spatial resolution that is dedicated to correction and cloud screening (443 nm for aerosol retrieval, 945 nm for moisture retrieval and 1380 nm for cirrus cloud detection). In the Sentinel-2a Imagery Level-1C for imagery geometric correction process is not done because this imagery data has a high resolution spectral that is 10 meters and not only has been corrected geometric but also has been corrected radiometric [11]. In this research, there are several steps, starting from pre-imageryprocessing, imagery processing to the result of comparison of seagrass distribution using algorithm and without Lyzenga's algorithm using high resolution imagery (Resolution 10 m).

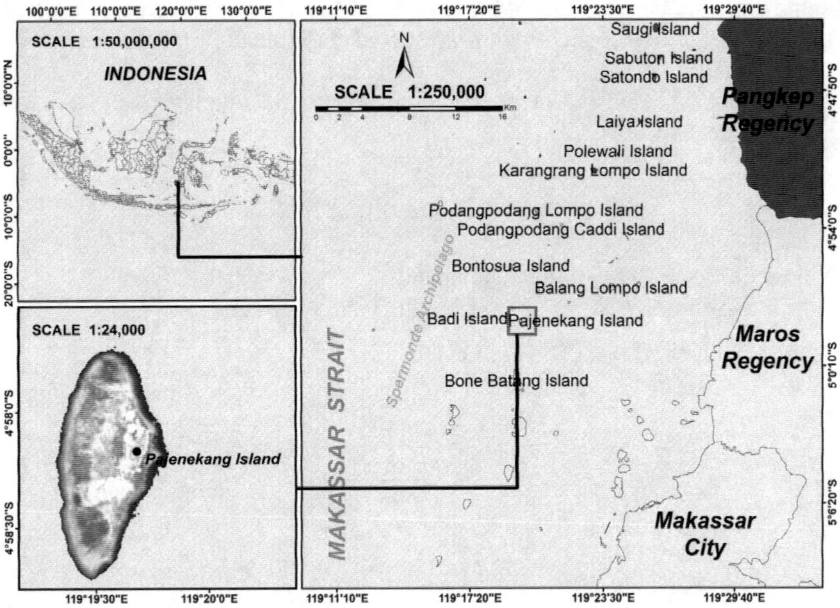

Figure 1. Research site, red circle is Pajenekang Island, one of small Island on Spermonde Archipelago

2.3.1. Atmosferic correction. Broadly speaking the flow of this research is depicted in the flow diagram on Figure. 2. The first stage to perform is the imagery correction to eliminate atmospheric effects on the imagery during recording. Corrections made by the Dark Object Subtraction (DOS) method. This correction is intended because every imagery that is processed always looks its visual been unclear, this atmospheric correction will clarify the appearance of the object, in order to facilitate to recognize the object when it was interpreted. The atmospheric correction is part of the pre-processing of imagery data which was used to eliminate radiance errors recorded on images as a result of atmospheric scattering (path radiance).

The rationale in the "Dark Object Subtraction (DOS)" method is that the number of pixels drawn from the deep water and the Digital Number (DN) value is then subtracted with each band. The formula can be seen in Equation.1[14].

Atmospherically corrected radiance :

$$\text{Li} - \text{Lsi} \tag{1}$$

Where Li is the pixel of the tire radiance i and Lsi is the average value of light for the water column in band i. But if atmospheric correction has taken place then the pixel value has been converted to a surface reflectance or surface value. The value of this reflectance is the value that when the wave is about and / or leaving the object, assuming that the information delivered from the aquatic object to be treated is free of the atmospheric effect. So the value of this surface becomes an initial data to perform data entry to the Lyzenga's algorithm but firstly separates the shore and ocean region so that the processing focuses on the ocean area only (as the seagrass habitat).

But before doing the necessary cutting of images for the study area of research. Cropping imagery or commonly called cropping which isa process of image processing used to minimize the area of observation of a study. It aims to minimize the capacity of the files to be processed and speed up the processes in the processing software used when compared with processing a full-scene data. At this scane Pajenekang Island is located at 4 ° 55'41.22 "LS and 119 ° 19'14.63" BT.

2.3.2. Composite band . The composite imagery in the image without using the Lyzenga's algorithm used is combining 3 natural bands (RGB: 432) from band 4 (red), band 3 (green) and band 2 (blue) for Sentinel-2A imagery data. Then conducting unsupervised classification (Unsupervised-K_Means).

2.3.3. Depth Invariant Index (DII) .The next method is water column correction using the method of "Depth Invariant Index (DII)" [15]:

$$\text{Xi} = \ln(\text{Li}) \tag{2}$$

$$\text{Ln}(\text{Li}) = -[(\text{ki}/\text{kj}) \ln(\text{lj})] \tag{3}$$

Where:

Li dan Lj : the reflectance value of the i and j bands.
Ki/Kj : the attenuation coefficient ratio from the i and j bands

While to find the value of coefficient attenuation (Ki / Kj) is:

$$Ki/Kj = a + a\sqrt{(a^2 + 1)} \tag{4}$$

$$a = \frac{(varXi - varXj)}{2\,covXiXj} \tag{5}$$

Where:

Xi : The variance of band i,
Xj : Variant of band j and
XiXj : Covariance from band ij.

In Equation (2) is used to find the reflection relationship with exponential attenuation for each depth. Therefore we need the value of two spectral data channel data to get attenuation coefficient value (Ki / Kj) (In equation 4). The value is the attenuation coefficient ratio of the blue channel (490 nm) and the green channel (560 nm). The algorithm calculation is influenced by channel pair i (blue channel) and j (green channel) used. According to Suwargana [16], the value of the wavelength for the channel used has an influence on how deep a channel is in detecting a watershed, so the blue channel and the green channel have the wavelength with the best penetration among other channels that is good to use.

The application of the Lyzenga's algorithm to find the coefficient of aquatic attenuation (ki / kj) first needs to find the value, the value of a is obtained by looking at the digital values in the blue channel and the green channel at the same geographical position through the region-making process or the sample area training in the polygon form . In this study used sample area training as much as 35 regions taken in shallow water areas. Then calculated statistically to get the value of variety (variance) and covariance for the result of training of sample area on blue channel and green channel, so that the value of a and ratio of attenuation coefficient (ki/kj) based on equation (4) and (5) . Furthermore, the imagery is classified as Unsupervised refers to the data of the processed imagery while connected with the results of the observation when the field data retrieval with. The classification of satellite imagery in this way has been widely used to produce basic shallow terrestrial habitat maps [17].

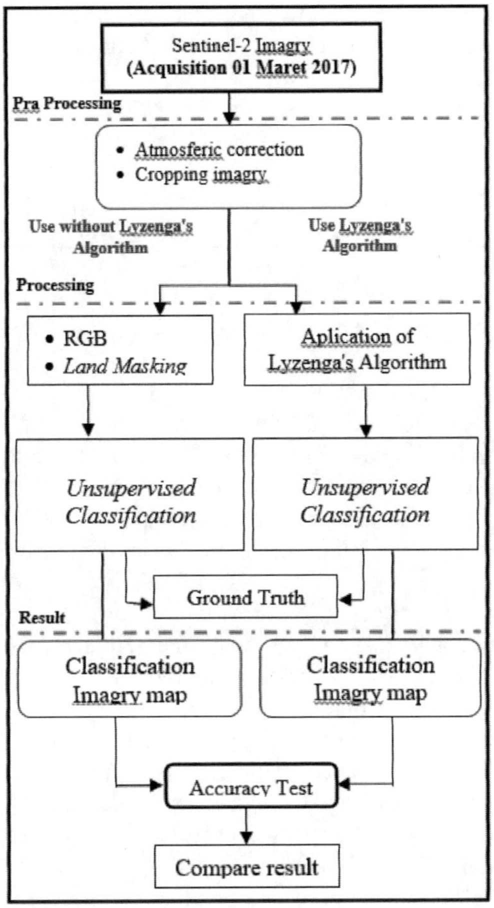

Figure 2. Flow Chart of research methods

The results of the two were then compared from the results of the accuracy test. This test is important to know how big the map can be trusted. The accuracy test used is using matrix error (error matrix) by comparing the actual class of field observation with the result of imagery classification. Here is a flow chart of research that can be seen in Figure. 2.

3. Results and discussion

This section discusses are the atmospheric correction; the land masking; the resulta imagery classificatin without Lyzenga's algorithm and Lyzenga's algorithm; the type objects of the seagrass dominant species found in the field; and accuracy test.

3.1. Atmospheric correction

The study was conducted in March-August 2017 in shallow waters of Pajenekang Island-Spermonde Islands of Pangkep, Indonesia with an area of 104.57 Ha geographically located at 119° 19'39.92 "E longitude and 4° 58'09, 42 "S (Figure. 1).

The result of atmospheric correction using "Dark Object Subtraction (DOS)" method can be seen from changing pixel value. The previous pixel value in the form of a reflectant value on a satellite is then converted to a reflectance value of the surface (surface reflectance). So the surface reflectance value with a range between 0-1. Where are according that the basic assumption in the picture of some complete shadow pixels and radiances received by satellites is also caused by atmospheric scattering (path radiance) [14]. This assumption is combined with the fact that very few targets on the surface of the earth are absolutely black, so it is assumed that one percent of the minimum reflectance is better than zero percent. Band values of corrected band_2, band_3 and band_4 can be seen in Table. 2.

Table. 2. Tables of band values resulting from Dark Object Subtraction (DOS) correction

Band	Reflectance Value Minimum	Percent (%)
Band_2	0,000000	1
Band_3	0,000000	1
Band_4	0,000000	1

3.2. Land Masking

The separation (masking) of land and sea boundaries needs to be done to facilitate during the transformation and classification process. The method used is the digitization of the mainland on the imagery. So that the sea area is allowed to continue to bring the reflectance value of the surface (surface reflectance), while the land area is left to have a value of 0 (null) so that when doing the transformation and classification there is no interference from the land value.

3.3. Imagery classification without Lyzenga's algorithm

When compared to the composite imagery of Red-Green-Blue (RGB = 432), the Red-Green-Blue imagery composite is further classified by the same method of Unsupervised Classification-K_Means The classification results are shown in Figure .3.

Classification results without the use of algorithms for any reflection of aquatic benthic objects. In the shallow areas of Pajenekang Island the reflection of objects from every pixel is the result of various objects as well as some objects that are reflections of dominant objects such as: sand class, live coral, seagrass categories are very good, good and rather good. Objects mixed in one class such as dark blue pixels are the result of reflection from live coral mix with Dead Coral Algae (DCA), algae, seagrass & sand and the light blue image is the result of a reflection of live coral mix with DCA (Figure 3). Based on the search results without using the Lyzenga algorithm, the seagrass category only contained three classes: very good category:> 75%, good: 50,5-75,4% and rather good: 25,5-50,4%. From the results that this category very good (> 75%) consists of 96 pixels, the good category (50.5-75.4%) consists of 255 pixels and rather category (25.5-50.4%) of 1872 pixels. For seagrass cover in each category of seagrass cover can be seen in Table. 3.

Table 3. Percentage coverage of Seagrass area without Lyzenga's algorithm.

Percent Cover (%)	Coverage area (Ha)	Coverage area (%)
>75 %	0.96	4.32
50,5-75,4 %	2.55	11.47
25,5-50,4 %	18.72	84.21
Total	22.23	100.00

The extent value for each category of seagrass cover classification class in the above Table.3 is obtained from the multiplication of the number of pixels produced from each seagrass cover category with the imagery spatial resolution value used is 10 x 10 m (100 m^2). Different area of each category produced is the result of imagery analysis that has been adjusted to the conditions in the field, the number shows the difference of cover area of each station. The total area of the seagrass is 22.23 Ha, this condition is obtained from the classification by calculating the classification area for each class on the imagery recording at low tide [18].

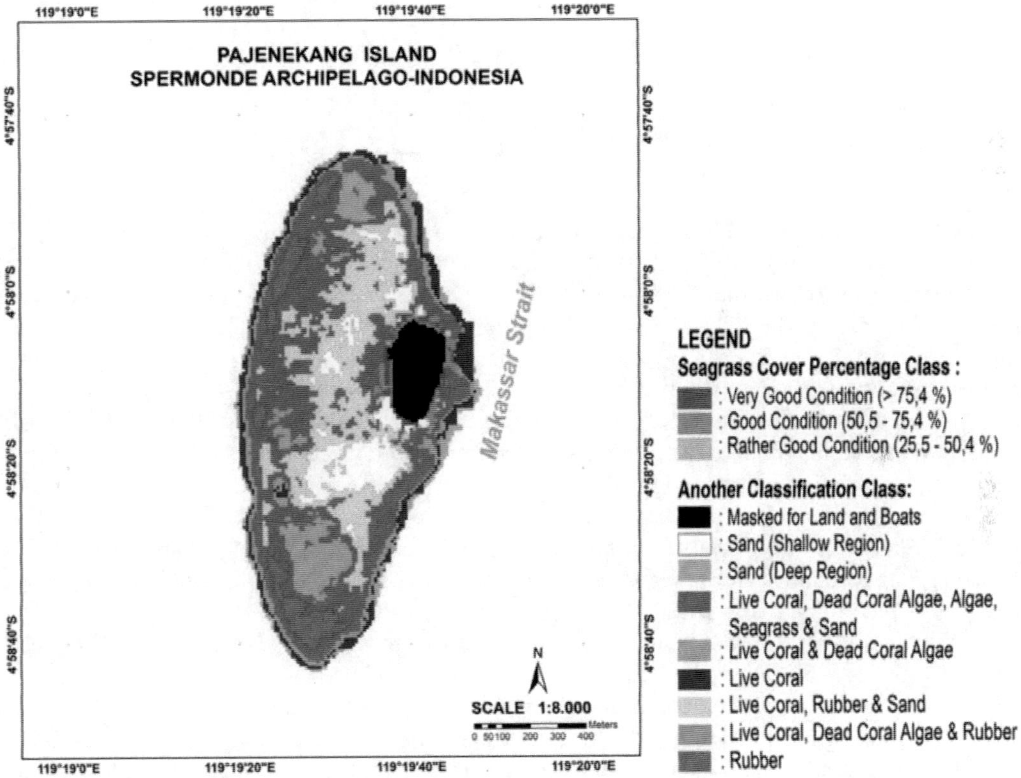

Figure 3. Classification Imagry without the Lyzenga's algorithm.

3.4 Imagery classification with Lyzenga's algorithm

In the process with using the Lyzenga's algorithm, the imagery used is a NIR-Green-Blue composite (RGB = 832). The composite imagery 421 shows better basic visibility, as of the NIR1, Green and Blue channels, has good spectral sensitivity to distinguish parameters in the ecosystem of a coastal waters [13]. The sample area training process carried out as many as 35 regions taken in shallow water areas with different depths for the bottom substrate of the waters. This is confirmed by Lyzenga [15] that in order to obtain the correct linear correlation value in the process of Depth Invariant Index (DII), then in the process of training sample on substrate of water base must be on homogeneous area and have different depth. Where the channel chosen for the calculated ratio is the channel with the highest penetration of water depth, they are the blue and green channels. Based on the sample training, we get the value of ki/kj of 0.7656. Then apply the algorithm in the process of getting a new imagery. Furthermore, the imagery is classified by unsupervised classification-K_Means method used to see the number of classes generated and perform the class naming based on ground truth.

Based on the results of classification by using the Lyzenga's algorithm visible color difference for each reflection of the basic object of the waters. In the shallow waters of Pajenekang Island the reflection of the objects of each pixel is the result of the reflection of several objects and also some other objects which are reflections of a dominant object such as: sand class, live coral, seagrass category very good, good, rather, etc. Mixed objects in one class such as yellow pixels show the result of the reflection of living coral mix with DCA and algae, seagrass and sand (Figure 9).

Table 4. Percentage of Seagrass coverage area with Lyzenga's algorithm.

Percent Cover (%)	Coverage area (Ha)	Coverage area (%)
>75 %	6.44	20.02
50,5-75,4 %	12.17	37.84
25,5-50,4 %	8.77	27.27
5,5-25,4 %	3.21	9.98
<5,5 %	1.57	4.48
Total	32.16	100.00

The value of the area for each seagrass objects of the 5 seagrass coverage area is obtained, is Very good:> 75%, Good: 50,5-75,4%, Rather Good: 25,5-50,4%, Bad: 5,5-25,4% and Very Bad: <5.5%. From the results it is known that the category> 75% consists of 644 pixels, the 50.5-75.4% category consists of 1217 pixels, the 25.5-50.4% category consists of 877 pixels, the categories 5.5-25.4 % consists of 321 pixels and the <5.5% category consists of 157 pixels. As for the value of seagrass cover in each category of seagrass cover can be seen in Table. 4 and Figure. 4-8.

Value of area for each class category of seagrass cover classification in the above Table. 4 is obtained from multiplication of pixel number resulting from each seagrass cover category with imagery spatial resolution value used is 10 x 10 m (100 m^2). Different area of each category produced is the result of imagery analysis that has been adjusted to the conditions in the field, where the number shows the difference of cover area from each station. The total area of the seagrass is 32.16 Ha (Figure 9), this condition is obtained from the classification by calculating the classification area for each class on the imagery recording at low tide.

However, the difference is the increase of two classes for the seagrass cover category are very bad condition categories (<5.5%) and bad condition categories (5.5-25,5%), also live corals that are in the mixed class of objects from living corals and DCA more clearly show the existence of life coral pixels on with Lyzenga's algorithm (descriptions in Fig. 9 that the red square on the zoom in results in the mixed object class area of the living coral and DCA is to clarify the pixels for live corals not found in the imagery image without the Lyzenga algorithm). The loss of the two classes is due to the area of the object is close to the mainland so it gets anthropogenic disturbance causing the waters in the area becomes turbid and different depth variations as well. Turbidity and depth variation is what affects the use of the algorithm in correction of the water column. So the emergence of a new class for the seagrass cover category increasing from 3 classes to 5 classes on the classification results using the imagery of the Lyzenga algorithm. This is confirmed by [15] who said that correction of the water column needs to be done to improve the imagery quality by reducing the disturbance in the water column. Therefore, to get the basic object of good water, preferably in the process of imagery processing before the classification , its required to do a correction of the water column first.

IOP Conf. Series: Earth and Environmental Science **165** (2018) 012028 doi:10.1088/1755-1315/165/1/012028

Figure 4. Percent cover of Seagrass on the field (Very good: > 75%)

Figure 5. Percent cover of Seagrass on the field (Good: 50,5-75,4%).

Figure 6. Percent cover of Seagrass on the field (Rather Good: 25,5-50,4%)

Figure 7. Percent cover of Seagrass on the field (Bad: 5,5-25,4%)

Figure 8. Percent cover of Seagrass on the field (Very Bad: <5.5%

Classification result using the Lyzenga algorithm, the seagrass with the bad condition category has an area of 3,21 ha and with very bad condition category has an area of 1,57 ha, also the live coral class mixed with DCA on the result of image classification using the Lyzenga's algorithm also shows a clearer life coral. The magnitude of the sand and the seagrass of the good category in the image without using the Lyzenga algorithm due to the absence of the influence of the water column in the spectral reflection process of sand and seagrass objects, so that the emergence of a new class for seagrass cover does not appear. This is different from the image of the result of the Lyzenga's algorithm to reduce the extent of the good category seagrass and there are two new classes for the seagrass cover category, which bad condition and very bad condition categories not found in the image classification result without the Lyzenga's algorithm, as shown in Figure. 9 on the legend marked with a red circle.

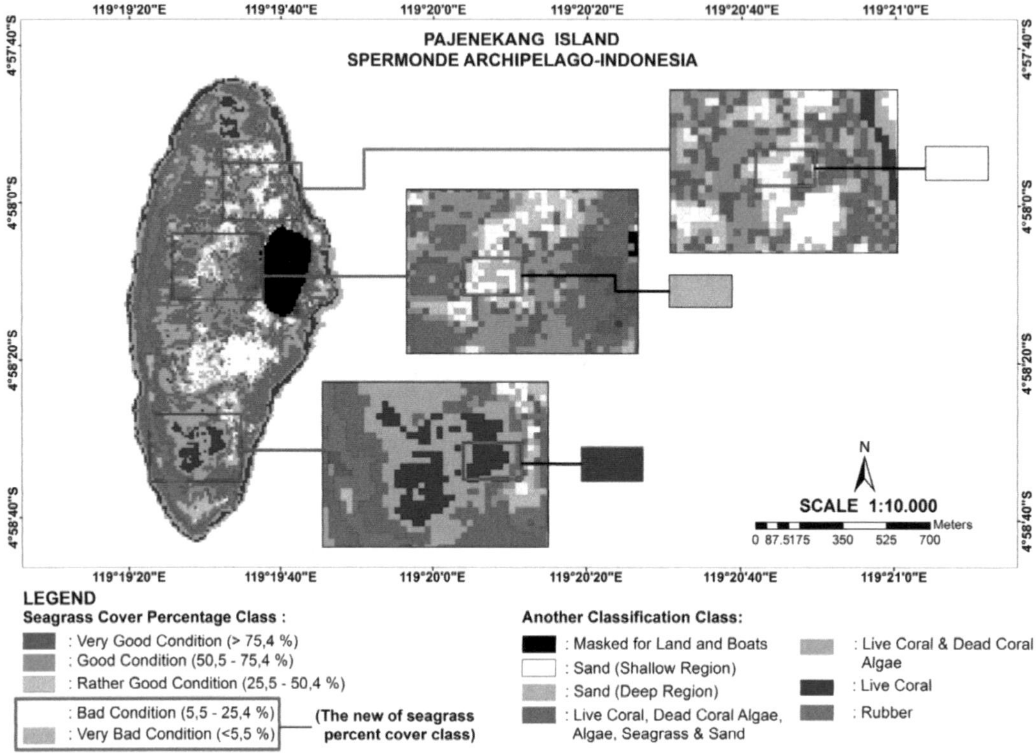

Figure 9. Result classification with the Lyzenga's algorithm. The description on the legend marked with a red square is a new class of seagrasspercents cover class not found in the image classification result without the Lyzenga's algorithm

3.5. The type objects of seagrass dominant species found in the field

In tropical regions such as Indonesia, the types of seagrasses generally grow together, forming a mixed meadows community of 2-8 species. While in temperate, Seagrass beds are often dominated by only one single species [3]. Type objects of the seagrass are 5 species dominant namely *Cymodocea rotundata, Enhalus acoroides, Halodule uninervis, Syringodium isoetifolium* and *Thalasia hemprichii.* Types of dominant seagrass, as shown in Figure. 10-14. Another objects some basic classes of shallow water in Pajenekang Island, Spermonde-Indonesia Islands, namely class are Sand; Dead Corals Algae (DCA) mixed with life corals, seagrass and sand; life corals mixed Dead Corals Algae (DCA); life corals; and rubber.

3.6. Accuracy test

Based on the results of accuracy tests using matrix error on the results of the classification of different methods on both images where obtained a value that states that 85.29% of the classification using algorithm is in accordance with the data or conditions in the field (ground truth 68 point coordinate), then Sentinel-2A's imagery is eligible for this study.

Anderson says that the thoroughness results of the classification must have a minimum value of 85% [19].The comparison of the classification results between the imagery and the Lyzenga's algorithm and the imagery without the Lyzenga's algorithm indicates that the imagery with the application of the algorithm has an effect on the process of detecting the appearance of the basal object in this case is better seagrass. Differences generated when the classification process by comparing the two results, it can be selected imagery classification results from the application of Lyzenga's algorithm. The imagery proved to provide a better baseline water substrate appearance with the addition of new classes for the category and seagrass slightly and very few of the 11 classes produced are shallow sand classes in shallow water; deep sand; DCA with algae mixed with living corals, seagrass and sand; live corals mixed with DCA; live corals; rubber and some classes of seagrass cover (Very good:> 75%, Good: 50,5-75,4%, Rather Good: 25,5-50,4%, Bad: 5,5-25,4% and Very Bad: <5.5%).

The error of 25% in the classification results without using the algorithm is caused by the change of the type of bottom closure of the waters where the imagery is classified as a good seagrass category whereas the validation results in the field show different types of objects is the area there are still two new categories namely a bad and very bad category, because in shallow waters with high depth variations and turbid water conditions will result in various classes due to correction of the water column. But when the research area has a variation in the depth that is less diverse (not correction of water column) then the resulting class will be less. So the extent of the object is different. This is what underlies the different seagrass and sand of the two images. Therefore in this study, the Lyzenga's algorithm gives effect to the result of classification due to turbidity variation and depth at the research location.

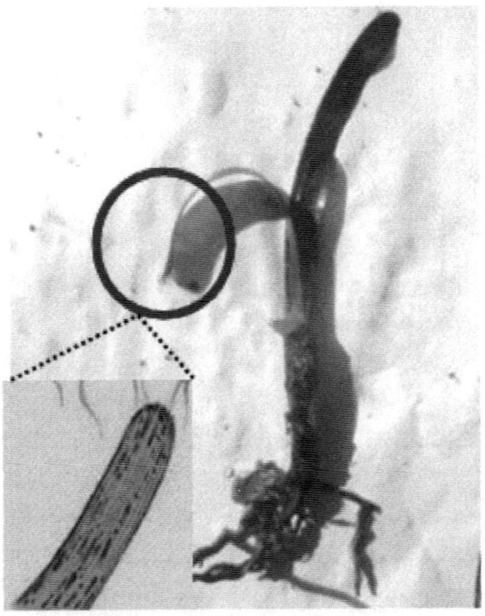

Figure 10. *Thalasia hemprichii*

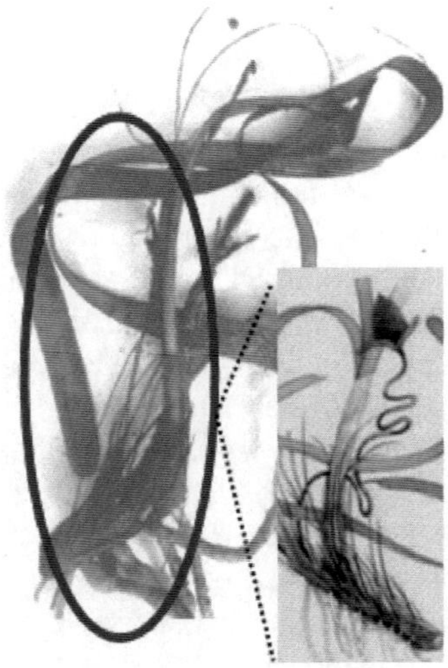

Figure 11. *Enhalus acoroides*

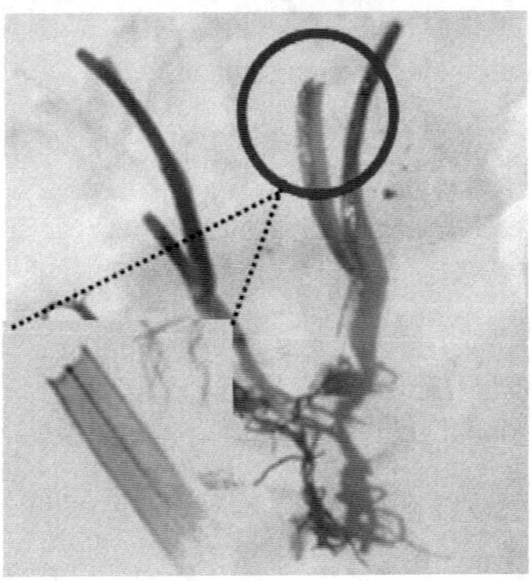

Figure 12. *Halodule uninervis*

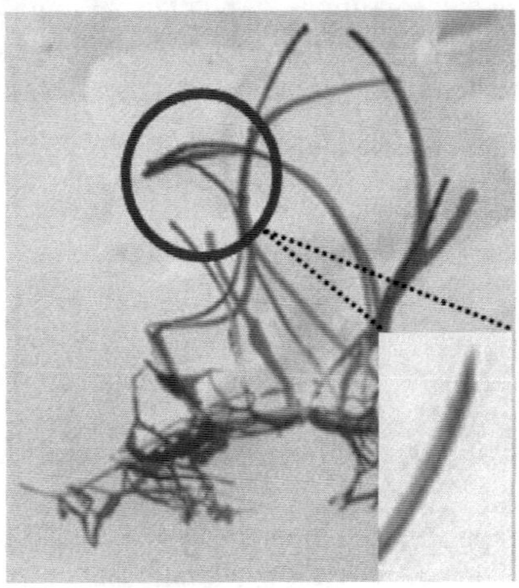

Figure 13. *Syringodium isoetifolium*

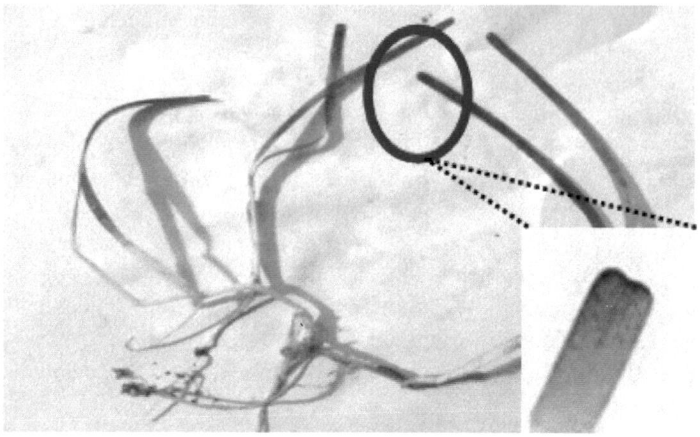

Figure 14. *Cymodocea rotundata*

4. Conclusions

Based on the comparison of imagery classification results with the Lyzenga's algorithm and without the Lyzenga's algorithm indicates that with the use of the Lyzenga's algorithm has an influence in detecting better baseline water features. Classes resulting from this research that is the class of shallow sand waters; deep sand; DCA and algae mixed with living corals, seagrass and sand; live corals mixed with DCA; live corals; rubber and some classes of seagrass cover (Very good:> 75%, Good: 50,5-75,4%, Rather Good: 25,5-50,4%, Bad: 5,5-25,4% and Very Bad: <5.5%), where the seagrass cover area is 30,75%, and there are 4 types of dominant seagrass namely *Cymodocea rotundata, Enhalus acoroides, Halodule uninervis, Syringodium isoetifolium* and *Thalasia hemprichi.*

Obstacles faced when the research is done that is the lack of reference about the results of imagery processing using Sentinel-2A Imagery, especially for seagrass beds so that the image of the interpretation for this type of imagery is needed by the author.

For further research, it is advisable to deepen the processing of this imagery data type and it is expected to apply the Lyzenga's algorithm for guided classification, especially for other substrate type of waters such as corals and more specific study to observe the condition of substrate base in Pajenekang Island not only look at the species but rather the texture of sediments and other oceanographic conditions. And the condition of the extent of seagrass beds large enough to need further research on the development and improvement of ecosystem of the area. So there is always the latest information that is expected to be the next research material on this island.

References

[1] Hemminga M A and Duarte C M 2000. *Seagrass Ecology*. Cambridge University Press.
[2] Kordi K M G H 2011 *EkosistemLamun (Seagrass).* Jakarta : PT. RinekaCipta.
[3] Vonk J A 2008 Shrimp burrow in tropical seagrass Meadow: an important sink for litter. *Estu.Coas.Shelf. Sci.* **79**: 79-8.
[4] Arifin and Supriadi 2006 Kondisi Padang Lamun Di Perairan ulau Sabangko, Salemo dan Sagara Kabupaten Pangkep. *Torani*, Vol. **16** (2) EdisiJuni 2006: 99 – 106.
[5] Faizal A and Amri K 2008 Digital Elevation Model for Mapping of Seagrass Bed Habitats in Cape of Bira, South Sulawesi. *Torani*, Volume. (**5**) December 2008:354-361.
[6] Aris A, Nurdin N and Baja S 2012 Modeling Spatial Dynamic of Coral Reef on The Small Island Case Study: BarrangLompo Island, Spermonde Archipelago, Indonesia.
[7] Nurdin N, Komatsu T, Agus A , Akbar M AS, Rasyid, A D and Amri K 2015 Multisensor and Multitemporal Data from Landsat Images to Detect Damage to Coral Reefs, Small Islands in

the Spermonde Archipelago, Indonesia. *Ocean Science Journal: Ocean Sci. J.* (2015) **50** (2):317-325.

[8] Nurdin N, Komatsu T, Rani C, Supriadi, Fakhriyyah S and Agus A 2016 Coral reef destruction of Small island in 44 years and destructive fishing in Spermonde Archipelago, Indonesia. *The 2nd International Conference of Indonesia Society for Remote Sensing (ICOIRS)* 2016.

[9] Nurdin N, Hidayatulah T and Akbar M A S 2009 Analisis Klasifikasi Obyek Penutupan Dasar Perairan laut Dangkal Menggunakan citra ALOS AVNIR-2. Volume **11** No. 1: Juni 2009 8-17.

[10] Sawayama S, Komatsu T and Nurdin N 2012 Coral reef habitats mapping of Spermonde Archipelago uding remote sensing, compared with in situ survey of fish abundance. SPIE (International Society for Optics and photonics.

[11] European Space Agency (ESA) 2012 ESA's Optical High-Resolution Mission for GMES Operation Service. ESA Communication: *ESA/ESTEC* (Frascati, Italy) and *ESA/ESRIN* (Noordwijk, The Netherlands). ESA SP-1322/2 March 2012, ISBN: 978-92-92221-419-7, ISSN: 0379-6566).

[12] Siregar Vincentius 1996 Pengembangan Algoritma Pemetaan perairan Dangkal (TerumbuKarang) dengan menggunakan Citra Satelit: Aplikasi pada Daerah Benoa, Bali *Kumpulan Makalah Seminar Konvensi Nasional Pembangunan benua Maritim Indonesia.*

[13] Jaelani L M, Laili N and Marini Y 2015 Pengaruh Algoritma Lyzenga dalam Pemetaan Terumbu Karang Menggunakan Worldview-2, Studi Kasus: Perairan PLTU Paiton Probolinggo (The Effect of Lyzenga's Algorithm on Coral Reef Mapping Using Worldview-2, A Case Study: Coastal Waiters of Paiton Probolinggo). Penelitian: Institusi Tekhnologi Sepuluh Nopember.

[14] Green E P, Mumbay P J, Edwards A J and Clark C D 2000 *Remote Sensing Handbook for Tropical Coastal Management.* Coastal Management Sourcebooks 3, UNESCO, Paris.

[15] Lyzenga D R 1981 Remote Sensing of Bottom Reflectance and Water Attenuation Parameters in Shallow Water Using Aircraft and Landsat Data. *International Journal Remote Sensing.*Vol **2** No.171 – 172.

[16] Suwargana N 2014 Analisis Citra ALOS AVNIR-2 Untuk Pemetaan Terumbu Karang (Studi Kasus: Banyu putih, Kab. Situbondo). Seminar Nasional Penginderaan Jauh, Deteksi Parameter Geobiofisik dan Diseminasi Penginderaan Jauh (pp. 588–9).

[17] Andréfouët S, Frank E, Muller-Karger, Eric J, Hochberg, Chuanmin Hu, Kendall L and Carder 2003 Change detection in shallow coral reef environments using Landsat 7 ETM+ data *Remote Sensing of Environmental* vol **78** pp 150-162.

[18] Dinas Hidrografi-Oseanografi (DISHIDROS) 2017 Daftar Pasang Surut Kepulauan Indonesia Tahun 2017. Dinas Hidrografi-Oseanografi TNI AngkatanLaut: Indonesia.

[19] Anderson J R 1976 A Land Use Cover Classification System for Use With Remote Sensor Data. *Geological Survei Profesional* Paper 946. Washington.

Acknowledgments

The authors express their gratitude to the head of Research Center for Regional Development and Spatial Information, Hasanuddin University and staffs for his help to allow authors to process data in their Department which not only provide much knowledge, but also provide direction and input for this research. Thanks also to **Mansyur Talibe, SE, ANT. II** and **Marwah Thalib, S.Pd** which has helped writers in improving grammar in writing this paper. Another thanks for **M. Akbar As, M. Sc, Turissa Pragunanti, S. Kel** and **Muhammad Afif Fauzan, S.Si** which has helped for data processing.

Strategy to Maintain Food Security in the Area of Flood Hazard in Karawang Regency

Bambang Riadi[1*]

[1]Geospatial Information Agency, Jalan Raya Bogor, Cibinong

bambang.riadi@big.go.id

Abstract. Rice yield in Java Island as national rice supplier in order to support food security. Java has a dense population, causing very high pressure on land use. The need for housing, roads and industrial infrastructure encourages the conversion of agricultural land for other land uses. This study aims to formulate strategies for the purpose of maintaining food security in flood-prone areas. Karawang Regency is a rice barn of West Java Province and is also a frequent area of flooding. Therefore, it is necessary to map the spread of flood-prone areas and to analyze land use linkages, rice productivity, population and food needs in the study area. Strategies for maintaining spatial food security are by mapping land cover multi years and mapping the flood hazard areas using high-resolution DEM data. Land cover mapping using RBI Map Scale 1: 25,000 the Year 2000 and Citra Landsat-8 and Spot-6 Year 2015. Wetland Area Year 2000 is 120,371 (62.83%), land area Year 2015 is 98,462 ha (51.40%). The average rice field area caused by flooding reaches 219.84 ha/year, if the productivity of 6.9 tons/ha annually loses rice production on average 1,517 tons. Food security in the research area is still maintained from various pressures of land use. There is unfulfilled support for national food security with the obligation to provide 1.5 million tons of paddy stock per year. The selection of strategies in maintaining food security in the research area is to tighten supervision in licensing in favor of the protection of paddy fields. The professionalism of agricultural extension resources will be able to increase rice productivity.

Keywords: DEM, Food security, multi years, spatial

1. Introduction

In general, flooding is caused by high rainfall above normal, so water runs from river to low land by riverside. The flood caused by river water flows into the surrounding environment as a result of high rainfall. Rainfall affects the production of food of agricultural products if the rainfall is very high can cause flooding, this event can reduce the production of food crops. The area of rice farming experiencing the immersion stress due to flood is expected to increase due to increased rainfall and rising sea levels due to global warming [1]. The submerged plants are inhibited by the process of photosynthesis, and in general, the rice plants will die if all parts of the plant are flooded for seven days [2]. And in flood-prone areas, where rice cultivation is often completely submerged for several days at different age crop conditions, causing crop damage and also decreased yield [3]. Floods often occur in the lowlands which are usually rice production centers, so the impact is quite significant related to the availability of food, especially rice. Wetland is generally located on the fluvial plains and its location is relatively lower than the road or settlement so that the paddy field is physically a flood-prone area and potentially experience deeper and longer pools in case of flooding.

The threat of flooding needs to get serious attention because this region contributes to the national rice demand on average reaching 865,000 tons of rice per year [4]. In the planting season of 2013/2014, floods caused many rice fields that failed to harvest and caused 7,700 ha of paddy fields to fail to harvest [4]. Another problem that can reduce rice production is reduced irrigated rice field, land conversion is permanent, which means after the rice field switch function cannot be returned to the original rice field again. Conversion of land that affects food production is the disappearance of effective stretches to contain excess runoff that can help reduce flooding. Increasing the area of rice fields affected by the flood is directly proportional to the area of crop failure [5], this condition is worrying because it can lead to decreased rice production.

Landform in the coastal area of Karawang regency is alluvial plain, an area with landforms from the alluvial plain land system is flood-prone area because it is low area or basin. For coastal areas with slope characteristics <2% and has an average rainfall of ≥ 200 mm included in flood-prone class [6]. The floodplain landform can be identified as a wetland area which is also a lowland form of the basin. To identify the wetland area in this study will utilize the map of the index of DEM TerraSAR-X in 2011. Food security is a multi-dimensional issue that requires analysis of production and food availability. Although there is no specific way to measure food security, the complexity of food security is simplified by focusing on the food availability dimension.

The objectives of this research are (1) to explain the factors needed to the maintain food security in the region with the danger of the flood, and to collect information about the impact of flood on rice production. (2) By utilizing advances in mapping technology to identify wetland changes and flooding incidents in rice fields. (3) wetland maintenance strategy for agricultural land, with anticipatory measures in the face of flood disaster by building and strengthening the forecasting system for flooding. The results of this study are expected to be useful for researchers and local governments in preparing the improvement of spatial plans to maintain food security.

2. Method and materials

2.1. Location and time of research

The research location is Karawang Regency and lies in the north part of Province of West Java, which is geographically located between 107º02 '- 107º40' East Longitude and 5º56 '- 6º34' South Latitude. The research and data processing activities were conducted from October 2016 to March 2017. Data analysis was conducted at the Geospatial Information Agency. Climatologically monthly average rainfall in Karawang Regency shows that rainfall at low-intensity level [7].

2.2. Types and data sources

The data used in the implementation of this research consist of primary data and secondary data. Primary data were obtained from field surveys to check the data on flood events, cover, and documentation and analysis results. While the secondary data in the form of thematic maps, statistical data, and other data required. The basic data types used in the study were Landsat 8 imagery October 2015 recording, topographic map of scale 1: 25,000 in 2000 [8], TerraSAR-X TWI map of 2011 (analysis results), land cover map of 2000, the land cover map of 2006 and the land cover map of 2015. Average monthly rainfall data from 2010 to 2017 [9]. Map Plan of Spatial Pattern Karawang regency 2011-2031 scale 1: 50,000.

2.3. Data Analysis Procedure

This research is conducted through several stages as follows:

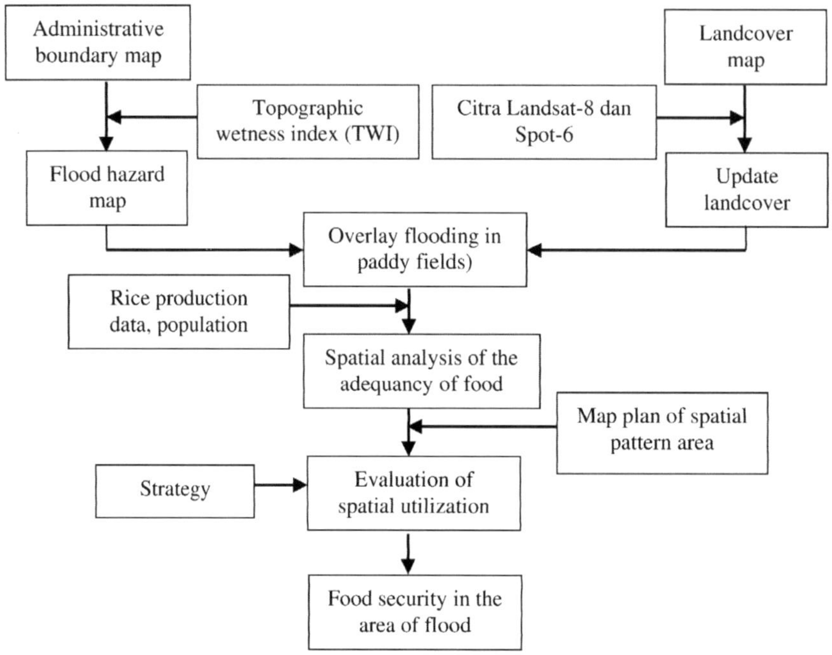

Figure.1. Flowchart of research methods

The expected output from this research is the strategy in maintaining food security in agriculture area with the danger of flood. The software used to process spatial data is Global Mapper V.10, ER Mapper 7, Ilwis 3.85 and ArcGIS 10.2, while for processing data attribute used Ms. Office Excel 2010.

2.4. Topographic Wetness Index Analysis(TWI)

Index calculation with DEM input fill sink and flow accumulation produce wetness index. TWI analysis is arranged sequentially with the Stream Networks analysis to construct a network of streams that exist in a particular region using flow accumulation information. Flow accumulation is explained by determining the value of flow accumulation considered as a river.

In this research, the high index value (> 25) has a sloping slope so that the potential for high water puddle. The index value of 15 in this study as an indication of boundary inundation area. The wetness index <15 indicates that a region has a steep slope and has a low TWI index value, with a low potency of water puddle. Delineation of boundary areas indicating the danger of flooding is done by converting raster to format vector format and/or through visual interpretation showing the distribution of flood hazard areas [10].

2.5. Analysis of Land Cover and Flooded Land

Land cover data extracted from topography map scale 1: 25,000 in 2000, grouped into 7 kinds of land cover (open area/fields, plantation, settlement/land woke up, forest, pond, water body, rice field). According to [11], spatial analysis of land use from satellite imagery is done by classifying, for this study using Landsat Satellite Image 2006 and Landsat-8 Image 2015. Furthermore, the classification result is detailed using a Spot-6 image with main concern on the use of wetland, changes in the use of wetland for other uses greatly affect rice production. To see the relation of existing land cover with potential flood vulnerability done by spatial analysis of overlay of hazard map of flood result of analysis with a land cover map.

Flood control programs have been widely implemented but floods (frequency, duration, intensity, the extent of inundation) continue to increase. Spatial or land use changes have more influence or contribution to flooding than the physical construction of flood controllers [12]. Production data, population data and historical data of flooding using secondary data from the Department of Agriculture and the Central Bureau of Statistics which will be dispatched and made its own attributes in the units of the district administration.

2.6. Spatial analysis for food sufficiency

The food adequacy analysis method used is analyzing food security policy and program done descriptively with content analysis method. This method is a method used to examine the content of food security policies and programs whose calculations use the formula (1):

$$FS = \frac{NC\ (kg/capita/year)}{A\ (kg/capita/year)} \quad (1)$$

FS : Food Sufficiency
NC : Normative Consumption
A : Availability

If the value of sufficiency is more than 1, then the area is food deficit and if the value of sufficiency is less than 1, then the area is food surplus.

Space utilization system basically contains two main components namely the component provider of space (supply), and components of space users (demand). Spatial components include natural and physical resources, while the userspace component (demand) includes people with activities, both production and consumption activities.

Supply components include agricultural land, analysis by identifying existing rice fields. This analysis begins with topographic extracts for making thematic land cover basic maps. Followed by updating the basic thematic map with a Landsat-8 image, increasing the depth of information of land cover analysis in update using Spot-6 image resolution 1,5 m and producing existing land cover map 2015. This existing map is used as the basis for calculating rice production, rice productivity 6.9 tons/ha and intensity of cropping 2 times per year [13].

The demand component includes the population with its activities, both production and consumption activities. Based on the data of population in 2010 - 2015, it can be seen that the average population growth rate (r) for 5 years (2010 - 2015) is at 1.948%. Predicted population and rice needs can be done for the coming years.

$$Pt = Po\,\{\,1 + (r.t)^n\} \quad (2)$$

Explanation :
Pt = Number of residents in the final year
Po = Population in the base year (2010).
t = Time period (e.g.: t = 2016 - 2010)
r = Population growth rate (1,948 %)

To be able to know the needs of population consumption of rice per year using the formula:

$$KK = SK\,.\,Pt \quad (3)$$

Conversion of grain to rice is 62,74% [1]

$$KG = KK\,(\frac{100}{62,74}) \quad (4)$$

Explanation:
KK = Needs of rice consumption of the population
SK = Standard consumption(kg/capita/year)
Pt = Population year (Projected population t year)
KG = Grain Consumption

2.7. Strategy
SWOT analysis is a strategic planning method used to evaluate the strengths, weaknesses, opportunities, and threats in a project. The four factors that form the acronym SWOT [14]. This process involves determining specific objectives and identifying internal and external factors that support and which are not in achieving that goal. SWOT analysis can be applied to analyzing and sorting things that affect the four factors. In the application is how strength is able to take advantage of existing opportunities. How to overcome the weaknesses that prevent the advantage of existing opportunities. How strengths are able to deal with existing threats. And how to overcome the weaknesses that can make the threat real or create a new threat.

3. Result and discussion
3.1. Topographic wetness index (TWI)
Topographic Wetness Index (TWI) is a wetness index that can be used as an indicator of an area that has the potential for flooding. This index is calculated from the surface variables formulated as TWI = ln (As / tan β) where As is Flow Accumulation and β is the slope. High Index values indicate sloping slopes that have a high potential for water puddles, low TWI index values, the low potency of water puddles. This index provides a wetland potential pattern in the catchment area, but the accuracy of the spatial prediction depends heavily on the quality of DEM used [15]. Delineation of boundary areas indicating the danger of flooding is done by converting the format of raster to format vector and or through a visual interpretation which shows the distribution of flood hazard areas (Figure.2.).

The study of flood hazard areas of most of the flood hazard areas is located in paddy fields (Figure.3.). From the results of the study in a certain period, it is found that the Karawang Regency with an area of 191,577 hectares in 2000 has paddy fields with an area of 120,371 ha (62.83%). In 2006 result of analysis of wetland area of 112.878 ha (58.92%), there was degradation of rice field of 7,493 ha. Results of existing land analysis of 2015 land use Landsat-8 image data and Citra Spot-6 obtained wetland area of 98,462 ha (51.41%). So in the period of 2000 to 2015 degradation of paddy fields with an area of 21,909 ha (11.42%) [16].

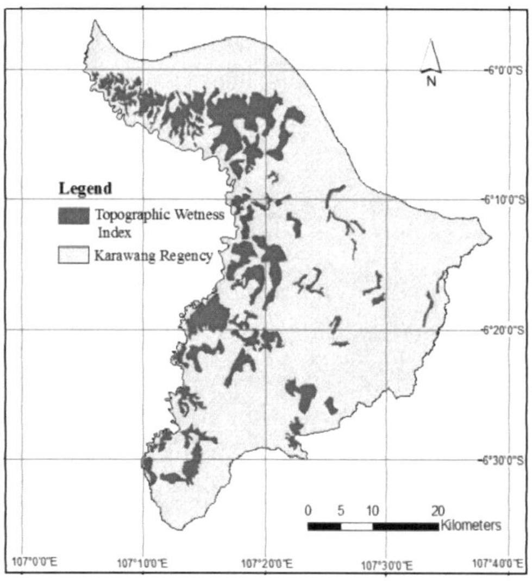

Figure.2. TWI Map

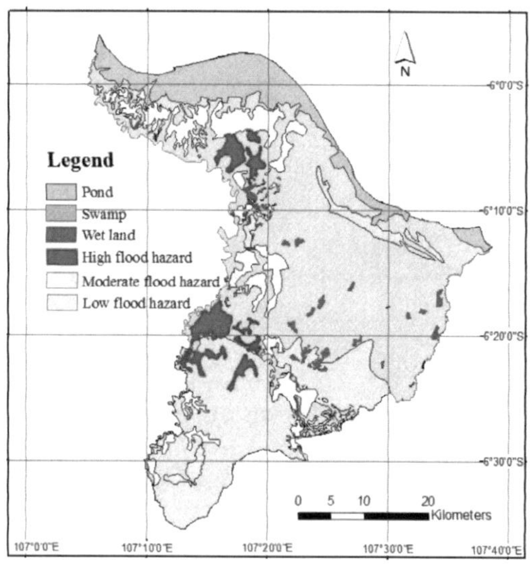

Figure.3. TWI analysis with landcover

Based on the calculation, the amount of reduction of rice field in Karawang regency is 1.88% per year [17]. As a result, it has reduced the contribution of rice out of the region beyond the internal consumption of 59.75% of production in 2000, to 49.82% in 2011 [17]. The difference in the results of the degraded rice field counts from previous research is the use of high-resolution image data in this

IOP Conf. Series: Earth and Environmental Science **165** (2018) 012029 doi:10.1088/1755-1315/165/1/012029

study. The real pressure of land use change on paddy fields is very high. The land transfer function is basically unavoidable in the implementation of development, population growth and demands of settlement land often conflicts and can be different land use with the designation plan. Limited land in urban areas also causes the city to develop physically towards the outskirts of the city, the periphery is a region that many experienced changes in land use from agricultural land to non-agricultural due to the influence of the development of nearby cities.

The heavy pressure on land use in the research area due to geographical proximity to the national capital, the need for residential and the industrial land increases the pressure on land conversion due to the low land rent of agricultural land compared to land use for industrial and residential land. Based on the plan of spatial pattern Karawang regency in 2011-2031 identified land use rice field remaining 95,556 ha or 49,88%.

To maintain food security Government of Karawang regency needs to focus on increasing aspect of productivity in food agriculture sector and preserving rice fields. The procedure for measuring food security is based on the availability category, particularly the adequacy/supply indicator. Therefore, production level becomes a very important element in maintaining food security, technological innovation is needed which can increase rice productivity by 6.9 ton/ha. Using the Formula (1) until Formula (4) can be projected for the research area of rice needs until 2031 will still be fulfilled with the remaining 95,556 ha (49.88%), but the surplus obligation of 1.5 million tons of grain for national stocks cannot be met if the level of rice productivity has not increased (Table.1.).

Flood hazard area with an area of 28,366 ha in Karawang regency mostly inundated rice fields. Overlay of the land cover map with TWI map (Figure.3.) Shows the distribution of flood hazard areas, the area of flood hazard with high flood hazard criteria with an area of 7,489 ha (7.61%), medium flood hazard level 19,188 ha (19.49%) and low flood hazard with an area of 1,689 ha (1.72%). The non-hazard level of flooding in paddy fields is 70,095 ha or 71.19% of rice area of 98,462 ha.

Areas with high flood hazard levels have characteristics close to industrial and urban areas, since there has been a lot of conversion of agricultural land to industrial land, resettlement or built up area (Figure.3.). This condition makes high surface flow and causes flooding in the area, special handling needs to be done so that in this area there is no failure of harvest so that rice production is maintained. Flood events that will result in crop failure is a flood that inundated rice fields for more than 7 days, which makes the rice cannot do photosynthesis.

Table.1. Population and Projection, Consumption of rice, Rice of grain per year, Area of rice field area, Grain production, Surplus of grain

Year	Population and Projection	Rice Consu mption (ton / year)	Needs of grain (ton / year)	Area of rice field (ha)	Grain production	Surplus of grain (ton)
2000	*1,787,319*	*176,051*	*280,604*	*120,371*	*1,661,120*	*1,380,516*
2006	*2,009,647*	*197,950*	*315,509*	*112,878*	*1,557,716*	*1,242,208*
2015	*2,273,579*	*223,948*	*356,945*	*98,462*	*1,358,776*	*1,001,830*
2020	*2,437,497*	*240,093*	*382,680*	*97,552*	*1,346,218*	*963,538*
2025	*2,596,832*	*255,788*	*407,695*	*96,642*	*1,333,660*	*925,964*
2030	*2,766,582*	*272,508*	*434,345*	*95,732*	*1,321,102*	*886,756*
2031	*2,801,841*	*275,981*	*439,881*	*95,556*	*1,318,673*	*878,792*

Source: Result of analysis and calculation

The average area of paddy fields that failed to harvest due to flooding reached 219.84 ha/year, the estimated value of economic losses due to flooding in paddy rice farming is a tangible loss consisting of direct and indirect. Direct losses include damage to agricultural land estimated through a cost recovery approach. The indirect losses include decreasing productivity and loss of farmer income using productivity change approach. Direct losses as a result of physical damage to farmland, are estimated from farmers' expenditure to improve the quality of agricultural land damaged by floodwater during one period. The cost of these improvements will be incorporated into production costs after the floods to calculate the income of farmers lost.

For the development of strategies to maintain food security, analyzed by using SWOT analysis based on the data obtained, internal and external factors owned and strategies that can be done referring to the SWOT matrix (Table.2.), Are as follows.

Table.2. Further arranged matrix in the following table:

Internal factors / **External factors**	**Strength (S)** *1. Food availability: Agricultural land available* *2. Access to food: Presence Departement of Food security* *3. Utilization and absorbent -an: Available means and agricultural infrastructure* *4. Food vulnerability: Area land failure due to flooding the minimum*	**Weakness (W)** *1. Food supply: high population growth, the difficulty of getting the means and production infrastructure* *2. Food access: Lack of quality and a strength of human resources extension workers* *3. Utilization and absorption: Low tech support and capital* *4. Food vulnerability: High overland function, low land ownership, disaster flood*
Opportunities (O) *1. Food availability: Open marketing access* *2. Access to food: means good road infrastructure* *3. Utilization and absorption: Expansion food Technology* *4. Food vulnerability: Increased facilities agricultural production*	**Strategy Opportunities (SO)** *1.Increased productivity and marketing of rice (S1, O1)* *2. Prepare management Effective food security (S2, O2, T2)*	**Strategy Weakness Oppor. (WO)** *1. Professional human resources extension workers enhanced quality and number (W1, W2, O3)* *2. Allocated development funding as needed (W3, O1, O3)* *3. Assistance, facilities supervision Production (W3, O2, T3)*
Threats (T) *1. Needs of food: Its high Growth population* *2. Food access: road and agricultural infrastructure* *3. Utilization and Absorp- tion: low public awareness of health* *4. Food vulnerability: Rice field function transfer, floods*	**Strategy Strengths Treats (ST)** *1.Making a system policy licensing and supervision deep area layout protecting farmland (S1, T1, T4)* *2. Improving infrastructure road and agriculture (S2, T2)*	**Strategy Weakness Threats (WT)** *1.In cooperation with financial or agricultural institutions (BUMN) in order availability of food (W3, T3)* *2. Cooperate with insurance for Protection plants due to flood (W4, T4)*

Table.3. Development of SWOT elements based on their importance

Strengths (S)		**Weakness** (W)		**Opportunities** (O)		**Threats** (T)	
Components	Values	Components	Values	Components	Values	Components	Values
S1	*3*	*W1*	*3*	*O1*	*2*	*T1*	*3*
S2	*2*	*W2*	*2*	*O2*	*2*	*T2*	*2*
S3	*3*	*W3*	*2*	*O3*	*3*	*T3*	*2*
S4	*2*	*W4*	*3*	*O4*	*3*	*T4*	*3*

Nilai 3 = Important 2 = Medium importance 1 = Not important

Table.4. The alternative selection strategy

SWOT element	Linkages	Weight	Rank
SO1	*S1, O1*	*3+2 = 5*	*7*
SO2	*S2,O3,T2*	*2+3+2=7*	*3*
WO1	*W1, W2, O3*	*3+2+3=8*	*2*
WO2	*W2, O1, O3*	*2+2+3=7*	*4*
WO3	*W3, O2, T3*	*2+ 2+2 = 6*	*6*
ST1	*S1, T1, T4*	*3+3+3 = 9*	*1*
ST2	*S2, T2*	*2+2 = 4*	*8*
WT1	*W3, T3*	*2+2 = 4*	*9*
WT2	*W4, T4*	*3+3 = 6*	*5*

Based on the value and weighting that has been done, the determination of food security strategy is the strategy ST1, WO1, and SO2, which include:

1. To make a policy of permit system and supervision of territorial scope in protecting agricultural land (S1/ Agricultural land available, T1/ High Growth population, T4/ Rice field degradation, floods).
2. The professionalism of extension human resources, quality improvement and a number of the productive workforces to increase food production (W1/ high population growth, the difficulty of getting the means and production infrastructure, W2/ Lack of quality and expertise of human resources agricultural extension, O3/ Expansion food Technology).
3. Preparing effective food security management (S2/ Presence Departement of Food security, O2/ Access to food: means good road infrastructure, T2/ road and agricultural infrastructure), controlling and maintaining agricultural land in order to avoid land conversion which resulted in widespread flooding in paddy fields

4. Conclusion

Strong pressure on land use in the research area due to the geographical location close to the state capital, the need for residential land and industrial land increases the pressure on land conversion. This condition occurs due to low land rent of paddy fields compared with land use for the industrial and residential land. Wetland land use to other uses from 2000 to 2015 reached 21,909 ha (18.2%). Another pressure is the danger of flooding in paddy fields that can reduce rice production. The results of the analysis obtained the area of flood hazard with high flood hazard criteria with an area of 7,489 ha (7.61%), medium flood hazard with 19.188 ha (19.49%) and low flood hazard with 1,689 ha (1.72%). The average rice field area caused by flooding reaches 219.84 ha/year if the productivity of 6.9 tons/ha each year average loss of rice production is on average 1,517 tons.

Food security in the research area is still in good condition, but it cannot support national food security as an obligation to provide 1.5 million tons of paddy stock per year. The selection of strategies in maintaining food security in the research area is to tighten the supervision in licensing in favor of the protection of paddy fields. The professionalism of agricultural extension resources will be able to increase rice productivity from the current average rice productivity of 6.9 tons/ha.

References

[1] Widiarta I N 2009 Pengaruh perubahan iklim global terhadap pertumbuhan hama tanaman padi dan musuh alaminya Hal 325-335 *Prosiding Seminar Nasional Padi* 2008 Balai Besar Penelitian Tanaman Padi Sukamandi

[2] Ito O, E Ella and N Kawano 1999 Physiological basis of submergence tolerance in rainfed lowland rice ecosystem *Field Crops*

[3] Ikhwani 2013 Ketahanan Varietas Padi Toleran Rendaman dan Responnya terhadap Pemupukan. *Jurnal Lahan Suboptimal.* ISSN: 2252-6188 Vol **2** No 1 April 2013

[4] Department of Agriculture Karawang Regency (Dinas Pertanian Kehutanan Perkebunan dan Peternakan/DPKPP) 2016 Potensi pertanian di Kabupaten Karawang

[5] Irianto G 2010 Kebijakan Sektor Pertanian Dalam Penguatan Adaptasi Perubahan Iklim Workshop I-MHERE B2 C IPB "Adaptasi Pertanian Dalam Merespon Perubahan Ikim Global Menuju Ketahanan dan Kedaulatan Pangan" Bogor 11 Maret 2010

[6] Standar Nasional Indonesia (SNI) 2015 Metode pemetaan bahaya banjir skala 1:50,000 dan 1:25,000.

[7] Badan Meterologi Klimatologi and Geofisika 2015 Peta distribusi curah hujan bulan April 2015 Kabupaten Karawang

[8] Badan Informasi Geospasial/BIG 2015 Peta Kejadian Banjir Tahun 2015

[9] The Tropical Rainfall Measuring Mission (TRMM) 2017 http://trmm.gsfc.nasa.gov Akses Agustus 1st 2017

[10] Riadi B, Barus B, Widiatmaka M, Yanuar J P and Pramudya B 2017 Identification and Delineation of Flood Hazard with Spatial Modeling Using High-accuracy DEM Data in Karawang *The 4th International Symposium On Lapan-IPB Satellite (Lisat)* (Bogor: Agricultural University)

[11] Parsa I M 2014 Uji Coba Model Pemetaan Lahan Sawah Berbasis Perubahan Penutup Lahan Citra Landsat Mosaik Tahunan di Jawa Barat *Jurnal Penginderaan Jauh* Vol.**11** No.1 Juni 2014: 15-28

[12] Rasyidie A 2013 Banjir Fakta dan Dampaknya Serta Pengaruh dari Perubahan Guna Lahan *Jurnal Perencanaan Wilayah dan Kota* Vol. **24** No.3 Desember 2013

[13] Badan Pusat Statistik/BPS 2016 https://karawangkab. bps.go.id

[14] Rangkuti Freddy 2002 *Analisis SWOT Teknik Membedah Kasus Bisnis* (Jakarta: PT.Gramedia Pustaka Utama)

[15] Merot P, Moy L H, Odoux C G, Clement B and Thenail J B C 2006 A Method for Improving the Management of Controversial Wetland *Environmental Management* Vol. **37** No. 2 pp. 258–*270* DOI: 10.1007/s00267-004-0391-4

[16] Riadi B, Barus B, Widiatmaka, Yanuar MJP and Pramudya B 2017 Review of Land use Changes Based on Space Pattern to Maintain Food Security in Karawang Regency- West Java Provinsi. *The 4th International Conference of Earth Sciences and Enginnering* (*ICEE 2017*) (Padang: Andalas University)

[17] Widiatmaka, Ambarwulan W and Munibah K 2013 Landuse Change During A Decade as Determined By Landsat Imagery of A Rice Production Region and Its Implication to Regional Contribution to Rice Self Sufficiency: Case Study of Karawang Regency West Java Indonesia Bridging Sustainable Asia *Proceeding of 34th. Asian Conference on Remote Sensing* (Bali: Bali) 20-24Oct 2013 pp. 834-840

Acknowledgments

Thank the Geospatial Information Agency for assistance TerraSAR-X DTM data and USGS Website for the ease of data access Landsat and The Faculty of Geo-Information Science and Earth Observation (ITC) for open source ILWIS 3.8 software, so this article can be resolved.

Modeling Deforestation and Green Houses Gas Emissions in Morowali Utara District, Central Sulawesi Province, Indonesia

Irmadi Nahib[1*] **and Yatin Suwarno**[1]

[1]Geospatial Information Agency, Jl Raya Jakarta Bogor KM 46 Cibinong, Jawa Barat, 16911, Indonesia

irmnahib@gmail.com, irmadi.nahib@big.go.id

Abstract. Land use cover conversion from forest to another land was constributing for total green houses gas (GHS) emission, especially from forestry sector. The Central Sulawesi Province has forest area about 4.258 million ha. The deforestation in Central Sulawesi Province for period 2000-2009 reached 432,111.50 Ha (10.15 %). Morowali Utara District is one of district in Central Sulawesi Province, which contribution 15,20 % total deforestation in Central Sulawesi province. This study aims to calculate the rate of deforestation and simulation carbon emission in Morowali Utara District by using REDD$^+$ model. The main data used in this study are land use or cover map (2003, 2009 and 2015), driver variable (distance from : forest edge, roads, streams, and slope) as the explanatory variables of deforestation. The results show that the deforestation on the period 2003-2015 reached 62,017.47 (9,19 %) or the rate of deforestation 5,168.12 ha year^{-1}. Logistic regression model for prediction of deforestation is $Y_1 = -2.2649 + 0.001126\ X_1 - 0.000152\ X_2 - 0.000225\ X_3 + 0.010704\ X_4$ (ROC = 0,7408), meanwhile the amount of carbon that the REDD project will protect given a departure from the business as usual deforestation scenario up to 18.868 million tCO_2e.

1. Introduction

Tropical deforestation is a major contributor to green house gas emissions in developing countries. Incentive mechanisms, such as reducing emissions from deforestation and forest degradation (REDD+), are currently being considered as a possible emissions reduction and offset solution [1].

Tropical forests and other vegetated landscapes like grasslands and wooded savannahs play a major role in the global carbon sequestration process. Their conservation and protection offers immense potential for reducing greenhouse gas emissions and global warming [2].

In 2008, The United Nations launched REDD (United Nations Collaborative Programme on Reducing Emissions from Deforestation and Forest Degradation in Developing Countries) to provide a mechanism to mitigate climate change by sequestering forest carbon. REDD also promotes the secondary ecosystem service benefits associated with this forest conservation, including protection of biodiversity and water quality [3].

Indonesia has the highest deforestation rates in the world, exceeding even Brazil while having only a quarter of Brazil's forest area [4]. Refer [5] the average annual deforestation in Indonesia for the period 2000-2012 was 671,420 hectares, accounts for 525,516 ha of deforestation in mineral land and 145,904 ha of deforestation in peat land. During this period, 9 percent of deforestation occurred in

Sulawesi island. The other research, refers to [6] the rate of deforestation Sulawesi Island in period 2000-2009 for was 1.667 million ha (15.58 %). In 2009, forest cover in Sulawesi Island was 9.039 million ha or approximately 46,65 % of the total area. It has been decreased by 1.667 compared to year 2000.

The Central Sulawesi province has forest about 4.258 million ha, so it have a strategic role in the implementation of REDD +. The deforestation occurred central Sulawesi Province in period 2000-2009, reached 432,111.50 ha (10.15 %), so forest cover in 2009 up to 3.826 ha. Morowali Utara District is one of district in Central Sulawesi Province, wich contribution of 15.20 % total deforestation in Central Sulawesi Province. The forest cover in Morowali Utara district in 2011 was 632,756.70 ha or approximately 73.84 % of the total area. It has been decreased by 33,013.80 ha (4.96%) compared to 2000. The rate of dorestation in period 2000-2011 adalah 3,001.25 ha year^{-1}.

Refer [7], land use cover conversion from forest to another land was constributing for total GHS emission, especially from forestry sector. In order to implement the REDD model, the constraints and incentives defined to produce the sustainable development scenario were used to create the input spatial data required. This study aims to calculate the rate of deforestation and simulation carbon emission in Morowali Utara District by using REDD+ model. The REDD project location for this study is conservation forest area (assumed) in Morowali Utara District.

2. Materials and method

2.1. Tools and Data

Tools use in this study area ArcGIS 10.2 and Terrset software. Refer [7], Terrset is an integrated geospatial software system for monitoring and modeling the earth system for sustainable development. The TerrSet System incorporates the IDRISI GIS Analysis and IDRISI Image Processing tools along with a constellation of vertical applications.

2.2. Data Used

Data used in this study:

- Indonesian topographic map, Scale 1: 50,000, Geospatial Information Agency
- Land cover map of Morowali Utara District, Scale 1: 250,000 (2003, 2009 and 2015), Directorate General of Forestry Planning - Ministry of Enviromental and Forestry **(Figure 1)**
- Environmental variable (dependent variable): distance from forest edge , distance from roads, distance from stream, and slopes, Indonesian Topographic Map, Scale 1: 50,000, Geospatial Information Agency **(Figure 2)**

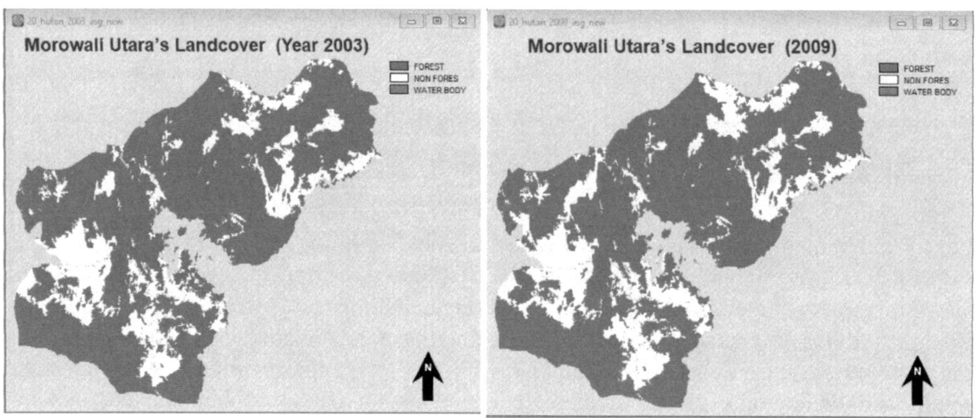

(a) Land Cover 2003 (b) Land Cover 2009

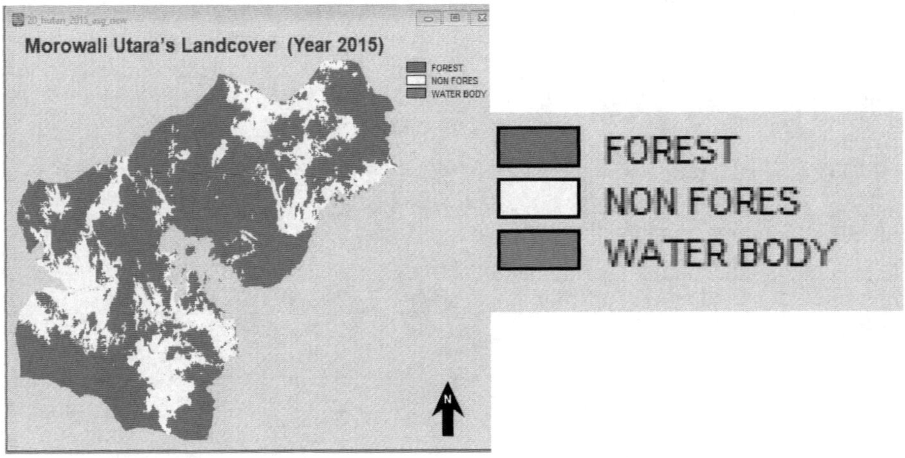

(c) Landcover 2015 Legend

Figure 1. Morowali Utara's Landcover : a) 2003, b) 2009 c) 2015

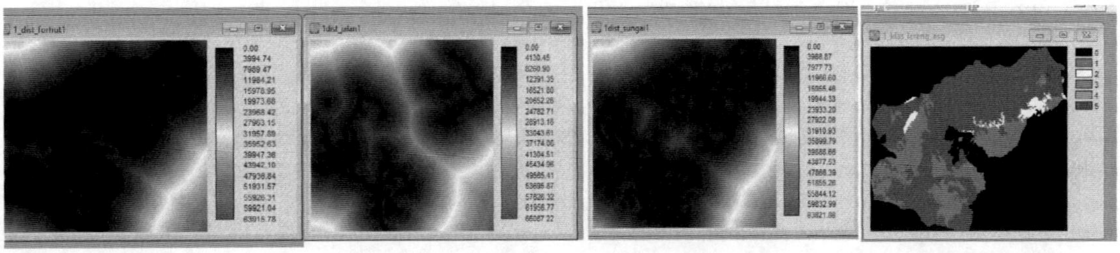

Distance from edge forest edge (m)	Distance from road (m)	Distance from streams (m)	Slopes(degrees)

Figure 2. Enviromental variable (dependent variable)

- The Bio-Carbon Fund Project (BioCF) methodology requires three basic spatial or geographical data inputs, the project area, the leakage area, and the reference area **(Figure 3).**

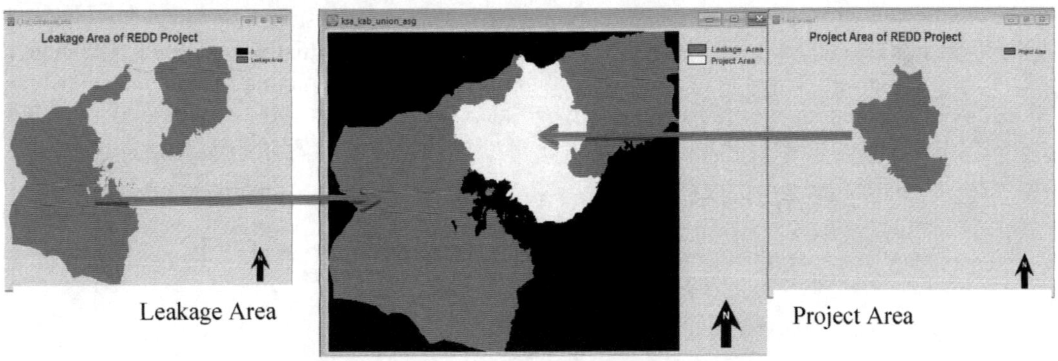

Leakage Area Project Area

Reference area = Project area + Leakage area

Figure 3 : Spatial data inputs for REDD project

Refer [7], the project area is the geographic extent of the area under consideration for a REDD project; the leakage area refers to the area around the project area that may experience impacts as a result of the creation of the protected area, and the reference area is the entire area of study encompassing the project and leakage areas.

- Refer [8] in central Sulawesi province, the carbon density for forest area which is 163,5 Mg/ha and non forest area is 16 Mg/ha. In this studi, only two carbon pools were counted (ei : above ground : constant, below ground ; Cairns),
- Calculated net GHG emission with the anticipated effectiveness assumption of the REDD of the life of project

Reporting interval	Stage 1	Stage 2	Stage 3	Stage 4	Stage 5	Stage 6
Success Rate	66	80	90	90	90	90
Leakage Rate	20	20	10	10	10	10

2.3 Landuse analysis and prediction model.

Land use maps of 2003, 2009 and 2015 were produced from the LANDSAT images. It consist of five land cover categories namely: dry forest, wet forest, estate crop, agricultural, non forest, and water bodies. Land use change analysis done by Land change modeler module in TerrSet sotware. Land change modeler is one of many tools which concern for enviromental spatial analyst. We can predict future land cover based on actual and previus land cover, the module also consider some driver variable such as road, river, settlement (distance), eleveation, slopes, etc.

We was conducted the study in the area by cross tabulating the derived land cover map of 2003 as the land cover early input image and the 2015 land cover map as land cover the later input image. Refer **[7, 9],** flowchart of the general procedure used in forest cover change modelling were shown in **Figure 4.**

Potential transition modeling was used to group land use transitions into a set of sub-models and then utilized to explore the potential power of the chosen explanatory variables. In the current study, the change analysis showed that there is only one set of transition, from forest to non forest category.

The explanatory variables chosen for this study are distance from existing forest edge, distance from road, distance from streams, and slope. In order to predict change, the transition was empirically modelled using multi-layer perception (MLP) neural network and logistic regression

Logistic regression model (LRM) was used to model and analyze the lancover change in IDRISI TerrSet. The binary presence or absence is the dependent variable for the periods 2003–2009. The predicted landcover of 2015 was validated using ROC / AUC (Relative Operating Characteristic/Area Under Curve) module of IDRISI TerrSet.

The ROC module compares a suitability image by depicting the likelihood of that class occurrance (the input image) and a boolean image showing where that class actually exists (the reference image). Refer [7, 11] the ROC module compares suitability image by depicting the likelihood of that class occurance (the input image) and a boolean image showing where that class actually exists (the reference image). The ROC curve is the true vs false positive fraction and the AUC is a measurement of overall performance.

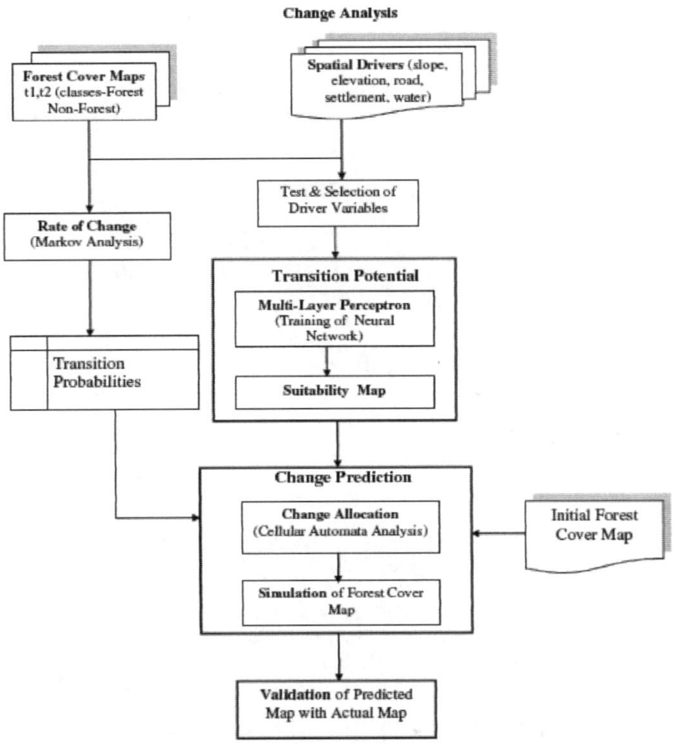

Figure 4. Flowchart of the general procedure used in forest cover change modelling (Source: Refer [9])

2.4 Calculating gas emission

Greenhouse gas emission reductions were calculated by taking the estimated carbon loss without a REDD project intervention and subtracting the estimated carbon that would be saved through a REDD project intervention, along with the estimated carbon loss through leakage. The difference is called additionality, as the net of GHG emissions that are reduced as a result of the REDD project [8]. Refer [2, 7] stage for calculating gas emisision :

- The REDD model utilizes a methodology for calculating and evaluating net anthropogenic greenhouse gas (GHG) emission reductions due to the implementation of a REDD project refer to World Bank's Bio-Carbon Fund Project (BioCF) Methodology for Estimation
- The REDD projects are modelled based on a 30 year projection with intermittent assessments every 5 years, that 6 prediction maps will be produced over the 30 year prediction between 2015 and 2045. The carbon reporting is done based on these intervals.
- Reductions in GHG emissions are calculated by subtracting the estimated carbon that would be saved through REDD project intervention, along with the estimated carbon loss through leakage from the estimated carbon loss (in this case through land use and cover change) without the implementation of a REDD project intervention.
- The formula employed in the calculation is stated by:

$$C - REDD = (C - Baseline) - (C - Actual) - (C - Leakage) \quad (1)$$

Where,

C-Baseline = Baseline greenhouse gas emissions within the project area; tCO2e
C-Actual = Actual greenhouse gas emissions within the project area; tCO2e
C-Leakage = Leakage greenhouse gas emissions; tCO2e

C-REDD = Net anthropogenic greenhouse gas emission reduction attributable to the REDD project activity; tCO2e

In this study, the conservation forest area is assumed to be a REDD project area, as wide as 205,151.49 ha.

3. Results and discussion

3.1 Landcover Change

Landcover change analysis was done for Morowali Utara district data time series comparison data from 2003 until 2015. Table 1, Table 2, and Figure 5 show the land cover and its changes of Morowali Utara District.

Table 1. Morowali Utara district land cover from 2003 to 2015

No	Landcover	2003		2009		2015	
		Ha	Percent	Ha	Percent	Ha	Percent
1	Dry Forest	664,482.24	75.97	638,846.01	73.04	603,161.37	68.96
2	Wet Forest	10,007.73	1.14	10,063.80	1.15	9,311.13	1.06
	Forest	*674,489.97*	*77.11*	*648,909.81*	*74.17*	*612,472.50*	*69.71*
3	Estate Crops	13,783.68	1.58	5,553.27	0.63	22,454.91	2.57
4	Agricultural	100,859.85	11.53	101,665.62	11.62	100,966.41	11.54
5	Non Forest	81,210.06	9.28	114,214.86	13.06	134,449.74	15.37
	Non Forest	*195,853.59*	*22.39*	*221,615.73*	*25.33*	*261,838.26*	*29.8*
6	Water Body	4,321.08	0.49	4,321.08	0.49	4,321.08	0.49
	Total	874,664.64	100.00	874,664.64	100	874,664.64	100.00

The forest cover area in Morowali Utara district in 2015 was 612,472.50 ha or approximately 69.71 % of the total area. It decreased by 62,017.74 ha (9.19%) compared to 2003. The rate of deforestation 5,168.14 ha year^{-1}. The rate of deforestation in this area is lower than deforestation in central sulawesi province. this condition is in accordance with Tumudi's research. Refer [10] Central Sulawesi province has forest area of 4.477 ha (year 2000) and 4.360 ha (year 2011). The rate of deforestation of Central Sulawesi Province in the period 2000-2011 amounted to 117,430 ha or 10.675 ha per year. The largest deforestation occurred in the Tojo Una-Una District up to 29,170 ha (25.01%) and the second, Morowali with 17. 850 ha. This condition shows deforestation in Morowali Utara district contributes for about 15.20% of all deforestation in Central Sulawesi Province.

Dryland forest conditions in Morowali Utrara District in 2003 covered 664,482.24 ha and reduced into 603,161.37 ha in 2015. The reduction of 61,320.87 ha or approximately 9.22 % over the 12 years. The average deforestation of dryland forest occurred in Morowali utara was 0.76 % per year or about 5,110.07 ha per year. This reduction was caused by deforestation which has changed dryland forest into a non forest.

In 2003, wet forest area in Moroowali Utra District was 10,007.73 ha and decreased to 9,311.13 ha in 2015. The reduction of wet forest area was 696.60 ha, or approximately 6.96 % over the 12 years. The average decrease of wet forest was 0.58 % per year, or about 58.05 ha per year. This reduction of wet forest was caused by the deforestation, which has changed the wet forest into non forest. Mostly occur in mangrove forests, converted into ponds.

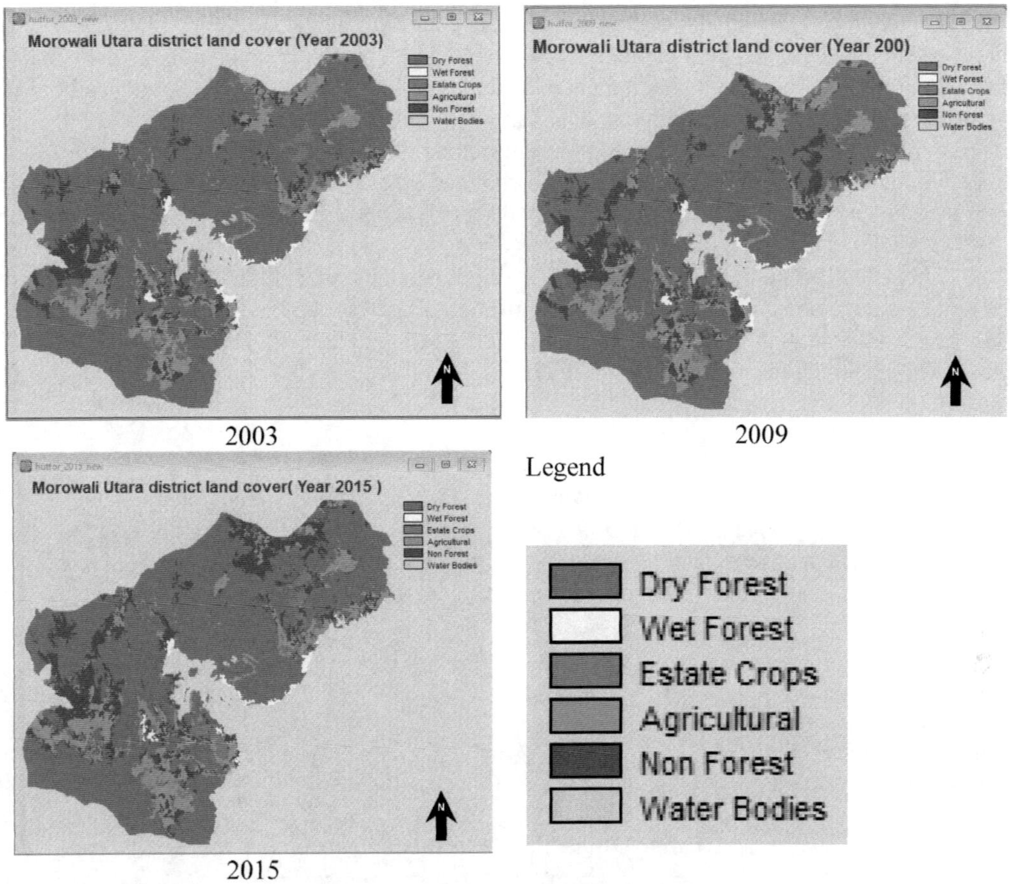

Figure 5. Morowali Utara District Landcover Map of years 2003, 2009 and 2015

Table 2. Morowali Utara district land cover change from 2003 to 2015

No	Land Cover	2003 – 2009		2009 – 2015		2003 – 2015	
		Hectare	Percent	Hectare	Percent	Hectare	Percent
1	Forest	25,580.16	3.79	36,437.31	5.62	62,017.47	9.19
2	Non Forest	-25,762.14	-13.15	-36,437.31	-16.44	-62,017.47	-31.66
3	Water bodies	0.00	0.00	0.00	0.00	0.00	0.00

Source : Result of Analysis of Landcover Map from 2003 to 2015

Over 62,017.47 ha were lost between 2003 and 2015 inside the study area (25,590.16 ha between 2003-2009 and (36,437.31 ha between 2009-2015). This roughly corresponds to 9.19% of the forest area that existed in the year 2003 (674,489.97 ha). Annual deforestation occurred at the rate of 0.76 % between 2003-2015. Between 2003 and 2009 the deforestation gross rate was 0.63%, whereas between 2009-2015 reaching 0.93.

Over the 12 years (2003-2015) the increase of non-forest area was 65,984.67 ha or 33.69 % The cause of deforestation is mining activity. Starting in 2009, there is a significant addition of mining company area. Up to 2016, in morowali uatara district there are as many as 51 nickel mining companies, 29 of the company has production license and just 22 company has exploration license. Mining activities with open pit systems making forest cover changes into non-forest areas. Besides that also in Morowali Utara district there are about 30 plantation companies. Plantation activity also causes the change of forest to non-forest (plantation) cover compared to the condition of the land cover in 2003. The rate of change was 5,498.72 ha per year or the reduction was 2.81% per year compared to non-forest areas condition in 2003. The increase of non-forest areas was caused by the activity of forest land conversion into non-forest areas (other uses).

Meanwhile, there were no change on the water bodies. The water bodies category recorded neither increase nor decrease. Refer [12] conveys that no changes in the body of water in certain period of time indicate that the changes of land cover are mostly oriented on agriculture and new settlements.

Meanwhile [13] declare that the changes of forest cover were caused by the deforestation, either planned or not. Planned deforestation is usually in the form of changes planned by the government for the benefit of forest land for plantations, agricultural or residential development, which is carried out lawfully in accordance with the legislation. Uunplanned deforestation is a deforestation through illegal activities. The forest degradation can be caused by illegal or unauthorized activities, such as harvesting and illegal logging.

3.2 Greenhouse Gas Emission

Based on the results of the land use c not equalhange analysis for the period between 2003 and 2009, the evaluation of the 4 explanatory variables in the transition potential sub-model and the transition potential surface that was subsequently created, the model was calibrated and employed to predict and simulate deforestation an 6-year period from 2009 until the year 2015. The model for predicting the forest cover change,

$$Y1 = -2.2649 + 0.001126 * X_1 - 0.000152 * X_2 - 0.000225 * X_3 + 0.010704 * X_4 \quad (2)$$

where :

X_1 = distance from edge forest

X_2 = distance from roads

X_3 = distance from streams

X_4 = slope

Table 3. Statistic of logistic regression

Statistic of logistic regression	Based on forest changes 2003- 2015
Number of total observations	715,440
-2logL0	230,240.0843
-2log(likelihood)	08,156.8311
Goodness of fit	632,710.3433
Pseudo R^2	0. 5946
Chi-square (*df=5*)	22,083.2531
ROC	0.7408
AUC	0.7516

The validation of the model was undertaken by computing the ROC statistic using a map of actual change between 2003 and 2009 as the reference image and the simulated land cover map of 2015 as the input image. The ROC statistic was obtained at 0.7408, a reasonably strong value that is good enough to make the results of the simulation acceptable. Pseudo R2 value greater (0.5946 and 0.3750) than 0.2 indicates that the model is a relatively good fit for the data. The simulated map is shown in figure 6. The result of that simulation shows that the forest category further decreased from 674,489.97 ha (77.11 %) to 612,472.50 Ha (69.71) invariably reducing the size of the non forest category by the same measure. The rate of deforestation 5,168.12 ha year^{-1}. This is the future business as usual scenario (figure 6). Deforestation has been observed mainly along the main road, in mining areas and close to previously deforested areas. The result of that simulation shows that the forest category further decreased from 674,489.97 ha (77.11 %) to 612,472.50 Ha (69.71) invariably reducing the size of the non forest category by the same measure.

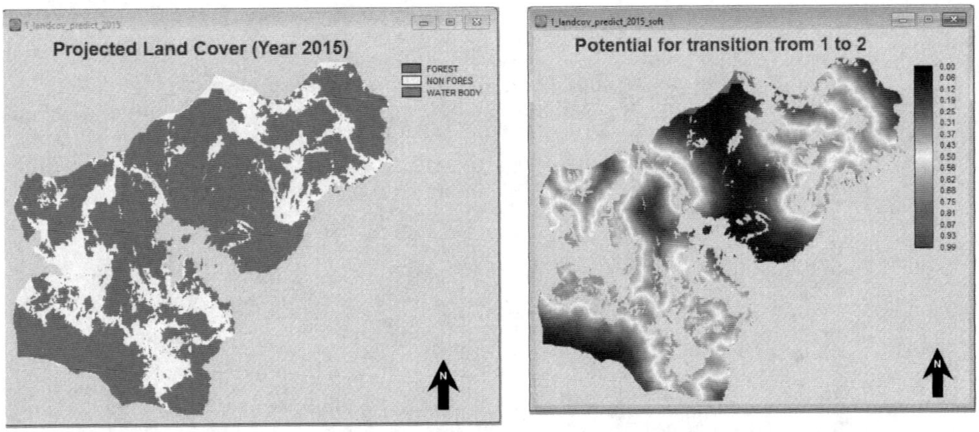

Figure 6: Simulated map and Transition potential Deforested of 2015

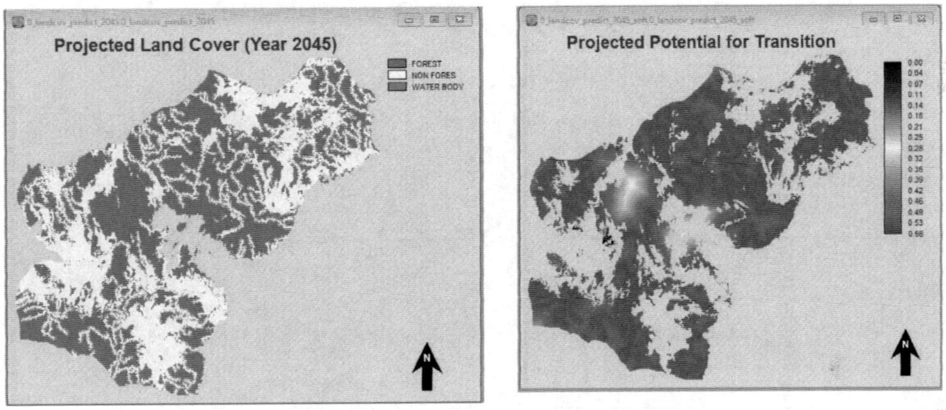

Figure 7. Future BAU and Suitable Deforestation

The result of REDD modeling that the scenario of future sustainable development with the protection of conservation forest area in Morowali Utara District as wide as 205,151.49 ha. For the next 30 years, the cumulative carbon is 18,868,706.00 tCO_2e will be saved or prevented from being released into the atmosphere as the consequences of reduction conversion of forest cover categories to Non-Forest categories until 2045. While continuing usual business scenarios without REDD project intervention, only 26,533,552,00 tCO2e will be saved by 2045. Detail of deforestation and carbon emissions estimates base on time series presented in Table 4 and Figure 8.

Table 4. Ex ante net anthropogenic carbon emission reduction (million t CO2E)

Project Years	Year	Carbon Baseline	Carbon Actual	Carbon Leakage	Carbon REDD
1	2016	- 0.906	- 0.308	- 0.181	- 0.417
6	2021	- 5.349	- 1.704	- 1.070	- 2.575
11	2026	- 9.485	- 2.445	- 1.810	- 5.230
16	2031	- 13.828	- 2.879	- 2.245	- 8.704
21	2036	- 18.256	- 3.322	- 2.687	- 12.247
26	2041	- 22.894	- 3.786	- 3.151	- 15.957

Project Years	Year	Carbon Baseline	Carbon Actual	Carbon Leakage	Carbon REDD
30	2045	- 26.534	- 4.150	- 3.515	- 18.869

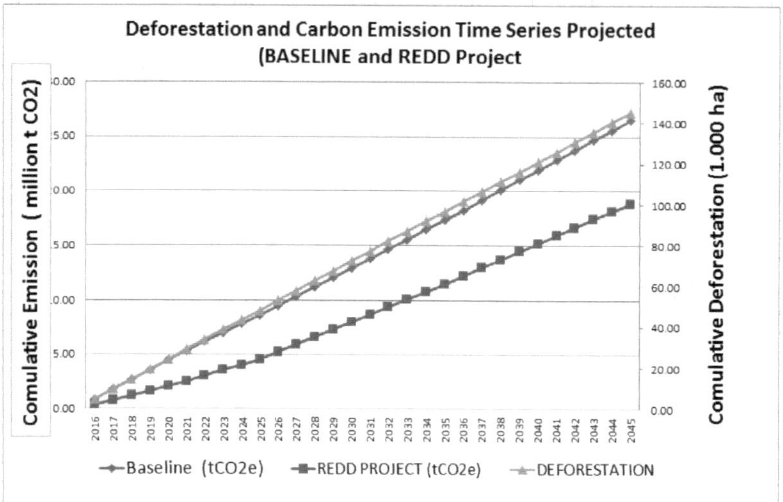

a. Deforestation and Carbon Emission

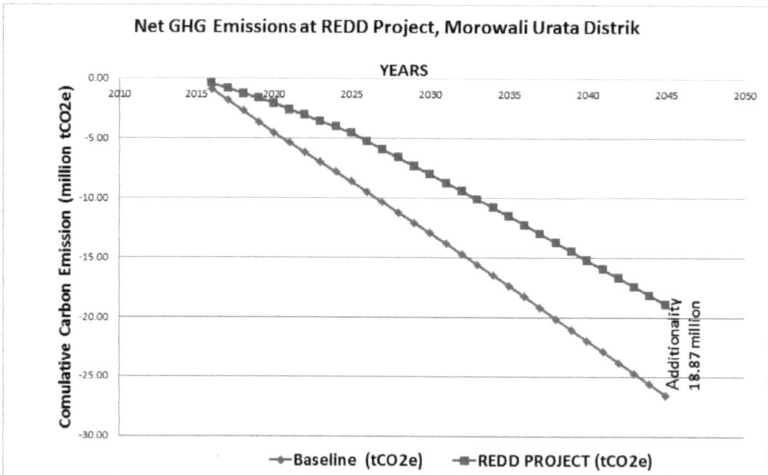

b. Net GHC Emission

Figure 8. Deforestation and Net Emisision at REDD Project, Morawali Utara District

4. Conclusion

Results show that the deforestation on the period 2003-2015 reached 62,017.47 (9,19 %) or the rate of deforestation 5,168.12 ha year^{-1}. The logistic model regression for prediction of deforestation is $Y_1 = -2.2649 + 0.001126X_1 - 0.000152\ X_2 - 0.000225\ X_3 + 0.010704\ X_4$ (ROC = 0,7408), while the amount of carbon that the REDD project will protect from the business as usual deforestation scenario up to 18.868 million tCO2e.

References

[1] Sangermano F, Toledano J and Eastman, J R 2012 Land covers change in the Bolivian Amazon and its implications for REDD+ and endemic biodiversity. *Landscape Ecology*. **27**(4), 571-584. DOI 10.1007/s10980-012-9710-y

[2] Bununu Y A, Ludin A N M and Hosni N Modelling Vegetation Loss And Greenhouse Gas Emissions In Kaduna, Nigeria. 10th SEATUC Symposium 22nd – 24th February 2016, Shibaura Institute of Technology, Tokyo, Japan

[3] Margono B A et al 2014 Primary forest covers loss in Indonesia over 2000- 2012 *Nature Climate Change* Volume **4**. pp. 730-735

[4] Government of Indonesia 2014 National Forest Reference Emission Le,el for Deforestation and Forest Degradation in the Context of the Activities Referred to Decision 1/CP.16. Paragraph 70 (REDD+) Under the UNFCCC. (Encourages developing country Parties to contribute to mitigation actions in the forest sector), Directorate General of Climate Change. The Ministry of Environment and Forestry. Indonesia

[5] Sumargo W, Naggara S G, Nainggolan F A and Apriani I 2011 Potret Keadaan Hutan Indonesia Periode Tahun 2000-2009 Edisi I *Forest Watch Indonesia* Nomor ISBN, 978-979

[6] Eastman J R 2014 Idrisi 18.2. The TerrSet Edition Clark Labs. Worchester, MA

[7] Suryadi I, Rauf A and Maulida M 2012 Provisional Reference Emission Level Provinsi Sulawesi Tengah Tecnical Report. UN-REDD Indonesia Program

[8] Reddy C S, Singh S, Dadhwal V K, Jha C S, Rao N R and Diwakar P G 2017 Predictive modelling of the spatial pattern of past and future forest cover changes in India. *Journal of Earth System Science*. **126**(1), 8

[9] Turmudi and Nahib I 2015 *Potret Hutan Sulawesi Tengah Berdasarkan Data Geospasial*. Buku Mosaik Informasi Geospasial Wilayah Sulawesi Tengah. IPB Press.

[10] Kumar R, Nandy S, Agarwal R and Kushwaha S P S 2014 Forest cover dynamics analysis and prediction modeling using logistic regression model. *Ecological Indicators*. **45,** 444-455.

[11] Nahib I and Widjojo S 2016 The Impact of Landcover Changes on Carbon Stock: A Study Case In Central Kalimantan Forest *36th Asian Conference on Remote Sensing 2015* (ACRS 2015) Quezon City, Metro Manila Philippines 19-23 October 2015 Volume 1 of 6. ISBN: 978-1-5108-1721-0.

[12] Chust G, D Ducrot and J L I Pretus 2004 Land Cover Mapping with Patch-Derived Landscape Indices. *Landscape and Urban Planning*. **67**: 45-53. www.elsevier.com/locate/landurbplan

Acknowledgements
This paper is the development of the activities of " Data and Spatial Information Colllection for Natural Resources of Central Sulawesi Province", which is the cooperation of Development Planning Board of Central Sulawesi Province with Geospatial Information Agency. We are grateful to Head of Centers for Research, Promotion and Cooperation, Geospasial Information Agency (BIG) for the data and laboratory facilities, and and to Head of the Regional Development Planning Board of Central Sulawesi Province, which has financed part of its funding activities.

IOP Conf. Series: Earth and Environmental Science **165** (2018) 012031 doi:10.1088/1755-1315/165/1/012031

UAV Photogrammetry Implementation Based on GNSS CORS UDIP to Enhance Cadastral Surveying and Monitoring Urban Development (Case Study: Ngresep Semarang)

B D Yuwono[1*], **A Suprayogi**[1], **R Azeriansyah**[1] **and D Nukita**[1]

[1]Geodesy Department, Faculty of Engineering, Diponegoro University, Tembalang-Semarang, Postcode 50277, Indonesia

bdyuwono92@gmail.com

Abstract. Unmanned Aerial Vehicle is unique to the field of mechanic which has the ability to fly through the air without a human operator on board, and perform complex and viable option for a number of different tasks, monitoring, coastal environment, cadastre, earthwork analysis. Land Registration is a major issue encountered in developing countries such as Indonesia. Registration of land that had been done in the form of terrestrial measurements expected to be addressed in the next few decades. Monitoring of urban development became a critical problem for urban planning. UAV technology is obtained is used as an alternative to accelerate both processes. The method used in this UAV is photogrammetry to obtain ortho map. The aim of this research is to evaluate the use of UAVs photogrammetry based on GNSS CORS UDIP to enhance cadastral survey and monitoring urban development.

Keyword: Land Registration, UAV Photogrammetry, Orthomap

1. Introduction

The land Surveying which related to the law and land ownership and the definition of property of boundary called cadastral surveying. It consists of interpreting and advising on boundary locations, on the status of land ownership and on the rights, restrictions, and interests in property, as well as the recording of such information for use on plans, maps, etc. It does the physical delineation of property boundaries and determination of dimensions, areas and certain rights associated with properties, whether they are on land, water or defined by natural or artificial features.

Cadastral surveys are generally performed to subdivide the land into parcels for ownership under a land title and to re-establish boundaries of previously surveyed properties to determine the physical extent of ownership or to facilitate the transfer of the property title. Cadastral survey became a part of land registration. Acceleration of registration of land is an urgent issue to be addressed urgently. [1] concluded the latest technology in the field of measurement and mapping and the Community involvement is needed to solve the problem. The method used for measuring the terrestrial mapping exercise with a number of 48 million packages will still be certified [2]. As a result, at the current rate, Indonesia required about 33 years to complete the mapping. Driven by the increased need for information on land in various fields, as well as a number of projects related to increasing land registration agencies with regard to the problem of land became the focus of improvements that are consistent with cost and time savings.

Land administration function can put the needs of the different and varied depending on geography and density of land use [4]. Therefore, when determining the technology and investment choices, the design approach must be "fit-for-purpose" that meets the needs of today's society and can gradually be increased from time to time - not follow the standard exists [5]. The use of UAV orthoimage is also for updating cadastral maps [7].

This research is going to seek to develop a new approach using UAV to support the cadastral boundary data acquisition and enhancing the application of photogrammetry UAV for identifying the hazard and urban development.

2. Data and Methods

The new phenomenon in the technology of photogrammetry applied in positioning is UAV. Unmanned aerial vehicles (UAVs) reached popularity in remote sensing because of they provide a rapid, low-cost and flexible acquisition system and produce a high-resolution data including digital especially of Digital surface models (DSMs), orthoimages and point clouds [3]. UAV can be applied in form 2D and 3D information from acquired images. UAV for 3D Data Recording shown in Figure 1.

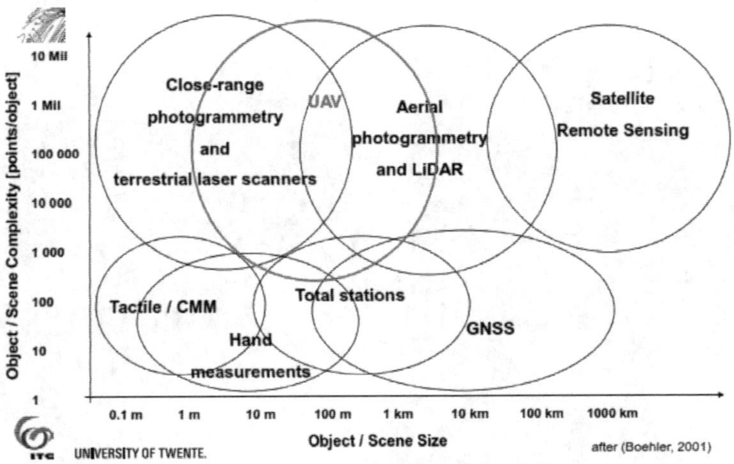

Figure 1. UAV Data for 3D Recording [8]

Cadastral mapping has become a part of the application for UAVs [6]. Cadastral maps show the extent, value and ownership of land, are combinable with a corresponding register and are considered crucial for a continuous and sustainable recording of land rights Using Manual delineation of visible cadastral boundary.is the next level after UAV Processing. An overview of case studies investigating the potential of UAVs for cadastral mapping and their approaches for boundary delineation is provided in

The methodology research framework that has been conceptualized around the central aim of automated feature extraction for cadastral boundary delineation and monitoring urban development. The framework includes four research objectives of which the first one is based on a literature review and the remaining three consist of the technical design and realization of the boundary delineation tool taking into account different spatial data sources.

2.1. Data Capture

The Study used a DJI Phantom 3 Professional equipped with a compact Sony EXMOR 12,4 M camera with a FOV 94° 20 mm lens. The platform has a single-frequency GPS receiver and IMU, which enables registration of approximate components of the external reference for each image. The UAV can fly at the altitude of 75-750 m and the design overlap of pictures it can take ranges from 60 to 90 percent. For control point will be conducted using GPS Geodetic Surve which GNSS CORS UDIP as Base Station. To observed control point using GPS Geodetic Topcon Hiper II and Hiper Gb.

Table 1. Test package

No	Parameter	Description
1.	Number of rows	4
2.	Camera / focal length [mm]	3,61
3.	Overlap (along/cros track direction)[%]	65/50
4.	Flight Altitude [m]	182m
5.	Camera Matrix pixel	1,56 μm

Reseach conducted with test package consisted of 1.387 digital photos taken by 4 independent flights within a single mission in Ngresep, Tembalang shown in table 1. The study measured coordinates of the 15 benchmark (12 for control and 3 for checkpoints. The coordinate of Benchmark was measured with Static GPS survey with two hours observation with the longest baseline about 3,181 km. The root means square error of coordinate was approx. 0.0058 m. Can be seen in Figure 2.

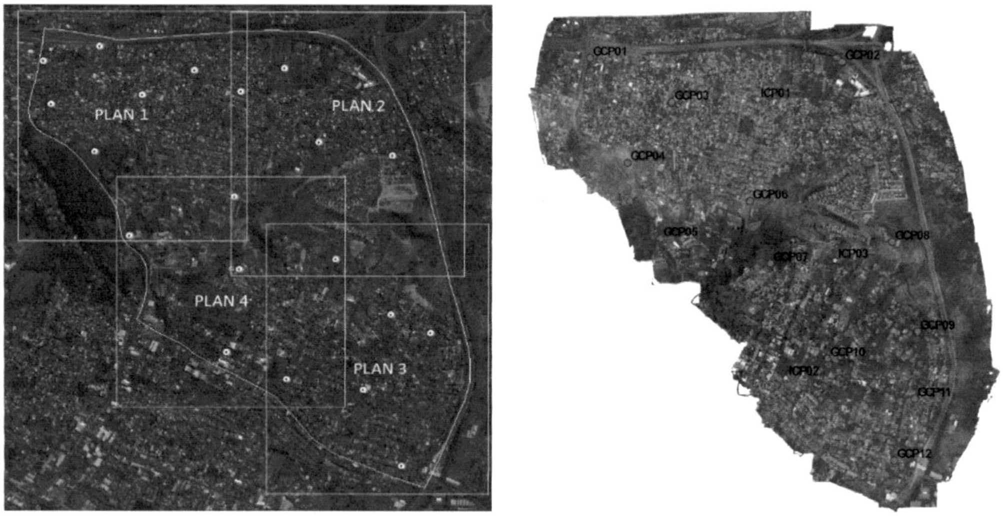

Figure 2. Design Flight and distribution of GCP [10]

2.2. Data Processing

After processing shown the mean error of a typical observation was 1.56 μm / pixel. The root means square errors in X Y positioning of the control points ranges from 0.0019 to 0.0032 And for the independent control point RMS errors ranged from 0.0372 to 0.0607 seen in Table 2.

Table 2. Accuracy of UAV

Items	Decription
σ0 [μm] /[pix]	1,56
Number of Control Points	12
Number of Check Points	3
RMS error for control point (X,Y,Z} [m]	0,0032; 0,0048; 0,0019

Items	Decription
RMS error for control check point (X,Y,Z) [m]	0,0607; 0,0411; 0,0372

2.3. Ground Sample Desain

The digitized results obtained in this study used baseline data from orthophoto obtained from the aerial photography. This aerial photography using UAV aircraft with the value of GSD (Ground Sampling Distance) 7,8648 cm/pixel. GSD is a pixel size value on the ground surface that is used as reference specifications of photographic images. This GSD calculation is based on the SNI aerial photogrammetry which can be calculated as follows:

$$GSD = \frac{Flight\ Height}{Focus\ Lenght} x\ pixel \tag{1}$$

Based on the above formula obtained GSD value of equal to 7.8648 cm.

Geometric precision is done by calculating the value of CE90 and LE90. The formula to find CE90 can be seen in the formula (2) the RMSe used in accordance with RMSe Report results checkpoints on the results orthorectification. The geometry accuracy testing referring to Geospatial Agency Regulation No. 15 the year 2014. About the Technical Guidelines, Basic Map Accuracy can be seen in Table 3.

$$CE90 = 1.5175\ X\ RMSE_r \tag{2}$$

$$LE90 = 1.5175\ X\ RMSE_z \tag{3}$$

Base on equation 2 and 3, It can be conclude that both vertical and horizontal accuracy reach Map scale 1: 1000 class 1. For detail can be seen at Table 3.

Table 3 Accuracy of UAV

RMS_E (m)	**CE90 (m)**	**Maps Scale**	**class**
0.0735	0.1212	1 :1000	1
RMS_Z(m)	**LE90 (m)**	**Maps Scale**	**class**
0.0373	0.0566	1 :1000	1

3. Result and Discussion

3.1. Comparison Land Registration Map and Orthophoto UAV

Characteristic of Land Registration Map is a thematic map, the map that informs about the shape, boundary, location, field number of each parcel and used for field accounting purposes.. The various of cadaster map scale are 1: 1000, 1: 2500 and 10.000. The boundaries of parcels identified on the map should be measured in the field (Regulation of The State Minister of Agrarian Affairs /BPN No.3 Year 1997). One of several the cadastral map functions is as the accounting of land parcels and preventing double registration. Any change or addition of parcel will be on registered map sheet and shall be drawn on the map. It's not necessary to put a building on the cadastral map unless they are used for the reconstruction of the plot boundary.

The analysis is done by comparing Land Registered Map and The delineation boundary from Orthophoto UAV. In this study, we used Land registration map with scale 1: 2500. The stereo plotting process is based on the appearance of the visible object on the aerial photograph. The digitized objects from the aerial photograph are a settlement, housing, vacant lot.

Figure 3. Orthoimage based on UAV imagery with vector layers of boundary vacant lot

To analyze the quality of the geometric orthoimage, accuracy was analyzed in absolute terms by reference to an independent control measurement. Accuracy can be expressed as the root mean square errors:

$$m_{ortho} = \sqrt{\left(\frac{d}{n}\right)} \tag{4}$$

d = deviation means the area base on Land Registration Map and area on the orthoimages to area determined in Orthophoto
n = number point

The analysis taken after comparing the area and the field distance that has been registered, there was delineation method with fix boundary and general boundary methods. Fix boundary is presented by the Land Registration Map of the data plots conducted from the registration measurement land in the field. Besides the general boundary method performed on orthophoto UAV Image. Reference data is the data area of the result of the fixed boundary. The large value of the residuals illustrated the disparities of the boundaries and areas of the parcel which might have caused by the systematic error of the method to delineate the boundaries. The analysis data divided in type vacant lot and residential area. The vacant lot means that there are not have the boundary at surrounding parcels. Base on equation (4) geometric accuracy of orthophoto at vacant area reaches 0.04. Delineation of the Vacant lot can be seen in figure 3. The percentage of area difference average between land registration map and orthophoto is 17 %. The Detail of comparison the parcels of the vacant lot can be seen at Table 4.

Table 4 Vacant lot Comparison

Parcel	Orthophoto (m^2)	Land Registered (m^2)	Differences (%)
1644/1643	379,594	396,594	17
2375	3033,324	3016,317	17

Table 5 shows there are some significant differences exist in the parcel boundary with the parcels Id; 1648, and 1646 with a different area of 8.26 m2 and 7,84m2. The causes significant difference area among others; (1) the one side of the corner boundary grow large trees that pre-mark is not visible, (2) some of the boundary monument has been move. Base on equation (4) geometric of orthophoto at residential area reaches 0.35 [11]. Based on the tolerance of National Land Agency, among 19 parcels studied, there are parcels that fulfill the tolerances and three others do not fulfill tolerances. The percentage of area difference average between land registration map. and orthophoto is 4,10%. Comparative analysis of area to know the difference of area is done by using paired t-test (paired sample t-test). In table 6 shown t_{obs} = 2,324 and t_{crit} = 2,10. *Since* $t_{obs} > t_{crit}$ we reject the null hypothesis and conclude with 95% confidence that the difference in area orthophoto and Land Registered Map has a significant difference seen at Table 6.

Table 5 Comparison UAV Photogrametry and Land Registration Map

Parcel	Orthophoto (m^2)	Land Registered (m^2)	Diffrences %	Parcel	Orthophoto (m^2)	Land Registered (m^2)	Differences %
2491	215,166	220,000	5	1646	182,599	190,436	8
2490	216,909	220,000	3	1645	181,023	186,254	5
2489	221,143	220,000	1	2260	260,987	266,987	6
2488	223,848	220,000	4	2257	198,465	200,000	2
2487	216,769	220,000	3	2256	198,969	200,000	1
2486	219,787	220,000	0	2254	203,003	200,000	3
2485	266,639	264,000	3	2253	196,956	202,259	5
2117	320,337	316,693	4	2250	195,195	203,044	8
1648	220,656	228,934	8	2249	193,976	200,000	6
1647	186,406	183,229	3				

Figure 4. Orthoimage base on UAV imagery with vector layer of residential area

Table 6 t-test (t paired sample test)

	Land Registered	*Orthophoto*
Mean	219,044	216,7806842
Variance	1062,385781	1158,377585
Observations	19	19
Pearson Correlation	0,992815365	
Hypothesized Mean Difference	0	
df	18	

	Land Registered	*Orthophoto*
t Stat	2,324300506	
P(T<=t) one-tail	0,016005735	
t Critical one-tail	1,734063607	
P(T<=t) two-tail	0,032011469	
t Critical two-tail	2,10092204	

3.2. Hazard dan Urban Development Monitoring

The need of land use map due to rapid growth in urbanization and industrialization, there is pressure on land, water, and environment. Urban sprawl may be found everywhere in the major city. The principle difference between land cover and land use. Land cover is the that which covers the surface of the earth and land use describe how the land cover is modified. For urban and spatial planning the land use is indispensable. A map that describes the information on the distribution of land utilization Area. Land use is a tangible manifestation of the influence of human activity on some physical surface of the earth.

Identification of land use and land cover is made easier by using orthophoto UAV image. Caused among others the visualization of the study object becomes clearer and has a large map scale. The results of orthophoto UAV analysis indicate that there are housing, residential area, trades, landslides, and open space was detected clearly. The figure 5 shown comparison between land use map and orthophoto image. The map of Orthophoto shows some changes land use. Changes in the land use include conservation became settlement and conservation area became trade area.

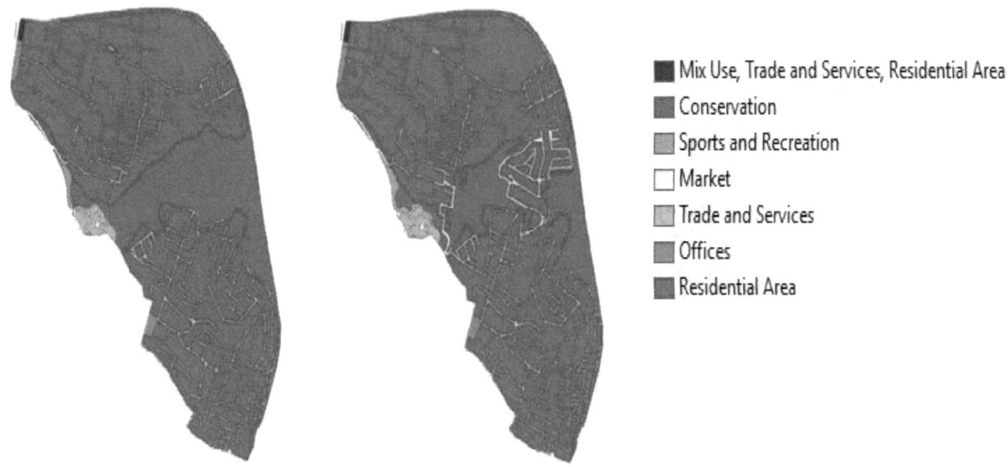

Figure 5. Comparison of landuse area observed by (a) UAV and (b) land use map

The Detail of land use can be seen in figure 6. (a) and (b) Shown the changes area from conservation became the residential area. The figure (c) Increasing of trade and service area.

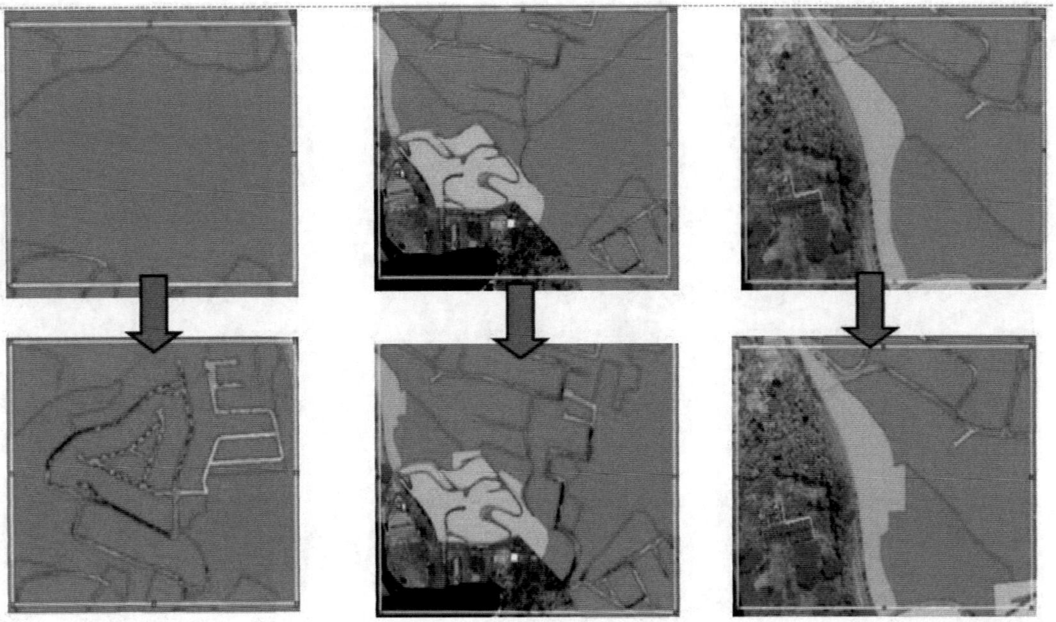

(a) Residential Area (b) Residential Area) (c) Trade and Service

Figure 6. Orthoimage base on UAV imagery

The detail of each area of land use map has been evaluated using Orthophoto. Change for settlement increased by 15.94 ha, while for trade area increased by 0.2 ha.Table 7 shown comparison between orthophoto map and land use map. The map is consist of Mix Use, Trade and service, Residential Area, Conservation, Sport and Recreation, Market, Trade and Services, Offices and Residential Area.Resume of the calculated area can be seen at Table 7.

Table. 7 Comparation between Orthophoto digitation and Land Use Map

Area	Land Use Map (Ha)	Orthophoto (Ha)
Mix Use, Trade and Service,Residential Area	0,395	0,395
Conservation	48,776	32,834
Sport and Recreation	0,893	1,012
Market	0,012	0,012
Trade and Service	2,656	2,863
Office	1,109	1,109
Residential Area	175,312	187,367

Figure 7. Landslide identification base on UAV imagery [10]

Figure 7 show that there are several differences between area observed by UAV and landuse map. On the UAV orthophoto looks that the area is a housing estate area whereas land use map is a consevation area. One of the advantages of the orthophoto observed by UAV map is to identify landslide. Figure 7 about 2 landslide area which located in ngesrep.

4. Conclusion

This research has been conducted to prove that UAV photogrammetry could provide the accuracy for horizontal scale1: 1000. By using UAV can be generated an orthophoto which can be used for delineation of parcels boundaries. This technology can achieve an accuracy of GSD 7.86 cm. The results have sufficient standards to be used as a base map. The result represents the residuals between observed and referenced value from the model fitness toward the data, the smaller the value, the closer the result is to the data. The difficulty encountered from data acquisition with UAV to enhance cadastral mapping is is the delineation of borders to field boundaries. Because actually in field observation not all boundaries of the object clearly visible. UAV platform has good ability in terms of area identification and visualization, especially for land use and landslide hazard.

References.

[1] Anshari R 2007 The Indonesia Land Administration System Decision Makers Meeting : Good Administration of Land in Asia and the Pacific: Land Administration for Poverty Reduction and Economic Growth: Tokyo Japan 21-22 June 2007 1–6

[2] Abidin, H Z, Haroen, T S, Adiyanto F H, Andreas H, Gumilar I, Mudita I and Soemarto I 2015 On the Establishment and Implementation of GPS CORS for Cadastral Surveying and Mapping in Indonesia Survey Review 47(340) 61–70 doi:10.1179/1752270614Y.0000000094

[3] Colomina I and Molina P 2014 Unmanned aerial systems for photogrammetry and remote sensing: A review *ISPRS J Photogrammetri Remote Sensing* **92** 79–97

[4] Enemark S 2013 Fit-for-Purpose: Building Spatial Frameworks for Sustainable and Transparent Land. Governance. *Land and Poverty Conference 2013 The World Bank*. Retrieved from

http://vbn.aau.dk/ws/files/77941985/Washington_2013._Paper._Enemark._WB_Land_and_P over ty_Conf._docx.pdf
[5] Enemark S, Bell K C, Robin C L and McLaren R 2014 Fit-For-Purpose Land Administration ISBN 978-87-92853-11-0 2015 *International Federation of Surveyors (FIG) Geospasial Agency Regulation* No. 15 year 2014
[6] Manyoky M, Theiler P, Steudler D and Eisenbeiss H 2011 Unmanned aerial vehicle in cadastral *applications. Int. Arch. Photogramm. Remote Sens. Spat. Inf. Sci* **63** 1–6
[7] Kedzierski M, Fryskowska A, Wierzbicki D and Delis P 2015 Cadastral Mapping Based on UAV Imagery *15th International Scientifi c and Technical Conference* "From imagery to map: digital photogrammetric technologies"
[8] Nex F and Remondino F 2014 UAV for 3D mapping applications a review https :// www.researchgate.net /publication /260529522
[9] Regulation of The State Minister of Agrarian Affairs / Head of National Land Agency on Provision of Land Registration Pub L No. 3 1997
[10] Reyhan A 2017 Identification Analysis Impat of Landslide hazard using UAV (Aerial Vehicle (UAV) (Case Study: Ngesrep, Semarang) Under gradute Geodetic Department Diponegoro University
[11] Rokhmana C A and Utomo S 2016 The Low-Cost UAV-Based Remote Sensing System Capabilities for Large Scale Cadaster Mapping. *IOP Conference Series: Earth and Environmental Science* numbered section of the paper

Acknowledgment
Authors would like thank Faculty of Engineering Diponegoro University for supporting the research funding, and also BPN Kota Semarang who has permitted field data collection at research area.

IOP Conf. Series: Earth and Environmental Science **165** (2018) 012032 doi:10.1088/1755-1315/165/1/012032

Using Remote Sensing and Geographic Information System (GIS) for Peak Discharge Estimating in Catchment of Way Ratai, Pesawaran District, Lampung Province

Risky Yanuar S[1*], Wahyu Nurbandi[1], Raden Ramadhani Yudha A[1], Brada Irmaning T[1], Artanti Prisma Z[1], Rosyita Alifiya[1], Wisudawan Putra D[1] and Sudaryatno[1]

[1]Departement of Geography Information Science, Universitas Gadjah Mada, Sleman, Special Region of Yogyakarta, 55281, Indonesia

risky.yanuar@gmail.com

Abstract. The catchment of Way Ratai is located in District of Pesawaran, Lampung Province. In the period of the last twenty years, there has occurred three times of flood disasters (1986, 1996, and 2017) in this catchment. Flood in March 2017 gave big impact for the Way Ratai village with the inundation was 1.0 – 1.7 meters. This study aims to estimate runoff coefficient (C) and peak discharge (Qp) using remote sensing and geographic information system (GIS). The research methodology includes cook method to estimate C value, rational method to estimate Qpr, and manning method to measure actual Qpm in field. Landsat 8 OLI imagery, digital elevation model (DEM) of SRTM, and rainfall data of 2014 – 2016 are used for estimate C and Qpr. Qpm is obtained from measure in the field. Qpr is compared with Qpm to analyze potential of flood. The results of this research are the catchment of Way Ratai has C: 0.615, Qpr: 543.45 m^3/s, and Qpm: 194.10 m^3/s. Qpm show river flow capacity is 543.45 m^3/s. Its value is greater than Qpr rational method. It shows that the catchment of Way Ratai has potential of flood.

1. Introduction

Catchment area is a land unit bounded by hills, where the rain that fell above this area received by the stream or river system and streamed through a single outlet channel [1]. Catchment area is a whole ecosystems and hydrological areas that have specific characteristics [2]. One of the characteristics of the catchment area has peak discharge values that can be used as indicator to determine the level of health of a system of river basin, catchment area planning, management, monitoring, evaluation, and to find out the potential of flooding in the catchment area [3].

Various techniques have been used to find out the value of peak discharge, either directly or indirectly [4]. Estimation of peak discharge indirectly can be done using remote sensing data and analysis of geographic information systems (GIS). Remote sensing as one technique to find out the characteristics of the catchment area by image data analysis and interpretation related to hydrological studies, especially in the estimation of peak discharge. Rational methods as one method to determine the value of peak discharge indirectly can be done using remote sensing data.

The rational method in determining peak discharge surface flow considering the time of concentration. Runoff in catchment area has reached disposal point and discharge the peak has been reached if rain fall has been going on during the time of concentration [5]. Concentration time is the time required by the water flowing in the ground from the furthest place in the catchment area to reach the exit point on the area. Rational methods assume the frequency of falling rain and surface flow is the same [6]. The Rational Method is an empirical relation between rainfall intensity and peak discharge flow, the runoff coefficient C is a dimensionless empirical coefficient related to the abstractive and diffusive properties of the basin [7]. Rational methods better use in small catchment area, maximal 800 Ha [8].

Another factor that needs to be taken into account in estimating peak discharge using rational method is the value of surface runoff coefficient (C). It is defined as the ratio between the peak surface runoff against rain intensity. The main factors that affect the value of the surface runoff coefficient is the rate of infiltration of land, land cover, and the intensity of the rain. Prior to the determination of surface runoff coefficient (C), it is necessary to determine the interval of rainfall, the width of the catchment area, type of land use, topography, and the soil properties [6]. Values of surface runoff coefficient can be estimated either by using the cook.

Estimation of discharge with rational method only as describing the discharge peaks in a catchment area in case of maximum intensity at a given time interval, but that value has not been able to describe the capacity of the major rivers accommodating water [9]. The capacity of the River in the catchment area can being estimated by performing a measurement on the field, one of them by using manning methods. The capacity of the river illustrates the culmination of a peak discharge flow of the main stream, where that value as the threshold to determine a peak discharge can cause flooding or not [10]. The manning method estimates the peak discharge value without having to use flow discharge data, but rather by identifying peak flood marks to find out the main river crossings, hydraulic gradients, and channel roughness factors [11].

Recently, a new technique based on the integration of Remote Sensing (RS) and Geographic Information Systems (GIS) has been developed [12]. Remote sensing can used to identify parameters affecting peak discharge such as rainfall, surface runoff coefficient, and broad basin areas, while GIS serves to assist the process of analysis and processing of peak discharge data [8].

This study aims to compare the results of the estimation of peak discharge using remote sensing and geographic information systems with rational methods and measurement results of peak discharge in the field by the method of manning. The results of the comparison of the value of peak discharge from both methods can be used to find out the potential of flooding in the area of study as well as knowing the level of the health catchment area system with surface runoff coefficient values analysis.

2. Material and Method

2.1. Study Site

This research was conducted in catchment area of Way Ratai, Pesawaran District, Lampung Province. Geographically catchment area of Way Ratai is located at position 05°25'12,294" – 05°39'45,064" south and 104°59'53,322" – 105°15'59,444" east. It has an area of 197.074251 km^2. It is a tropical region with an average rainfall of 158.4 mm/month with a wind speed of 2 knots in 2014 and it has a minimum average air temperature of 23.08°C, the average temperature air maximum 32.84°C, average air temperature of 26.83°C, with an average relative air humidity of 81.19%. The average number of rainy days in the Way Ratai catchment is 12.1 days/month. The Way Ratai catchment has a varied topography, ranging from lowland to highland areas, some of which are hilly to mountainous areas with varying heights of sea level. The elevation class found is about 500 to 600 meters above sea level. Based on the slope, it is about >40%. It appears that the Way Ratai catchment has a characteristic rounded basin shape resembling a bowl shape. This indicates that the potential for surface runoff accumulation in the main river with short distances. The analysis results show that the total length of the main river is calculated using tools in ArcGIS and get the value of 27.84 km, while the average length of the main river in each channel of 0.73 km.

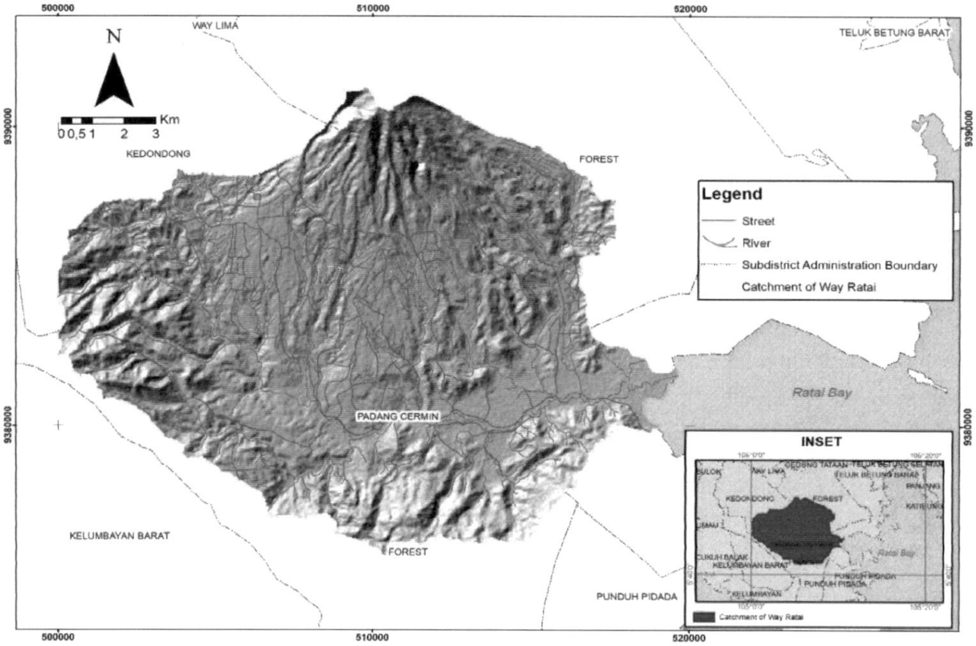

Figure 1. Study site in Way Ratai catchment area

2.2. Data Collecting

2.2.1. SRTM Imagery. SRTM imagery is obtained from the website usgs.glovis.gov is used to determine the topographical condition of the study area. This imagery is selected as the basic data making slope because the ability of radar that can penetrate the vegetation, so that the height data obtained from this image is the actual surface height of the soil [13]. SRTM image is derived into slope and shape information. A land distribution map was used to estimate texture conditions and soil infiltration in the study area. The SRTM image has a spatial resolution of 30 m.

2.2.2. Landsat-8 OLI/TIRS Imagery. The Landsat 8 OLI image is obtained from the usgs.glovis.gov website with 123 rows 64 path recording date June 16, 2015. Landsat 8 OLI imagery is used to obtain land cover information by visual interpretation. Land cover information is used to estimate the percentage of vegetation of each type of land cover [11].

2.2.3. Rainfall Data. Rainfall data from 2014 to 2016 is obtained from the Meteorology, Climatology and Geophysics Agency of Lampung Province. This data is used to analysis the meteorological condition of the study area and determine the intensity value of the design rain in estimating the peak discharge value in the rational method [14].

2.2.4. Field Survey. Field measurement data is measurement of peak discharge by manning method [9]. This data is used to determine the value of the main river capacity in the study area and compare with the estimation data of peak discharge indirectly using the rational method so that it can be known the potential of flood in the study area [15]. The variables measured in the field include the cross-sectional area of the river, the value of the hydraulic radius, the river hydraulic gradient, and the estimated roughness of the river surface.

2.3. SRTM Imagery Slope

The SRTM image is transformed into slope. SRTM image transformation to slope is using ArcGIS 10.3 software. Slope classification is divided into 4 classes i.e. 0-5%, 5-10%, 10% -30% and >30%. The size of the slope pixel follows the SRTM image size of 30 m. slope inclination information is used to determine the effect of surface flow factors.

2.4. Landsat-8 OLI/TIRS Pre-processing

The aim of pre-processing data to make pixel value in ideal condition so that it can be used for both visual and mathematical analysis [16]. Landsat-8 OLI/TIRS images should be performed radiometric calibration and atmospheric correction. The raw image value calibrated pixel value or digital number (DN) of the multispectral sensor must be converted into Top of Atmosphere (TOA) reflectance.

$$L\lambda\ (Landsat - 8) = (ML * Qcal) + AL \qquad (1)$$

Where $L\lambda$ is the TOA spectral radiance (Watts/(m2*srad*μm)), ML is the band-specific multiplication factor rescaling from metadata, AL is the band-specific additive rescaling factor from metadata, and $Qcal$ is the the value of standard pixel products quantized and calibrated (DN) [17]. From TOA spectral radiance, then converted to TOA reflected value

The image processed into required atmospheric correction to the level of at-sensor reflectance. Atmospheric correction of Landsat-8 OLI/TIRS image using histogram adjustment. The histogram evaluated is a histogram of reflected values on the sensor in the form of fractional numbers [18]. The method used in histogram adjustment is dark subtraction [19].

Radiometric corrected satellite imagery is done geometric correction process to improve geometric position accuracy and reduce distortion of earth curvature effect [18]. The first thing to do is to determine the GCP (Ground Control Points) and then done the process of geometric correction. This control point is an object visible to the image as well as visible on the reference map used in geometric correction. This control point can be an object visible to the image as well as seen on the reference map used in geometric correction, such as crossing between rivers and road or crossroads and some other objects clearly visible in the image and reference map [20]. The reference map in this study is a map of the Geospatial Information Agency with a projection of the WGS 1984 Transverse Mercator 49M with a UTM grid. The points taken as much as five points for each image with RMS less than 0.5.

2.5. Maximum Likelihood Classification

The Landsat-8 OLI image is classified as land cover using maximum likelihood algorithm. Maximum likelihood algorithm using base probability calculation or homogeneous object will always display normal distributed histogram [18]. Pixels are described as objects through the shape, size and orientation of samples in feature space [21]. Detailed maximum likelihood algorithm procedure can be found in [11]. The number of land cover classes is divided into 8 classes, namely forest, built-up land, bare land, mangrove forest, plantation, irrigated paddy field, shrubs and sea. This land cover classification uses 200 samples per class of land cover.

2.6. Data Analysis

2.6.1. Flow Surface Coefficient (C). The flow surface coefficient (C) is a number that shows the ratio of the surface flow to the amount of rainfall. Mathematically, the value of C is formulated in equation (2) which refers to the Cook method [22]. C is an indicator showing the physical condition of a catchment area. It has number ranges from 0 to 1. The larger the value of C or near 1 then the more rainfall that becomes the flow of the surface or the amount of water infiltrated less. A large C value indicates a significant probability of erosion. Associated with the level of watershed health, the greater the value of C or more than 0.5 indicates an unhealthy catchment area.

$$\text{Flow coefficient (C)} = \text{surface flow (mm) / rainfall (mm)} \qquad (2)$$

Surface flow coefficient values were obtained using slope parameters, infiltration, vegetation percentage, and flow score. The technique used in calculating the C value is quantitative modelling (overlay) with GIS. The result of modelling is the value of flow coefficient in each unit of land.

Scoring results of each parameter then summed to get the total score. To determine the value of C per unit of land, the total score of each unit of land is multiplied by the weighting factor of each unit of land. The weighting factor is obtained by equation (3). The value of C DAS is the total of the total C of each unit of land. **Table 1. – 4.** shows some parameters and its score [23].

$$\text{Weight factor per unit of land} = \text{Area of Land} / \text{Width of catchment} \tag{3}$$

Table 1. Slope parameter

Slope class	Relief configuration	Score
0 – 5 %	Flat	10
5 – 10%	Wavy	20
10 – 30%	Hills	30
>30%	Rugged and Crude	40

Table 2. Infiltration parameter

Infiltration	Terrain characteristic	Score
Fast	Sand in or other land that can absorb faster	5
Medium	Deep clay with infiltration approximately setipe with prairie ground	10
Slow	Clay / other soils with low infiltration capacity	15
Very slow (ignored)	No effective ground cover, thin, ground layer	20

Table 3. Land Cover parameter

Land cover	Score
Approximately 90% covered either by wood	5
Approximately 50% covered both by trees and grasses	10
Crop cover little, no crops and natural cover little	15
No effective cover or similar	20

Table 4. Flow Density parameter

Class	Flow density (mil/mil2)	Score
High	>5	5
Normal	2 – 5	10
Low	1 – 2	15
Ignored	<1	20

2.6.2. Rational Method Peak Discharge. Estimate the peak discharge by rational method using the following equation.

$$Qp = 0.278\ C \times I \times A \tag{4}$$

where Qp: peak discharge (m^3/s), A: area of river basin (km^2), I: maximum rain intensity for the same time concentration (mm/h), C: surface flow coefficient, and 0.278: coefficient.

Rain intensity values are obtained through the following equation.

$$I = (R / 24) \times (24 / Tc)\ 0.67 \tag{5}$$

where I: intensity of rain plan during concentration time (mm/h), R: daily rain plan (mm), Tc: concentration time (hours).

The rain calculation plan uses the Log Pearson Type III technique with the following equation.

$$\text{Log Ptr} = R + KS \tag{6}$$

where Ptr: rain plan with return period Tr, R: average rainfall based on population sample, K: frequency factor, S: standard deviation.

Calculation time of concentration using equation as follows.

$$Tc = L / V \quad (7)$$

Where L: long river main river basin (m), V: 72 x (H x L) 0.6 with H being the height difference of the watershed (m).

2.6.3. Manning Method Peak Discharge

$$Qp = \frac{1}{n} x A x R^{2/3} x S^{1/2} \quad (8)$$

where Qp: peak discharge (m^3/s), n: roughness coefficient of river channel surface, A: sectional area of the river on the former flood (m), R: hydraulic radius (m), S: hydraulic gradient.
The value of the hydraulic radius is obtained by the following equation.

$$R = A / P \quad (9)$$

where A: sectional area of the river on the former flood (m^2), P: wet perimeter (m)
The value of the hydraulic gradient is obtained by the following equation.

$$S = H / L \quad (10)$$

where H: height difference of river surface (m), L: length of measurement (m)
The coarse coefficient value is obtained by equation (9). The amount of coefficient value based on field observation with value determination is shown in **Table 5.** [10].

$$.n = (n0 + n1 + n2 + n3 + n4)\, m5 \quad (11)$$

where .n: value of the roughness of the channel, n0: basic material, n1: level of channel uniformity, n2: cross section variation of channels, n3: effect of narrowing on transverse sections, n4: plants, and n5: level of meander.

Table 5. Manning Roughness Coefficient

Channel condition	Characteristic	n	Coefficient
Basic material	Soil	.n0	0.02
	Rock		0.024
	Smooth gravel		0.025
	Rough gravel		0.028
Channel uniformity	Smooth	.n1	0
	Low smooth		0.005
	Medium		0.01
	Coarse		0.02
Cross sectional variation	Very slow	.n2	0
	Slow		0.005
	Medium		
	Fast		0.01 – 0.015
Effect of narrowing	Ignored	.n3	0
	Low influence		0.01 – 0.015
	Medium influence		0.02 – 0.03
	High influence		0.04 – 0.06
Plants	Low	.n4	0.0005 – 0.01
	Medium		0.01 – 0.025
	High		0.025 – 0.05
	Very high		0.05 – 0.1
Meander	Low	.n5	1
	Medium		1.15
	High		1.3

3. Result and Discussion

3.1. Flow Surface Parameters

There are four parameters considered in determining the runoff coefficient using Cook method of slope, soil texture, land cover, and drainage density. **Figure 2d.** Represents slope information in Way Ratai catchment area. It appears that the study area has a gradient of slope that varies with the dominance of slope >10%. There is a difference in the value of the slope between the flat area and the surrounding area. The flat area or lowest slope (0 - 5%) is located in the main stream to the downstream. A firm slope divergence indicates that surface runoff will accumulate rapidly in the region with the lowest slope.

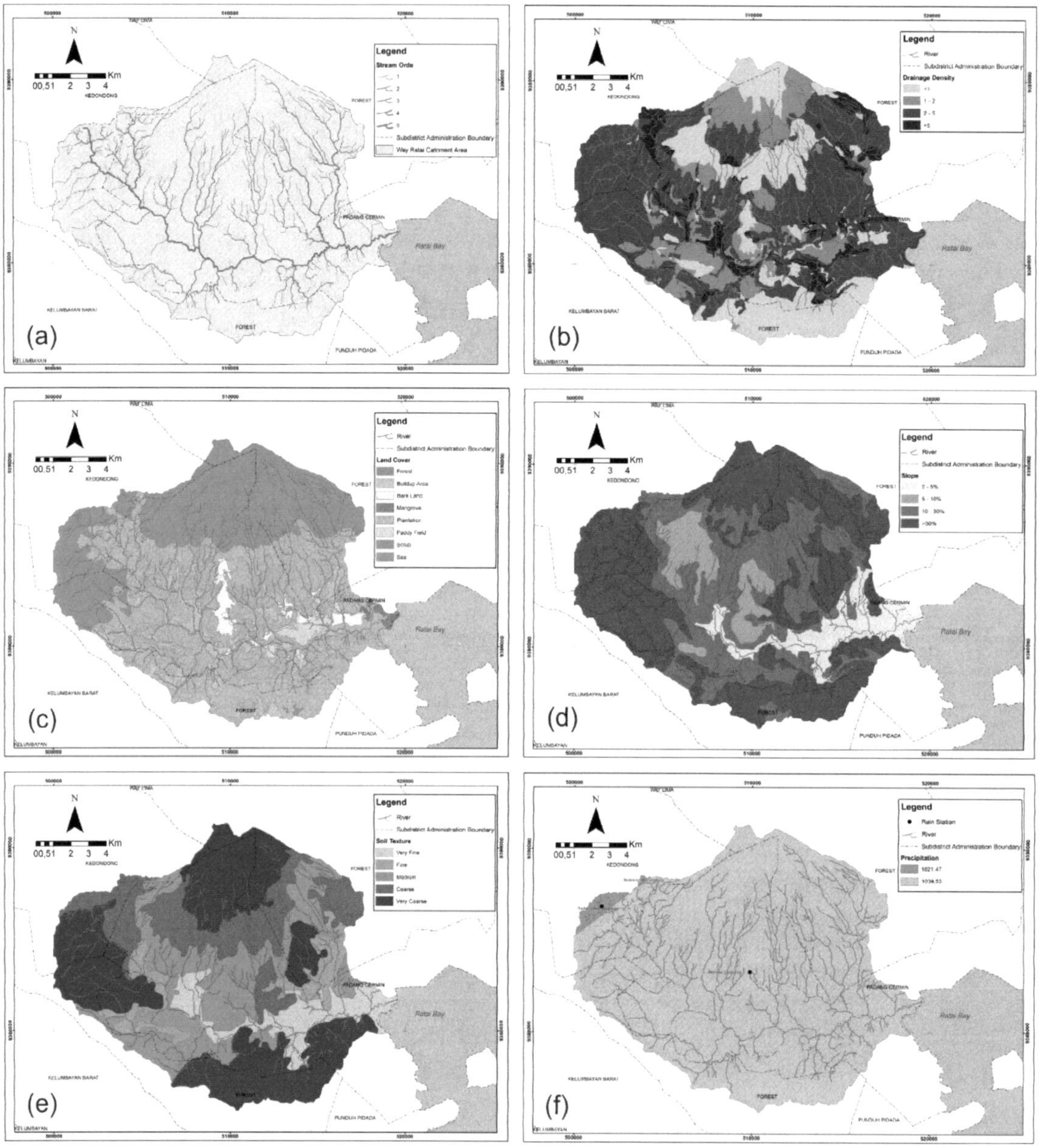

Figure 2. Way Ratai catchment (a) stream order, (b) drainage density, (c) land cover, (d) slope, (e) soil texture and (f) polygon thiessen map. These six paramaters for flow surface coefficient variable using Cook method.

The slope conditions of the Way Ratai catchment are influenced by the landform formed by volcanic and fluvial processes. The northern and western regions of the Way Ratai catchment have a slope of >30%. This is due to the inactive volcanoes of Mount Ratai in the northern part of Way Ratai catchment and the volcanic hills degraded in the western part of Way Ratai catchment. The central region of Way Ratai catchment has a slope of 0-5%. This is due to the presence of alluvial plains that are the main stream of Way Ratai. Based on **Figure 2d.** the direction of the Way Ratai stream flows leads to the eastern part of the catchment due to the lower slopes in the eastern part of the catchment.

Based on **Figure 2d.** the drainage density of Way Ratai catchment is influenced by the slope characteristic. It is shown that the steeper the slope will be the higher the drainage density. The northern and western part of the Way Ratai catchment has a high drainage density level proportional to the slope level, while the middle of the Way Ratai catchment has low drainage density with sloping slopes.

Based on **Figure 2e**. the soil texture in the north, west and south of Way Ratai catchment is dominated by very rough soil texture. This is due to the formation of depressed volcanic hills and volcanic mountains and steep slopes. The soil texture in the middle of the Way Ratai River catchment is dominated by a fine and very fine soil texture. The texture of the soil is very fine and fine it is in the landform of alluvial plains and flat slope. **Figure 2e.** shows the closer to the main river Way Ratai then the texture of the soil will be finer. This geomorphological process has influenced the soil texture characteristics of Way Ratai stream by contrast.

Distribution information of relatively soil texture obtained by considering the parameters of landform and slope. There are four relatively soil texture classes that are soft, coarse, and very coarse. Based on Figure 3. Way Ratai catchment area is dominated by relatively coarse texture - very coarse. There is an area with a fine texture is a region with a relatively flat. The relative texture value of the soil is reduced to the information of the soil infiltration rate. The assumption is built that the more coarse soil texture the higher the infiltration rate, the more fine the texture of the soil the lower the infiltration rate. Referring to the relatively soil texture distribution in **Figure 2e.** Then most Way Ratai catchment area have high infiltration rates - very high, while low infiltration rates are present in areas of low slope and fine texture.

Land cover that has a high density vegetation presentation in the form of timber will have an impact on low surface runoff. Land cover classes that have low surface runoff impacts are plantations and forests except for mangrove forests due to flooded areas, while land cover that has a high impact of surface runoff i.e. bare land, irrigated paddy fields and built-up land. The absorption capacity of the surface flow to the cover is influenced by the presence of vegetation which can absorb the surface flow so that no surface runoff will occur.

Figure 4. Represents conditions of Way Ratai catchment area dominated by land cover of plantation. The northern region of the Way Ratai Catchment area has bush cover and forest cover. This shrub cover is located on the lower slopes and the slopes of Mount Pesawaran. The southern region of Way Ratai catchment area has a land cover that is plantation and built up area. The central region of the Way Ratai Catchment area consists of built-up land, irrigated paddy fields and bare land. The built up area in the middle of the Way Ratai Catchment area follows the main stream of Way Ratai. There is a mangrove forest in the eastern part of the vast Way Ratai catchment area. This mangrove forest is located in the upstream of Way Ratai Catchment area or in the form of alluvial coastal plains.

Based on **Figure 2c.** the area with the land cover has a high impact of surface runoff i.e. the middle of Way Ratai catchment. In the middle of Way Ratai catchment there is built-up land, bare land and irrigated paddy fields which are potential for surface runoff, while the northern and western part of Way Ratai catchment is plantation and forest so there is no potential for surface runoff.

Drainage density is used as one of the determinants of the runoff coefficient Cook method. Drainage density is obtained by comparing the length of the stream in each unit of land with the area of the land. Assumption of this measurements is more dense that drainage, the area is considered to have a high runoff. Units of high density are considered high if their drainage density is above 5 mil/mil^2 so that it is given a score of 20, whereas for units of land whose drainage density below 1 mile/mil^2 is considered to have very low value and given a score of 5. In units of land with normal drainage density values of 2-5 mil/mil^2 are given a score of 15 and that have a low value of 1-2 miles/mil^2 given a score of 10.

Drainage density in the study area has a very diverse value. In the area around Way Ratai tends to have high drainage density value, it is caused by the number of tributaries that go to Way Ratai. Visually the study area is dominated by normal drainage density classes in the range 2-5 mil/mil^2. The total area of this class is 99.490419 km^2 while the total area of study is 197.074246 km^2 or almost half of it is an area with normal drainage density. Based on this it can be assessed if the Way Ratai catchment area has a normally tended surface flow, only a few areas of high or low drainage density.

3.2. Flow Surface Coefficient (C)

The runoff coefficient of the cook method obtained based on the above input parameters shows a considerable result that is 61.5% or 0.615. The value indicates that 61.5% of the water falling in the study area is likely to become a runoff. The above runoff coefficient indirectly expresses the characteristic of the Way Ratai Catchment area which tends to show that falling rainwater will form a surface stream compared to infiltrate into ground water. If analysis based on four input parameters, slope parameter is analysis as the most influential parameter to high runoff coefficient value. This is in accordance with the prediction because the study area is surrounded by hills and has a high slope of the slope it will produce a high surface flow.

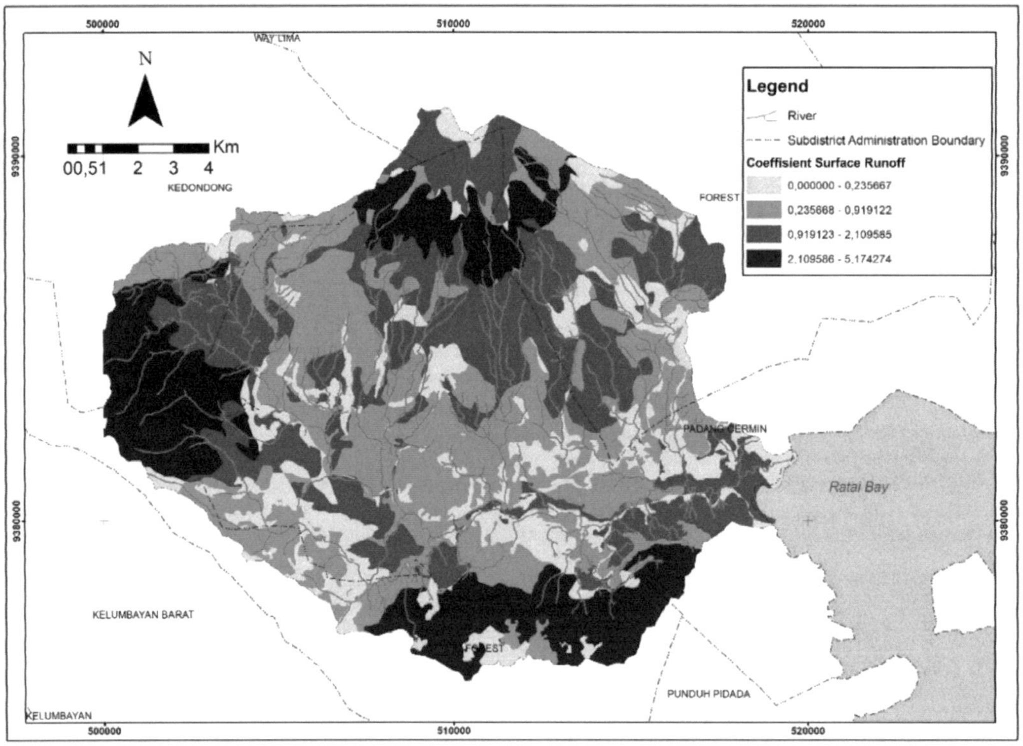

Figure 3. Way Ratai catchment flow surface coefficient map

The coefficient of runoff in each unit of land is in the range 0 to 5.174%. Based on its distribution, units of land with high flow coefficient values or in the range of values of 2.11 - 5.174% are located on the part of Pesawaran Mount and structural hills on the west and south sides of the study area. Compared to slope gradient maps in areas where high flow coefficient values lie in high slope classes, this further strengthens the assumption that high slope gradients are the reason for the large surface runoff coefficients.

Figure 3. shows that the middle section is dominated by a moderate to low-grade runoff coefficient class, due either to a low slope or in grades 0 to 10%. When compared to other parameters, in the middle of the catchment area basically has a high score or influence on large surface flows. For example, in infiltration parameters dominated by moderate to very low infiltration rates, as well as land cover parameters dominated by open-field and built up area that are considered to have no effective vegetation cover or the like, increasing the likelihood of surface runoff. However, with a small slope inclination, the movement of the water tends to be slow and ultimately remains infiltrated, although other parameters show a large impact on the formation of surface flow.

3.3. Rational and Manning Method Peak Discharge

The result of runoff coefficient value with the Cook method is 0.615 (**Table 6**.). C values greater than 0.5 indicate that catchment area conditions in the study area already have high surface flow potential. A value of 0.615 indicates that 61.5% of water derived from rainfall flows in the form of a runoff. This means that the amount of water absorbed into the soil is less than the amount of water that becomes runoff. Judging from the level of catchment area, the value of coefficient 0.615 indicates the existence of damage catchment area management. The result of the analysis shows that the slope parameter has a considerable influence on the value of C obtained. This is characterized by a rounded catchment area shape such as a bowl with a fairly firm slope between the downstream with the center and the upstream. The High slope enough to cause rain water quickly to be runoff than infiltrated. The watershed-like watershed condition results in accumulated rapid water downstream or the lowest slope.

Table 6. Rational Method Result

C	I (mm/hour)	A (km^2)	Qp (m^3/s)
0.615	58.06	197.07	543.45

(Source: Data Processing, 2017)

Table 7. Manning Method Result

n	A (m^2)	R (m)	S	Qp (m^3/s)
0.083	163.91	5.33	0.0014	194.10

(Source: Data Processing, 2017)

The peak discharge measurements with rational method resulted in values of 543.45 m^3/s and Manning method of 194.10 m^3/s. The peak discharge value of the Manning method signifies the value of the stream capacity in holding water. The result of measuring the value of stream capacity is smaller than the peak discharge of the rational method. Higher peak discharge of the stream capacity indicates that the study area has potential flooding. Based on the results of field observations, there have been floods in other areas of study in March 2017. This indicates that there is alignment between the results of flood potential analysis using remote sensing with actual flood events. Based on interviews, the flood that occurred in 2017 was influenced by the increasingly shallow stream conditions. The silting of the river causes the river capacity to decrease, so as to reduce the potential for flooding, one of the efforts that can be done is to rehabilitate the river by dredging the river sediment material to increase the river capacity.

4. Conclusion

The Way Ratai catchment area has a surface coefficient of 0.615 which indicates the condition of the Way Ratai Catchment area is unhealthy. Based on four parameters used, slope parameter is the most influential parameter to high surface coefficient value. The coefficient of surface flow in each unit of land is in the range 0 - 5,174%. The peak discharge measurements with rational method resulted in values of 543.45 m^3/s and Manning method of 194.10 m^3/s. The result of the measurement of stream capacity value is smaller than the peak discharge of rational method result. Higher peak discharge of the stream capacity indicates that the study area has potential flooding.

References

[1] Triadmodjo Bambang 2010 *Applied Hydrology: Second Edition* (Yogyakarta: Beta Offset Yogyakarta) p 47.

[2] Maryono Agus 2005 *Handling Flood, Drought, and Environment* (Yogyakarta: Gadjah Mada University Press) p 23.

[3] Pramono I B, Wahyuningrum N, and Wuryanta, A 2010 Application of Rational Methods for Peak Discharge Estimated on Several Sub-Basin Areas *Jurnal Penelitian Hutan dan Konservasi Alam* vol **7**(2) pp 161-176.

[4] Chow V T, Maidment D R and Mays L W 1988 *Applied Hydrology* (New York: Mc. Graw Hill International Edition. Civil Engineering Series) p 45.

[5] Haridjaja O, Murtilaksono K, Sudarmono, and Rachman L M 1990 *Hydrology Agriculture* (Bogor: Departement of Soil Science, University of Agriculture Bogor (IPB)) p 72.

[6] Arsyad S 2010 *Soil and Water Conservation* (Bogor: University of Agriculture Bogor (IPB)) p 14.

[7] Hayes D and Young R 2006 *Comparison of Peak Discharge and Runoff Characteristic Estimates from The Rational Method to Field Observations for Small Basins in Central Virginia* (US Geological Survey: Scientific Investigations Report 2005-5254).

[8] Sigit Heru Murti, Ach Firyal Wijdani, Aisya Jaya D, Andika Putri F, Assyria Fahsya U, Dian Prabantoro, Dzimar A R P, Nila Ratnasari 2015 Estimation of Peak Discharge for Identification of Potential Flood in Jangkok Basin using Remote Sensing and GIS *Proc. Geoinformation Science National Symposium* vol **1** pp 176-186.

[9] Gunawan T 1991 Application of Remote Sensing Technique to Assume Peak Discharge: Case Study in Bengawan Solo Upper Basin of Central Java *Dissertation* (Bogor: University of Agriculture Bogor (IPB)) p 79.

[10] Rahim S E 2006 *Control of Soil Erosion: In the Framework of Environmental Conservation* (Jakarta: PT. Bumi Aksara) p 34.

[11] Gao Jay 2010 *Digital Analysis of Remotely Sensed Imagery* (New York: McGrawHill) p 125.

[12] Kants Meshrams, Doharer and Singhs 2015 Morphometric Analysis of Sonar Sub-basin Using SRTM Data and Geographical Information System (GIS) *African journal agriculture resolution* vol **10** pp 1401–1406.

[13] USGS 2015 *Shuttle Radar Topography Mission (SRTM) 1 Arc-Second Global* Available online: (https://lta.cr.usgs.gov/SRTM1Arc) (accessed: 20 September 2017).

[14] Pratisto Arif and Danoedoro Projo 2008 Impact of Land Use Change on Debit Response and Flood Hazard (Case Study in Gesing River Basin, Purworejo) *Pertemuan Ilmiah Tahunan MAPIN XII ITB Bandung* vol **1** pp 1-11.

[15] Lee R 1980 *Forest Hydrology* (Yogyakarta: Gadjah Mada University Press) p 36.

[16] Campbell James B and Wynne Randolph H 2011 *Introduction to Remote Sensing* (New York: The Guilford Press) p 305.

[17] USGS 2006 Multi-resolution Land Characteristic 2001 (MRLC2001) Image Processing Procedure Available online: (http://landcover.usgs.gov/pdf/image_preprocessing.pdf) (accessed: 5 September 2017).

[18] Danoedoro Projo 2012 *Introduction to Digital Image Processing* (Yogyakarta: Andi Press) p 168 and 186.

[19] Xu H 2008 A New Indice for Delineating Built-Up Land Features in Satellite Imagery *Int. Journal of Remote Sensing* vol **29**(14) pp 4269-4276.

[20] Lillesand Thomas M, Kiefer Ralph W, and Chipman Jonathan W 2003 *Remote sensing and Image Interpretation 5th ed.* (New York: John Wiley & Sons, Inc) p 498.

[21] Shresta D P 1991 *Digital Image Processing* (Enschede: International Institute for Aerospace

Survey and Earth Sciences) p 103.
[22] Asdak C 1995 *Hydrology and Watershed Management* (Yogyakarta: Gadjah Mada University Press) p 20.
[23] Gunawan T, Sudaryatno, Gisarsih S 2012 *Guidelines for Watershed Characteristics of Cooperative Research Report* (Jakarta: Ministry of Forestry of the Republic of Indonesia) p 15.

Acknowledgments
We are thankful to Mr. Sudaryatno, Cartography and Remote Sensing field work study (KKL 3) Comitte, and Departement of Geography Information Science, Faculty of Geography, Universitas Gadjah Mada whose support to this research. Many thanks are also goes to US Geological Survey (USGS) to give opportunity for the authors to access remote sensing data, such as Landsat-8 OLI/TIRS and SRTM imagery, Climatology Meterology and Geophysic Agency (BMKG) for rainfall data and also to Geospatial Information Agency (BIG) for map of Rupabumi Indonesia.

Analysis of Vegetation Density Effect In Bengawan Solo Watershed To The Total Suspended Solid (TSS) In Gajah Mungkur Reservoir

Bambang Sudarsono[1,2*], **Abdi Sukmono**[1,2] **and Aziz Anjar Santoso**[1]

[1]Department of Geodesy Engineering, Faculty of Engineering, Diponegoro University Semarang, Indonesia

[2]Center for Spatial Data Infrastructure Development (PPIDS) of Diponegoro University, Indonesia

bambangf22@gmail.com

Abstract. Gajah Mungkur Reservoir located in Wonogiri Regency Central Java Province. This is a multipurpose reservoir with an area of approximately 8,800 ha. This reservoir can irrigate 23.600 ha of rice fields in Sukoharjo, Klaten, Karanganyar and Sragen districts. The amount of sedimentation increased rapidly resulting in degradation of water quality of Gajah Mungkur Reservoir. High sedimentation can be shown from the high value of Total Suspended Solid (TSS) in the waters. TSS is a solid material, including organic and inorganic matter suspended in aquatic area. TSS Dynamics in the waters of Gajah Mungkur Reservoir can not be separated from the dynamics of vegetation density change in Watershed area (DAS). The watershed that flows into the Gajah Mungkur Reservoir is Bengawan solo Watershed.Total Suspended Solid (TSS) concentration can be identified using Total Suspended Solid (TSS) algorithm in Landsat 8 Satellite imagery. This research is monitored by TSS development in 2013, 2015, and 2017 which then compared with TSS level in the field. Vegetation density was identified using the Normalized Difference Vegetation Index (NDVI) method.Based on the results obtained the most suitable algorithm in the waters of Gajah Mungkur Reservoir is TSS Syarif Budhiman algorithm with 92% regression coefficient. The concentration of TSS in the waters of Gajah Mungkur Reservoir in the period of 2013-2017 has increased. Then the result of vegetation density in Bengawan Solo watershed showed that dominates is a fairly tight class. While the density of vegetation in the watershed that flows into the Gajah Mungkur Reservoir in 2013-2017 significant changes also some Sub watershed that decreased vegetation density occurred in the sub watershed Keduang, Solo Hulu and Wuryantoro. The vegetation density of several sub watersheds has an effect on the content of TSS in Gajah reservoir at estuary of the sub-watershed.

1. Introduction

Siltation is a common problem in reservoir buildings in Indonesia. One of the reservoirs in Indonesia that has serious problems with silting is the Gajah Mungkur Reservoir of Wonogiri Regency of Central Java Province. One of the factors causing the silting problem in Gajah Mungkur Reservoir is the increasing amount of sedimentation. This high sedimentation is due to the high erosion charge of the rivers flowing into the Gajah Mungkur reservoir. The rapidly increasing and uncontrolled sedimentation resulted water quality degradation in Gajah Mungkur Reservoir. One of them is increasing Total

Suspended Solid (TSS) content in reservoir area from year to year. Although there has been a normalization of reservoirs every year but high sedimentation continues in this reservoir [1].

TSS Dynamics in the waters of Gajah Mungkur Reservoir cannot be separated from the dynamics of soil surface erosion, erosion of cliff, slope avalanche, river bank erosion, and erosion of side of road. Many locations of cliffs and landslides in the catchment area, the slopes (cliffs) of the Watershed (DAS) area. The Watershed (DAS) is the total land and water surface bounded by a topographical water boundary and one way to contribute to the discharge of a river in a certain cross section. Watershed catchments are not only run-off, but also the erosion and transportation of chemicals.

Bengawan Solo watershed (DAS) is a watershed that has a stream to the Gajah Mungkur reservoir. There are at least ten Sub watersheds (Sub DAS) that flows and empties into the Gajah Mungkur reservoir. The sub watersheds include Sub-Watershed of Alang, Sub-Watershed of Durensewu, Sub-Watershed of Keduang, Sub-Watershed of Kedung Guling, Sub-Watershed of Kepuh, Sub-Watershed of Pondok, Sub-Watershed of Solohulu, Sub-Watershed of Temon, Sub-Watershed of Wiroko and Sub-Watershed of Wuryantoro. They supply water discharge and sedimentation to the Gajah Mungkur reservoir.The level of erosion hazard can be affected by several factors, one of which is the vegetation density. Decreasing of vegetation density in watershed can increase erosion hazard in the watershed. Because functionally vegetation serves as an erosion barrier in the watershed area [2].

Several sub-watersheds (Sub-DAS) areas that flows into the Gajah Mungkur Reservoir are fairly developed areas. The change of land use in this Sub DAS area triggers a decrease in vegetation density which can result in increased erosion hazard in Gajah Mungkur reservoir. In this research will be studied the development of sedimentation in Gajah Mungkur reservoir in multi temporal (2013, 2015, 2017) so that the pattern of sedimentation will be seen. Sedimentation observation method using remote sensing method of TSS Landsat-8 image algorithm from several models (syarif budiman, parwati, Woerd pastercam). Some of these models will be tested for the best approach compared to in situ TSS. The result of this best algorithm model is then used to observe the development of TSS in Gajah Mungkur Reservoir in multi temporal.

Then the development of the pattern of vegetation density change in the Bengawan Solo watershed was also observed using the NDVI Landsat-8 image technique in the same year (2013, 2015, and 2017). This observation was conducted on each sub-basin unit.Both are then linked to determine the relationship of vegetation density changes in each sub-catchment to erosion runoff results in sedimentation at the Sub-Basin estuary found in the Mungkur elephant reservoir. The result of this research is expected to give direction of sedimentation handling solution in Gajah Mungkur reservoir. The purpose and purpose of this study are: 1. Mapping of vegetation density of the Watershed which leads to Gajah Mungkur Reservoir 2. Determine the best Suspended Solid Total Algorithm for the decline in Total Suspended Solid value in the waters of the Gajah Mungkur Reservoir 3. Mapping the distribution of Total Suspended Solid (TSS) concentrations in the Gajah Mungkur Reservoir in 2013, 2015, and 2017. Then from these results the objective of this research is knowing the influence of vegetation density in every sub-watershed to the Total Suspended Solid (TSS) concentration changing in Gajah mungkur reservoir using Landsat-8 imagery.

2. Methodology

2.1. Description of Study Area

Gajah Mungkur Reservoir is a reservoir located 6 km south of Wonogiri Regency, Central Java Province. The waters of this artificial lake is made by stem the longest river in Java Island is Bengawan Solo River. It was built in the late 1970s and started operations in 1978. The reservoirs covering an area of approximately 8,800 ha in 7 districts can irrigate 23.600 ha of paddy fields in Sukoharjo, Klaten, Karanganyar and Sragen areas. In addition to supplying drinking water Wonogiri City also generates electricity from hydropower (PLTA) of 12.4 MegaWatt. The location of study area can be shown in figure 1.

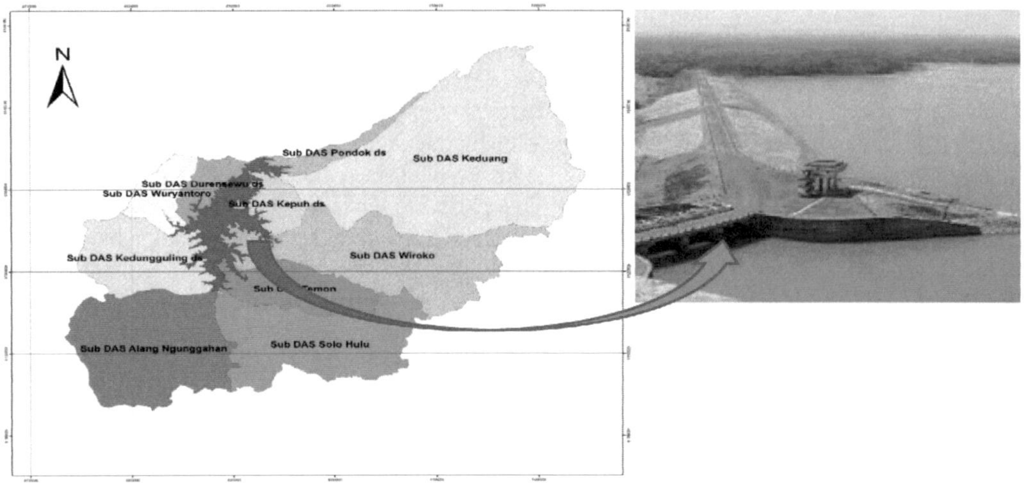

Figure 1. Location of study area (Gajah Mungkur Reservoir and several sub watershed around)

2.2. Data And Equipment

The tools used in this research are:

a. Software

- Windows 8 Ultimate Operating System
- GIS Software
- Image Processing Software

b. Field Measurement Equipment

- GPS Handheld
- Bottle where water is sampled
- Motorboat
- Camera

c. Data

The data used in this research are:

- Landsat Image 8 year 2013, 2015, 2017 (source: USGS)
- Map of Bengawan Solo Watershed (source: BPDAS)
- Indonesian topographic Map (RBI) scale 1: 25,000 Wonogiri regency (source: BIG)
- TSS in situ data and coordinate sampling in field for validation (source: Field sampling)

2.3. Methodology

The method of this study was followed the flowchart in Figure 2. In this study is using 4 images multi-temporal with reference to Rice Planting Time 1st in Demak 2013/2014, i.e. the image on October 30, December 1, 2013, January 2 and March 7, 2014.

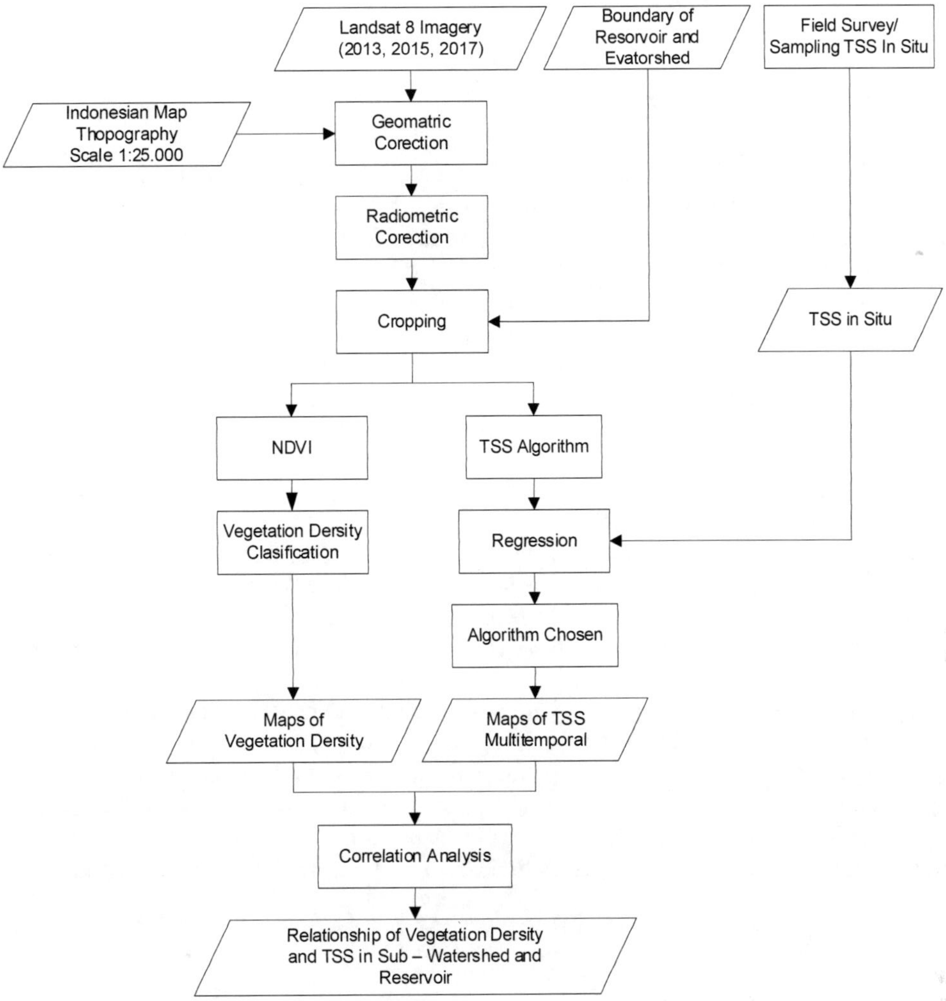

Figure 2. Flowchart of the research

2.3.1. Geometric correction. Geometric correction geo-referenced image to real-world coordinates. The geometric correction method of the image used a image to map rectification with reference Indonesian Topographical map (RBI) of Demak, scale 1: 25,000. Ground control point (GCP) on this research is taken as many as 16 points of each image. Root mean square error (RMSE) in this process were required maximum 1 pixel.

2.3.2. Radiometric correction. This stage corrected image using the conversion of a digital number to Top of Atmosphere (TOA) Reflectance. OLI band data of Landsat 8 can be converted to TOA planetary reflectance using reflectance rescaling coefficients provided in the product metadata file (MTL file). MTL is the format of metadata file for Landsat 8 imagery. The following equation is used to convert a digital number (DN) or pixel values to TOA reflectance for operational land imager (OLI) data as follows [3]:

$$\rho\lambda' = M\rho Qcal + A\rho \quad (1)$$

TOA reflectance with a correction for the sun angle is then [6]:

$$\rho\lambda = \frac{\rho\lambda'}{\cos(\theta SZ)} = \frac{\rho\lambda'}{\sin(\theta SE)} \quad (2)$$

where:

$\rho\lambda'$ = TOA planetary reflectance, without correction for the solar angle. Note that $\rho\lambda'$ does not contain a correction for the sun angle
M_ρ = Band-specific multiplicative rescaling factor from the metadata (REFLECTANCE_MULT_BAND_x, where x is the band number)
A_ρ = Band-specific additive rescaling factor from the metadata (REFLECTANCE_ADD_BAND_x, where x is the band number)
Q_{cal} = Quantized and calibrated standard product pixel values (DN)
$\rho\lambda$ = TOA planetary reflectance
θSE = Local sun elevation angle. The scene centre sun elevation angle in degrees is provided in the metadata (SUN_ELEVATION)
θ_{SZ} = Local solar zenith angle; $\theta_{SZ} = 90° - \theta_{SE}$

2.3.3. Image Cropping. Cropping image was used to concentrate the area study. It reduced the file size of the image so that the processing of data to be lighter and faster. It is made in accordance with the required area or region of interest (ROI). Moreover, in this study uses data multi-temporal for classification process, so cropping of the image is needed in order to accelerate the image processing.

2.3.4. Estimating Vegetation Density. The value of vegetation density in this study was obtained from the transformation of Normalized Difference Vegetation Index (NDVI). The value of this vegetation index is calculated as the ratio between the measured reflections of the red band (RED) and the infrared band (approached by the NIR band). The use of these two bands was chosen as the vegetation index parameter because the band size results were influenced by chlorophyll absorption, sensitive to vegetation biomass, and facilitated the distinction between vegetated, soil, bare land and water.

$$NDVI = \frac{NIR - RED}{NIR + RED} \quad (3)$$

where:
NIR = band near infrared
RED = band red

The resulting coupling between the red and infrared bands results in the maximum difference between vegetation and soil. NDVI generated original values always range from -1 to +1 [4]. The original values between -1 and +1 resulting from NDVI transformations have different presentations on land use. NDVI values around 0, usually represent land use with little vegetation to no vegetation at all.

Table 1. Classification of vegetation Density Based NDVI [4]

Class value NDVI	Classification Vegetation Density	Type of Land Cover
-1 to 0	Water Body	Lake, Sea, River
0 to 0,25	Bare Vegetation	Settlement, Bare Soil
0,25 to 0,55	Quite dense vegetation	Rice field, Moor
0,55 to 0,78	Dense vegetation	Rice field, Shrubs
0,78 to 1	Very dense vegetation	Tree shrubs, Forest

2.3.5. Estimating TSS. Sedimentation observation method using remote sensing method of TSS Landsat-8 image algorithm from several models (Syarif Budiman, Parwati, Woerd Pastercam). Some of these models will be tested for the best approach compared to in situ TSS. The result of this best algorithm model is then used to observe the development of TSS in Gajah Mungkur Reservoir in multi-temporal. The algorithm is:

- Parwati Algorithm

Algorithm which is the result of research conducted by Ety Parwati in order to make efforts to optimize the supervision of the quality of coastal waters environment due to industrial waste by using remote sensing technology.

These water quality observations use reflected values that are converted into TSS values [5]. The algorithm equation is as follows:

$$TSS\ (mg/l) = 0{,}6211(7{,}9038 \times \exp(23{,}942 \times Redband)) \times 0.9645 \tag{4}$$

Where :

TSS = *Total Suspended Solid*

Red Band = reflectance value of the *Red band*

- Woerd and Patercamp algorithm

The algorithm developed by Hans van der Woerd and Reinold Pasterkamp in 2004 is the result of his research in charting suspended material in a conservation area in the northern seas of Europe [6].

The algorithm equation is as follows:

If the value of X ≤ 2.76 or the reflectance value of band 2 ≤ 0.0282 then

$$TSS\ (mg/l) = 1{,}0585e^{1{,}3595x} \tag{5}$$

If the value of X> 2.76 or the reflectance value of band 2> 0.0282 then:

$$TSS\ (mg/l) = 32{,}918x - 46{,}616 \tag{6}$$

Where X is:

$$X = (-0{,}53 * RefB2) + \frac{0.001}{0{,}03 * RefB2} - 0{,}0059 \tag{7}$$

Where,

TSS = *Total Suspended Solid*

RefB2 = reflectance value of band 2

- Syarief Budiman Algorithm

The algorithm was developed in the waters of the Mahakam Delta with a method developed based on bio-optical modelling to analyze a distribution of material suspended through Remote Sensing technology. Syarif Budhiman's algorithm used the constants A and S to complete the algorithm to be built [7].

The algorithm equation is as follows:

$$TSS(mg/l) = A * \exp(S * R(0\ -)Redband) \tag{8}$$

where,

TSS = *Total Suspended Solid*

R(0-)*redband* = *irradiant* of reflectance

In this research, the TSS in Situ survey is used for comparative validation. This validation result is used to get the best TSS algorithm model. This will be used in TSS mapping in Gajah Mungkur reservoir. Validation was done on 26 spots that are evenly distributed in the reservoir area. The distribution of TSS in situ validation point spots can be seen in Figure 3. The results of water sampling at the validation point were tested in the laboratory.

2.3.6. Detection of Vegetation Density Effect to TSS changing .In this study, the result of Vegetation density and TSS multi years were calculated the different or the changing year to year. Then the vegetation density changing in the Bengawan Solo Watersheds were correlated to the the TSS changing in Gajah Mungkur reservoir. The correlated method is description correlation. The result of vegetation changing in every sub-watershed was correlated using see the TSS changing in the estuary of the main river in Gajah Mungkur reservoir. This result can detect how are it effect to the TSS in Gajahmungkur reservoir and Which are the sub-watershed most have effect to TSS in Gajah Mungkur reservoir.

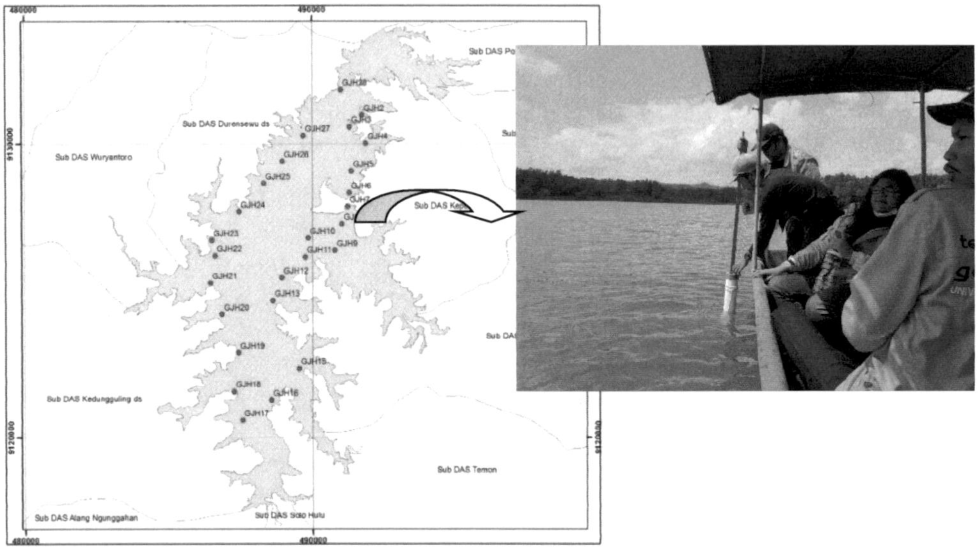

Figure 3. Point Map of TSS in Situ Validation

3. Result and Discussion

3.1. TSS Distribution in Gajah Mungkur Reservoir

3.1.1. Determination the best TSS algoritm in Gajah Mungkur Reservoir

The determination of the best algorithm result is intended to perform the selection of algorithm that is suitable with the condition of TSS in situ distribution, done by regressing the result of TSS algorithm calculation from the previous research with TSS of water sample test which amounted to 26 sample points. The algorithm used is the algorithm of Syarif Budhiman, Parwati Algorithm, Woerd and Pastercamp algorithm based on the reflectance value of the band. One of the three algorithms were chosen to analyse the multiplicity of TSS distribution and the effect of vegetation density to the TSS. The result of linear regression between in situ TSS and TSS algorithm is shown in Fig. 4.

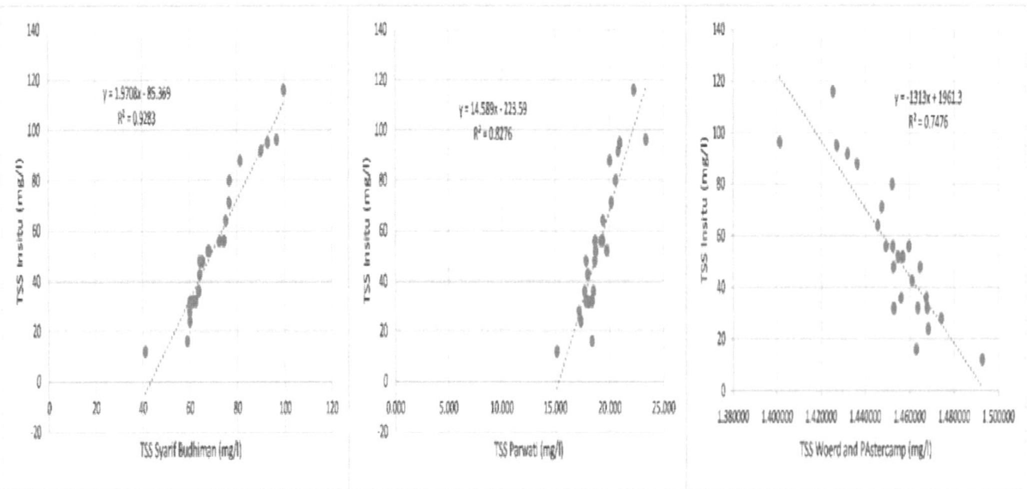

Figure 4. Graph of Linear Regression between TSS algorithm Syarif Budhiman, TSS Parwati, and TSS Woerd and Pastercamp with TSS insitu

Comparison of TSS value from Landsat 8 image processing data acquisition on April 25, 2017 with three TSS algorithms namely Syarif Budhiman algorithm, Parwati algorithm, and Woerd and Pastercamp Algorithm with TSS value of insitu taken on April 25, 2017 which has coefficient value of determination close to 1 is Syarif Budhiman's algorithm. Shariah alhiwah algorithm has a determination value of 0.92 or 92%. TSS value obtained from the calculation Syarif Budhiman algorithm almost have the same pattern with the value of insitu TSS scattered in Gajah Mungkur Reservoir as much as 26 points. Then it can be concluded that the best Syarif Budhiman algorithm and will be used in the process of TSS mapping in Gajah Mungkur Reservoir.

3.1.2. Distribution Analysis of Total Suspended Solid (TSS) Concentration in Gajah Mungkur Reservoir
Based on the results of the best algorithm determination Syarif Budhiman algorithm is obtained with the value of in situ TSS. Furthermore, this Syarif Budhiman algorithm is used as input for mapping of TSS distribution in Gajah Mungkur Reservoir waters in 2013, 2015 and 2017. The map of TSS distribution in Gajah Mungkur Reservoir waters in 2013, 2015 and 2017 is presented in figure 5.

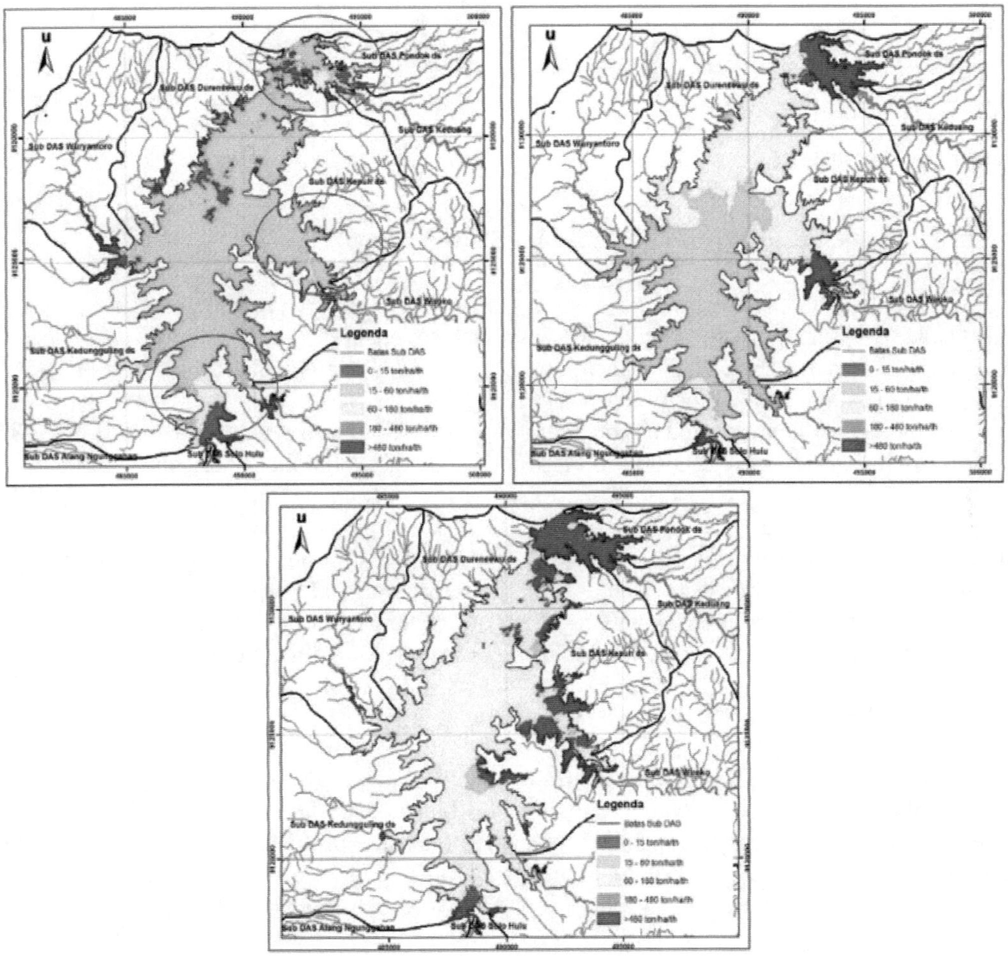

Figure 5. Map of concentration distribution of TSS Gajah Mungkur Reservoir in 2013, 2015, and 2017

From the results obtained by comparison of the distribution of TSS Reservoir of Gajah Mungkur in 2013, 2015, and 2017 is presented in the graph in Table 2 and Figure 6.

Table 2. Comparison of TSS distribution area 2013, 2015, and 2017

No	Year	Area (ha)			
		0-20 (mg/L)	20-50 (mg/L)	50-100 (mg/L)	>100 (mg/L)
1	2013	97,956	4.779,848	616,404	503,78
2	2015	51,521	2.880,392	2.526,905	539,171
3	2017	51,521	103,573	4.656,853	1.186,042

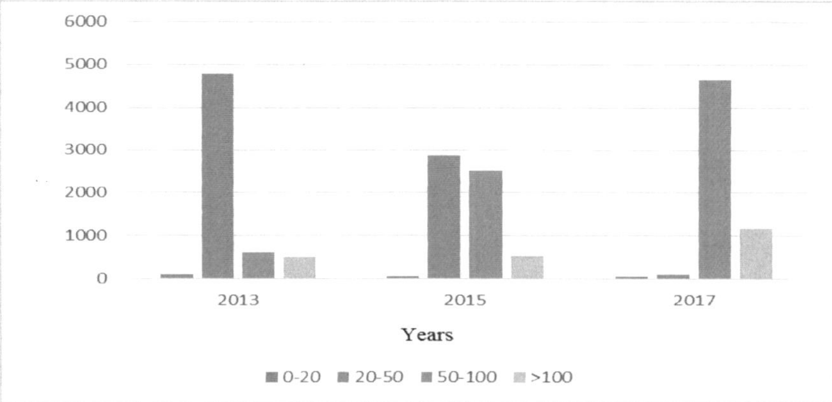

Figure 6. Comparison diagram of TSS distribution 2013, 2015, and 2017

Based on the calculation of TSS in the waters of Gajah Mungkur Reservoir in 2013 the spread area is found at concentration 20-50 mg / l which is 4,779,848 ha or equal to 79,690%. Followed concentration 50-100% has an area of 616,404 ha. Concentrations> 100 mg / l possess an area of 503,780 ha. And the concentration of 0-20 mg / l has an area of 97.956 ha.

In 2015 the largest area is found at concentrations of 20-50 mg / l with an area of 2,880,392 ha or 48,022%. At concentrations of 50-100 mg / l have an area of almost the same concentration of 20-50 mg / l, i.e. 2526,905 ha or equal to 42.129%. Concentrations> 100 mg / l have an area of 539,171 ha and a concentration of 0-20 mg / l has an area of 51,521 ha.

In 2017 the extent of spread is concentrated at 50-100 mg / l of 4,656,853 ha or 77,640%. Followed concentration> 100 has an area of 1,186,042 ha. Concentrations of 20-50 mg / l possess an area of 103,573 ha. And the concentration of 0-20 mg / l has an area of 51,521 ha. Based on the TSS map results in Figure 5, seen in estuary of the Keduang Sub-Watersheds, Solo Hulu Sub-Watersheds and Wiroko Sub-Watersheds (indicated by red circle) have an increasing TSS value from 2013, 2015 and 2017. This may indicate that these three Sub-Watersheds have high erosion.

3.2. Vegetation Density in Bengawan Solo watersheds

The results of the vegetation density process using NDVI method are shown in Figure 7. From the results it is seen that there is a significant change in vegetation density in 2013, 2015 and 2017. Although in general the vegetation classes are still dominated, bare vegetation classes are annually increasing. This can increase the amount of erosion on each Sub-DAS. Moreover, most bare vegetation classes located close to the slope.

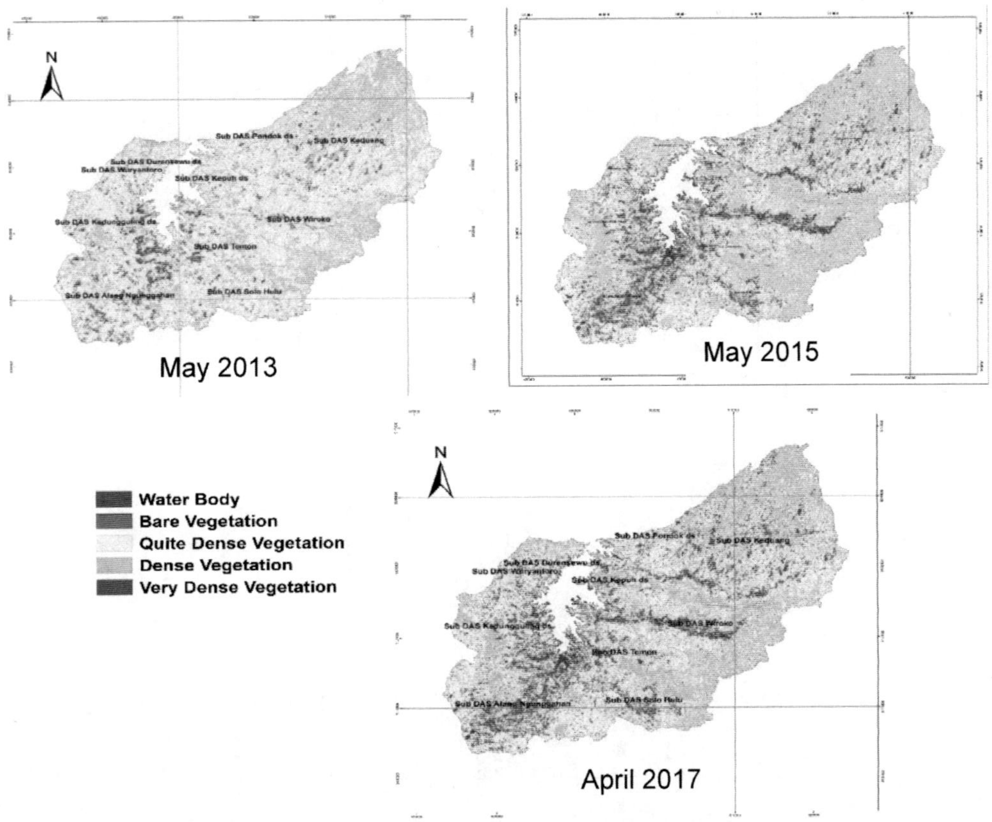

Figure 7. Distribution of vegetation Density in Bengawan Solo Watershed multitemporal

From the results obtained, the vegetation density each Sub-Watershed in 2013, 2015, and 2017 is presented in the graph in Figure 8.

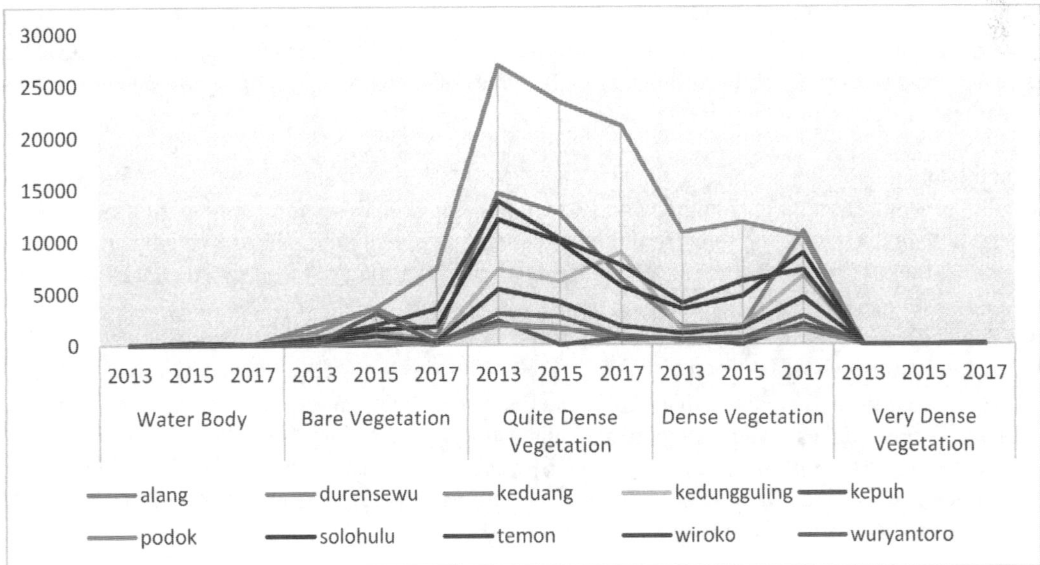

Figure 8. Graph of changes in the area (Ha) of vegetation density classes of each Sub-Watershed in 2013, 2015, 2017

From these data can be seen on the Kedung Sub-watershed there is a decrease in the vegetation class is quite dense vegetation and there is an increase in the bare vegetation. The same thing happened Solo Hulu sub-watershed and Wiroko Sub-Watershed. The class quite dense vegetation in Kedung Guling sub-watershed decrease in 2013 to 2015. And there is an increase in 2015 to 2017. But in the vegetation dense class increased. Figure 9.

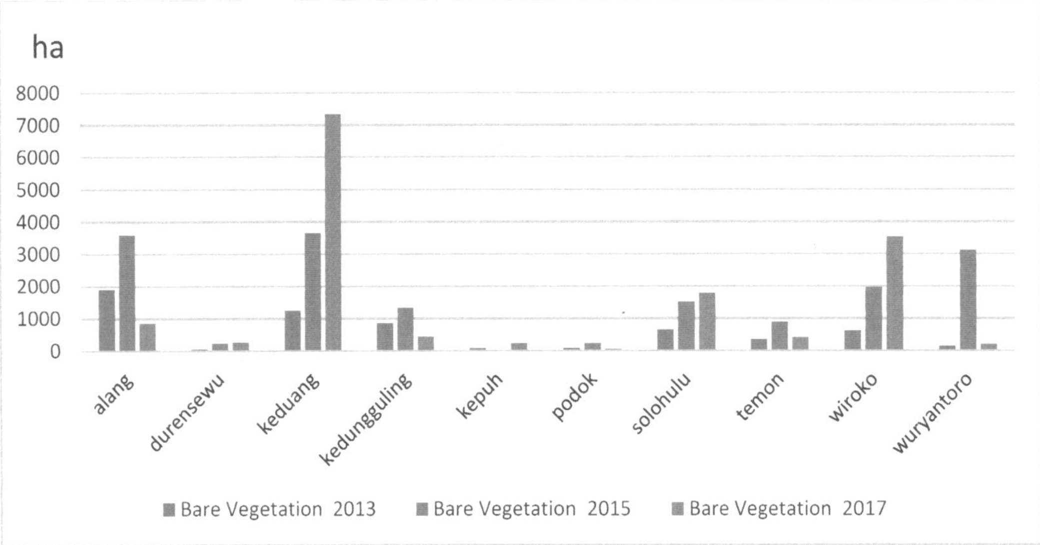

Figure 9. The bare vegetation class change in 2013, 2015, 2017

3.3. Correlation Vegetation Density in Several Sub-Watersheds and TSS Distribution in Gajah Mungkur Reservoir

Based on the results of the analysis on TSS in Gajah Mungkur Reservoir and Vegetation Density Change on each sub-watershed can be seen that both have a real relationship. This can be seen from the result of TSS change at sub watershed estuary that at sub watershed estuary Kedung, Solo Hulu and Wiroko have TSS value which always increases from year to year. On the other hand, the value of vegetation density in the bare vegetation class on these three sub-basins has a significant increase. This means land degradation occurs in these three sub-catchments. Or in other words a decrease in vegetation density causes erosion increase in them

4. Conclusion

- The concentration of TSS in the waters of Gajah Mungkur Reservoir in the period of 2013-2017 were increased. It can be seen from the largest area in each year, 2013 at concentrations of 20-50 mg / l, by 2015 at concentrations of 20-50 mg / l and 50-100 mg / l, and by 2017 at concentrations of 50-100 mg/l.
- The vegetation density in some sub-watershed in Bengawan Solo watershed was decreased such as sub-watersheds of keduang, Sub Watershed of Solo Hulu and Sub watershed of Wiroko.
- The vegetation density level changing in the Bengawan Solo watershed influences the distribution pattern and changes in the TSS value in Gajah Mungkur reservoir. The relationship of both shows a direct relationship. Sub watershed that has vegetation density change significantly and has an influence on increasing TSS in the Gajah Mungkur reservoir are sub watershed of Kedung, Sub Watershed of Solo Hulu and Sub watershed of Wiroko.

References

[1] Wirosoedarmo R 2011 Sedimentation Behavior and Its Influence on Channel Performance On Waru Tegal Irrigation Network Kili Kediri. Department Agriculture Engineering, Faculty of Agriculture, Brawijaya University.

[2] Asdak C 2015 *Hydrology and Watershed Management.* Gadjah Mada University Press Yogyakarta.

[3] USGS 2013 *Using the USGS Landsat 8 Product.* http://landsat.usgs.gov. 10 April 2017.

[4] Danoedoro P 1996 *Teoritical Digital Image Processing and It's Application in Remote Sensing.* Gadjah Mada University Press

[5] Parwati E 2006 Modeling Spatial Dynamics of Coastal Land Management Kabupaten Berau, East Kalimantan using Remote Sensing. *Final Research Report of LAPAN Aerospace Independence*

[6] Pastercamp P, S W M Peters, M. Rijkeboer and A G Dekber 1999 RESTWES: Retrieval of total suspended matter concentration from SPOT image. *Report Number W-99/33*, Institute for Environmental Studies, Vrije University, Amsterdam, The Netherlands.

[7] Budhiman S 2005 Total Suspended Matter Spreading Mapping Using Aster Data with Bio-Optical Model Approach. *Prosiding PIT MAPIN XIV*, Surabaya

Acknowledgment

We extend our thanks to remote sensing and photogrammetry laboratory- Diponegoro University for all their help in data processing equipment